ELECTRODYNAMICS OF SOLIDS
OPTICAL PROPERTIES OF ELECTRONS IN MATTER

The authors of this book present a thorough discussion of the optical properties of solids, with a focus on electron states and their response to electrodynamic fields. A review of the fundamental aspects of the propagation of electromagnetic fields, and their interaction with condensed matter, is given. This is followed by a discussion of the optical properties of metals, semiconductors, and collective states of solids such as superconductors.

Theoretical concepts, measurement techniques, and experimental results are covered in three inter-related sections. Well established, mature fields are discussed, together with modern topics at the focus of current interest. The substantial reference list included will also prove to be a valuable resource for those interested in the electronic properties of solids.

The book is intended for use by advanced undergraduate and graduate students, and researchers active in the fields of condensed matter physics, materials science, and optical engineering.

MARTIN DRESSEL received his Doctor of Sciences degree in 1989 from the Universität Göttingen where he subsequently worked as a postdoctoral research fellow. Since then he has held positions in the University of British Columbia at Vancouver; the University of California, Los Angeles; the Technische Universität, Darmstadt; and the Center of Electronic Correlations and Magnetism at the Universität Augsburg. Professor Dressel is now Head of the 1. Physikalisches Institut at the Universität Stuttgart.

GEORGE GRÜNER obtained his Doctor of Sciences degree from the Eötvös Lorand University, Budapest, in 1972, and became Head of the Central Research Institute of Physics in Budapest in 1974. In 1980 he took up the position of Professor of Physics at the University of California, Los Angeles, and later became Director of the Solid State Science Center there. Professor Grüner has been a distinguished visiting professor at numerous institutions worldwide and is a consultant for several international corporations and advisory panels. He is a Guggenheim Fellow and is also a recipient of the Alexander Humboldt Senior American Scientist Award.

Electrodynamics of Solids

Optical Properties of Electrons in Matter

Martin Dressel

Stuttgart

and

George Grüner

Los Angeles

CAMBRIDGE UNIVERSITY PRESS

PUBLISHED BY THE PRESS SYNDICATE OF THE UNIVERSITY OF CAMBRIDGE
The Pitt Building, Trumpington Street, Cambridge, United Kingdom

CAMBRIDGE UNIVERSITY PRESS
The Edinburgh Building, Cambridge CB2 2RU, UK
40 West 20th Street, New York, NY 10011-4211, USA
477 Williamstown Road, Port Melbourne, VIC 3207, Australia
Ruiz de Alarcón 13, 28014, Madrid, Spain
Dock House, The Waterfront, Cape Town 8001, South Africa

http://www.cambridge.org

First published 2002

Printed in the United Kingdom at the University Press, Cambridge

Typeface Times 11/14pt.   *System* LATEX 2_ε [DBD]

A catalogue record of this book is available from the British Library

Library of Congress Cataloguing in Publication data

Dressel, Martin, 1960–
Electrodynamics of solids: optical properties of electrons in matter /
Martin Dressel and George Grüner.
p.   cm.
includes bibliographical references and index.
ISBN 0 521 59253 4 – ISBN 0 521 59726 9 (pb.)
1. Solids – Optical properties. 2. Solids – Electric properties.
I. Grüner, George. II. Title. QC176.8.O6 D74 2002
530.4$'$12–dc21   2001025962

ISBN 0 521 59253 4   hardback
ISBN 0 521 59726 9   paperback

Contents

Preface

This book has its origins in a set of lecture notes, assembled at UCLA for a graduate course on the optical studies of solids. In preparing the course it soon became apparent that a modern, up to date summary of the field is not available. More than a quarter of a century has elapsed since the book by Wooten: Optical Properties of Solids – and also several monographs – appeared in print. The progress in optical studies of materials, in methodology, experiments and theory has been substantial, and optical studies (often in combination with other methods) have made definite contributions to and their marks in several areas of solid state physics. There appeared to be a clear need for a summary of the state of affairs – even if with a somewhat limited scope.

Our intention was to summarize those aspects of the optical studies which have by now earned their well deserved place in various fields of condensed matter physics, and, at the same time, to bring forth those areas of research which are at the focus of current attention, where unresolved issues abound. Prepared by experimentalists, the rigors of formalism are avoided. Instead, the aim was to reflect upon the fact that the subject matter is much like other fields of solid state physics where progress is made by consulting both theory and experiment, and invariably by choosing the technique which is most appropriate.

'A treatise expounds, a textbook explains', said John Ziman, and by this yardstick the reader holds in her or his hands a combination of both. In writing the book, we have in mind a graduate student as the most likely audience, and also those not necessarily choosing this particular branch of science but working in related fields. A number of references are quoted throughout the book, these should be consulted for a more thorough or rigorous discussion, for deeper insight or more exhaustive experimental results.

There are limits of what can be covered: choices have to be made. The book focuses on 'mainstream' optics, and on subjects which form part of what could be termed as one of the main themes of solid state physics: the electrodynamics or (to choose a more conventional term) the optical properties of electrons in matter. While we believe this aspect of optical studies will flourish in future years, it is also evolving both as far as the techniques and subject matter are concerned. Near-field optical spectroscopy, and optical methods with femtosecond resolution are just two emerging fields, not discussed here; there is no mention of the optical properties of nanostructures, and biological materials – just to pick a few examples of current and future interest.

Writing a book is not much different from raising a child. The project is abandoned with frustration several times along the way, only to be resumed again and again, in the hope that the effort of this (often thankless) enterprise is, finally, not in vain. Only time will tell whether this is indeed the case.

Acknowledgements

Feedback from many people was essential in our attempts to improve, correct, and clarify this book, for this we are grateful to the students who took the course. Wolfgang Strohmaier prepared the figures. The Alexander von Humboldt and the Guggenheim Foundations have provided generous support; without such support the book could not have been completed.

Finally we thank those who shared our lives while this task was being completed, Annette, Dani, Dora, and Maria.

<h1 style="text-align:center">1</h1>

Introduction

Ever since Euclid, the interaction of light with matter has aroused interest – at least among poets, painters, and physicists. This interest stems not so much from our curiosity about materials themselves, but rather to applications, should it be the exploration of distant stars, the burning of ships of ill intent, or the discovery of new paint pigments.

It was only with the advent of solid state physics about a century ago that this interaction was used to explore the properties of materials in depth. As in the field of atomic physics, in a short period of time optics has advanced to become a major tool of condensed matter physics in achieving this goal, with distinct advantages – and some disadvantages as well – when compared with other experimental tools.

The focus of this book is on optical spectroscopy, defined here as the information gained from the absorption, reflection, or transmission of electromagnetic radiation, including models which account for, or interpret, the experimental results. Together with other spectroscopic tools, notably photoelectron and electron energy loss spectroscopy, and Raman together with Brillouin scattering, optics primarily measures charge excitations, and, because of the speed of light exceeding substantially the velocities of various excitations in solids, explores in most cases the $\Delta\mathbf{q} = 0$ limit. While this is a disadvantage, it is amply compensated for by the enormous spectral range which can be explored; this range extends from well below to well above the energies of various single-particle and collective excitations.

The interaction of radiation with matter is way too complex to be covered by a single book; so certain limitations have to be made. The response of a solid at position $\mathbf{r}$ and time t to an electric field $\mathbf{E}(\mathbf{r}', t')$ at position $\mathbf{r}'$ and time t' can be written as

$$D_i(\mathbf{r}, t) = \int \int \bar{\bar{\epsilon}}_{ij}(\mathbf{r}, \mathbf{r}', t, t') E_j(\mathbf{r}', t') \, \mathrm{d}t' \, \mathrm{d}\mathbf{r}' \qquad (1.0.1)$$

1

where i and j refer to the components of the electric field $\mathbf{E}$ and displacement field $\mathbf{D}$; thus $\bar{\bar{\epsilon}}_{ij}$ is the so-called dielectric tensor. For homogeneous solids, the response depends only on $\mathbf{r} - \mathbf{r}'$ (while time is obviously a continuous variable), and Eq. (1.0.1) is reduced to

$$D_i(\mathbf{r}, t) = \int \int \bar{\bar{\epsilon}}_{ij}(\mathbf{r} - \mathbf{r}', t - t') E_j(\mathbf{r}', t') \, dt' \, d\mathbf{r}' \quad . \tag{1.0.2}$$

We further assume linear response, thus the displacement vector $\mathbf{D}$ is proportional to the applied electric field $\mathbf{E}$. In the case of an alternating electric field of the form

$$\mathbf{E}(\mathbf{r}, t) = \mathbf{E}_0 \exp\{i(\mathbf{q} \cdot \mathbf{r} - \omega t)\} \tag{1.0.3}$$

the response occurs at the same frequency as the frequency of the applied field with no higher harmonics. Fourier transform then gives

$$D_i(\mathbf{q}, \omega) = \bar{\bar{\epsilon}}_{ij}(\mathbf{q}, \omega) E_j(\mathbf{q}, \omega) \tag{1.0.4}$$

with the complex dielectric tensor assuming both a wavevector and frequency dependence. For $\bar{\bar{\epsilon}}_{ij}(\mathbf{r} - \mathbf{r}', t - t')$ real, the $\mathbf{q}$ and ω dependent dielectric tensor obeys the following relation:

$$\bar{\bar{\epsilon}}_{ij}(\mathbf{r} - \mathbf{r}', t - t') = \bar{\bar{\epsilon}}_{ij}^{*}(\mathbf{r} - \mathbf{r}', t - t') \quad ,$$

where the star (*) refers to the complex conjugate. Only cubic lattices will be considered throughout most parts of the book, and then $\hat{\epsilon}$ is a scalar, complex quantity.

   Of course, the response could equally well be described in terms of a current at position $\mathbf{r}$ and time t, and thus

$$J(\mathbf{r}, t) = \int \int \hat{\sigma}(\mathbf{r}, \mathbf{r}', t, t') E(\mathbf{r}', t') \, dt' \, d\mathbf{r}' \tag{1.0.5}$$

leading to a complex conductivity tensor $\hat{\sigma}(\mathbf{q}, \omega)$ in response to a sinusoidal time-varying electric field. The two response functions are related by

$$\hat{\epsilon}(\mathbf{q}, \omega) = 1 + \frac{4\pi i}{\omega} \hat{\sigma}(\mathbf{q}, \omega) \quad ; \tag{1.0.6}$$

this follows from Maxwell's equations.

   Except for a few cases we also assume that there is a local relationship between the electric field $\mathbf{E}(\mathbf{r}, t)$ and $\mathbf{D}(\mathbf{r}, t)$ and also $\mathbf{j}(\mathbf{r}, t)$, and while these quantities may display well defined spatial dependence, their spatial variation is identical; with

$$\frac{\mathbf{J}(\mathbf{r})}{\mathbf{E}(\mathbf{r})} = \hat{\sigma} \quad \text{and} \quad \frac{\mathbf{D}(\mathbf{r})}{\mathbf{E}(\mathbf{r})} = \hat{\epsilon} \tag{1.0.7}$$

two spatially independent quantities. This then means that the Fourier transforms of $\hat{\epsilon}$ and $\hat{\sigma}$ do not have $\mathbf{q} \neq 0$ components. There are a few notable exceptions when some important length scales of the problem, such as the mean free path ℓ in metals or the coherence length ξ_0 in superconductors, are large and exceed the length scales set by the boundary problem at hand. The above limitations then reduce

$$\hat{\sigma}(\omega) = \sigma_1(\omega) + i\sigma_2(\omega) \qquad \text{and} \qquad \hat{\epsilon}(\omega) = \epsilon_1(\omega) + i\epsilon_2(\omega) \qquad (1.0.8)$$

to scalar and $\mathbf{q}$ independent quantities, with the relationship between $\hat{\epsilon}$ and $\hat{\sigma}$ as given before. We will also limit ourselves to non-magnetic materials, and will assume that the magnetic permeability $\mu_1 = 1$ with the imaginary part $\mu_2 = 0$.

We will also make use of what is called the semiclassical approximation. The interaction of charge e_i with the radiation field is described as the Hamiltonian

$$\mathcal{H} = \frac{1}{2m} \sum_i \left[\mathbf{p}_i - \frac{e_i}{c} \mathbf{A}(\mathbf{r}_i) \right]^2 \quad , \qquad (1.0.9)$$

and while the electronic states will be described by appropriate first and second quantization, the vector potential $\mathbf{A}$ will be assumed to represent a classical field. We will also assume the so-called Coulomb gauge, by imposing a condition

$$\nabla \cdot \mathbf{A} = 0 \quad ; \qquad (1.0.10)$$

this then implies that $\mathbf{A}$ has only transverse components, perpendicular to the wavevector $\mathbf{q}$.

Of course one cannot do justice to all the various interesting effects which arise in the different forms of condensed matter – certain selections have to be made, this being influenced by our prejudices. We cover what could loosely be called the electrodynamics of electron states in solids. As the subject of what can be termed electrodynamics is in fact the response of charges to electromagnetic fields, the above statement needs clarification. Throughout the book our main concern will be the optical properties of electrons in solids, and a short guide of the various states which may arise is in order.

In the absence of interaction with the underlying lattice, and also without electron–electron or electron–phonon interactions, we have a collection of free electrons obeying – at temperatures of interest – Fermi statistics, and this type of electron liquid is called a Fermi liquid. Interactions between electrons then lead to an interacting Fermi liquid, with the interactions leading to the renormalization of the quasi-particles, leaving, however, their character unchanged. Under certain circumstances, notably when the electron system is driven close to an instability, or when the electronic structure is highly anisotropic, this renormalized Fermi-liquid picture is not valid, and other types of quantum liquids are recovered. The – not too appealing – notion of non-Fermi liquids is usually adopted when deviations from a Fermi liquid are found. In strictly one dimension (for example) the nature of the quantum liquid, called the Luttinger liquid, with all of its implications, is well known. Electron–phonon interactions also lead to a renormalized Fermi liquid.

If the interactions between the electrons or the electron–phonon interactions are of sufficient strength, or if the electronic structure is anisotropic, phase transitions to what can be termed electronic solids occur. As is usual for phase transitions, the ordered state has a broken symmetry, hence the name broken symmetry states of metals. For these states, which are called charge or spin density wave states, translational symmetry is broken and the electronic charge or spin density assumes a periodic variation – much like the periodic arrangement of atoms in a crystal. The superconducting state has a different, so-called broken gauge symmetry. Not surprisingly for these states, single-particle excitations have a gap – called the single-particle gap – a form of generalized rigidity. As expected for a phase transition, there are collective modes associated with the broken symmetry state which – as it turns out – couple directly to electromagnetic fields. In addition, for these states the order parameter is complex, with the phase directly related to the current and density fluctuations of the collective modes.

Disorder leads to a different type of breakdown of the Fermi liquid. With increasing disorder a transition to a non-conducting state where electron states are localized may occur. Such a transition, driven by an external parameter (ideally at $T = 0$ where only quantum fluctuations occur) and not by the temperature, is called a quantum phase transition, with the behavior near to the critical disorder described – in analogy to thermal phase transitions – by various critical exponents. This transition and the character of the insulating, electron glass state depend on whether electron–electron interactions are important or not. In the latter case we have a Fermi glass, and the former can be called a Coulomb glass, the two cases being distinguished by temperature and frequency dependent excitations governed by different exponents, reflecting the presence or absence of Coulomb gaps.

A different set of states and properties arises when the underlying periodic lattice leads to full and empty bands, thus to semiconducting or insulating behavior. In this case, the essential features of the band structure can be tested by optical experiments. States beyond the single-electron picture, such as excitons, and also impurity states are essential features here. All this follows from the fundamental assumption about lattice periodicity and the validity of Bloch's theorem. When this is not relevant, as is the case for amorphous semiconductors, localized states with a certain amount of short range order are responsible for the optical properties.

The response of these states to an electromagnetic field leads to dissipation, and this is related to the fluctuations which arise in the absence of driving fields. The relevant fluctuations are expressed in terms of the current–current or density–density correlation functions, related to the response through the celebrated fluctuation-dissipation theorem. The correlation functions in question can be derived using an appropriate Hamiltonian which accounts for the essential features of the particular electron state in question. These correlations reflect and the dissipation occurs through the elementary excitations. Single-particle excitations, the excitation of the individual quasi-particles, may be the source of the dissipation, together with the collective modes which involve the cooperative motion of the entire system governed by the global interaction between the particles. Electron–hole excitations in a metal are examples of the former, plasmons and the response of the broken symmetry ground state are examples of the latter. As a rule, these excitations are described in the momentum space by assuming extended states and excitations with well defined momenta. Such excitations may still exist in the case of a collection of localized states; here, however, the excitations do not have well defined momenta and thus restrictions associated with momentum conservation do not apply.

Other subjects, interesting in their own right, such as optical phonons, dielectrics, color centers (to name just a few) are neglected; and we do not discuss charge excitations in insulators – vast subjects with interesting properties. Also we do not discuss the important topic of magneto-optics or magneto-transport phenomena, which occur when both electric and magnetic fields are applied.

The organization of the book is as follows: underlying theory, techniques, and experimental results are discussed as three, inter-relating parts of the same endeavor. In Part 1 we start with the necessary preliminaries: Maxwell's equations and the definition of the optical constants. This is followed by the summary of the propagation of light in the medium, and then by the discussion of phenomena which occur at an interface; this finally brings us to the optical parameters which are measured by experiment. The three remaining chapters of Part 1 deal with the optical properties of metals, semiconductors, and the so-called broken symmetry states of metals. Only simple metals and semiconductors are dealt with here, and only the conventional broken symmetry states (such as BCS superconductors) will be covered in the so-called weak coupling limit. In these three chapters three different effects are dominant: dynamics of quasi-free electrons, absorption due to interband processes, and collective phenomena.

In Part 2 the experimental techniques are summarized, with an attempt to bring out common features of the methods which have been applied at vastly different spectral ranges. Here important similarities exist, but there are some important differences as well. There are three spectroscopic principles of how the response in a wide frequency range can be obtained: measurements can be performed in the frequency domain, the time domain, or by Fourier transform technique. There are also different ways in which the radiation can interact with the material studied:

simply transmission or reflection, or changes in a resonance structure, can be utilized.

In Part 3 experimental results are summarized and the connection between theory and experiment is established. We first discuss simple scenarios where the often drastic simplifications underlying the theories are, in the light of experiments, justified. This is followed by the discussion of modern topics, much in the limelight at present. Here also some hand-waving arguments are used to expound on the underlying concepts which (as a rule) by no means constitute closed chapters of condensed matter physics.

Part one

Concepts and properties

Introductory remarks

In this part we develop the formalisms which describe the interaction of light (and sometimes also of a test charge) with the electronic states of solids. We follow usual conventions, and the transverse and longitudinal responses are treated hand in hand. Throughout the book we use simplifying assumptions: we treat only homogeneous media, also with cubic symmetry, and assume that linear response theory is valid. In discussing various models of the electron states we limit ourselves to local response theory – except in the case of metals where non-local effects are also introduced. Only simple metals and semiconductors are treated; and we offer the simple description of (weak coupling) broken symmetry – superconducting and density wave – states, all more or less finished chapters of condensed matter physics. Current topics of the electrodynamics of the electron states of solids are treated together with the experimental state of affairs in Part 3. We make extensive use of computer generated figures to visualize the results.

After some necessary preliminaries on the propagation and scattering of electromagnetic radiation, we define the optical constants, including those which are utilized at the low energy end of the electrodynamic spectrum, and summarize the so-called Kramers–Kronig relations together with the sum rules. The response to transverse and longitudinal fields is described in terms of correlation and response functions. These are then utilized under simplified assumptions such as the Drude model for metals or simple band-to-band transitions in the case of semiconductors.

Broken symmetry states are described in their simple form using second quantized formalism.

General books and monographs

A.A. Abrikosov, *Fundamentals of the Theory of Metals* (North-Holland, Amsterdam, 1988)

M. Born and E. Wolf, *Principles of Optics*, 6th edition (Cambridge University Press, Cambridge, 1999)

J. Callaway, *Quantum Theory of Solids*, 2nd edition (Academic Press, New York, 1991)

R.G. Chambers, *Electrons in Metals and Semiconductors* (Chapman and Hall, London, 1990)

W.A. Harrison, *Solid State Theory* (McGraw-Hill, New York, 1970)

H. Haug and S.W. Koch, *Quantum Theory of the Optical and Electronic Properties of Semiconductors*, 3rd edition (World Scientific, Singapore, 1994)

J. D. Jackson, *Classical Electrodynamics*, 2nd edition (John Wiley & Sons, New York, 1975)

C. Kittel, *Quantum Theory of Solids*, 2nd edition (John Wiley & Sons, New York, 1987)

L.D. Landau, E.M. Lifshitz, and L.P. Pitaevskii, *Electrodynamics of Continuous Media*, 2nd edition (Butterworth-Heinemann, Oxford, 1984)

I.M. Lifshitz, M.Ya. Azbel', and M.I. Kaganov, *Electron Theory of Metals* (Consultants Bureau, New York, 1973)

G.D. Mahan, *Many-Particle Physics*, 2nd edition (Plenum Press, New York, 1990)

D. Pines, *Elementary Excitations in Solids* (Addison-Wesley, Reading, MA, 1963)

D. Pines and P. Noizières, *The Theory of Quantum Liquids*, Vol. 1 (Addison-Wesley, Reading, MA, 1966)

J.R. Schrieffer, *Theory of Superconductivity*, 3rd edition (W.A. Benjamin, New York, 1983)

F. Stern, *Elementary Theory of the Optical Properties of Solids*, in: *Solid State Physics* **15**, edited by F. Seitz and D. Turnbull (Academic Press, New York, 1963), p. 299

J. Tauc (ed.), *The Optical Properties of Solids*, Proceedings of the International School of Physics 'Enrico Fermi' **34** (Academic Press, New York, 1966)

F. Wooten, *Optical Properties of Solids* (Academic Press, San Diego, CA, 1972)

P.Y. Yu and M. Cardona, *Fundamentals of Semiconductors* (Springer-Verlag, Berlin, 1996)

J.M. Ziman, *Principles of the Theory of Solids*, 2nd edition (Cambridge University Press, London, 1972)

2

The interaction of radiation with matter

Optics, as defined in this book, is concerned with the interaction of electromagnetic radiation with matter. The theoretical description of the phenomena and the analysis of the experimental results are based on Maxwell's equations and on their solution for time-varying electric and magnetic fields. The optical properties of solids have been the subject of extensive treatises [Sok67, Ste63, Woo72]; most of these focus on the parameters which are accessible with conventional optical methods using light in the infrared, visible, and ultraviolet spectral range. The approach taken here is more general and includes the discussion of those aspects of the interaction of electromagnetic waves with matter which are particularly relevant to experiments conducted at lower frequencies, typically in the millimeter wave and microwave spectral range, but also for radio frequencies.

After introducing Maxwell's equations, we present the time dependent solution of the equations leading to wave propagation. In order to describe modifications of the fields in the presence of matter, the material parameters which characterize the medium have to be introduced: the conductivity and the dielectric constant. In the following step, we define the optical constants which characterize the propagation and dissipation of electromagnetic waves in the medium: the refractive index and the impedance. Next, phenomena which occur at the interface of free space and matter (or in general between two media with different optical constants) are described. This discussion eventually leads to the introduction of the optical parameters which are accessible to experiment: the optical reflectivity and transmission.

2.1 Maxwell's equations for time-varying fields

To present a common basis of notation and parameter definition, we want to recall briefly some well known relations from classical electrodynamics. Before we consider the interaction of light with matter, we assume no matter to be present.

9

2.1.1 Solution of Maxwell's equations in a vacuum

The interaction of electromagnetic radiation with matter is fully described by Maxwell's equations. In the case of a vacuum the four relevant equations are

$$\nabla \times \mathbf{E}(\mathbf{r}, t) + \frac{1}{c}\frac{\partial \mathbf{B}(\mathbf{r}, t)}{\partial t} = 0 \quad , \tag{2.1.1a}$$

$$\nabla \cdot \mathbf{B}(\mathbf{r}, t) = 0 \quad , \tag{2.1.1b}$$

$$\nabla \times \mathbf{B}(\mathbf{r}, t) - \frac{1}{c}\frac{\partial \mathbf{E}(\mathbf{r}, t)}{\partial t} = \frac{4\pi}{c}\mathbf{J}(\mathbf{r}, t) \quad , \tag{2.1.1c}$$

$$\nabla \cdot \mathbf{E}(\mathbf{r}, t) = 4\pi\rho(\mathbf{r}, t) \quad . \tag{2.1.1d}$$

$\mathbf{E}$ and $\mathbf{B}$ are the electric field strength and the magnetic induction, respectively; $c = 2.997\,924\,58 \times 10^8$ m s^{-1} is the velocity of light in free space. The current density $\mathbf{J}$ and the charge density ρ used in this set of equations refer to the total quantities including both the external and induced currents and charge densities; their various components will be discussed in Section 2.2. All quantities are assumed to be spatial, $\mathbf{r}$, and time, t, dependent as indicated by $(\mathbf{r}, t)$. To make the equations more concise, we often do not explicitly include these dependences. Following the notation of most classical books in this field, the equations are written in Gaussian units (cgs), where $\mathbf{E}$ and $\mathbf{B}$ have the same units.[1]

The differential equations (2.1.1a) and (2.1.1b) are satisfied by a vector potential $\mathbf{A}(\mathbf{r}, t)$ and a scalar potential $\Phi(\mathbf{r}, t)$ with

$$\mathbf{B} = \nabla \times \mathbf{A} \tag{2.1.2}$$

and

$$\mathbf{E} + \frac{1}{c}\frac{\partial \mathbf{A}}{\partial t} = -\nabla\Phi \quad . \tag{2.1.3}$$

The first equation expresses the fact that $\mathbf{B}$ is an axial vector and can be expressed as the rotation of a vector field. If the vector potential vanishes ($\mathbf{A} = 0$) or if $\mathbf{A}$ is time independent, the electric field is conservative: the electric field $\mathbf{E}$ is given by the gradient of a potential, as seen in Eq. (2.1.3). Substituting the above expressions into Ampère's law (2.1.1c) and employing the general vector identity

$$\nabla \times (\nabla \times \mathbf{A}) = -\nabla^2\mathbf{A} + \nabla(\nabla \cdot \mathbf{A}) \tag{2.1.4}$$

gives the following equation for the vector potential $\mathbf{A}$:

$$\nabla^2\mathbf{A} - \frac{1}{c^2}\frac{\partial^2\mathbf{A}}{\partial t^2} = -\frac{4\pi}{c}\mathbf{J} + \frac{1}{c}\nabla\frac{\partial\Phi}{\partial t} + \nabla(\nabla \cdot \mathbf{A}) \quad ; \tag{2.1.5}$$

a characteristic wave equation combining the second time derivative and the second

---

[1] For a discussion of the conversion to rational SI units (mks), see for example [Bec64, Jac75]. See also Tables G.1 and G.3.

spatial derivative. As this equation connects the current to the scalar and vector potentials, we can find a corresponding relationship between the charge density and the potentials. Substituting Eq. (2.1.3) into Coulomb's law (2.1.1d) yields

$$\nabla \cdot \mathbf{E} = 4\pi\rho = -\nabla^2\Phi - \frac{1}{c}\frac{\partial}{\partial t}(\nabla \cdot \mathbf{A}) \quad .$$

Using the Coulomb gauge[2]

$$\nabla \cdot \mathbf{A} = 0 \quad , \tag{2.1.6}$$

the last term in this equation vanishes; the remaining part yields Poisson's equation

$$\nabla^2\Phi = -4\pi\rho \quad , \tag{2.1.7}$$

expressing the fact that the scalar potential $\Phi(\mathbf{r}, t)$ is solely determined by the charge distribution $\rho(\mathbf{r}, t)$. From Eq. (2.1.1c) and by using the definition of the vector potential, we obtain (in the case of static fields) a similar relation for the vector potential

$$\nabla^2\mathbf{A} = -\frac{4\pi}{c}\mathbf{J} \quad , \tag{2.1.8}$$

connecting only $\mathbf{A}(\mathbf{r}, t)$ to the current density $\mathbf{J}(\mathbf{r}, t)$. Employing the vector identity $\nabla \cdot (\nabla \times \mathbf{B}) = 0$, we can combine the derivatives of Eqs (2.1.1c) and (2.1.1d) to obtain the continuity equation for electric charge

$$\frac{\partial\rho}{\partial t} = -\nabla \cdot \mathbf{J} \quad , \tag{2.1.9}$$

expressing the fact that the time evolution of the charge at any position is related to a current at the same location.

Equation (2.1.7) and Eq. (2.1.9) have been obtained by combining Maxwell's equations in the absence of matter, without making any assumptions about a particular time or spatial form of the fields. In the following we consider a harmonic time and spatial dependence of the fields and waves. A monochromatic plane

---

[2] Another common choice of gauge is the Lorentz convention $\nabla \cdot \mathbf{A} + \frac{1}{c}\frac{\partial\Phi}{\partial t} = 0$, which gives symmetric wave equations for the scalar and vector potentials:

$$\nabla^2\Phi - \frac{1}{c^2}\frac{\partial^2\Phi}{\partial t^2} = -4\pi\rho \quad ,$$

$$\nabla^2\mathbf{A} - \frac{1}{c^2}\frac{\partial^2\mathbf{A}}{\partial t^2} = -\frac{4\pi}{c}\mathbf{J} \quad ;$$

or – in the case of superconductors – the London gauge, assuming $\nabla^2\Phi = 0$. For more details on the properties of various gauges and the selection of an appropriate one, see for example [Jac75, Por92]. Our choice restricts us to non-relativistic electrodynamics; relativistic effects, however, can safely be neglected throughout the book.

electric wave of frequency $f = \omega/2\pi$ traveling in a certain direction (given by the wavevector $\mathbf{q}$) can then be written as

$$
\begin{aligned}
\mathbf{E}(\mathbf{r}, t) &= \mathbf{E}_{01} \sin\{\mathbf{q} \cdot \mathbf{r} - \omega t\} + \mathbf{E}_{02} \cos\{\mathbf{q} \cdot \mathbf{r} - \omega t\} \\
&= \mathbf{E}_{03} \sin\{\mathbf{q} \cdot \mathbf{r} - \omega t + \psi\} \quad ,
\end{aligned}
\tag{2.1.10}
$$

where $\mathbf{E}_{0i}$ ($i = 1, 2, 3$) describe the maximum amplitude; but it is more convenient to write the electric field $\mathbf{E}(\mathbf{r}, t)$ as a complex quantity

$$
\mathbf{E}(\mathbf{r}, t) = \mathbf{E}_0 \exp\{i(\mathbf{q} \cdot \mathbf{r} - \omega t)\} \quad .
\tag{2.1.11}
$$

We should keep in mind, however, that only the real part of the electric field is a meaningful quantity. We explicitly indicate the complex nature of $\mathbf{E}$ and the possible phase factors only if they are of interest to the discussion. A few notes are in order: the electric field is a vector, and therefore its direction as well as its value is of importance. As we shall discuss in Section 3.1 in more detail, we distinguish longitudinal and transverse components with respect to the direction of propagation; any transverse field polarization can be obtained as the sum of two orthogonal transverse components. If the ratio of both is constant, linearly polarized fields result, otherwise an elliptical or circular polarization can be obtained. Similar considerations to those presented here for the electric field apply for the magnetic induction $\mathbf{B}(\mathbf{r}, t)$ and other quantities. The sign in the exponent is chosen such that the wave travels along the $+\mathbf{r}$ direction.[3] We assume that all fields and sources can be decomposed into a complete continuous set of such plane waves:

$$
\mathbf{E}(\mathbf{r}, t) = \frac{1}{(2\pi)^4} \int \int_{-\infty}^{\infty} \mathbf{E}(\mathbf{q}, \omega) \exp\{i(\mathbf{q} \cdot \mathbf{r} - \omega t)\} \, d\omega \, d\mathbf{q} \quad ,
\tag{2.1.12}
$$

and the four-dimensional Fourier transform of the electric field strength $\mathbf{E}(\mathbf{r}, t)$ is

$$
\mathbf{E}(\mathbf{q}, \omega) = \int \int_{-\infty}^{\infty} \mathbf{E}(\mathbf{r}, t) \exp\{-i(\mathbf{q} \cdot \mathbf{r} - \omega t)\} \, dt \, d\mathbf{r} \quad .
\tag{2.1.13}
$$

Analogous equations and transformations apply to the magnetic induction $\mathbf{B}$. Some basic properties of the Fourier transformation are discussed in Appendix A.1. Equation (2.1.12), which assumes plane waves, requires that a wavevector $\mathbf{q}$ and a frequency ω can be well defined. The wavelength $\lambda = 2\pi/|\mathbf{q}|$ should be much smaller than the relevant dimensions of the problem, and therefore the finite sample size is neglected; also the period of the radiation should be much shorter than the typical time scale over which this radiation is applied. We assume, for the moment, that the spatial dependence of the radiation always remains that of a plane wave.

---

[3]　For $\exp\{i(\mathbf{q} \cdot \mathbf{r} + \omega t)\}$ and $\exp\{-i(\mathbf{q} \cdot \mathbf{r} + \omega t)\}$ the wave travels to the left ($-\mathbf{r}$), while $\exp\{i(\mathbf{q} \cdot \mathbf{r} - \omega t)\}$ and $\exp\{-i(\mathbf{q} \cdot \mathbf{r} - \omega t)\}$ describe right moving waves ($+\mathbf{r}$). With the convention $\exp\{-i\omega t\}$ the optical constants $\hat{\epsilon}$ and $\hat{N}$ have positive imaginary parts, as we shall find out in Section 2.2. The notation $\exp\{i\omega t\}$ is also common and leads to $\hat{N} = n - ik$ and $\hat{\epsilon} = \epsilon_1 - i\epsilon_2$.

Using the assumption of harmonic waves of the form (2.1.12), the spatial and time derivatives of the electric field strength $\mathbf{E}(\mathbf{r}, t)$ can be calculated as

$$
\begin{aligned}
\nabla \times \mathbf{E}(\mathbf{r}, t) &= i\mathbf{q} \times \mathbf{E}(\mathbf{r}, t) \\
\nabla \cdot \mathbf{E}(\mathbf{r}, t) &= i\mathbf{q} \cdot \mathbf{E}(\mathbf{r}, t) \\
\frac{\partial \mathbf{E}(\mathbf{r}, t)}{\partial t} &= -i\omega \mathbf{E}(\mathbf{r}, t) \quad ,
\end{aligned}
\tag{2.1.14}
$$

with the same relations found for the magnetic induction $\mathbf{B}(\mathbf{r}, t)$. The spatial and time periodicity of the radiation can now be utilized to write Maxwell's equations in Fourier transformed form:

$$
\mathbf{q} \times \mathbf{E}(\mathbf{q}, \omega) - \frac{\omega}{c}\mathbf{B}(\mathbf{q}, \omega) = 0 \quad ,
\tag{2.1.15a}
$$

$$
\mathbf{q} \cdot \mathbf{B}(\mathbf{q}, \omega) = 0 \quad ,
\tag{2.1.15b}
$$

$$
i\mathbf{q} \times \mathbf{B}(\mathbf{q}, \omega) + i\frac{\omega}{c}\mathbf{E}(\mathbf{q}, \omega) = \frac{4\pi}{c}\mathbf{J}(\mathbf{q}, \omega) \quad ,
\tag{2.1.15c}
$$

$$
i\mathbf{q} \cdot \mathbf{E}(\mathbf{q}, \omega) = 4\pi\rho(\mathbf{q}, \omega) \quad .
\tag{2.1.15d}
$$

The vector and scalar potentials, $\mathbf{A}$ and Φ, can also be converted to Fourier space, and from Eqs (2.1.2), (2.1.3), and (2.1.7) we then obtain

$$
\mathbf{B}(\mathbf{q}, \omega) = i\mathbf{q} \times \mathbf{A}(\mathbf{q}, \omega) \quad ,
\tag{2.1.16}
$$

$$
\mathbf{E}(\mathbf{q}, \omega) = i\frac{\omega}{c}\mathbf{A}(\mathbf{q}, \omega) - i\mathbf{q}\Phi(\mathbf{q}, \omega) \quad ,
\tag{2.1.17}
$$

$$
q^2\Phi(\mathbf{q}, \omega) = 4\pi\rho(\mathbf{q}, \omega) \quad .
\tag{2.1.18}
$$

As mentioned above, the assumption of harmonic waves is not a restriction for most cases; since the transformation to Fourier space turns out to be particularly convenient, we have utilized these relations heavily.

2.1.2 Wave equations in free space

In the absence of matter and in the absence of free current and external charge ($\mathbf{J} = 0$ and $\rho = 0$), the combination of Faraday's induction law (2.1.1a) and Coulomb's law (2.1.1d) yields

$$
\frac{1}{c}\frac{\partial}{\partial t}(\nabla \times \mathbf{B}) = \nabla^2\mathbf{E} \quad ;
$$

using Eq. (2.1.1c), we obtain the following relation for the electric field:

$$
\nabla^2\mathbf{E} - \frac{1}{c^2}\frac{\partial^2\mathbf{E}}{\partial t^2} = 0 \quad .
\tag{2.1.19a}
$$

This is a wave equation in its simplest form without dissipation or other complications; the second spatial derivative is equal to the second derivative in time with

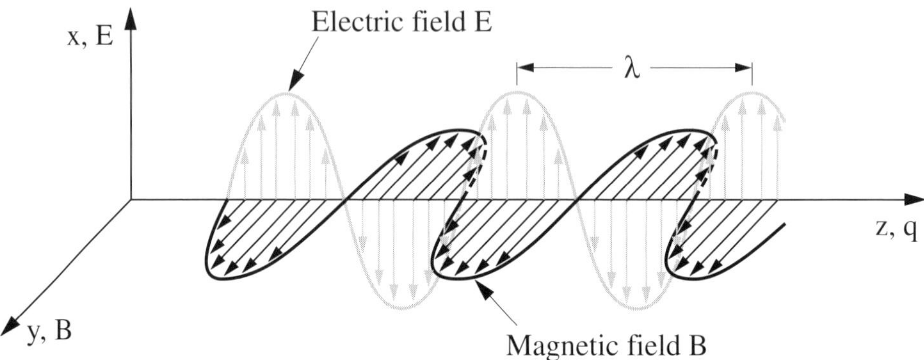

Fig. 2.1. Linearly polarized electromagnetic radiation of wavelength λ propagating along the z direction. Assuming that the electric field $\mathbf{E}$ is along the x direction, the magnetic field $\mathbf{B}$ points in the y direction and the wavevector $\mathbf{q}$ is directed along the z axis.

the square of the velocity as proportionality factor. One possible solution of this differential equation is given by a harmonic wave $\mathbf{E}(\mathbf{r}, t) = \mathbf{E}_0 \exp\{\mathrm{i}(\mathbf{q} \cdot \mathbf{r} - \omega t)\}$ as suggested in Eq. (2.1.11). Similarly, the value of the wavevector $\mathbf{q}$ is simply ω/c. The corresponding wave equation for the $\mathbf{B}$ field can easily be derived from Maxwell's equations and has the same form:

$$\nabla^2 \mathbf{B} - \frac{1}{c^2} \frac{\partial^2 \mathbf{B}}{\partial t^2} = 0 \quad . \tag{2.1.19b}$$

A linearly polarized electromagnetic wave is shown in Fig. 2.1, with the notations and conventions adopted in this book. According to Eqs (2.1.15a) and (2.1.15c) the three vectors $\mathbf{E}$, $\mathbf{B}$, and $\mathbf{q}$ are perpendicular to one another. The electric and magnetic fields are oriented in the x and y directions, respectively; the wave propagates along the z axis. Note that the electric and magnetic fields are oscillating in phase as seen from Eq. (2.1.15a). At any given $\mathbf{r}$ and t the magnetic field is zero when the electric field vanishes; because $\mathbf{J} = 0$ in free space, we see from Eq. (2.1.15c) that $\mathbf{E}$ is also zero when there is no $\mathbf{B}$ field. Since the equation of an undamped wave contains only the second spatial and time derivatives, the choice of the particular sign in the exponent of Eq. (2.1.11) has no significance: left- and right-going waves satisfy the wave equation.

There is an energy associated with the creation of electric and magnetic fields; the time average density of the electric field energy is $u_{\mathrm{e}} = \frac{1}{4\pi} \langle \mathbf{E} \rangle_t^2$ and of the magnetic field $u_{\mathrm{m}} = \frac{1}{4\pi} \langle \mathbf{B} \rangle_t^2$. A traveling electromagnetic wave is associated with transport of electromagnetic energy. There will be no dissipation of energy (or attenuation of the waves) as long as we do not consider damping effects. The energy density, i.e. the energy in a unit volume, associated with a harmonic wave is written as

$$u = \frac{1}{8\pi} E^2 + \frac{1}{8\pi} B^2 \quad . \tag{2.1.20}$$

In the case of free space $|\mathbf{E}| = |\mathbf{B}|$, and the energies associated with the electric and magnetic fields are equal ($u_{\mathrm{e}} = u_{\mathrm{m}}$). Let us define the vector

$$\mathbf{S} = \frac{c}{4\pi} \mathbf{E} \times \mathbf{B} \quad , \tag{2.1.21}$$

called the Poynting vector, which is directed along the direction of propagation. It describes the energy transported, i.e. the flow of energy per unit area and per unit time, as can be easily seen by

$$\nabla \cdot \mathbf{S} = -\frac{du}{dt} \quad , \tag{2.1.22}$$

an expression of energy conservation. We will use these relations in Section 2.3.1 when the energy dissipated in the medium is discussed.

2.2 Propagation of electromagnetic waves in the medium

Because we are interested in the optical properties of solids, we now have to discuss the influence of matter on the wave propagation. First we will define some material parameters and see how they enter Maxwell's equations.

2.2.1 Definitions of material parameters

The presence of a medium in electric and magnetic fields may lead to electric dipoles and magnetic moments, polarization charges, and induced current. Clearly the electric and magnetic fields will not be uniform within the material but fluctuate from point to point reflecting the periodicity of the atomic lattice. For wavelengths appreciably larger than the atomic spacing we can nevertheless consider an average value of the electric and magnetic fields. These fields, however, are different compared with the fields in vacuum; consequently the electric displacement $\mathbf{D}$ instead of the electric field strength $\mathbf{E}$, and the magnetic induction $\mathbf{B}$ associated with the magnetic field strength $\mathbf{H}$, are introduced to account for the modifications by the medium.

The total charge density $\rho = \rho_{\mathrm{total}}$ used in Coulomb's law (2.1.1d) now has two components

$$\rho_{\mathrm{total}} = \rho_{\mathrm{ext}} + \rho_{\mathrm{pol}} \quad , \tag{2.2.1}$$

an external charge ρ_{ext} added from outside and a contribution due to the spatially varying polarization

$$\rho_{\mathrm{pol}} = -\nabla \cdot \mathbf{P} \quad . \tag{2.2.2}$$

For a homogeneous polarization the positive and negative charges cancel everywhere inside the material, leading to no net charge ρ_{pol} in our limit at large enough wavelengths.

Let us assume that there is no external current present: $\mathbf{J}_{\text{ext}} = 0$. The total current density $\mathbf{J} = \mathbf{J}_{\text{total}}$ entering Maxwell's equation then consists of a contribution $\mathbf{J}_{\text{cond}}$ arising from the motion of electrons in the presence of an electric field and of a contribution $\mathbf{J}_{\text{bound}}$ arising from the redistribution of bound charges:

$$\mathbf{J}_{\text{total}} = \mathbf{J}_{\text{cond}} + \mathbf{J}_{\text{bound}} \quad . \tag{2.2.3}$$

Ohm's law, i.e. current proportional to electric field $\mathbf{E}$, is assumed to apply to this conduction current:

$$\mathbf{J}_{\text{cond}} = \sigma_1 \mathbf{E} \quad . \tag{2.2.4}$$

σ_1 is the conductivity of the material; since we limit ourselves to the linear response, σ_1 does not depend on the electric field strength. We also assume it should not be changed by an external magnetic field, implying that we do not consider magnetoresistive effects. The time dependent polarization $\partial \mathbf{P}/\partial t$ or spatially dependent magnetization $\nabla \times \mathbf{M}$ contribute to what is called the displacement current or bound current density $\mathbf{J}_{\text{bound}}$.

Let us now include these terms explicitly into Maxwell's equations (2.1.1). The first two equations do not change because they contain neither current nor charge densities. Upon substitution, Eq. (2.1.1c) becomes

$$\nabla \times \mathbf{B}(\mathbf{r}, t) - \frac{1}{c} \frac{\partial \mathbf{E}(\mathbf{r}, t)}{\partial t} = \frac{4\pi}{c} \mathbf{J}_{\text{total}}(\mathbf{r}, t)$$

$$= \frac{4\pi}{c} \left[\mathbf{J}_{\text{cond}}(\mathbf{r}, t) + \frac{\partial \mathbf{P}(\mathbf{r}, t)}{\partial t} + c\nabla \times \mathbf{M}(\mathbf{r}, t) \right]$$

and Eq. (2.1.1d) changes to

$$\nabla \cdot \mathbf{E}(\mathbf{r}, t) = 4\pi \rho_{\text{total}}(\mathbf{r}, t) = 4\pi \left[\rho_{\text{ext}}(\mathbf{r}, t) - \nabla \cdot \mathbf{P}(\mathbf{r}, t) \right] \quad .$$

These reformulations of Maxwell's equations suggest the definitions of the electric displacement $\mathbf{D}$ and magnetic field strength $\mathbf{H}$. The electric field strength $\mathbf{E}$ and the electric displacement (or electric flux density) $\mathbf{D}$ are connected by the dielectric constant (or permittivity) ϵ_1:

$$\mathbf{D} = \epsilon_1 \mathbf{E} = (1 + 4\pi \chi_e)\mathbf{E} = \mathbf{E} + 4\pi \mathbf{P} \quad , \tag{2.2.5}$$

where χ_e is the dielectric susceptibility and $\mathbf{P} = \chi_e \mathbf{E}$ is the dipole moment density or polarization density. The dielectric constant ϵ_1 can be either positive or negative. Similarly, the magnetic field strength $\mathbf{H}$ is connected to the magnetic induction $\mathbf{B}$

by the permeability μ_1:

$$\mathbf{B} = \mu_1 \mathbf{H} = (1 + 4\pi\chi_{\mathrm{m}})\mathbf{H} = \mathbf{H} + 4\pi\mathbf{M} \quad , \tag{2.2.6}$$

where χ_{m} is the magnetic susceptibility and $\mathbf{M} = \chi_{\mathrm{m}}\mathbf{H}$ is the magnetic moment density, or magnetization. The quantities ϵ_1, χ_{e}, μ_1, and χ_{m} which connect the fields are unitless. The magnetic susceptibility χ_{m} is typically four to five orders of magnitude smaller (except in the case of ferromagnetism) than the dielectric susceptibility χ_{e}, which is of the order of unity. For this reason the dia- and para-magnetic properties can in general be neglected compared to the dielectric properties when electromagnetic waves pass through a medium. Throughout the book we shall not discuss the properties of magnetic materials and therefore we assume that $\mu_1 = 1$ (unless explicitly indicated otherwise), although we include it in some equations to make them more general. We assume furthermore that there are no magnetic losses and consequently that the imaginary part of the permeability is zero.

2.2.2 Maxwell's equations in the presence of matter

With these definitions we can rewrite Maxwell's equations (2.1.1) in the presence of matter. The rearrangement of the terms and some substitution yield

$$\nabla \times \mathbf{E} + \frac{1}{c}\frac{\partial \mathbf{B}}{\partial t} = 0 \quad , \tag{2.2.7a}$$

$$\nabla \cdot \mathbf{B} = 0 \quad , \tag{2.2.7b}$$

$$\nabla \times \mathbf{H} - \frac{1}{c}\frac{\partial \mathbf{D}}{\partial t} = \frac{4\pi}{c}\mathbf{J}_{\mathrm{cond}} \quad , \tag{2.2.7c}$$

$$\nabla \cdot \mathbf{D} = 4\pi\rho_{\mathrm{ext}} \quad . \tag{2.2.7d}$$

Using Eqs (2.2.5) and Ohm's law (2.2.4) and recalling that there is no external current, Eq. (2.2.7c) can be written as

$$c\nabla \times \mathbf{H} = -\mathrm{i}\omega\epsilon_1\mathbf{E} + 4\pi\sigma_1\mathbf{E} = -\mathrm{i}\omega\hat{\epsilon}\mathbf{E} \quad ,$$

where we have assumed a harmonic time dependence of the displacement term $\partial\mathbf{D}/\partial t = -\mathrm{i}\omega\mathbf{D}$, and we have defined the complex dielectric quantity[4]

$$\hat{\epsilon} = \epsilon_1 + \mathrm{i}\frac{4\pi\sigma_1}{\omega} = \epsilon_1 + \mathrm{i}\epsilon_2 \quad . \tag{2.2.8}$$

By writing $\mathbf{D} = \hat{\epsilon}\mathbf{E}$, the change in magnitude and the phase shift between the

---

[4] Whenever we need to distinguish complex quantities from real ones, we indicate complex quantities with a circumflex, and their real and imaginary parts by subscripts 1 and 2, except if they have well established names.

displacement $\mathbf{D}$ and the electric field $\mathbf{E}$ are conveniently expressed. According to Eq. (2.1.10) with $\psi = 0$ the electric field reads $\mathbf{E} = \mathbf{E}_0 \sin\{\mathbf{q} \cdot \mathbf{r} - \omega t\}$, and we can write for the displacement

$$
\begin{aligned}
\mathbf{D} &= \mathbf{D}_0 \sin\{\mathbf{q} \cdot \mathbf{r} - \omega t + \delta(\omega)\} \\
&= \mathbf{D}_0 \left[\sin\{\mathbf{q} \cdot \mathbf{r} - \omega t\} \cos\{\delta(\omega)\} + \cos\{\mathbf{q} \cdot \mathbf{r} - \omega t\} \sin\{\delta(\omega)\}\right] \\
&= \epsilon_1 \mathbf{E}_0 \sin\{\mathbf{q} \cdot \mathbf{r} - \omega t\} + \epsilon_2 \mathbf{E}_0 \cos\{\mathbf{q} \cdot \mathbf{r} - \omega t\}
\end{aligned}
\tag{2.2.9}
$$

to demonstrate that ϵ_1 and ϵ_2 span a phase angle of $\pi/2$. Here ϵ_1 is the in-phase and ϵ_2 is the out-of-phase component; we come back to the loss angle δ in Eq. (2.3.25). The notation accounts for the general fact that the response of the medium can have a time delay with respect to the applied perturbation. Similarly the conductivity can be assumed to be complex

$$
\hat{\sigma} = \sigma_1 + i\sigma_2
\tag{2.2.10}
$$

to include the phase shift of the conduction and the bound current, leading to a more general Ohm's law

$$
\mathbf{J}_{\mathrm{tot}} = \hat{\sigma}\mathbf{E} \quad ;
\tag{2.2.11}
$$

and we define[5] the relation between the complex conductivity and the complex dielectric constant as

$$
\hat{\epsilon} = 1 + \frac{4\pi i}{\omega}\hat{\sigma} \quad .
\tag{2.2.12}
$$

Besides the interchange of real and imaginary parts in the conductivity and dielectric constant, the division by ω becomes important in the limits $\omega \to 0$ and $\omega \to \infty$ to avoid diverging functions. Although we regard the material parameters ϵ_1 and σ_1 as the two fundamental components of the response to electrodynamic fields, in subsequent sections we will use a complex response function, in most cases the conductivity $\hat{\sigma}$, when the optical properties of solids are discussed.

A number of restrictions apply to this concept. For example, the dielectric constant ϵ_1 introduced by Eq. (2.2.5) is in general not constant but a function of both spatial and time variables. It may not just be a number but a response function or linear integral operator which connects the displacement field $\mathbf{D}(\mathbf{r}, t)$ with the electric field $\mathbf{E}(\mathbf{r}', t')$ existing at all other positions $\mathbf{r}'$ and times t earlier than t'

$$
\mathbf{D}(\mathbf{r}, t) = \int \int_{-\infty}^{t} \epsilon_1(\mathbf{r}, \mathbf{r}', t') \mathbf{E}(\mathbf{r}', t')\, \mathrm{d}t'\mathrm{d}\mathbf{r}' \quad .
\tag{2.2.13}
$$

The consequence of this stimulus response relation is discussed in more detail in

---

[5] Sometimes the definition $\sigma_2 = -\omega\epsilon_1/(4\pi)$ is used, leading to $\hat{\epsilon} = 4\pi i\hat{\sigma}/\omega$. At low frequencies the difference between the two definitions is negligible. At high frequencies σ_2 has to be large in order to obtain $\epsilon_1 = 1$ in the case of a vacuum.

Section 3.2. In the general case of an anisotropic medium, the material parameters ϵ_1, μ_1, and σ_1 may be direction dependent and have to be represented as tensors, leading to the fact that the displacement field does not point in the same direction as the electric field. For high electric and magnetic fields, the material parameters ϵ_1, μ_1, and σ_1 may even depend on the field strength; in such cases higher order terms of a Taylor expansion of the parameters have to be taken into account to describe the non-linear effects.[6] The dielectric properties $\hat{\epsilon}$ can also depend on external magnetic fields [Lan84, Rik96], and of course the polarization can change due to an applied magnetic field (Faraday effect, Kerr effect); we will not consider these magneto-optical phenomena. We will concentrate on homogeneous[7] and isotropic[8] media with $\mu_1 = 1$, ϵ_1, and σ_1 independent of field strength and time.

2.2.3 Wave equations in the medium

To find a solution of Maxwell's equations we consider an infinite medium to avoid boundary and edge effects. Furthermore we assume the absence of free charges ($\rho_{\text{ext}} = 0$) and external currents ($\mathbf{J}_{\text{ext}} = 0$). As we did in the case of a vacuum we use a sinusoidal periodic time and spatial dependence for the electric and magnetic waves. Thus,

$$\mathbf{E}(\mathbf{r}, t) = \mathbf{E}_0 \exp\{i(\mathbf{q} \cdot \mathbf{r} - \omega t)\} \qquad (2.2.14a)$$

and

$$\mathbf{H}(\mathbf{r}, t) = \mathbf{H}_0 \exp\{i(\mathbf{q} \cdot \mathbf{r} - \omega t - \phi)\} \qquad (2.2.14b)$$

describe the electric and magnetic fields with wavevector $\mathbf{q}$ and frequency ω. We have included a phase factor ϕ to indicate that the electric and magnetic fields may be shifted in phase with respect to each other; later on we have to discuss in detail what this actually means. As we will soon see, the wavevector $\mathbf{q}$ has to be a complex quantity: to describe the spatial dependence of the wave it has to include a propagation as well as an attenuation part. Using the vector identity (2.1.4) and with Maxwell's equations (2.2.7a) and (2.2.7d), we can separate the magnetic and electric components to obtain

$$\frac{1}{c}\frac{\partial}{\partial t}(\nabla \times \mathbf{B}) = \nabla^2 \mathbf{E} - \nabla\left(\frac{4\pi\rho_{\text{ext}}}{\epsilon_1}\right) \ . \qquad (2.2.15)$$

By substituting the three materials equations (2.2.4)–(2.2.6) into Ampère's law (2.2.7c), we arrive at $\nabla \times \mathbf{B} = (\epsilon_1\mu_1/c)(\partial\mathbf{E}/\partial t) + (4\pi\mu_1\sigma_1/c)\mathbf{E}$. Combining

---

[6] A discussion of non-linear optical effects can be found in [Blo65, But91, Mil91, She84].

[7] The case of inhomogeneous media is treated in [Ber78, Lan78].

[8] Anisotropic media are the subject of crystal optics and are discussed in many textbooks, e.g. [Gay67, Lan84, Nye57].

this with Eq. (2.2.15) eventually leads to the wave equation for the electric field

$$\nabla^2 \mathbf{E} - \frac{\epsilon_1 \mu_1}{c^2} \frac{\partial^2 \mathbf{E}}{\partial t^2} - \frac{4\pi \mu_1 \sigma_1}{c^2} \frac{\partial \mathbf{E}}{\partial t} = 0 \quad , \tag{2.2.16a}$$

if $\rho_{\text{ext}} = 0$ is assumed. In Eq. (2.2.16a) the second term represents Maxwell's displacement current; the last term is due to the conduction current. Similarly we obtain the equation

$$\nabla^2 \mathbf{H} - \frac{\epsilon_1 \mu_1}{c^2} \frac{\partial^2 \mathbf{H}}{\partial t^2} - \frac{4\pi \mu_1 \sigma_1}{c^2} \frac{\partial \mathbf{H}}{\partial t} = 0 \quad , \tag{2.2.16b}$$

describing the propagation of the magnetic field. In both equations the description contains an additional factor compared to that of free space, which is proportional to the first time derivative of the fields. Of course, we could have derived an equation for $\mathbf{B}$. As mentioned above, we neglect magnetic losses. From Eq. (2.2.7b) we can immediately conclude that $\mathbf{H}$ always has only transverse components. The electric field may have longitudinal components in certain cases, for from Eq. (2.2.7d) we find $\nabla \cdot \mathbf{E} = 0$ only in the absence of a net charge density.

If Faraday's law is expressed in $\mathbf{q}$ space, Eq. (2.1.15a), we immediately see that for a plane wave both the electric field $\mathbf{E}$ and the direction of the propagation vector $\mathbf{q}$ are perpendicular to the magnetic field $\mathbf{H}$, which can be written as

$$\mathbf{H}(\mathbf{q}, \omega) = \frac{c}{\mu_1 \omega} \mathbf{q} \times \mathbf{E}(\mathbf{q}, \omega) \quad . \tag{2.2.17}$$

$\mathbf{E}$, however, is not necessarily perpendicular to $\mathbf{q}$. Without explicitly solving the wave equations (2.2.16), we already see from Eq. (2.2.17) that if matter is present with finite dissipation ($\sigma_1 \neq 0$) – where the wavevector is complex – there is a phase shift between the electric field and magnetic field. This will become clearer when we solve the wave equations for monochromatic radiation. Substituting Eq. (2.2.14a) into Eq. (2.2.16a), for example, we obtain the following dispersion relation between the wavevector $\mathbf{q}$ and the frequency ω:

$$\mathbf{q} = \frac{\omega}{c} \left[\epsilon_1 \mu_1 + i \frac{4\pi \mu_1 \sigma_1}{\omega} \right]^{1/2} \mathbf{n_q} \quad , \tag{2.2.18}$$

where $\mathbf{n_q} = \mathbf{q}/|\mathbf{q}|$ is the unit vector along the $\mathbf{q}$ direction. Note that we have made the assumption that no net charge ρ_{ext} is present: i.e. $\nabla \cdot \mathbf{E} = 0$. A complex wavevector $\mathbf{q}$ is a compact way of expressing the fact that a wave propagating in the $\mathbf{n_q}$ direction experiences a change in wavelength and an attenuation in the medium compared to when it is in free space. We will discuss this further in Section 2.3.1. The propagation of the electric and magnetic fields (Eqs (2.2.16)) can now be written in Helmholtz's compact form of the wave equation:

$$(\nabla^2 + \mathbf{q}^2)\mathbf{E} = 0 \quad \text{and} \quad (\nabla^2 + \mathbf{q}^2)\mathbf{H} = 0 \quad . \tag{2.2.19}$$

It should be pointed out that the propagation of the electric and magnetic fields is described by the same wavevector $\mathbf{q}$; however, there may be a phase shift with respect to each other ($\phi \neq 0$).

In the case of a medium with negligible electric losses ($\sigma_1 = 0$), Eqs (2.2.16) are reduced to the familiar wave equations (2.1.19) describing propagating electric and magnetic fields in the medium:

$$\nabla^2 \mathbf{E} - \frac{\epsilon_1 \mu_1}{c^2}\frac{\partial^2 \mathbf{E}}{\partial t^2} = 0 \quad \text{and} \quad \nabla^2 \mathbf{H} - \frac{\epsilon_1 \mu_1}{c^2}\frac{\partial^2 \mathbf{H}}{\partial t^2} = 0 \quad . \quad (2.2.20)$$

There are no variations of the magnitude of $\mathbf{E}$ and $\mathbf{H}$ inside the material; however, the velocity of propagation has changed by $(\epsilon_1 \mu_1)^{1/2}$ compared to when it is in a vacuum. We immediately see from Eq. (2.2.18) that the wavevector $\mathbf{q}$ is real for non-conducting materials; Eq. (2.2.17) and the corresponding equation for the electric field then become

$$\mathbf{H} = \left(\frac{\epsilon_1}{\mu_1}\right)^{1/2} \mathbf{n_q} \times \mathbf{E} \quad \text{and} \quad \mathbf{E} = \left(\frac{\mu_1}{\epsilon_1}\right)^{1/2} \mathbf{n_q} \times \mathbf{H} \quad , \quad (2.2.21)$$

indicating that both quantities are zero at the same time and at the same location and thus $\phi = 0$ as sketched in Fig. 2.1. The solutions of Eqs (2.2.20) are restricted to transverse waves. In the case $\sigma_1 = 0$, both $\mathbf{E}$ and $\mathbf{H}$ are perpendicular to the direction of propagation $\mathbf{n_q}$; hence, these waves are called transverse electric and magnetic (TEM) waves.

2.3 Optical constants

2.3.1 Refractive index

The material parameters such as the dielectric constant ϵ_1, the conductivity σ_1, and the permeability μ_1 denote the change of the fields and current when matter is present. Due to convenience and historical reasons, optical constants such as the real refractive index n and the extinction coefficient k are used for the propagation and dissipation of electromagnetic waves in the medium which is characterized by the wavevector (2.2.18). Note that we assume the material to extend indefinitely, i.e. we do not consider finite size or surface effects at this point. To describe the optical properties of the medium, we define the complex refractive index as a new response function

$$\hat{N} = n + \mathrm{i}k = \left[\epsilon_1 \mu_1 + \mathrm{i}\frac{4\pi \mu_1 \sigma_1}{\omega}\right]^{1/2} = [\hat{\epsilon}\mu_1]^{1/2} \quad ; \quad (2.3.1)$$

the value of the complex wavevector $\mathbf{q} = \hat{q}\mathbf{n_q}$ then becomes

$$\hat{q} = \frac{\omega}{c}\hat{N} = \frac{n\omega}{c} + \mathrm{i}\frac{k\omega}{c} \quad , \quad (2.3.2)$$

where the real refractive index n and the extinction coefficient (or attenuation index) k are completely determined by the conductivity σ_1, the permeability μ_1, and the dielectric constant ϵ_1:[9]

$$n^2 = \frac{\mu_1}{2}\left\{\left[\epsilon_1^2 + \left(\frac{4\pi\sigma_1}{\omega}\right)^2\right]^{1/2} + \epsilon_1\right\} \tag{2.3.3}$$

$$k^2 = \frac{\mu_1}{2}\left\{\left[\epsilon_1^2 + \left(\frac{4\pi\sigma_1}{\omega}\right)^2\right]^{1/2} - \epsilon_1\right\} \quad. \tag{2.3.4}$$

These two important relations contain all the information on the propagation of the electromagnetic wave in the material and are utilized throughout the book. The optical constants describe the wave propagation and cannot be used to describe the dc properties of the material. For $\omega = 0$ only ϵ_1, σ_1, and μ_1 are defined. The dielectric constant, permeability, and conductivity are given in terms of n and k:

$$n^2 - k^2 = \epsilon_1\mu_1 \quad, \tag{2.3.5}$$

$$2nk = \frac{4\pi\mu_1\sigma_1}{\omega} \quad; \tag{2.3.6}$$

and Eq. (2.3.1) can be written as

$$\hat{N}^2 = \mu_1\left[\epsilon_1 + i\frac{4\pi\sigma_1}{\omega}\right] = \mu_1\hat{\epsilon} \approx \frac{4\pi i\mu_1\hat{\sigma}}{\omega} \quad, \tag{2.3.7}$$

where the approximation assumes $|\epsilon_1| \gg 1$. In Table 2.1 we list the relationships between the various response functions. In the following chapter we show that the real and imaginary components of $\hat{N}$, $\hat{\epsilon}$, and $\hat{\sigma}$, respectively, are not independent but are connected by causality expressed through the Kramers–Kronig relations (Section 3.2). If $\hat{N}$ is split into an absolute value $|\hat{N}| = (n^2 + k^2)^{1/2}$ and a phase ϕ according to $\hat{N} = |\hat{N}|\exp\{i\phi\}$, then the phase difference ϕ between the magnetic and dielectric field vectors introduced in Eqs (2.2.14) is given by

$$\tan\phi = k/n \quad. \tag{2.3.8}$$

In a perfect insulator or free space, for example, the electric and magnetic fields are in phase and $\phi = 0$ since $k = 0$. In contrast, in a typical metal at low frequencies $\sigma_1 \gg |\sigma_2|$, leading to $n \approx k$ and hence $\phi = 45°$.

Propagation of the electromagnetic wave

Let us now discuss the meaning of the real part of the refractive index n attributed to the wave propagation; the dissipations denoted by k are then the subject of the

---

[9] We consider only the positive sign of the square root to make sure that the absorption expressed by k is always positive.

Table 2.1. *Relationships between the material parameters and optical constants*
$$\hat{\epsilon}, \hat{\sigma}, \text{ and } \hat{N}.$$

The negative sign in the time dependence of the traveling wave $\exp\{-i\omega t\}$ was chosen (cf. Eqs (2.2.14)).

	Dielectric constant $\hat{\epsilon}$	Conductivity $\hat{\sigma}$	Refractive index $\hat{N}$
$\hat{\epsilon}$	$\hat{\epsilon} = \epsilon_1 + i\epsilon_2$	$\epsilon_1 = 1 - \dfrac{4\pi\sigma_2}{\omega}$ $\epsilon_2 = \dfrac{4\pi\sigma_1}{\omega}$	$\epsilon_1 = \dfrac{n^2-k^2}{\mu_1}$ $\epsilon_2 = \dfrac{2nk}{\mu_1}$
$\hat{\sigma}$	$\sigma_1 = \dfrac{\omega\epsilon_2}{4\pi}$ $\sigma_2 = (1-\epsilon_1)\dfrac{\omega}{4\pi}$	$\hat{\sigma} = \sigma_1 + i\sigma_2$	$\sigma_1 = \dfrac{nk\omega}{2\pi\mu_1}$ $\sigma_2 = \left(1 - \dfrac{n^2-k^2}{\mu_1}\right)\dfrac{\omega}{4\pi}$
$\hat{N}$	$n = \left\{\dfrac{\mu_1}{2}\left[\epsilon_1^2 + \epsilon_2^2\right]^{1/2} + \dfrac{\epsilon_1\mu_1}{2}\right\}^{1/2}$ $k = \left\{\dfrac{\mu_1}{2}\left[\epsilon_1^2 + \epsilon_2^2\right]^{1/2} - \dfrac{\epsilon_1\mu_1}{2}\right\}^{1/2}$	$n = \left\{\dfrac{\mu_1}{2}\left[\left(1 - \dfrac{4\pi\sigma_2}{\omega}\right)^2 + \left(\dfrac{4\pi\sigma_1}{\omega}\right)^2\right]^{1/2} + \dfrac{\mu_1}{2} - \dfrac{2\pi\mu_1\sigma_2}{\omega}\right\}^{1/2}$ $k = \left\{\dfrac{\mu_1}{2}\left[\left(1 - \dfrac{4\pi\sigma_2}{\omega}\right)^2 + \left(\dfrac{4\pi\sigma_1}{\omega}\right)^2\right]^{1/2} - \dfrac{\mu_1}{2} + \dfrac{2\pi\mu_1\sigma_2}{\omega}\right\}^{1/2}$	$\hat{N} = n + ik$

following subsections. If the wavevector given by Eq. (2.3.2) is substituted into the equations for the electromagnetic waves (2.2.14), we see the real part of the wavevector relate to the wavelength in a medium by $\text{Re}\{\mathbf{q}\} = q = 2\pi/\lambda$. If $\mu_1 = 1$, the wavelength λ in the medium is given by

$$\lambda = \frac{\lambda_0}{n} \quad ; \tag{2.3.9}$$

except in the vicinity of a strong absorption line (when $n < 1$ is possible in a narrow range of frequency), it is smaller than the wavelength in a vacuum λ_0, and the reduction is given by the factor n.

The phase velocity is simply the ratio of frequency and wavevector

$$v_{\text{ph}} = \frac{\omega}{q} \quad , \tag{2.3.10}$$

while the group velocity is defined as

$$v_{\text{gr}} = \frac{\partial \omega}{\partial q} \quad . \tag{2.3.11}$$

v_{ph} describes the movement of the phase front, v_{gr} can be pictured as the velocity of the center of a wavepackage; in a vacuum $v_{\text{ph}} = v_{\text{gr}} = c$. In general, $v_{\text{ph}} = c/n(\omega)$ can be utilized as a definition of the refractive index n. Experimentally the wavelength of standing waves is used to measure v_{ph}.

Both n and k are always positive, even for $\epsilon_1 < 0$; but as seen from Eq. (2.3.3) the refractive index n becomes smaller than 1 if $\epsilon_1 < 1 - (\pi\sigma_1/\omega)^2$. For materials with $\sigma_1 = 0$, the wavevector $\mathbf{q}$ is real and we obtain $\mathbf{q} = \frac{\omega}{c}(\epsilon_1\mu_1)^{1/2}\mathbf{n_q}$, with the refractive index n given by the so-called Maxwell relation

$$n = (\epsilon_1\mu_1)^{1/2} \tag{2.3.12}$$

(a real quantity and $n \geq 1$). From Eq. (2.3.4) we immediately see that in this case the extinction coefficient vanishes: $k = 0$. On the other hand, for good metals at low frequencies the dielectric contribution becomes less important compared to the conductive contribution $\sigma_1 \gg |\sigma_2|$ (or $|\epsilon_1| \ll \epsilon_2$) and thus

$$k \approx n \approx \left[\frac{2\pi\sigma_1\mu_1}{\omega}\right]^{1/2} = \left[\frac{\epsilon_2\mu_1}{2}\right]^{1/2} \quad . \tag{2.3.13}$$

Attenuation of the electromagnetic wave

Substituting the complex wavevector Eq. (2.3.2) into the expression (2.2.14a) for harmonic waves and decomposing it into real and imaginary parts yields

$$\mathbf{E}(\mathbf{r}, t) = \mathbf{E}_0 \exp\left\{i\omega\left(\frac{n}{c}\mathbf{n_q}\cdot\mathbf{r} - t\right)\right\} \exp\left\{-\frac{\omega k}{c}\mathbf{n_q}\cdot\mathbf{r}\right\} \quad . \tag{2.3.14}$$

Now it becomes obvious that the real part of the complex wavevector $\mathbf{q}$ expresses a traveling wave while the imaginary part takes into account the attenuation. The first exponent of this equation describes the fact that the velocity of light (phase velocity) is reduced from its value in free space c to c/n. The second exponent gives the damping of the wave, $\mathbf{E}(\mathbf{r}) \propto \exp\{-\alpha r/2\} = \exp\{-r/\delta_0\}$. It is the same for electric and magnetic fields because their wavevectors $\mathbf{q}$ are the same. The amplitudes of the fields are reduced by the factor $\exp\{-2\pi k/n\}$ per wavelength λ in the medium (Fig. 2.2). We can define a characteristic length scale for the attenuation of the electromagnetic radiation as the distance over which the field decreases by the factor $1/e$ (with $e = 2.718$):

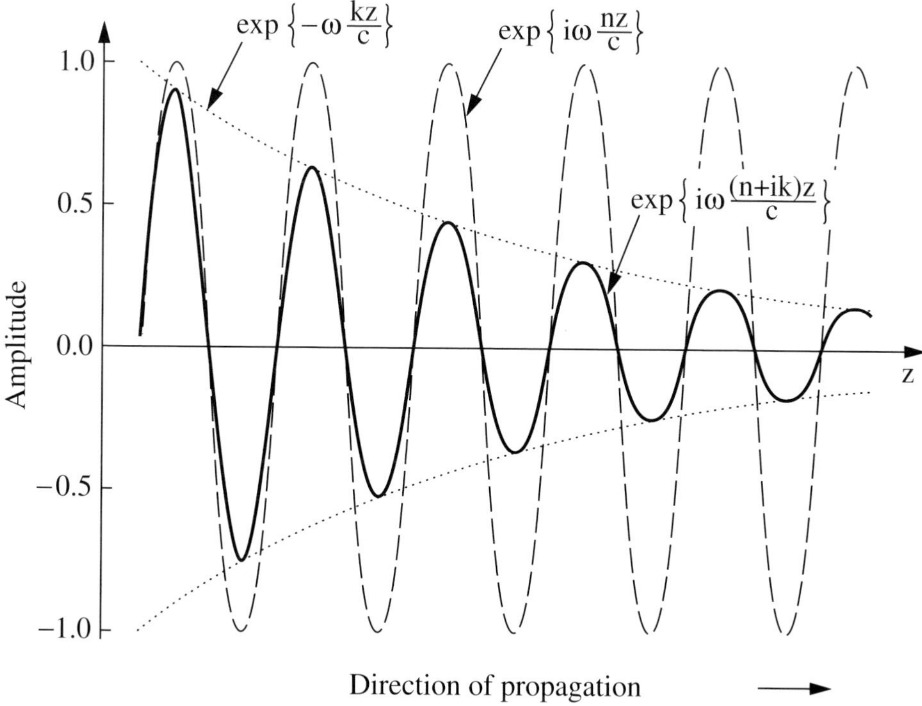

Fig. 2.2. Spatial dependence of the amplitude of a damped wave as described by Eq. (2.3.14) (solid line). The envelope $\exp\{-\frac{\omega k}{c}z\}$ is shown by the dotted lines. The dashed lines represent the undamped harmonic wave $\exp\{i\frac{\omega n}{c}z\}$.

$$\delta_0 = \left(\frac{\alpha}{2}\right)^{-1} = \frac{1}{\mathrm{Im}\{\hat{q}\}} = \frac{c}{\omega k} \tag{2.3.15a}$$

$$= \frac{c}{(2\pi\omega\mu_1)^{1/2}}\left\{\left[\sigma_1^2 + \sigma_2^2\right]^{1/2} + \sigma_2\right\}^{-1/2} \quad , \tag{2.3.15b}$$

where we have used Eq. (2.3.4) and Table 2.1 for the transformation in the limit $\epsilon_1 \approx -4\pi\sigma_2/\omega$. In the limit of $\sigma_1 \gg |\sigma_2|$ the previous expression simplifies to

$$\delta_0 = \left[\frac{c^2}{2\pi\omega\mu_1\sigma_1}\right]^{1/2} \quad , \tag{2.3.16}$$

the so-called classical skin depth of metals. Note that by definition (2.3.15a) δ_0 is the decay length of the electric (or magnetic) field and not just a surface property as inferred by its name. The skin depth is inversely proportional to the square root of the frequency and the conductivity; thus high frequency electromagnetic waves interact with metals only in a very thin surface layer. As seen from the definition of the skin depth, the field penetration depends on the imaginary part of the wavevector $\mathbf{q}$ inside the material; δ_0 is therefore also a measure of the phase shift caused by the material. While for metals the expression (2.3.16) is sufficient, there are cases such as insulators or superconductors where the general formulas (2.3.15) have to be used.

Complementary to the skin depth δ_0, we can define an absorption coefficient $\alpha = 2/\delta_0$ by Lambert–Beer's law[10]

$$\alpha = -\frac{1}{I}\frac{dI}{dr} \tag{2.3.17}$$

to describe the attenuation of the light intensity $I(r) = I_0 \exp\{-\alpha r\}$ propagating in a medium of extinction coefficient k:

$$\alpha = \frac{2k\omega}{c} = \frac{4\pi k}{\lambda_0} = 2\,\mathrm{Im}\{\hat{q}\} \quad . \tag{2.3.18}$$

The power absorption coefficient has the units of an inverse length; α is not a fundamental material parameter, nevertheless it is commonly used because it can be easily measured and intuitively understood. With $k = 2\pi\sigma_1\mu_1/(n\omega)$ from Eq. (2.3.6), we obtain for the absorption coefficient

$$\alpha = \frac{4\pi\sigma_1\mu_1}{nc} \quad , \tag{2.3.19}$$

implying that, for constant n and μ_1, the absorption $\alpha \propto \sigma_1$; i.e. highly conducting materials attenuate radiation strongly. Later we will go beyond this phenomenological description and try to explain the absorption process.

Energy dissipation

Subtracting Eq. (2.2.7c) from Eq. (2.2.7a) after multiplying them by $\mathbf{H}$ and $\mathbf{E}$, respectively, leads to an expression of the energy conservation in the form of

$$\frac{1}{4\pi}\left(\mathbf{E}\cdot\frac{\partial\mathbf{D}}{\partial t} + \mathbf{H}\cdot\frac{\partial\mathbf{B}}{\partial t}\right) + \frac{c}{4\pi}\nabla\cdot(\mathbf{E}\times\mathbf{H}) + \mathbf{J}_{\mathrm{cond}}\cdot\mathbf{E} = 0 \tag{2.3.20}$$

in the absence of an external current $\mathbf{J}_{\mathrm{ext}}$, where we have used the vector relation $\nabla\cdot(\mathbf{E}\times\mathbf{H}) = \mathbf{H}\cdot(\nabla\times\mathbf{E}) - \mathbf{E}\cdot(\nabla\times\mathbf{H})$. The first terms in the brackets of Eq. (2.3.20) correspond to the energy stored in the electric and magnetic fields. In the case of matter being present, the average electric energy density in the field u_{e} and magnetic energy density u_{m} introduced in Eq. (2.1.20) become

$$u = u_{\mathrm{e}} + u_{\mathrm{m}} = \frac{1}{8\pi}(\mathbf{E}\cdot\mathbf{D}) + \frac{1}{8\pi}(\mathbf{H}\cdot\mathbf{B}) \quad . \tag{2.3.21}$$

---

[10] Sometimes the definition of the attenuation coefficient is based on the field strength $\mathbf{E}$ and not the intensity $I = |E|^2$, which makes a factor 2 difference.

While in the case of free space the energy is distributed equally ($u_e = u_m$), this is not valid if matter is present. We recall that the energy transported per area and per time (energy flux density) is given by the Poynting vector $\mathbf{S}$ of the electromagnetic wave:

$$\mathbf{S} = \frac{c}{4\pi}(\mathbf{E} \times \mathbf{H}) \quad ; \tag{2.3.22}$$

for plane waves the Poynting vector is oriented in the direction of the propagation $\mathbf{q}$. The conservation of energy for the electromagnetic wave can now be written as

$$\frac{\mathrm{d}u}{\mathrm{d}t} + \nabla \cdot \mathbf{S} + \mathbf{J}_{\mathrm{cond}} \cdot \mathbf{E} = 0 \quad ; \tag{2.3.23}$$

the limiting case valid for free space with $\mathbf{J}_{\mathrm{cond}} = 0$ has already been derived in Eq. (2.1.22). Hence, the energy of the electromagnetic fields in a given volume either disperses in space or dissipates as Joule heat $\mathbf{J}_{\mathrm{cond}} \cdot \mathbf{E}$. The latter term may be calculated by

$$
\begin{aligned}
\mathbf{J}_{\mathrm{cond}} \cdot \mathbf{E} &= \hat{\sigma} \mathbf{E} \cdot \mathbf{E} \approx \frac{-\mathrm{i}\omega}{4\pi\mu_1} \hat{N}^2 \mathbf{E} \cdot \mathbf{E} \\
&\approx \left[\frac{2nk\omega}{4\pi\mu_1} - \mathrm{i}\frac{\omega}{4\pi\mu_1}(n^2 - k^2) \right] E_0^2 \quad ,
\end{aligned}
\tag{2.3.24}
$$

where we have made use of Eqs (2.2.11) and (2.3.7). The real part of this expression, $P = \sigma_1 E_0^2$, describes the loss of energy per unit time and per unit volume, the absorbed power density; it is related to the absorption coefficient α by $\mathbf{J}_{\mathrm{cond}} \cdot \mathbf{E} = \frac{c}{4\pi}\frac{n}{\mu_1}\alpha E_0^2$. The phase angle between the current density $\mathbf{J}$ and the electric field strength $\mathbf{E}$ is related to the so-called loss tangent already introduced in Eq. (2.2.9) and also defined by $\hat{N}^2 = (n^2 + k^2)\exp\{\mathrm{i}\delta\}$ leading to

$$\tan\delta = \frac{\epsilon_2}{\epsilon_1} = \frac{2nk}{n^2 - k^2} \quad . \tag{2.3.25}$$

The loss tangent denotes the phase angle between the displacement field $\mathbf{D}$ and electric field strength $\mathbf{E}$. It is commonly used in the field of dielectric measurements where it relates the out-of-phase component to the in-phase component of the electric field.

Since the real part of the Poynting vector $\mathbf{S}$ describes the energy flow per unit time across an area perpendicular to the flow, it can also be used to express the energy dissipation. We have to substitute the spatial and time dependence of the electric field (Eq. (2.3.14)) and the corresponding expression for the magnetic field into the time average of Eq. (2.3.22) which describes the intensity of the radiation $\langle S \rangle_t = (c/16\pi)(E_0 H_0^* + E_0^* H_0)$. The attenuation of the wave is then calculated by the time averaged divergency of the energy flow

$$\langle \nabla \cdot \mathbf{S} \rangle_t = \frac{c}{4\pi} \frac{n}{2\mu_1} \frac{-2\omega k}{c} E_0^2 \exp\left\{ -\frac{2\omega k}{c} r \right\} \quad ,$$

leading to

$$\frac{4\pi}{c} \frac{\langle \nabla \cdot \mathbf{S} \rangle_t}{n \langle E \rangle_t^2} = -\frac{\omega k}{c\mu_1} = -\frac{2}{\mu_1} \alpha \quad .$$

The power absorption, as well as being dependent on k, is also dependent on n because the electromagnetic wave travels in the medium at a reduced velocity c/n. We see that the absorption coefficient α is the power absorbed in the unit volume, which we write as $\sigma_1 E^2$, divided by the flux, i.e. the energy density times the energy velocity

$$\alpha = \frac{\sigma_1 \mu_1 \langle \mathbf{E}^2 \rangle}{(\epsilon_1/4\pi)\langle \mathbf{E}^2 \rangle v} = \frac{4\pi \sigma_1 \mu_1}{\epsilon_1 v} \approx \frac{4\pi \sigma_1 \mu_1}{nc} = \frac{\omega \epsilon_2 \mu_1}{nc} \quad , \qquad (2.3.26)$$

where $v = c/n$ is the velocity of light within the medium of the index of refraction n.

2.3.2 Impedance

The refractive index $\hat{N}$ characterizes the propagation of waves in the medium; it is related to a modified wavevector $\mathbf{q}$ compared to the free space value ω/c. Let us now introduce the impedance as another optical constant in order to describe the relationship between the electric and magnetic fields and how it changes upon the wave traveling into matter.

The ratio of the electric field $\mathbf{E}$ to the magnetic field $\mathbf{H}$ at position z defines the load presented to the wave by the medium beyond the point z, and is the impedance

$$\hat{Z}_S = \frac{4\pi}{c} \frac{\hat{E}(z)}{\hat{H}(z)} \quad , \qquad (2.3.27)$$

with units of resistance. The impedance is a response function which determines the relationship between the electric and magnetic fields in a medium. Using Eqs (2.2.14) we can write the complex impedance as $\hat{Z}_S = |\hat{Z}_S| \exp\{i\phi\}$. The absolute value of $|\hat{Z}_S|$ indicates the ratio of the electric and magnetic field amplitudes, while the phase difference between the fields equals the phase angle ϕ. Since $\mathbf{E}$ and $\mathbf{H}$ are perpendicular to each other, using Eq. (2.2.17) we obtain

$$\hat{\mathbf{H}} = \frac{c}{i\omega\mu_1} \frac{\partial \hat{\mathbf{E}}}{\partial z} = \frac{\hat{N}}{\mu_1} \hat{\mathbf{E}}$$

if a harmonic spatial and time dependence of the fields is assumed as given by Eqs (2.2.14). This leads to the following expressions for the impedance:

$$\hat{Z}_S = \frac{4\pi}{c}\frac{\mu_1}{\hat{N}} = \frac{4\pi}{c^2}\frac{\mu_1\omega}{\hat{q}} = \frac{4\pi}{c}\left(\frac{\mu_1}{\hat{\epsilon}}\right)^{1/2} \quad . \tag{2.3.28}$$

If $\sigma_1 = 0$, the dielectric constant $\hat{\epsilon}$ and thus the impedance are real; in this case the electric displacement and magnetic induction fields are equal ($\hat{D} = \hat{B}$) and in phase ($\phi = 0$). The above equation is then reduced to $Z_S = (4\pi/c)(\mu_1/\epsilon_1)^{1/2}$. In the case of free space ($\mu_1 = 1, \epsilon_1 = 1$), we obtain the impedance of a vacuum:

$$Z_0 = \frac{4\pi}{c} = 4.19 \times 10^{-10} \text{ s cm}^{-1} = 377 \ \Omega \quad . \tag{2.3.29}$$

The impedance fully characterizes the propagation of the electromagnetic wave as we will discuss in more detail in Section 9.1. The presence of non-magnetic matter ($\mu_1 = 1$) in general leads to a decrease of the electric field compared with the magnetic field, implying a reduction of $|\hat{Z}_S|$; for a metal, $|\hat{Z}_S|$ is small because $|\hat{H}| \gg |\hat{E}|$. Depending on the context, $\hat{Z}_S$ is also called the characteristic impedance or wave impedance. For reasons we discuss in Section 2.4.4, the expression surface impedance is commonly used for $\hat{Z}_S$ when metals are considered.

In the case of conducting matter a current is induced in the material, thus the electric and magnetic fields experience a phase shift. The impedance of the wave has a non-vanishing imaginary part and can be evaluated by substituting the full definition (2.2.18) of the complex wavevector $\hat{q}$ into Eq. (2.3.28). The final result for the complex impedance is:

$$\hat{Z}_S = R_S + iX_S = \frac{4\pi}{c}\left(\frac{\mu_1}{\epsilon_1 + i\frac{4\pi\sigma_1}{\omega}}\right)^{1/2} = \frac{4\pi}{c}\left(\frac{\mu_1}{\hat{\epsilon}}\right)^{1/2} \quad . \tag{2.3.30}$$

The real part R_S is called the surface resistance and the imaginary part X_S the surface reactance. If we assume that $|\epsilon_1| \gg 1$, which is certainly true in the case of metals at low temperatures and low frequencies but may also be fulfilled for dielectrics, we obtain from Eq. (2.3.30) the well known relation for the surface impedance

$$\hat{Z}_S \approx \left(\frac{4\pi\omega\mu_1}{c^2 i(\sigma_1 + i\sigma_2)}\right)^{1/2} \quad ; \tag{2.3.31}$$

the expressions for the real and imaginary parts, R_S and X_S, are[11]

$$R_S = \left(\frac{2\pi\omega\mu_1}{c^2}\right)^{1/2} \left\{\frac{[\sigma_1^2 + \sigma_2^2]^{1/2} - \sigma_2}{\sigma_1^2 + \sigma_2^2}\right\}^{1/2} \tag{2.3.32a}$$

$$X_S = -\left(\frac{2\pi\omega\mu_1}{c^2}\right)^{1/2} \left\{\frac{[\sigma_1^2 + \sigma_2^2]^{1/2} + \sigma_2}{\sigma_1^2 + \sigma_2^2}\right\}^{1/2} . \tag{2.3.32b}$$

For highly conducting materials at low frequencies, $\epsilon_2 \gg |\epsilon_1|$ or $\sigma_1 \gg |\sigma_2|$, we find $R_S = -X_S$; meaning that the surface resistance and (the absolute value of) the reactance are equal in the case of a metal.

In cases where both the surface resistance and surface reactance are measured, we can calculate the complex conductivity $\hat{\sigma} = \sigma_1 + i\sigma_2$ by inverting these expressions:

$$\sigma_1 = -\frac{8\pi\omega}{c^2}\frac{R_S X_S}{(R_S^2 + X_S^2)^2} , \tag{2.3.33a}$$

$$\sigma_2 = \frac{4\pi\omega}{c^2}\frac{X_S^2 - R_S^2}{(R_S^2 + X_S^2)^2} . \tag{2.3.33b}$$

In Eq. (2.3.28) we have shown that the impedance is just proportional to the inverse wavevector and to the inverse complex refractive index

$$\hat{Z}_S = \frac{4\pi}{c}\frac{\mu_1}{\hat{N}} = \frac{4\pi\mu_1}{c}\frac{1}{n + ik} .$$

This allows us to write the surface resistance R_S and surface reactance X_S as :[12]

$$R_S = \frac{\mu_1 n}{n^2 + k^2}Z_0 \tag{2.3.34a}$$

$$X_S = \frac{-\mu_1 k}{n^2 + k^2}Z_0 \tag{2.3.34b}$$

where $Z_0 = 4\pi/c$ is the wave impedance of a vacuum as derived above. Obviously $X_S = 0$ for materials with no losses ($k = 0$). The ratio of the two components of the surface impedance

$$\frac{-X_S}{R_S} = \frac{k}{n} = \tan\phi \tag{2.3.35}$$

gives the phase difference ϕ between magnetic and electric fields, already expressed by Eq. (2.3.8).

---

[11] In the case of a vacuum, for example, $\sigma_1 = 0$ and the negative root of $\sqrt{\sigma_2^2}$ has to be taken, since σ_2 stands for $-\epsilon_1\omega/4\pi$. The impedance then is purely real ($X_S = 0$) but the real part is not zero, $R_S = Z_0 = 4\pi/c$. It becomes identical to the wave impedance derived in Eq. (2.3.29). In the case of metals at low frequencies, $\epsilon_1 < 0$ and σ_2 is positive.

[12] The negative sign of X_S is a result of the definition of the surface impedance as $\hat{Z}_S = R_S + iX_S$ and thus purely conventional; it is often neglected or suppressed in the literature.

2.4 Changes of electromagnetic radiation at the interface

Next we address the question of how the propagation of electromagnetic radiation changes at the boundary between free space and a medium, or in general at the interface of two materials with different optical constants $\hat{N}$ and $\hat{N}'$. Note, we assume the medium to be infinitely thick; materials of finite thickness will be discussed in detail in Appendix B. The description of the phenomena leads to the optical parameters, such as the reflectivity R, the absorptivity A, and the transmission (or transmissivity) T of the electromagnetic radiation. Optical parameters as understood here are properties of the interface and they depend upon the boundary for their definition. All these quantities are directly accessible to experiments, and it just depends on the experimental configuration used and on the spectral range of interest that one is more useful for the description than the other. In this section we define these parameters, discuss their applicability, and establish the relationship between them. In general, experiments for both the amplitude and phase of the reflected and transmitted radiation can be conducted, but most often only quantities related to the intensity of the electromagnetic radiation (e.g. $I = |E|^2$) are of practical interest or only these are accessible.

2.4.1 Fresnel's formulas for reflection and transmission

Let us consider the propagation of a plane electromagnetic wave from a vacuum ($\epsilon'_1 = \mu'_1 = 1, \sigma'_1 = 0$) into a material with finite ϵ_1 and σ_1. The surface lies in the xy plane while the z axis is positive in the direction toward the bulk of the medium; the surface plane is described by the unit vector $\mathbf{n_s}$ normal to the surface. We further suppose that the direction of propagation is the xz plane (plane of incidence), as displayed in Fig. 2.3. The incident waves

$$\mathbf{E}_i = \mathbf{E}_{0i} \exp\{i(\mathbf{q}_i \cdot \mathbf{r} - \omega_i t)\} \tag{2.4.1a}$$

$$\mathbf{H}_i = \mathbf{n_{q_i}} \times \mathbf{E}_i \tag{2.4.1b}$$

arrive at the surface at an angle ψ_i, which is the angle between wavevector $\mathbf{q}_i$ and $\mathbf{n_s}$ in the xz plane; $\mathbf{n_{q_i}}$ is the unit vector along $\mathbf{q}_i$. One part of the electric and magnetic fields enters the material, and we write this portion as

$$\mathbf{E}_t = \mathbf{E}_{0t} \exp\{i(\mathbf{q}_t \cdot \mathbf{r} - \omega_t t)\} \tag{2.4.2a}$$

$$\mathbf{H}_t = \left(\frac{\epsilon_1}{\mu_1}\right)^{1/2} \mathbf{n_{q_t}} \times \mathbf{E}_t \ ; \tag{2.4.2b}$$

the other part is reflected off the surface and is written as

$$\mathbf{E}_r = \mathbf{E}_{0r} \exp\{i(\mathbf{q}_r \cdot \mathbf{r} - \omega_r t)\} \tag{2.4.3a}$$

$$\mathbf{H}_r = \mathbf{n_{q_r}} \times \mathbf{E}_r \ . \tag{2.4.3b}$$

(a)

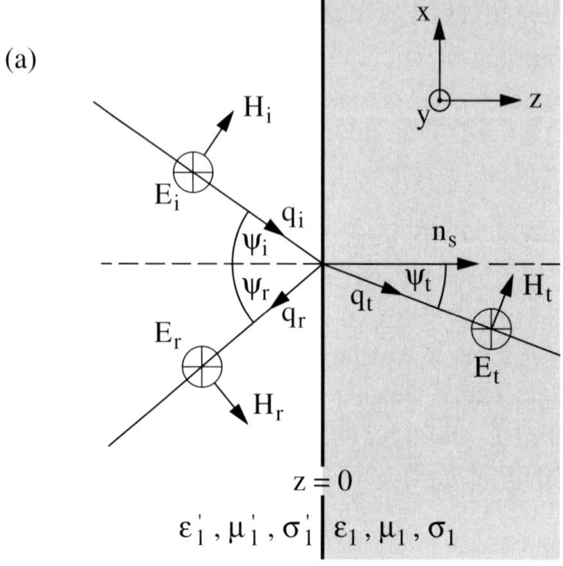

(b)

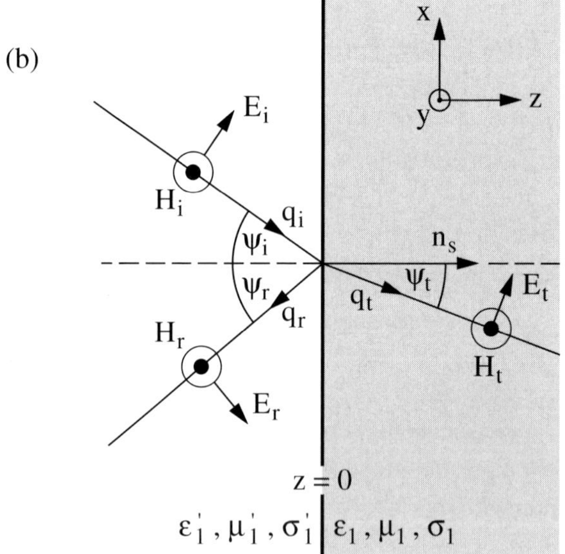

Fig. 2.3. (a) Reflection and refraction of an electromagnetic wave with the electric field **E** perpendicular to the plane of incidence; the magnetic field **H** lies in the plane of incidence. (b) Reflection and refraction of the electromagnetic wave with **E** parallel to the plane of incidence; the magnetic field **H** is directed perpendicular to the plane of incidence. In the xy plane at $z = 0$ there is an interface between two media. The first medium (left side) is characterized by the parameters ϵ_1', σ_1', and μ_1', the second medium by ϵ_1, σ_1, and μ_1. ψ_i, ψ_r, and ψ_t are the angles of incidence, reflection, and transmission, respectively.

By using the subscripts i, t, and r we make explicitly clear that all parameters may have changed upon interaction with the material. At the boundary, the normal components of **D** and **B**, as well as the tangential components of **E** and **H**, have to be continuous:

$$[\mathbf{E}_{0i} + \mathbf{E}_{0r} - \epsilon_1 \mathbf{E}_{0t}] \cdot \mathbf{n}_s = 0 \ , \qquad (2.4.4a)$$

$$[(\mathbf{q}_i \times \mathbf{E}_{0i}) + (\mathbf{q}_r \times \mathbf{E}_{0r}) - (\mathbf{q}_t \times \mathbf{E}_{0t})] \cdot \mathbf{n}_s = 0 \ , \qquad (2.4.4b)$$

$$[\mathbf{E}_{0i} + \mathbf{E}_{0r} - \mathbf{E}_{0t}] \times \mathbf{n}_s = 0 \ , \qquad (2.4.4c)$$

$$[(\mathbf{q}_i \times \mathbf{E}_{0i}) + (\mathbf{q}_r \times \mathbf{E}_{0r}) - \frac{1}{\mu_1}(\mathbf{q}_t \times \mathbf{E}_{0t})] \times \mathbf{n}_s = 0 \ . \qquad (2.4.4d)$$

At the surface ($z = 0$) the spatial and time variation of all fields have to obey these boundary conditions; the frequency[13] and the phase factors have to be the same for the three waves:

$$\omega_i = \omega_t = \omega_r = \omega$$

$$(\mathbf{q}_i \cdot \mathbf{r})_{z=0} = (\mathbf{q}_t \cdot \mathbf{r})_{z=0} = (\mathbf{q}_r \cdot \mathbf{r})_{z=0} \ .$$

From the latter equation we see that both the reflected and the refracted waves lie in the plane of incidence; furthermore $\hat{N}_i \sin \psi_i = \hat{N}_t \sin \psi_t = \hat{N}_r \sin \psi_r$. Since $\hat{N}_i = \hat{N}_r = \hat{N}'$ (the wave travels through the same material), we find the well known fact that the angle of reflection equals the angle of incidence: $\psi_r = \psi_i$. Setting $\hat{N}_t = \hat{N}$, Snell's law for the angle of refraction yields

$$\frac{\sin \psi_i}{\sin \psi_t} = \frac{\hat{N}}{\hat{N}'} \ , \qquad (2.4.5)$$

which states that the angle of the radiation transmitted into the medium ψ_t becomes smaller as the refractive index increases.[14] For $\sigma_1 \neq 0$, we see that formally $\sin \psi_t$ is complex because the refractive index $\hat{N}$ has an imaginary part, indicating that the wave gets attenuated. Although the angle ϕ_t is solely determined by the real part of the refractive indices, in addition the complex notation expresses the fact that the wave experiences a different attenuation upon passing through the interface. As mentioned above, we assume that $n' = 1$ and $k' = 0$, i.e. $\hat{N}' = 1$ (the wave travels in a vacuum before it hits the material), thus Snell's law can be reduced to

$$\frac{\sin \psi_i}{\sin \psi_t} = \hat{N} = (n + ik) = \left[\epsilon_1 \mu_1 + i\frac{4\pi \mu_1 \sigma_1}{\omega} \right]^{1/2} \ . \qquad (2.4.6)$$

There are two cases to be distinguished. First, the electric field is normal to the

---

[13] Non-linear processes may lead to higher harmonics; also (inelastic) Raman and Brillouin scattering cause a change in frequency. These effects are not considered in this book.

[14] The angle is measured with respect to normal incidence, in contrast to the Bragg equation where the angle is commonly measured with respect to the surface.

plane of incidence (Fig. 2.3a), and therefore parallel to the surface of the material; second, the electric field is in the plane of incidence[15] (Fig. 2.3b). We want to point out that we use the Verdet convention which relates the coordinate system to the wavevectors.[16] In the case when the electric vector is perpendicular to the plane of incidence, Eqs (2.4.4c) and (2.4.4d) give $E_{0i}+E_{0r}-E_{0t} = 0$ and $(E_{0i}-E_{0r})\cos\psi_i - (\epsilon_1/\mu_1)^{1/2}E_{0t}\cos\psi_t = 0$. This yields Fresnel's formulas for $\mathbf{E}_i$ perpendicular to the plane of incidence; the complex transmission and reflection coefficients are:

$$\hat{t}_\perp = \frac{E_{0t}}{E_{0i}} = \frac{2\mu_1\cos\psi_i}{\mu_1\cos\psi_i + \left(\hat{N}^2 - \sin^2\psi_i\right)^{1/2}} \quad , \tag{2.4.7a}$$

$$\hat{r}_\perp = \frac{E_{0r}}{E_{0i}} = \frac{\mu_1\cos\psi_i - \left(\hat{N}^2 - \sin^2\psi_i\right)^{1/2}}{\mu_1\cos\psi_i + \left(\hat{N}^2 - \sin^2\psi_i\right)^{1/2}} \quad . \tag{2.4.7b}$$

If the electric vector $\mathbf{E}_i$ is in the plane of incidence, the Eqs (2.4.4a), (2.4.4c), and (2.4.4d) lead to $(E_{0i} - E_{0r})\cos\psi_i - E_{0t}\cos\psi_t = 0$ and $(E_{0i} + E_{0r}) - (\epsilon_1/\mu_1)^{1/2}E_{0t} = 0$. We then obtain Fresnel's formulas for $\mathbf{E}_i$ parallel to the plane of incidence:

$$\hat{t}_\parallel = \frac{E_{0t}}{E_{0i}} = \frac{2\mu_1\hat{N}\cos\psi_i}{\hat{N}^2\cos\psi_i + \mu_1\left(\hat{N}^2 - \sin^2\psi_i\right)^{1/2}} \quad , \tag{2.4.7c}$$

$$\hat{r}_\parallel = \frac{E_{0r}}{E_{0i}} = \frac{\hat{N}^2\cos\psi_i - \mu_1\left(\hat{N}^2 - \sin^2\psi_i\right)^{1/2}}{\hat{N}^2\cos\psi_i + \mu_1\left(\hat{N}^2 - \sin^2\psi_i\right)^{1/2}} \quad . \tag{2.4.7d}$$

These formulas are valid for $\hat{N}$ complex. To cover the general case of an interface between two media (the material parameters of the first medium are indicated by a prime: $\epsilon_1' \neq 1, \mu_1' \neq 1, \sigma_1' \neq 0$; and the second medium without: $\epsilon_1 \neq 1, \mu_1 \neq 1, \sigma_1 \neq 0$), the following replacements in Fresnel's formulas are sufficient: $\hat{N} \rightarrow \hat{N}/\hat{N}'$ and $\mu_1 \rightarrow \mu_1/\mu_1'$.

2.4.2 Reflectivity and transmissivity by normal incidence

In the special configuration of normal incidence ($\psi_i = \psi_t = \psi_r = 0$), the distinction between the two cases, parallel and perpendicular to the plane of incidence, becomes irrelevant (Fig. 2.3). The wavevector $\mathbf{q}$ is perpendicular to the surface while $\mathbf{E}$ and $\mathbf{H}$ point in the x and y directions, respectively. In order to satisfy

---

[15] From the German words 'senkrecht' and 'parallel', they are often called s-component and p-component.

[16] A different perspective (Fresnel convention) relates the coordinate system to the sample surface instead of to the field vectors [Hol91, Mul69, Sok67].

the boundary conditions, i.e. the tangential components of **E** and **H** must be continuous across the boundary, the amplitudes of the incident (E_{0i}, H_{0i}), transmitted (E_{0t}, H_{0t}), and reflected (E_{0r}, H_{0r}) waves are

$$E_{0t} = E_{0i} + E_{0r} \tag{2.4.8a}$$

$$H_{0t} = H_{0i} - H_{0r}. \tag{2.4.8b}$$

Starting with Eqs (2.2.14) and Eq. (2.3.2), on the right hand side of the xy plane, the electric and magnetic fields in the dielectric medium ($\mu_1 = 1$) are given by

$$E_x(z, t) = E_{0t} \exp\left\{ i\omega \left(\frac{\hat{N} z}{c} - t \right) \right\} \tag{2.4.9a}$$

$$H_y(z, t) = \sqrt{\epsilon_1} E_{0t} \exp\left\{ i\omega \left(\frac{\hat{N} z}{c} - t \right) \right\}, \tag{2.4.9b}$$

while on the left hand side, in a vacuum ($\epsilon_1' = \mu_1' = 1, \sigma_1' = 0$)

$$E_x(z, t) = E_{0i} \exp\left\{ i\omega \left(\frac{z}{c} - t \right) \right\} + E_{0r} \exp\left\{ i\omega \left(-\frac{z}{c} - t \right) \right\} \tag{2.4.10a}$$

$$H_y(z, t) = E_{0i} \exp\left\{ i\omega \left(\frac{z}{c} - t \right) \right\} - E_{0r} \exp\left\{ i\omega \left(-\frac{z}{c} - t \right) \right\} \tag{2.4.10b}$$

From Eq. (2.2.7a) $\frac{\partial E_x}{\partial z} = -\frac{1}{c} \frac{\partial H_y}{\partial t}$, applied to Eqs (2.4.9b) and (2.4.10b), we obtain

$$\hat{N} E_{0t} = E_{0i} \quad E_{0r} \quad . \tag{2.4.11}$$

Combining this equation with Eq. (2.4.8a), we arrive at

$$\hat{r} = \hat{r}_\parallel = -\hat{r}_\perp = \frac{E_{0r}}{E_{0i}} = \frac{1 - \hat{N}}{1 + \hat{N}} = |\hat{r}| \exp\{i\phi_r\} \tag{2.4.12}$$

for the complex reflection coefficient $\hat{r}$, in agreement with Eqs (2.4.7b) and (2.4.7d) on setting $\psi = 0$. Note, we only consider $\mu_1' = \mu_1 = 1$. The phase shift ϕ_r is the difference between the phases of the reflected and the incident waves. More generally, at the interface between two media ($\hat{N} \neq 1$ and $\hat{N}' \neq 1$) we obtain

$$\hat{r} = \frac{\hat{N}' - \hat{N}}{\hat{N}' + \hat{N}} \tag{2.4.13}$$

with a phase change

$$\phi_r = \arctan\left\{ \frac{2(k'n - kn')}{n'^2 + k'^2 - n^2 - k^2} \right\} . \tag{2.4.14}$$

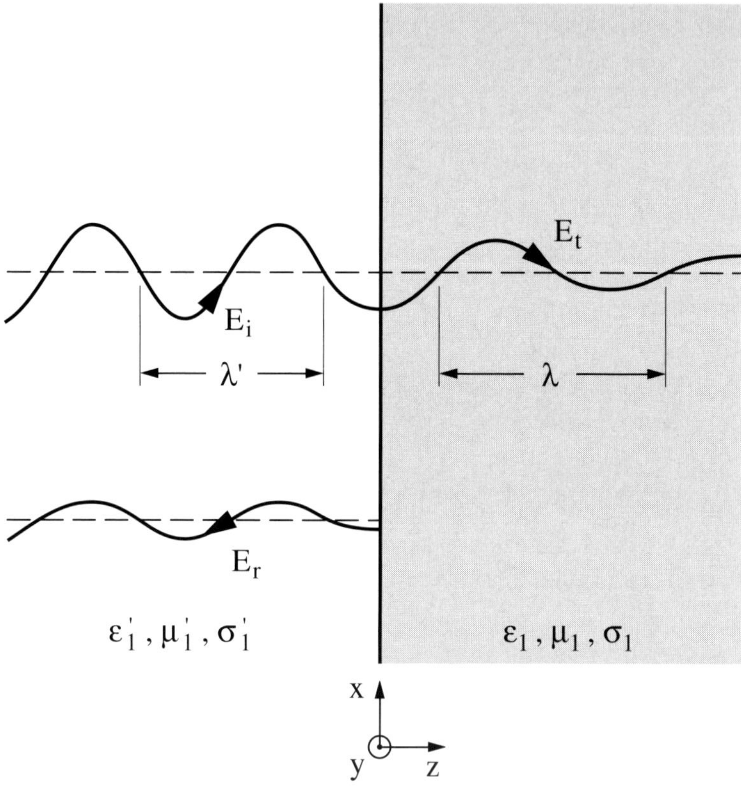

Fig. 2.4. Incident $\mathbf{E}_i$, reflected $\mathbf{E}_r$, and transmitted electric wave $\mathbf{E}_t$ traveling normal to the interface between two media. The first medium has the parameters ϵ'_1, σ'_1, and μ'_1, the second medium ϵ_1, σ_1, and μ_1; the longer wavelength $\lambda > \lambda'$ indicates that $\epsilon_1 < \epsilon'_1$ assuming $\mu_1 = \mu'_2$. The amplitudes of the reflected and transmitted waves as well as the phase difference depend on the optical properties of the media.

Here $\phi_r \to \pi$ if (crudely speaking) $\hat{N} > \hat{N}'$; this means that for normal incidence the wave suffers a phase shift of $180°$ upon reflection on an optically denser medium (defined by the refractive index); $\phi_r \to 0$ if $n'^2 + k'^2 > n^2 + k^2$.

During an experiment the reflected power is usually observed and the phase information is lost by taking the absolute value of the complex quantity $|\hat{\mathbf{E}}|^2 = E_0^2$. The reflectivity R is defined as the ratio of the time averaged energy flux reflected from the surface $S_r = (c/4\pi)|\mathbf{E}_{0r} \times \mathbf{H}_{0r}|$ to the incident flux $S_i = (c/4\pi)|\mathbf{E}_{0i} \times \mathbf{H}_{0i}|$. Substituting the electric and magnetic fields given by Eqs (2.4.1) and (2.4.3) into the definition of the Poynting vector Eq. (2.3.22) yields the simple expression

$$R = \frac{S_r}{S_i} = \frac{|E_{0r}|^2}{|E_{0i}|^2} = |\hat{r}|^2 = \left| \frac{1 - \hat{N}}{1 + \hat{N}} \right|^2 = \frac{(1-n)^2 + k^2}{(1+n)^2 + k^2} \tag{2.4.15}$$

for $\hat{N}' = 1$. Corresponding to the reflectivity, the phase change of the reflected wave ϕ_r is given by Eq. (2.4.14):

$$\tan \phi_r = \frac{-2k}{1 - n^2 - k^2} \quad .$$

For a dielectric material without losses ($k \to 0$), the reflectivity is solely determined by the refractive index:

$$R = \left(\frac{1 - n}{1 + n} \right)^2 \quad , \tag{2.4.16}$$

and it can approach unity if n is large; then $\tan \phi_r = 0$. Both parameters R and ϕ_r can also be expressed in terms of the complex conductivity $\hat{\sigma} = \sigma_1 + i\sigma_2$ by using the relations listed in Table 2.1; and then

$$R = \frac{1 + \frac{4\pi}{\omega} \left(\sigma_1^2 + \sigma_2^2 \right)^{1/2} - \left(\frac{8\pi}{\omega} \right)^{1/2} \left[\left(\sigma_1^2 + \sigma_2^2 \right)^{1/2} + \sigma_2 \right]^{1/2}}{1 + \frac{4\pi}{\omega} \left(\sigma_1^2 + \sigma_2^2 \right)^{1/2} + \left(\frac{8\pi}{\omega} \right)^{1/2} \left[\left(\sigma_1^2 + \sigma_2^2 \right)^{1/2} + \sigma_2 \right]^{1/2}} \tag{2.4.17}$$

$$\tan \phi_r = \frac{\left(\frac{8\pi}{\omega} \right)^{1/2} \left[\left(\sigma_1^2 + \sigma_2^2 \right)^{1/2} - \sigma_2 \right]^{1/2}}{1 + \frac{4\pi}{\omega} \left(\sigma_1^2 + \sigma_2^2 \right)^{1/2}} \quad . \tag{2.4.18}$$

If $\sigma_1 \gg |\sigma_2|$, the reflectivity is large ($R \to 1$) and the phase ϕ_r approaches π.

The transmission coefficient $\hat{t}$ for an electromagnetic wave passing through the boundary is written as

$$\hat{t} = \frac{E_{0t}}{E_{0i}} = \frac{2N'}{N + N'} = |\hat{t}| \exp\{i\phi_t\}; \tag{2.4.19}$$

$$\phi_t = \arctan \left\{ \frac{nk' - n'k}{nn' - n'^2 + kk' + k'^2} \right\} = \arctan \left\{ \frac{-k}{n + 1} \right\} \tag{2.4.20}$$

is the phase shift of the transmitted to incident wave; again we have assumed the case $\hat{N}' = 1$ for the second transformation. The power transmitted into the medium is given by the ratio of the time averaged transmitted energy flux $S_t = (c/4\pi)|\mathbf{E}_{0t} \times \mathbf{H}_{0t}|$ to the incident flux S_i. With Eqs (2.4.1) and (2.4.2) we obtain the so-called transmissivity (often simply called transmission)

$$T = \frac{S_t}{S_i} = \sqrt{\epsilon_1} \frac{|E_{0t}|^2}{|E_{0i}|^2} = \sqrt{\epsilon_1} |\hat{t}|^2 = \frac{4n}{(n + 1)^2 + k^2} = 1 - R \quad . \tag{2.4.21}$$

It is important to note that $|\hat{t}|^2 \neq 1 - |\hat{r}|^2$ due to the modified fields and thus energy density inside the material.

2.4.3 *Reflectivity and transmissivity for oblique incidence*

We now want to consider the general case of light hitting the interface at arbitrary angles. As derived in Fresnel's equations (2.4.7a)–(2.4.7d) the result depends upon whether the electric fields are oriented parallel or perpendicular to the plane of incidence. In general the applicability of Fresnel's equations is not restricted to certain angles of incidence ψ_i or limited in frequency ω. However, if the frequency ω is comparable to the plasma frequency ω_p, (longitudinal) plasma waves (so-called plasmons) may occur for $\psi_i \neq 0$ and parallel polarization [Bec64]. This is the case in thin metal films [McA63], anisotropic conducting materials [Bru75], but also in bulk metals [Mel70]. More details on oblique incidence can be found in several textbooks on optics [Bor99, Hec98, Kle86].

In Figs 2.5–2.7 we show the angular dependence of the reflection and transmission coefficients together with the change in phase for both polarizations parallel and perpendicular to the plane of incidence. Fig. 2.5a was calculated with $n = 1.5$ and $k = 0$ by using Eqs (2.4.7a)–(2.4.7d). The reflection coefficients are real and $|r_\parallel| \leq |r_\perp|$ in the entire range; $\hat{r}_\parallel$ shows a zero-crossing at the Brewster angle

$$\psi_B = \arctan n \quad , \tag{2.4.22}$$

where the phase shift $\phi_{r\parallel}$ jumps from 0 to π. For $\psi_i \rightarrow \pi/2$ the reflection coefficient of both polarizations are equal ($\hat{r}_\perp = \hat{r}_\parallel$), for $\psi_i = 0$ we obtain $\hat{r}_\perp = -\hat{r}_\parallel$; as mentioned before the change of sign is a question of definition. Although for ψ_B only one direction of polarization is reflected, the transmitted light is just slightly polarized. The features displayed in Fig. 2.5 are smeared out if the dielectric becomes lossy, i.e. if $k > 0$. The coefficients $\hat{r}_\parallel$ and $\hat{r}_\perp$ become complex quantities, indicating the attenuation of the wave. As an example, in Fig. 2.6 we plot the same parameters for $n = 1.5$ and $k = 1.5$. The absolute value $|\hat{r}_\parallel|$ still shows a minimum, but there is no well defined Brewster angle. The properties of oblique incident radiation are utilized in ellipsometry to determine the complex reflection coefficient (Section 11.1.4). Grazing incidence is also used to enhance the sensitivity of reflection measurements off metals because the absorptivity $A = 1 - |\hat{r}_\parallel|^2$ increases approximately as $1/\cos\psi_i$. Using oblique incidence with parallel polarization, the optical properties perpendicular to the surface can be probed.

By traveling from an optically denser medium to a medium with smaller n, as shown in Fig. 2.7, the wave is totally reflected if the angle of incidence exceeds $\psi_T > \psi_B$, the angle of total reflection ($\tan\psi_B = \sin\psi_r = n$). The example calculated corresponds to the case where the wave moves from a medium with $n' = 1.5$ to free space $n = 1$ (corresponding to $n' = 1$ and $n = 1/1.5$). Again $|r_\perp| \geq |r_\parallel|$ in the entire range. Interestingly the reflectivity for $\psi_i = 0$ is the same

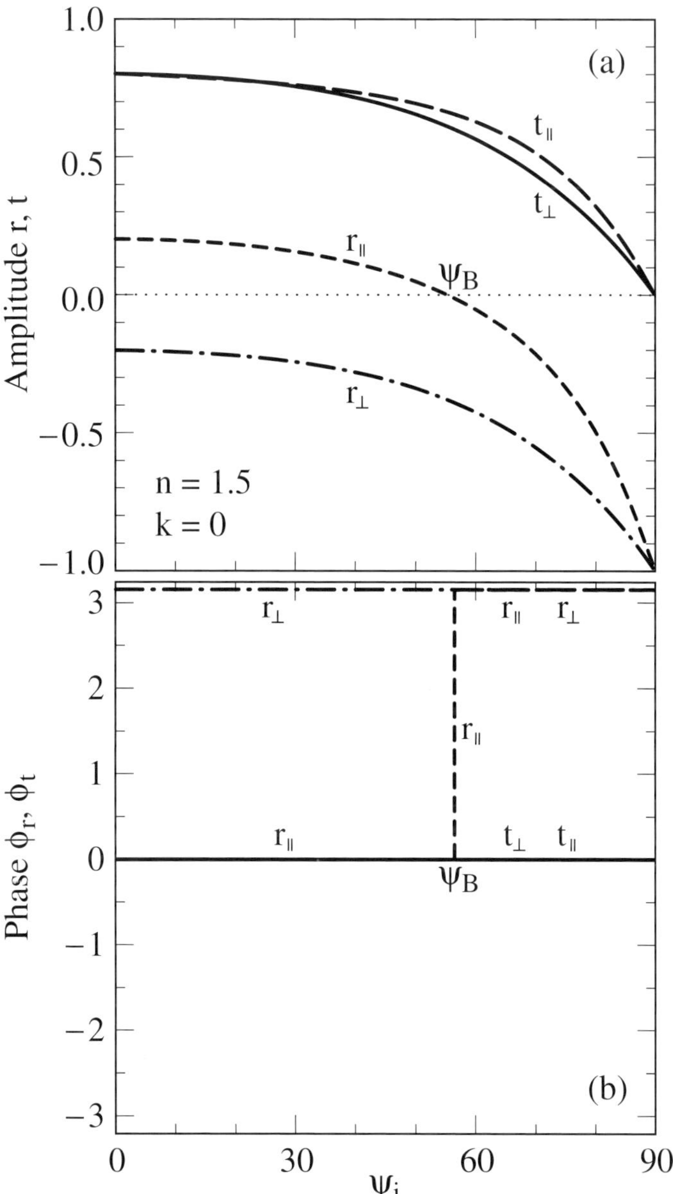

Fig. 2.5. (a) The (real) reflection and transmission coefficients, r and t (in both polarizations parallel and perpendicular to the plane of incidence) as a function of angle of incidence ψ_i for $n = 1.5$, $n' = 1$, $k = k' = 0$, and $\mu_1 = \mu_1' = 1$. The Brewster angle is defined as $r_\parallel(\psi_B) = 0$. (b) The corresponding phase shifts, ϕ_r and ϕ_t, of the reflected and transmitted waves; here $\psi_t = 0$ and $\psi_r = \pi$ for the electric field perpendicular to the plane of incidence (referred to as $r_\perp$ and $t_\perp$). In the case of **E** parallel to the plane of incidence ($r_\parallel$ and $t_\parallel$), ψ_t remains zero, while the phase ψ_r changes by π at the Brewster angle ψ_B.

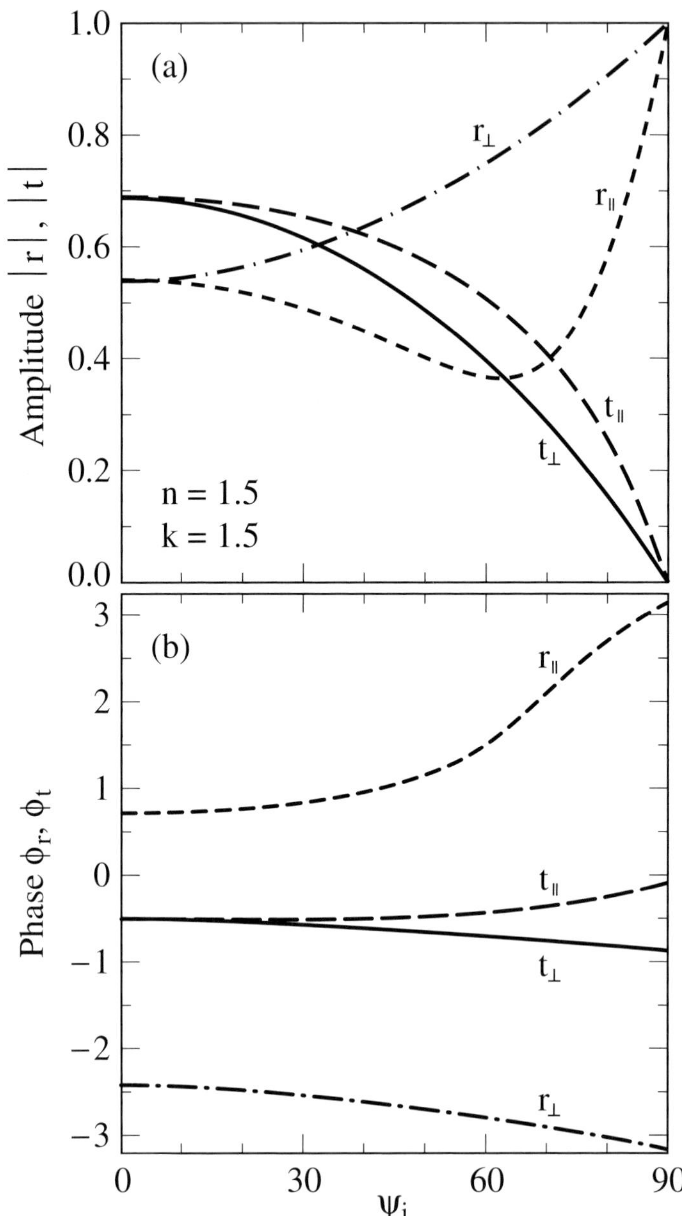

Fig. 2.6. (a) The absolute values of the reflection and transmission coefficients, $|r|$ and $|t|$, as a function of angle of incidence ψ_i in polarizations parallel and perpendicular to the plane of incidence. Besides the refractive index $n = 1.5$, the material also has losses described by the extinction coefficient $k = 1.5$; again $n' = 1$, $k' = 0$, and $\mu_1 = \mu'_1 = 1$. (b) The angular dependences of the corresponding phase change upon reflection, ϕ_r, and transmission, ϕ_t. The different cases are indicated by $r_\parallel$, $t_\parallel$, $r_\perp$, and $t_\perp$, respectively.

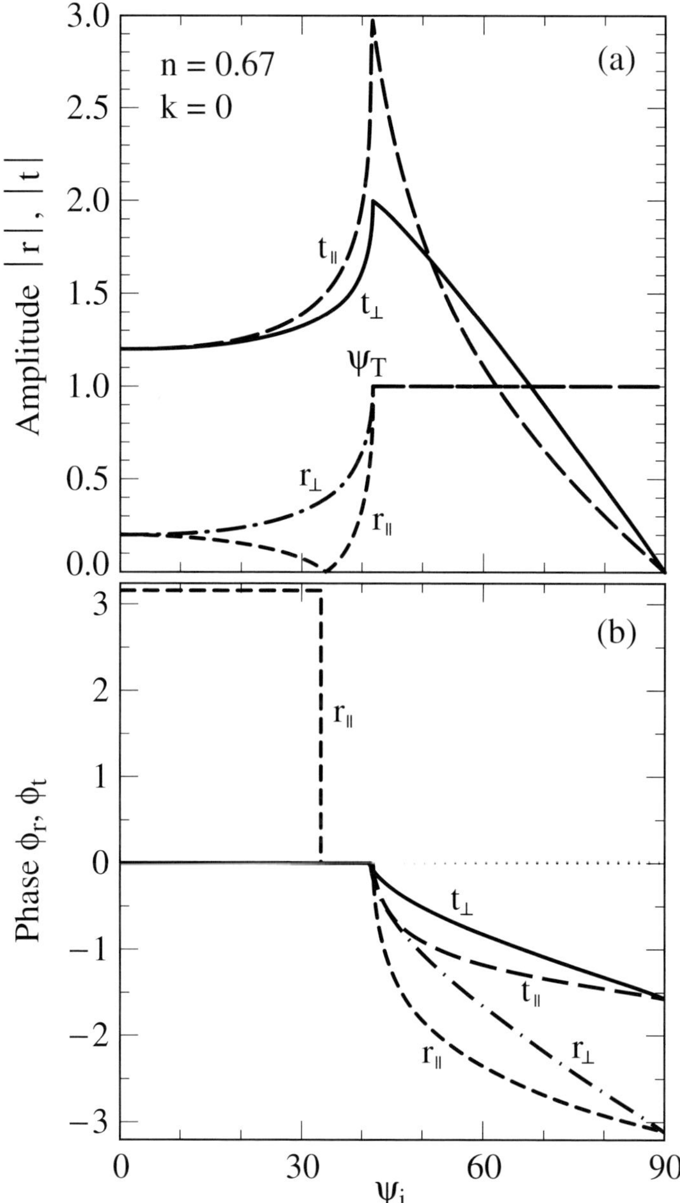

Fig. 2.7. (a) The absolute values of the reflection and transmission coefficients, $|r|$ and $|t|$, in polarizations parallel and perpendicular to the plane of incidence, as a function of angle of incidence ψ_i for $n/n' = 1/1.5$ and $\mu_1 = \mu'_1 = 1$. The Brewster angle $r_\parallel(\psi_B) = 0$ and the angle of total reflection ψ_T is clearly seen. (b) The phase angles ϕ_r and ϕ_t change significantly in the range of total reflection. The case of **E** parallel to the plane of incidence is referred to as $r_\parallel$ and $t_\parallel$, while $r_\perp$ and $t_\perp$ refer to **E** perpendicular to the plane, respectively.

for $n = 1.5$ and $n = 1/1.5$ due to the reciprocity of the optical properties if $k = 0$. For $\psi_i > \psi_T$, surface waves develop which lead to a spatial offset of the reflected wave [Goo47]. The range of total reflection is characterized by a phase difference Δ of the polarizations $\mathbf{E}_\parallel$ and $\mathbf{E}_\perp$ in time. If we write $\mathbf{E} = E_\parallel \mathbf{n}_\parallel + E_\perp \mathbf{n}_\perp$ with $E_\parallel = E \cos\phi \exp\{i\Delta\}$ and $E_\perp = E \sin\phi$ we find

$$E_x(t) = |E_\parallel| \cos\{\omega t + \Delta\} \qquad \text{and} \qquad E_y(t) = |E_\perp| \sin\{\omega t\} \quad ,$$

respectively, which characterize an elliptically polarized light. Although the light is totally reflected, the fields extend beyond the interface by a distance of the order of the wavelength and their amplitude decays exponentially (evanescent waves). This fact is widely used by a technique called attenuated total reflection, for instance to study lattice and molecular vibrations at surfaces [Spr91]. The fact that the transmission coefficient $\hat{t}$ is larger than 1 does not violate the energy conservation because the transmitted energy is given by $T = \left(\epsilon_1' \mu_1 / \mu_1' \epsilon_1\right)^{1/2} |\hat{t}|^2 (\cos\psi_t / \cos\psi_i)$ as a generalization of Eq. (2.4.21) for oblique incidence; this parameter is always smaller than unity.

In general, the angle of incidence is not very critical for most experimental purposes; if it is chosen sufficiently near to zero, the value of the reflectance coefficients for the two states of polarization $\hat{r}_\parallel$ and $\hat{r}_\perp$ will differ from $\hat{r}$ by an amount less than the precision of the measurement; e.g. less than 0.001 for angles of incidence as large as $10°$.

2.4.4 Surface impedance

As already mentioned above, the impedance of the wave is called surface impedance in the case of conducting matter because originally $\hat{Z}_S$ was defined as the ratio of the electric field $\mathbf{E}$ normal to the surface of a metal to the total current density $\mathbf{J}$ induced in the material:

$$\hat{Z}_S = R_S + iX_S = \frac{\hat{E}_{z=+0}}{\int_0^\infty \hat{J}\, dz} \quad , \tag{2.4.23}$$

where $\hat{J}(z)$ is the spatially dependent current per unit area which decays exponentially with increasing distance from the surface z. The electric field $\hat{E}_{z=+0}$ refers to the transmitted wave at the interface upon entering the material. Here we see that, for the definition of $\hat{Z}_S$, it is not relevant that the wave approaches the material coming from the vacuum because the material prior to $z = 0$ does not enter the constituent equation (2.4.23). The surface impedance only depends on the optical properties of the material beyond z; hence $\hat{Z}_S$ is defined at any point in a material, as long as it extends to infinity. In Appendix B we explain how even this drawback of the definition can be overcome.

The real part R_S, the surface resistance, determines the power absorption in the metal; the surface reactance X_S accounts for the phase difference between **E** and **J**. Since the ratio of electric field to current density corresponds to a resistivity, the surface impedance can be interpreted as a resistance. Although mainly used in the characterization of the high frequency properties of metals, the concept of surface impedance is more general and is particularly useful if the thickness of the medium d is much larger than the skin depth δ_0.

We can regain the above definition of the impedance (2.3.27) by using Maxwell's third equation (2.2.7c) which relates the magnetic field **H** and the current density **J**. For ordinary metals up to room temperature the displacement current $\omega \epsilon_1 E/(4\pi i)$ can be neglected for frequencies below the visible spectral range, and

$$\nabla \times \mathbf{H} = \frac{4\pi}{c}\mathbf{J}$$

remains. We can then apply Stokes's theorem and obtain

$$\hat{Z}_S = \frac{\hat{E}}{\int_0^\infty \hat{J}\,\mathrm{d}z} = \frac{4\pi}{c}\left.\frac{\hat{E}}{\hat{H}}\right|_{z=0} \quad , \tag{2.4.24}$$

where $\hat{E}$ and $\hat{H}$ refer to the electric field and the magnetic field as complex quantities defined in Eqs (2.2.14) with a phase shift ϕ between $\hat{E}$ and $\hat{H}$. The surface impedance $\hat{Z}_S$ does not depend on the interface of two materials but rather is an optical parameter like the refractive index. Another approach used to calculate the surface impedance from Eq. (2.4.23) utilizes Eq. (2.3.24) to obtain the current density

$$\hat{J}(z) = \frac{-c^2}{4\pi i\omega\mu_1}\frac{\partial^2\hat{E}}{\partial z^2}$$

leading to the total current in the surface layer

$$\int_0^\infty \hat{J}(z)\,\mathrm{d}z = \frac{c^2}{4\pi i\omega\mu_1}\left.\left(\frac{\partial\hat{E}}{\partial z}\right)\right|_0^\infty = \frac{c}{4\pi\mu_1}\hat{N}\hat{E} \quad . \tag{2.4.25}$$

This brings us back to Eq. (2.3.28) which connects the impedance $\hat{Z}_S$ to the complex refractive index and the wavevector.

Since the surface impedance relates the electric field to the induced current in a surface layer, R_S and X_S must also be connected to the skin depth δ_0. Combining Eq. (2.3.32b) with Eq. (2.3.15b), we find

$$-X_S = \frac{1}{\delta_0}\frac{1}{(\sigma_1^2 + \sigma_2^2)^{1/2}} \approx \frac{1}{\delta_0\sigma_1} \approx R_S \tag{2.4.26}$$

for $\sigma_1 \gg |\sigma_2|$, which is the low frequency limit of good conductors. The last approximation can be easily understood by replacing the integration in Eq. (2.4.23) by the characteristic length scale of the penetration. Also we can write

$$\delta_0 = \frac{c^2}{4\pi\omega\mu_1}\frac{R_S^2 + X_S^2}{R_S} \approx \frac{c^2}{2\pi\omega\mu_1}R_S = \frac{-\lambda_0 c}{(2\pi)^2}X_S \qquad (2.4.27)$$

if $R_S = -X_S$ upon penetrating the material. Since $2\pi c/\omega = \lambda_0$ is the wavelength of the radiation in a vacuum,

$$\hat{Z}_S = \frac{(2\pi)^2\mu_1}{c}\frac{\delta_0}{\lambda_0}(1+\mathrm{i}) \quad ; \qquad (2.4.28)$$

i.e. in the limit $\sigma_1 \gg |\sigma_2|$ the surface impedance is determined by the ratio of the skin depth to the wavelength. We again want to point out that, in spite of the common names, both the surface impedance and the skin depth have been defined without referring to the surface of the material. We just consider the decay of the **E** and **H** fields and the change in phase between them.

2.4.5 Relationship between the surface impedance and the reflectivity

Finally we want to present the relationships between the surface impedance and the optical properties such as the reflectivity or transmissivity. The specular reflection for normal incidence ($\psi_i = \psi_t = 0$) at the boundary between two media has been derived in Eq. (2.4.13) assuming a wave traveling from one medium ($\hat{N}', \mu_1'$) to another ($\hat{N}, \mu_1$). We can combine this with the relation for the surface impedance (2.3.28):

$$\hat{r} = \frac{\mu_1\hat{N}' - \mu_1'\hat{N}}{\mu_1\hat{N}' + \mu_1'\hat{N}} = \frac{\hat{Z}_S - \hat{Z}_S'}{\hat{Z}_S + \hat{Z}_S'} \quad . \qquad (2.4.29)$$

The reflectivity (i.e. reflected power) is then given by the very general expression

$$R = \left|\frac{\hat{Z}_S - \hat{Z}_S'}{\hat{Z}_S + \hat{Z}_S'}\right|^2 \quad . \qquad (2.4.30)$$

As we discuss in more detail in Chapter 9, a reflection of waves also occurs if the impedance of a guiding structure (waveguide, cable) changes; hence the previous equation holds for any impedance mismatch.

The reflectivity at the interface does not depend on whether the wave travels from medium 1 to medium 2 or from medium 2 to medium 1; the reflectivity of an interface is reciprocal. The phase ϕ_r, however, changes by π. In the case that the first medium is a vacuum ($\hat{Z}_S' = Z_0 = 377\ \Omega$) the reflectivity can be written as

$$R = \left|\frac{Z_0 - \hat{Z}_S}{Z_0 + \hat{Z}_S}\right|^2 = 1 - \frac{4R_S}{Z_0}\left(1 + \frac{2R_S}{Z_0} + \frac{R_S^2 + X_S^2}{Z_0^2}\right)^{-1} \quad . \qquad (2.4.31)$$

The transmission coefficient as a function of the surface impedances is

$$\hat{t} = \frac{2\hat{Z}_S}{\hat{Z}_S + \hat{Z}'_S} \quad . \tag{2.4.32}$$

In the special case of transmission from a vacuum into the medium with finite absorption:

$$T = 1 - R = \frac{4R_S}{Z_0}\left(1 + \frac{2R_S}{Z_0} + \frac{R_S^2 + X_S^2}{Z_0^2}\right)^{-1} \quad . \tag{2.4.33}$$

In the case of an infinite thickness, which can be realized by making the material thicker than the skin depth δ_0, all the power transmitted through the interface is absorbed, and the absorptivity $A = T$. From Eqs (2.4.31) and (2.4.33),

$$R = 1 - T = 1 - A \tag{2.4.34}$$

in this limit.

References

[Bec64] R. Becker, *Electromagnetic Fields and Interaction* (Bluisdell Publisher, New York, 1964)

[Ber78] D. Bergman, *Phys. Rep.* **43**, 377 (1978)

[Blo65] N. Bloembergen, *Nonlinear Optics* (Benjamin, New York, 1965)

[Bor99] M. Born and E. Wolf, *Principles of Optics*, 6th edition (Cambridge University Press, Cambridge, 1999)

[Bru75] P. Brüesch, *Optical Properties of the One-Dimensional Pt-Complex Compounds*, in: *One-Dimensional Conductors*, edited by H.G. Schuster, Lecture Notes in Physics **34** (Springer-Verlag, Berlin, 1975), p. 194

[But91] P.N. Butcher and D. Cotter, *The Elements of Nonlinear Optics* (Cambridge University Press, Cambridge, 1991)

[Gay67] P. Gay, *An Introduction to Crystal Optics* (Longman, London, 1967)

[Goo47] F. Goos and H. Hänchen, *Ann. Phys. Lpz.* **1**, 333 (1947); F. Goos and H. Lindberg-Hänchen, *Ann. Phys. Lpz.* **5**, 251 (1949)

[Hec98] E. Hecht, *Optics*, 3rd edition (Addison-Wesley, Reading, MA, 1998)

[Hol91] R.T. Holm, *Convention Confusions*, in: *Handbook of Optical Constants of Solids*, Vol. II, edited by E.D. Palik (Academic Press, Boston, MA, 1991), p. 21

[Jac75] J. D. Jackson, *Classical Electrodynamics*, 2nd edition (John Wiley & Sons, New York, 1975); 3rd edition (John Wiley & Sons, New York, 1998)

[Kle86] M.V. Klein, *Optics*, 2nd edition (John Wiley & Sons, New York, 1986)

[Lan84] L.D. Landau, E.M. Lifshitz, and L.P. Pitaevskii, *Electrodynamics of Continuous Media*, 2nd edition (Butterworth-Heinemann, Oxford, 1984)

[Lan78] R. Landauer, *AIP Conference Proceedings* **40**, 2 (1978)

[McA63] A.J. McAlister and E.A. Stern, *Phys. Rev.* **132**, 1599 (1963)

[Mel70] A.R. Melnyk and M.J. Harrison, *Phys. Rev. B* **2**, 835 (1970)

[Mil91] P.L. Mills, *Nonlinear Optics* (Springer-Verlag, Berlin, 1991)

[Mul69] R.H. Müller, *Surf. Sci.* **16**, 14 (1969)

[Nye57] J.F. Nye, *Physical Properties of Crystals* (Clarendon, Oxford, 1957)

[Por92] A.M. Portis, *Electrodynamics of High-Temperature Superconductors* (World Scientific, Singapore, 1992)

[Rik96] G.L.J.A. Rikken and B.A. van Tiggelen, *Nature* **381**, 54 (1996); *Phys. Rev. Lett.* **78**, 847 (1997)

[She84] Y.R. Shen, *The Principles of Nonlinear Optics* (Wiley, New York, 1984)

[Sok67] A.V. Sokolov, *Optical Properties of Metals* (American Elsevier, New York, 1967)

[Spr91] G.J. Sprokel and J.D. Swalen, *The Attenuated Total Reflection Method*, in: *Handbook of Optical Constants of Solids*, Vol. II, edited by E.D. Palik (Academic Press, Orlando, FL, 1991), p. 75

[Ste63] F. Stern, *Elementary Theory of the Optical Properties of Solids*, in: *Solid State Physics* **15**, edited by F. Seitz and D. Turnbull (Academic Press, New York, 1963), p. 299

[Woo72] F. Wooten, *Optical Properties of Solids* (Academic Press, San Diego, CA, 1972)

Further reading

[Bri60] L. Brillouin, *Wave Propagation and Group Velocity* (Academic Press, New York, 1960)

[Cle66] P.C. Clemmov, *The Plane Wave Spectrum Representation of Electrodynamic Fields* (Pergamon Press, Oxford, 1966)

[Fey63] R.P. Feynman, R.B. Leighton, and M. Sand, *The Feynman Lectures on Physics*, Vol. 2 (Addison-Wesley, Reading, MA, 1963)

[For86] F. Forstmann and R.R. Gerhardts, *Metal Optics Near the Plasma Frequency*, Springer Tracts in Modern Physics, **109** (Springer-Verlag, Berlin, 1987)

[Gro79] P. Grosse, *Freie Elektronen in Festkörpern* (Springer-Verlag, Berlin, 1979)

[Lek87] J. Lekner, *Theory of Reflection* (Martinus Nijhoff, Dordrecht, 1987)

3

General properties of the optical constants

In Chapter 2 we described the propagation of electromagnetic radiation in free space and in a homogeneous medium, together with the changes in the amplitude and phase of the fields which occur at the interface between two media. Our next objective is to discuss some general properties of what we call the response of the medium to electromagnetic fields, properties which are independent of the particular description of solids; i.e. properties which are valid for basically all materials. The difference between longitudinal and transverse responses will be discussed first, followed by the derivation of the Kramers–Kronig relations and their consequences, the so-called sum rules. These relations and sum rules are derived on general theoretical grounds; they are extremely useful and widely utilized in the analysis of experimental results.

3.1 Longitudinal and transverse responses

3.1.1 General considerations

The electric field strength of the propagating electromagnetic radiation can be split into a longitudinal component $\mathbf{E}^{\mathrm{L}} = (\mathbf{n_q} \cdot \mathbf{E})\mathbf{n_q}$ and a transverse component $\mathbf{E}^{\mathrm{T}} = (\mathbf{n_q} \times \mathbf{E}) \times \mathbf{n_q}$, with $\mathbf{E} = \mathbf{E}^{\mathrm{L}} + \mathbf{E}^{\mathrm{T}}$, where $\mathbf{n_q} = \mathbf{q}/|\mathbf{q}|$ indicates the unit vector along the direction of propagation $\mathbf{q}$. While $\mathbf{E}^{\mathrm{L}} \parallel \mathbf{q}$, the transverse part $\mathbf{E}^{\mathrm{T}}$ lies in the plane perpendicular to the direction $\mathbf{q}$ in which the electromagnetic radiation propagates; it can be further decomposed into two polarizations which are usually chosen to be normal to each other. Since by definition $\nabla \times \mathbf{E}^{\mathrm{L}} = 0$ and $\nabla \cdot \mathbf{E}^{\mathrm{T}} = 0$, we obtain

$$\nabla \times \mathbf{E} = \nabla \times \mathbf{E}^{\mathrm{T}} \qquad \text{and} \qquad \nabla \cdot \mathbf{E} = \nabla \cdot \mathbf{E}^{\mathrm{L}} \quad ,$$

implying that longitudinal components have no influence on the rotation, and transverse fields do not enter the calculation of the divergence. In a similar way the current density ($\mathbf{J} = \mathbf{J}^{\mathrm{L}} + \mathbf{J}^{\mathrm{T}}$), the magnetic induction ($\mathbf{B} = \mathbf{B}^{\mathrm{L}} + \mathbf{B}^{\mathrm{T}}$), and the

47

vector potential ($\mathbf{A} = \mathbf{A}^{\mathrm{L}} + \mathbf{A}^{\mathrm{T}}$) can also be decomposed into components parallel and perpendicular to the direction $\mathbf{q}$ of the propagating wave. We use the Coulomb gauge and assume $\nabla \cdot \mathbf{A} = 0$. Then $\mathbf{q} \cdot (\mathbf{A}^{\mathrm{T}} + \mathbf{A}^{\mathrm{L}})$ vanishes. As $\mathbf{q}$ points in the direction of propagation, $\mathbf{A}^{\mathrm{L}}$ must be zero; i.e. in this gauge the vector potential has only a transverse component.

First we notice that in the case of free space $\mathbf{E}^{\mathrm{L}} = \mathbf{B}^{\mathrm{L}} = 0$, i.e. the electromagnetic wave is entirely transverse.[1] From Eq. (2.1.15a) we already know that the two components of the electromagnetic radiation are normal to each other: $\mathbf{E}^{\mathrm{T}} \perp \mathbf{B}^{\mathrm{T}} \perp \mathbf{q}$.

If electrical charges and current are present, we can employ the above relations to reduce the electrical continuity equation (2.1.9) to $-\partial \rho / \partial t = \nabla \cdot (\mathbf{J}^{\mathrm{L}} + \mathbf{J}^{\mathrm{T}}) = \nabla \cdot \mathbf{J}^{\mathrm{L}}$. From Poisson's equation (2.1.7) we obtain

$$\nabla \frac{\partial \Phi}{\partial t} = 4\pi \mathbf{J}^{\mathrm{L}} \quad , \tag{3.1.1}$$

which, when substituted into the wave equation (2.1.5), leads to

$$\nabla^2 \mathbf{A} - \frac{1}{c^2} \frac{\partial^2 \mathbf{A}}{\partial t^2} = -\frac{4\pi}{c} \mathbf{J}^{\mathrm{T}} \quad .$$

From the previous two relations we see that the longitudinal current density $\mathbf{J}^{\mathrm{L}}$ is only connected to the scalar potential Φ, and the transverse current density $\mathbf{J}^{\mathrm{T}}$ is solely determined by the vector potential $\mathbf{A}$. Similar relations can be obtained for the electric field. Since $\mathbf{A}$ has no longitudinal component as just derived, Eq. (2.1.17) can be split into two components: the longitudinal part

$$\mathbf{E}^{\mathrm{L}} = -\mathrm{i}\mathbf{q}\Phi \quad , \tag{3.1.2}$$

which vanishes if $\rho_{\mathrm{ext}} = 0$ as a consequence of Poisson's equation (2.1.18), and the transverse electric field

$$\mathbf{E}^{\mathrm{T}} = \mathrm{i}\frac{\omega}{c}\mathbf{A} \quad . \tag{3.1.3}$$

Again, the longitudinal component of $\mathbf{E}$ is related to the electrical potential Φ, whereas the vector potential $\mathbf{A}$ determines the transverse part of the electric field. Looking at two of Maxwell's equations (2.2.7a and d)

$$\nabla \cdot (\hat{\epsilon}\mathbf{E}) = \nabla \cdot \hat{\epsilon}\mathbf{E}^{\mathrm{L}} = 4\pi \rho_{\mathrm{ext}} \quad ,$$

$$\nabla \times \mathbf{E} = \nabla \times \mathbf{E}^{\mathrm{T}} = -\frac{1}{c}\frac{\partial \mathbf{B}}{\partial t} \quad ,$$

---

[1] This statement is valid in the absence of boundaries, and it is not valid, for example, at the surface of a metal. There the propagation of electromagnetic waves with longitudinal electric field or magnetic field components is possible; these, for example, are the TE and TM modes which can propagate in waveguides and will be discussed in Section 9.1.

it becomes clear that the longitudinal component corresponds to the rearrangement of the electronic charge, whereas the transverse component is related to the induced electrical currents.

3.1.2 Material parameters

Our next goal is to describe the longitudinal and transverse responses in terms of the parameters which characterize the medium: the dielectric constant and the magnetic permeability. Let us confine ourselves to isotropic and homogeneous media, although some of the relations are more general. The longitudinal and transverse dielectric constant can be described using a tensor notation:

$$\mathbf{D}(\mathbf{q}, \omega) = \bar{\bar{\epsilon}}_1(\mathbf{q}, \omega)\mathbf{E}(\mathbf{q}, \omega) \quad , \tag{3.1.4}$$

where the tensor components $(\epsilon_1)_{ij}(\mathbf{q}, \omega)$ can in general be written as

$$(\epsilon_1)_{ij}(\mathbf{q}, \omega) = \epsilon_1^{\mathrm{L}}(\mathbf{q}, \omega)\frac{\mathbf{q}_i \circ \mathbf{q}_j}{q^2} + \epsilon_1^{\mathrm{T}}(\mathbf{q}, \omega)\left[\delta_{ij} - \frac{\mathbf{q}_i \circ \mathbf{q}_j}{q^2}\right] \quad , \tag{3.1.5}$$

with $\epsilon_1^{\mathrm{L}}(\mathbf{q}, \omega)$ the diagonal and $\epsilon_1^{\mathrm{T}}(\mathbf{q}, \omega)$ the off-diagonal components of the dielectric tensor. The dyad which projects out the longitudinal component is defined to be $\bar{\bar{\mathbf{1}}}^{\mathrm{L}} = \mathbf{q} \circ \mathbf{q}/q^2$; and the one which yields the transverse component is $\bar{\bar{\mathbf{1}}}^{\mathrm{T}} = \bar{\bar{\mathbf{1}}} - \bar{\bar{\mathbf{1}}}^{\mathrm{L}} = \bar{\bar{\mathbf{1}}} - \mathbf{q} \circ \mathbf{q}/q^2$. We decompose the electric displacement $\mathbf{D}(\mathbf{q}, \omega)$ into two orthogonal parts

$$\begin{aligned}\mathbf{D}(\mathbf{q}, \omega) \quad &- \quad [\mathbf{n_q} \cdot \mathbf{D}(\mathbf{q}, \omega)]\mathbf{n_q} + [\mathbf{n_q} \times \mathbf{D}(\mathbf{q}, \omega)] \times \mathbf{n_q} \\ &= \quad \frac{\mathbf{q} \cdot \mathbf{D}(\mathbf{q}, \omega)}{q^2}\mathbf{q} + \frac{\mathbf{q} \times \mathbf{D}(\mathbf{q}, \omega)}{q^2} \times \mathbf{q} \quad .\end{aligned}$$

Equation (3.1.4) can now be written as

$$\mathbf{D}(\mathbf{q}, \omega) = \epsilon_1^{\mathrm{L}}(\mathbf{q}, \omega)\frac{\mathbf{q} \cdot \mathbf{E}(\mathbf{q}, \omega)}{q^2}\mathbf{q} + \epsilon_1^{\mathrm{T}}(\mathbf{q}, \omega)\frac{\mathbf{q} \times \mathbf{E}(\mathbf{q}, \omega)}{q^2} \times \mathbf{q} \quad , \tag{3.1.6}$$

and we find the following relationship between the electric field strength $\mathbf{E}$ and electrical displacement $\mathbf{D}$ for the two components:

$$\begin{aligned}\mathbf{D}^{\mathrm{L}}(\mathbf{q}, \omega) &= \epsilon_1^{\mathrm{L}}(\mathbf{q}, \omega)\mathbf{E}^{\mathrm{L}}(\mathbf{q}, \omega) \tag{3.1.7}\\ \mathbf{D}^{\mathrm{T}}(\mathbf{q}, \omega) &= \epsilon_1^{\mathrm{T}}(\mathbf{q}, \omega)\mathbf{E}^{\mathrm{T}}(\mathbf{q}, \omega) \quad ; \tag{3.1.8}\end{aligned}$$

i.e. for an isotropic and homogeneous medium the longitudinal and transverse components of the dielectric constant are independent response functions which do not mix. From Eq. (3.1.2) we see that the longitudinal dielectric constant describes the response of the medium to a scalar potential Φ. This in general arises due to an additional charge, leading to the rearrangement of the initial electric charge distribution. Equation (3.1.3) states that the transverse dielectric constant describes the response of the medium to a vector potential $\mathbf{A}$, related to the presence of electromagnetic radiation.

Maxwell's equations (2.2.7), in Fourier transformed version, can now be written as

$$\mathbf{q} \times \mathbf{E}(\mathbf{q}, \omega) \;=\; \frac{\omega}{c}\mathbf{B}(\mathbf{q}, \omega) \tag{3.1.9a}$$

$$\mathbf{q} \cdot \mathbf{B}(\mathbf{q}, \omega) \;=\; 0 \tag{3.1.9b}$$

$$i\mathbf{q} \times \mathbf{H}(\mathbf{q}, \omega) \;=\; -\frac{i\omega}{c}\mathbf{D}(\mathbf{q}, \omega) + \frac{4\pi}{c}\mathbf{J}(\mathbf{q}, \omega) \tag{3.1.9c}$$

$$i\epsilon_1^{\mathrm{L}}(\mathbf{q}, \omega)\mathbf{q} \cdot \mathbf{E}(\mathbf{q}, \omega) \;=\; 4\pi\rho(\mathbf{q}, \omega) \quad, \tag{3.1.9d}$$

if we assume a harmonic spatial and time dependence of the fields, given in the usual way by Eqs (2.2.14). For simplicity, we do not include explicitly the wavevector and frequency dependence of μ_1. Comparing Eq. (3.1.9c) with the transformed Eq. (2.2.7c), $(i/\mu_1)\mathbf{q} \times \mathbf{B}(\mathbf{q}, \omega) = -(i\omega/c)\epsilon_1(\mathbf{q}, \omega)\mathbf{E}(\mathbf{q}, \omega) + (4\pi/c)\mathbf{J}(\mathbf{q}, \omega)$, we obtain for the relationship between displacement and electric field

$$\left(1 - \frac{1}{\mu_1}\right)\mathbf{q} \times [\mathbf{q} \times \mathbf{E}(\mathbf{q}, \omega)] - \frac{\omega^2}{c^2}\epsilon_1(\mathbf{q}, \omega)\mathbf{E}(\mathbf{q}, \omega) = -\frac{\omega^2}{c^2}\mathbf{D}(\mathbf{q}, \omega) \quad. \tag{3.1.10}$$

Utilizing Eq. (3.1.6) to substitute $\mathbf{D}(\mathbf{q}, \omega)$ yields

$$\left[\frac{q^2 c^2}{\omega^2}\left(1 - \frac{1}{\mu_1}\right) + \epsilon_1(\mathbf{q}, \omega)\right]\mathbf{q} \times [\mathbf{q} \times \mathbf{E}(\mathbf{q}, \omega)] - \epsilon_1(\mathbf{q}, \omega)[\mathbf{q} \cdot \mathbf{E}(\mathbf{q}, \omega)]\mathbf{q}$$
$$= \epsilon_1^{\mathrm{T}}(\mathbf{q}, \omega)\mathbf{q} \times [\mathbf{q} \times \mathbf{E}(\mathbf{q}, \omega)] - \epsilon_1^{\mathrm{L}}(\mathbf{q}, \omega)[\mathbf{q} \cdot \mathbf{E}(\mathbf{q}, \omega)]\mathbf{q} \quad; \tag{3.1.11}$$

which finally leads to

$$q^2 \left(1 - \frac{1}{\mu_1}\right) = \frac{\omega^2}{c^2}[\epsilon_1^{\mathrm{T}}(\mathbf{q}, \omega) - \epsilon_1^{\mathrm{L}}(\mathbf{q}, \omega)] \quad. \tag{3.1.12}$$

In the limit $\mathbf{q}c/\omega \to 0$, both components of the dielectric constant become equal: $\epsilon_1^{\mathrm{T}}(0, \omega) = \epsilon_1^{\mathrm{L}}(0, \omega)$. For long wavelength we cannot distinguish between a longitudinal and transverse electric field in an isotropic medium.

For $\mathbf{J}(\mathbf{q}, \omega) = 0$, Eq. (3.1.10) can be simplified one step further: $\frac{ic}{\omega}\mathbf{q} \times [\mathbf{q} \times \mathbf{E}(\mathbf{q}, \omega)] = -\frac{i\omega}{c}\mathbf{D}(\mathbf{q}, \omega)$, and the dispersion relation of the dielectric constant is then

$$\left[\frac{c^2}{\omega^2}q^2 - \epsilon_1^{\mathrm{T}}(\mathbf{q}, \omega)\right]\mathbf{q} \times [\mathbf{q} \times \mathbf{E}(\mathbf{q}, \omega)] + \epsilon_1^{\mathrm{L}}(\mathbf{q}, \omega)[\mathbf{q} \cdot \mathbf{E}(\mathbf{q}, \omega)]\mathbf{q} = 0 \quad, \tag{3.1.13}$$

after substitution of Eq. (3.1.6) for $\mathbf{D}(\mathbf{q}, \omega)$. Since this equation has to be valid for any $\mathbf{E}$, both the longitudinal and the transverse components have to vanish independently at certain $\mathbf{q}$ and ω values. Thus we find the frequencies ω_T and ω_L from the two conditions

$$q^2 - \frac{\omega_T^2}{c^2}\epsilon_1^T(\mathbf{q}, \omega_T) = 0 \qquad (3.1.14)$$

$$\epsilon_1^L(\mathbf{q}, \omega_L) = 0 \quad . \qquad (3.1.15)$$

At these two frequencies, oscillations in the transverse and in the longitudinal directions are sustained. The consequence of these equations will be discussed in detail in Section 5.4.4, where we see that the vanishing $\epsilon_1(\omega)$ can lead to collective excitations. The characteristic frequencies ω_T and ω_L are called transverse and longitudinal plasma frequencies, respectively, because they describe the resonance frequency of the electronic charges moving against the positive ions.

To find expressions for the longitudinal and transverse components of the optical conductivity, we separate Ampère's law (3.1.9c) into transverse and longitudinal components. Due to Eq. (3.1.9b) the magnetic induction is purely transverse ($\mathbf{B}^L = 0$); hence, the longitudinal part simplifies to

$$0 = -i\omega\epsilon_1^L \mathbf{E}_{\text{total}}^L(\mathbf{q}, \omega) + 4\pi \mathbf{J}_{\text{total}}(\mathbf{q}, \omega) \quad , \qquad (3.1.16)$$

and, after separating the total current $\mathbf{J}_{\text{total}} = \mathbf{J}_{\text{ind}} + \mathbf{J}_{\text{ext}}$, yields

$$i\omega\left(\epsilon_1^L + i\frac{4\pi\sigma_1}{\omega}\right)\mathbf{E}_{\text{total}}^L(\mathbf{q}, \omega) = \omega q \Phi - 4\pi \mathbf{J}_{\text{ind}}(\mathbf{q}, \omega) \qquad (3.1.17)$$

by using the Fourier transform of Eq. (3.1.1). Rearranging the terms and using Eqs (2.2.8) and (3.1.2), this eventually leads to

$$\mathbf{J}_{\text{ind}}^L(\mathbf{q}, \omega) = \frac{i\omega}{4\pi}(1 - \hat{\epsilon}^L)\mathbf{E}_{\text{total}}^L(\mathbf{q}, \omega) \quad . \qquad (3.1.18)$$

Similar considerations can be made for the transverse direction; however, $\mathbf{B}_{\text{total}}^T \neq 0$, and the solution to Eq. (3.1.9c) is more complicated. According to Eq. (3.1.12) we eventually arrive at

$$\mathbf{J}_{\text{ind}}^T(\mathbf{q}, \omega) = \frac{i\omega}{4\pi}\left[(1 - \hat{\epsilon}^L) - \frac{c^2}{\omega^2}q^2\left(1 - \frac{1}{\mu_1}\right)\right]\mathbf{E}_{\text{total}}^T(\mathbf{q}, \omega) \quad . \qquad (3.1.19)$$

For $\mu_1 = 1$ and $\mathbf{q} = 0$ these two equations can be combined in the general form

$$\mathbf{J}_{\text{ind}}^{L,T}(\mathbf{q}, \omega) = \frac{i\omega}{4\pi}(1 - \hat{\epsilon}^{L,T})\mathbf{E}_{\text{total}}^{L,T}(\mathbf{q}, \omega) \quad . \qquad (3.1.20)$$

In the limit of long wavelengths, the behavior for transverse and longitudinal responses becomes similar. We can also define a complex conductivity $\hat{\sigma} = (i\omega/4\pi)(1 - \hat{\epsilon})$ in agreement with Eq. (2.2.12), so that

$$\mathbf{J}_{\text{ind}}^{L,T}(\mathbf{q}, \omega) = \hat{\sigma}(\mathbf{q}, \omega)\mathbf{E}_{\text{total}}^{L,T}(\mathbf{q}, \omega) \quad . \qquad (3.1.21)$$

The current density as a response to an electric field is described by a transverse or a longitudinal conductivity. Our goal for the chapters to follow is to calculate this response.

3.1.3 Response to longitudinal fields

Let us apply an external longitudinal electric field $\mathbf{E}^{L}$ and assume that the material equation has the general form given in Eq. (2.2.13), with $\hat{\epsilon}$ the time and space dependent, or alternatively through the Fourier transformation the frequency and wavevector dependent, dielectric constant. The electric field is related to the external charge by Poisson's equation (2.2.7d), $\nabla \cdot \mathbf{D}^{L}(\mathbf{r}, t) = 4\pi \rho_{\text{ext}}(\mathbf{r}, t)$. The external displacement field $\mathbf{D}^{L}$ leads to a rearrangement of the charge density and, in turn, to a space charge field $\mathbf{E}_{\text{pol}}$. The total electric field is then given by

$$\mathbf{E}^{L}(\mathbf{r}, t) = \mathbf{D}^{L}(\mathbf{r}, t) + \mathbf{E}_{\text{pol}}(\mathbf{r}, t) \quad ,$$

where the space charge field $\mathbf{E}_{\text{pol}}(\mathbf{r}, t)$ is related to the polarization charge density by

$$\nabla \cdot \mathbf{E}_{\text{pol}}(\mathbf{r}, t) = 4\pi \rho_{\text{ind}}(\mathbf{r}, t) \quad .$$

By combining the previous equations, we obtain the Poisson relation

$$\nabla \cdot \mathbf{E}^{L}(\mathbf{r}, t) = 4\pi [\rho_{\text{ext}}(\mathbf{r}, t) + \rho_{\text{ind}}(\mathbf{r}, t)] \quad .$$

In summary, the spatial and temporal Fourier transforms of the longitudinal displacement, electric field, and space charge field lead to

$$
\begin{aligned}
i\mathbf{q} \cdot \mathbf{D}^{L}(\mathbf{q}, \omega) &= 4\pi \rho_{\text{ext}}(\mathbf{q}, \omega) \quad , & \text{(3.1.22a)}\\
i\mathbf{q} \cdot \mathbf{E}^{L}(\mathbf{q}, \omega) &= 4\pi [\rho_{\text{ext}}(\mathbf{q}, \omega) + \rho_{\text{ind}}(\mathbf{q}, \omega)] = 4\pi \rho(\mathbf{q}, \omega) \quad , & \text{(3.1.22b)}\\
i\mathbf{q} \cdot \mathbf{E}_{\text{pol}}(\mathbf{q}, \omega) &= 4\pi \rho_{\text{ind}}(\mathbf{q}, \omega) \quad , & \text{(3.1.22c)}
\end{aligned}
$$

where $\rho(\mathbf{q}, \omega) = \rho_{\text{ext}}(\mathbf{q}, \omega) + \rho_{\text{ind}}(\mathbf{q}, \omega) = \rho_{\text{total}}(\mathbf{q}, \omega)$, the total charge density.

Within the framework of the linear approximation, we can define a wavevector and frequency dependent dielectric constant $\hat{\epsilon}(\mathbf{q}, \omega)$ (which in the case of an isotropic medium is a scalar):

$$\mathbf{E}^{L}(\mathbf{q}, \omega) = \frac{\mathbf{D}^{L}(\mathbf{q}, \omega)}{\hat{\epsilon}^{L}(\mathbf{q}, \omega)} \quad . \tag{3.1.23}$$

Substituting the material equation (2.2.5), $\mathbf{D}^{L}(\mathbf{q}, \omega) = \mathbf{E}^{L}(\mathbf{q}, \omega) + 4\pi \mathbf{P}(\mathbf{q}, \omega)$, the polarization $\mathbf{P}$ is related to the space charge field $\mathbf{E}_{\text{pol}}$ by

$$\mathbf{P}(\mathbf{q}, \omega) = -\frac{1}{4\pi} \mathbf{E}_{\text{pol}}(\mathbf{q}, \omega) \quad . \tag{3.1.24}$$

In accordance with Eq. (2.2.5), the longitudinal dielectric susceptibility $\hat{\chi}_e^L(\mathbf{q}, \omega)$ is defined as the parameter relating the polarization to the applied electric field: $\mathbf{P}(\mathbf{q}, \omega) = \hat{\chi}_e^L(\mathbf{q}, \omega)\mathbf{E}^L(\mathbf{q}, \omega)$. Substituting this into Eq. (2.2.5) and comparing it with Eq. (3.1.23) yields a relationship between the dielectric susceptibility $\hat{\chi}_e^L$ and the dielectric constant $\hat{\epsilon}^L$:

$$\hat{\chi}_e^L(\mathbf{q}, \omega) = \frac{\hat{\epsilon}^L(\mathbf{q}, \omega) - 1}{4\pi} \quad . \tag{3.1.25}$$

Using Eqs (3.1.22), one can write $\hat{\chi}_e^L(\mathbf{q}, \omega)$ in terms of a scalar quantity such as the charge density. By substituting the definition of $\hat{\chi}_e^L(\mathbf{q}, \omega)$ into Eq. (3.1.24), we obtain

$$-\mathbf{E}_{\text{pol}}(\mathbf{q}, \omega) = 4\pi\,\hat{\chi}_e^L(\mathbf{q}, \omega)\mathbf{E}^L(\mathbf{q}, \omega) \quad .$$

If we now multiply both sides by $\mathbf{iq}$ and substitute Eqs (3.1.22a) and (3.1.22b), we obtain for the dielectric susceptibility

$$\hat{\chi}_e^L(\mathbf{q}, \omega) = -\frac{1}{4\pi}\frac{\rho_{\text{ind}}(\mathbf{q}, \omega)}{\rho(\mathbf{q}, \omega)} \quad . \tag{3.1.26}$$

Thus, the dielectric susceptibility is the ratio of the induced charge density to the total charge density, which by virtue of Eq. (3.1.25) can also be written in terms of the dielectric constant

$$\hat{\epsilon}^L(\mathbf{q}, \omega) = 1 - \frac{\rho_{\text{ind}}(\mathbf{q}, \omega)}{\rho(\mathbf{q}, \omega)} = \frac{\rho_{\text{ext}}(\mathbf{q}, \omega)}{\rho(\mathbf{q}, \omega)} \quad . \tag{3.1.27}$$

The dielectric constant is the ratio of external charge to the total charge. Alternatively, $\hat{\epsilon}^L(\mathbf{q}, \omega)$ can be related to the external and induced charge density since, by using Eq. (3.1.27), we find

$$\frac{1}{\hat{\epsilon}^L(\mathbf{q}, \omega)} = 1 + \frac{\rho_{\text{ind}}(\mathbf{q}, \omega)}{\rho_{\text{ext}}(\mathbf{q}, \omega)} \quad . \tag{3.1.28}$$

The imaginary part of $1/\hat{\epsilon}$ is the loss function, so-called for reasons explained later. Equations (3.1.27) and (3.1.28) are often referred to as the selfconsistent field approximation (SCF) or random phase approximation (RPA) [Mah90]. The major assumption in the derivation is that the electrons respond to the total charge density, which initially is not known, as it is the result of the calculation; we try to evaluate this function selfconsistently. If the interaction of an electron with the induced fields in the medium is weak and $\rho_{\text{ind}} \ll \rho$, we can assume $\rho \approx \rho_{\text{ext}}$ and find

$$\left(\frac{1}{\hat{\epsilon}^L(\mathbf{q}, \omega)}\right)_{\text{HF}} = 1 + \frac{\rho_{\text{ind}}(\mathbf{q}, \omega)}{\rho(\mathbf{q}, \omega)} \quad . \tag{3.1.29}$$

This approach is referred to as the Hartree–Fock (HF) approximation. One can also express the susceptibility $\hat{\chi}_e^L(\mathbf{q}, \omega)$ in terms of the scalar potential; the total quantity $\Phi(\mathbf{q}, \omega) = \Phi_{\text{ext}}(\mathbf{q}, \omega) + \Phi_{\text{ind}}(\mathbf{q}, \omega)$ is often referred to as the screened potential. Poisson's equation (2.1.7) can be transformed in Fourier form: $-q^2\Phi(\mathbf{q}, \omega) = -4\pi\rho(\mathbf{q}, \omega) = -4\pi[\rho_{\text{ext}}(\mathbf{q}, \omega) + \rho_{\text{ind}}(\mathbf{q}, \omega)]$. Thus we obtain

$$\hat{\chi}_e^L(\mathbf{q}, \omega) = -\frac{1}{q^2}\frac{\rho_{\text{ind}}(\mathbf{q}, \omega)}{\Phi(\mathbf{q}, \omega)} \quad , \tag{3.1.30}$$

and

$$\hat{\epsilon}^L(\mathbf{q}, \omega) = 1 + 4\pi\,\hat{\chi}_e^L(\mathbf{q}, \omega) = 1 - \frac{4\pi}{q^2}\frac{\rho_{\text{ind}}(\mathbf{q}, \omega)}{\Phi(\mathbf{q}, \omega)} \quad . \tag{3.1.31}$$

If we are interested in the screened potential $\Phi(\mathbf{q}, \omega)$, the appropriate response function[2] $\hat{\chi}^L(\mathbf{q}, \omega)$ is that which describes the charge $\rho_{\text{ind}}(\mathbf{q}, \omega)$ induced in response to the screened potential $\Phi(\mathbf{q}, \omega)$:

$$\hat{\chi}^L(\mathbf{q}, \omega) = \frac{\rho_{\text{ind}}(\mathbf{q}, \omega)}{\Phi(\mathbf{q}, \omega)} = \frac{q^2}{4\pi}[1 - \hat{\epsilon}^L(\mathbf{q}, \omega)] \quad , \tag{3.1.32}$$

or in the inverted form $\hat{\epsilon}^L(\mathbf{q}, \omega) = 1 - (4\pi/q^2)\hat{\chi}^L(\mathbf{q}, \omega)$. In the case of only free charge carriers, as in a metal, the real part χ_1 is always positive. The equation $\rho_{\text{ind}}(\mathbf{q}, \omega) = \hat{\chi}^L(\mathbf{q}, \omega)\Phi(\mathbf{q}, \omega)$ can be compared with Ohm's law, Eq. (2.2.11): in both cases, the response functions ($\hat{\sigma}$ and $\hat{\chi}$) relate the response of the medium [$\rho(\mathbf{q}, \omega)$ and $\mathbf{J}(\mathbf{q}, \omega)$] to the total (i.e. screened) perturbation ($\mathbf{E}$ and Φ, respectively). This was first derived by Lindhard [Lin54]; $\hat{\chi}$ is consequently called the Lindhard dielectric function.

From Eqs (3.1.27) and (3.1.22b) we can derive an expression for the dielectric constant in terms of the electric field and displacement field:

$$\hat{\epsilon}^L(\mathbf{q}, \omega) = 1 + \frac{4\pi\,i\rho_{\text{ind}}(\mathbf{q}, \omega)}{\mathbf{q}\cdot\mathbf{E}^L(\mathbf{q}, \omega)} \quad : \tag{3.1.33}$$

whereas using Eqs (3.1.28) and (3.1.22b) gives

$$\frac{1}{\hat{\epsilon}^L(\mathbf{q}, \omega)} = 1 - \frac{4\pi\,i\rho_{\text{ind}}(\mathbf{q}, \omega)}{\mathbf{q}\cdot\mathbf{D}^L(\mathbf{q}, \omega)} \quad . \tag{3.1.34}$$

The significance of the above equations is clear: if the response to an external field $\mathbf{E}(\mathbf{q}, \omega)$ has to be evaluated, $1/\hat{\epsilon}(\mathbf{q}, \omega)$ is the appropriate response function, whereas the response to a screened field is best described by $\hat{\epsilon}(\mathbf{q}, \omega)$.

We have also seen that the conductivity connects the local current density to the local electric field via Ohm's law (2.2.11), and therefore $\mathbf{J}^L(\mathbf{q}, \omega) = \hat{\sigma}^L(\mathbf{q}, \omega)\mathbf{E}^L(\mathbf{q}, \omega)$. The continuity equation (2.1.9) for charge and current is $\nabla\cdot\mathbf{J}^L(\mathbf{q}, \omega) + \partial\rho_{\text{ind}}/\partial t = 0$, and for a current which is described by the usual

---

[2] Commonly, the dielectric response function is called $\hat{\chi}$, which must not lead to confusion with the electric susceptibility $\hat{\chi}_e$ or the magnetic susceptibility $\hat{\chi}_m$. Sometimes χ is called generalized susceptibility.

harmonic wavefunction $\exp\{i(\mathbf{q} \cdot \mathbf{r} - \omega t)\}$, the above equation can be written as $\mathbf{q} \cdot \mathbf{J}^{L}(\mathbf{q}, \omega) = \omega \rho_{\mathrm{ind}}(\mathbf{q}, \omega)$. Using Eqs (3.1.33) and (2.2.11) finally leads to

$$\hat{\epsilon}^{L}(\mathbf{q}, \omega) = 1 + \frac{4\pi i}{\omega} \hat{\sigma}^{L}(\mathbf{q}, \omega) \quad . \tag{3.1.35}$$

This is the well known relationship between the dielectric constant and the conductivity.

3.1.4 Response to transverse fields

Similarly to longitudinal fields, a simple relationship between the electric field and the induced current can also be derived for transverse fields. In a similar way as the charge density ρ is connected to the scalar potential Φ by the Lindhard function, $\rho = \hat{\chi}\Phi$, now the current density $\mathbf{J}$ is related to the vector potential $\mathbf{A}$ by the conductivity; the proportionality factor is

$$\mathbf{J} = \hat{\sigma}\mathbf{E} = \frac{i\hat{\sigma}\omega}{c}\mathbf{A} \quad . \tag{3.1.36}$$

Thus we can define a dielectric constant for the transverse response according to Eq. (3.1.8); this gives the relationship between the transverse conductivity and the transverse dielectric constant:

$$\hat{\epsilon}^{T}(\mathbf{q}, \omega) = 1 + \frac{4\pi i}{\omega} \hat{\sigma}^{T}(\mathbf{q}, \omega) \quad , \tag{3.1.37}$$

the same as obtained previously for the longitudinal conductivity and dielectric constant.

The analogy between longitudinal and transverse fields ends here; it is not possible to define a transverse dielectric constant corresponding to the longitudinal case. Assuming a homogeneous isotropic medium, the divergence of the transverse fields vanishes by definition. With $\mathbf{q} \cdot \mathbf{E}^{T} = \mathbf{q} \cdot \mathbf{D}^{T} = 0$ the denominator in Eqs (3.1.33) and (3.1.34) is zero.

3.1.5 The anisotropic medium: dielectric tensor

As mentioned earlier, the strict separation of the longitudinal and transverse response as summarized in Eq. (3.1.20) only holds for non-magnetic ($\mu_1 = 1$) and isotropic materials. In anisotropic media the situation is different: the polarization and induced currents can generally flow in any direction that is different from that of the electric field, and it is possible to mix both components to induce, for example, a longitudinal current with a purely transverse electric field. The situation can be described by the dielectric tensor $\bar{\bar{\epsilon}}$. The nine components of $\bar{\bar{\epsilon}}$ are not all independent from each other.

In the absence of an external magnetic field, even for anisotropic media, the real part of the complex dielectric tensor is symmetric (the Onsager relation): it is always possible to find a set of principal dielectric axes such that the real dielectric tensor can be diagonalized. The imaginary part of the dielectric tensor (i.e. the conductivity tensor) is also symmetric and can be put into diagonal form. However, the directions of the principal axes of these two tensors are in general not the same. They do, however, coincide for crystals with symmetry at least as high as orthorhombic [Nye57].

A mixing of longitudinal and transverse responses also occurs if the medium is bounded, i.e. in the presence of surfaces. In particular if the frequency ω is comparable to the plasma frequency ω_p, longitudinal plasma waves may occur.

3.2 Kramers–Kronig relations and sum rules

The various material parameters and optical constants introduced in the previous section describe the response of the medium to applied electromagnetic radiation within the framework of linear response theory; the frequency dependence of the response is called dispersion. The complex dielectric constant $\hat{\epsilon}$ and the complex conductivity $\hat{\sigma}$ can be regarded as the prime response functions of the material, describing the electric polarization and current induced in response to the applied electric field. The change of the electromagnetic wave in the material was discussed in terms of the refractive index $\hat{N}$ and in terms of the complex surface impedance $\hat{Z}_S$, for example; these are also complex response functions. For the optical parameters, such as the amplitude of the electromagnetic wave which is transmitted through an interface or which is reflected off a boundary of two materials, the second components are the phase shifts ϕ_t and ϕ_r, respectively, which are experienced by the electromagnetic fields there; both components constitute a response function. Hence we always deal with complex response functions describing the response of the system to a certain stimulus; it always contained possible dissipation and some phase change. General considerations, involving causality, can be used to derive important relations between the real and imaginary parts of the complex response functions. They were first given by Kramers [Kra26] and Kronig [Kro26], and play an important role, not only in the theory of response functions [Bod45, Mac56].

These relations are also of great practical importance: they allow for the evaluation of the components of the complex dielectric constant or conductivity when only one optical parameter such as the reflected or absorbed power is measured. With $R(\omega)$ obtained over a broad frequency range, the dispersion relations can be utilized to evaluate $\phi_r(\omega)$. There are two restrictions of practical importance: the data have to cover a wide spectral range and the sample must not be transparent.

If just the transmitted power through a sample of finite thickness is measured, the Kramers–Kronig analysis does not allow the determination of both components without knowing the reflected portion.

These dispersion relations can also be used – together with physical arguments about the behavior of the response in certain limits – to derive what are called sum rules.

3.2.1 Kramers–Kronig relations

For the derivation of the general properties we assume a linear response to an external perturbation given in the form of

$$\hat{X}(\mathbf{r}, t) = \int \int_{-\infty}^{\infty} \hat{G}(\mathbf{r}, \mathbf{r}', t, t') \hat{f}(\mathbf{r}', t') \, d\mathbf{r}' \, dt' \quad . \tag{3.2.1}$$

This describes the response $\hat{X}$ of the system at time t and position $\mathbf{r}$ to an external stimulus $\hat{f}$ at times t' and locations $\mathbf{r}'$. The function $\hat{G}(\mathbf{r}, \mathbf{r}', t, t')$ is called the response function, and may be the conductivity, the dielectric constant, the susceptibility, or any other optical constant, such as the refractive index. Since the origin of the time scale should not be of physical significance, $\hat{G}(t, t')$ is a function of the difference of the argument $t - t'$ only. In the following we will neglect the spatial dependence of the external perturbation[3] and restrict ourselves to the local approximation: we assume that the response at a particular position $\mathbf{r}$ depends only on the field which exists at that particular place (described by the delta function: $\delta\{x\} - 1$ if $x = 0$):

$$\hat{G}(\mathbf{r}, \mathbf{r}', t, t') = \delta\{\mathbf{r} - \mathbf{r}'\}\hat{G}(t - t') \quad .$$

The validity of this approximation is discussed at length in later chapters. We also assume that the medium is isotropic and homogeneous, and thus $\hat{G}$ is a scalar. With these assumptions, Eq. (3.2.1) becomes

$$\hat{X}(t) = \int_{-\infty}^{\infty} \hat{G}(t - t') \hat{f}(t') \, dt' \quad , \tag{3.2.2}$$

where we suppress the spatial dependence for above reasons. If the system is required to be causal, then

$$\hat{G}(t - t') = 0 \qquad \text{for} \qquad t < t' \quad , \tag{3.2.3}$$

which basically means that there is no response prior to the stimulus and we can simplify $\int_{-\infty}^{\infty} dt'$ to $\int_{-\infty}^{t} dt'$. Further analysis is more convenient in Fourier space,

---

[3] For a discussion of non-local effects on the derivation of the Kramers–Kronig relations and sum rules, see [Mar67].

with the spectral quantities defined as

$$\hat{f}(\omega) \;=\; \int \hat{f}(t)\exp\{i\omega t\}\,dt \quad , \tag{3.2.4a}$$

$$\hat{X}(\omega) \;=\; \int \hat{X}(t)\exp\{i\omega t\}\,dt \quad , \tag{3.2.4b}$$

$$\hat{G}(\omega) \;=\; \int \hat{G}(t-t')\exp\{i\omega(t-t')\}\,dt \tag{3.2.4c}$$

leading to the convolution

$$\begin{aligned}
\hat{X}(\omega) \;&=\; \int dt\,\exp\{i\omega t\}\left[\int \hat{G}(t-t')\hat{f}(t')\,dt'\right] \\
&=\; \int dt'\,\hat{f}(t')\exp\{i\omega t'\}\left[\int \hat{G}(t-t')\exp\{i\omega(t-t')\}\,dt\right] \\
&=\; \hat{G}(\omega)\hat{f}(\omega) \quad .
\end{aligned} \tag{3.2.5}$$

$\hat{G}(\omega)$ is also often referred to as the frequency dependent generalized susceptibility. In general, it is a complex quantity with the real component describing the attenuation of the signal and the imaginary part reflecting the phase difference between the external perturbation and the response.

For mathematical reasons, let us assume that the frequency which appears in the previous equations is complex, $\hat{\omega} = \omega_1 + i\omega_2$; then from Eq. (3.2.4c) we obtain

$$\hat{G}(\hat{\omega}) = \int \hat{G}(t-t')\exp\{i\omega_1(t-t')\}\exp\{-\omega_2(t-t')\}\,dt \quad . \tag{3.2.6}$$

The factor $\exp\{-\omega_2(t-t')\}$ is bounded in the upper half of the complex plane for $t-t' > 0$ and in the lower half plane for $t-t' < 0$, and $\hat{G}(t-t')$ is finite for all $t-t'$. The required causality (Eq. (3.2.3)) hence limits $\hat{G}(\hat{\omega})$ to the upper half of the $\hat{\omega}$ plane. Let us consider a contour shown in Fig. 3.1 with a small indentation near to the frequency ω_0. Because the function is analytic (i.e. no poles) in the upper half plane, Cauchy's theorem applies:

$$\oint_C \frac{\hat{G}(\hat{\omega}')}{\hat{\omega}' - \hat{\omega}_0}\,d\hat{\omega}' = 0 \quad .$$

For a detailed discussion of the essential requirements and properties of the response function $\hat{G}(\hat{\omega})$, for example its boundedness and linearity, see [Bod45, Lan80], for instance. While the integral over the large semicircle vanishes[4] as $\hat{G}(\hat{\omega}') \to 0$ when $\hat{\omega}' \to \infty$, the integral over the small semicircle of radius η can be evaluated using $\hat{\omega}' = \hat{\omega}_0 - \eta\exp\{i\phi\}$; then, by using the general relation

---

[4] This is not a serious restriction since we can redefine $\hat{G}$ in the appropriate way as $\hat{G}(\omega) - \hat{G}(\infty)$.

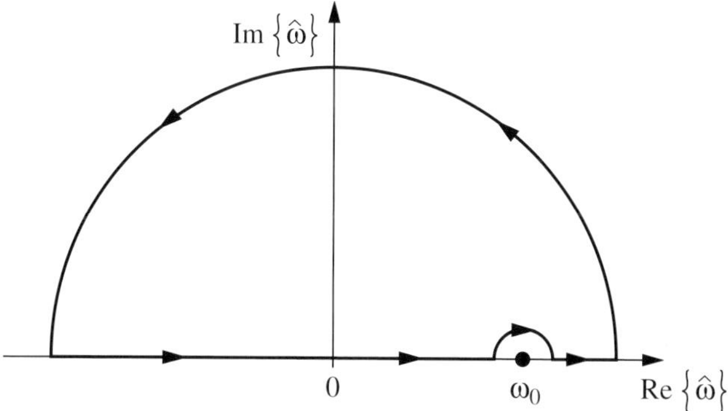

Fig. 3.1. Integration contour in the complex frequency plane. The limiting case is considered where the radius of the large semicircle goes to infinity while the radius of the small semicircle around ω_0 approaches zero. If the contribution to the integral from the former one vanishes, only the integral along the real axis from $-\infty$ to ∞ remains.

(sometimes called the Dirac identity)

$$\lim_{\eta \to 0} \frac{1}{x \pm i\eta} = \mathcal{P}\left\{\frac{1}{x}\right\} \mp i\pi\,\delta(x) \quad , \tag{3.2.7}$$

we obtain

$$\lim_{\eta \to 0} \oint_\eta \frac{\hat{G}(\hat{\omega}')}{\hat{\omega}' - \hat{\omega}_0}\,\mathrm{d}\hat{\omega}' = -i\pi\,\hat{G}(\hat{\omega}_0) \quad .$$

This integral along the real frequency axis gives the principal value $\mathcal{P}$. Therefore

$$\hat{G}(\omega) = \frac{1}{i\pi}\mathcal{P}\int_{-\infty}^{\infty} \frac{\hat{G}(\omega')}{\omega' - \omega}\,\mathrm{d}\omega' \quad , \tag{3.2.8}$$

where we have omitted the subscript of the frequency ω_0. In the usual way, the complex response function $\hat{G}(\omega)$ can be written in terms of the real and imaginary parts as $\hat{G}(\omega) = G_1(\omega) + iG_2(\omega)$, leading to the following dispersion relations between the real and imaginary parts of the response function:

$$G_1(\omega) = \frac{1}{\pi}\mathcal{P}\int_{-\infty}^{\infty} \frac{G_2(\omega')}{\omega' - \omega}\,\mathrm{d}\omega' \tag{3.2.9a}$$

$$G_2(\omega) = -\frac{1}{\pi}\mathcal{P}\int_{-\infty}^{\infty} \frac{G_1(\omega')}{\omega' - \omega}\,\mathrm{d}\omega' \quad ; \tag{3.2.9b}$$

i.e. G_1 and G_2 are Hilbert transforms of each other. Using these general relations we can derive various expressions connecting the real and imaginary parts of different optical parameters and response functions discussed earlier. The relationship between causality and the dispersion relations is also illustrated in Fig. 3.2. Thus

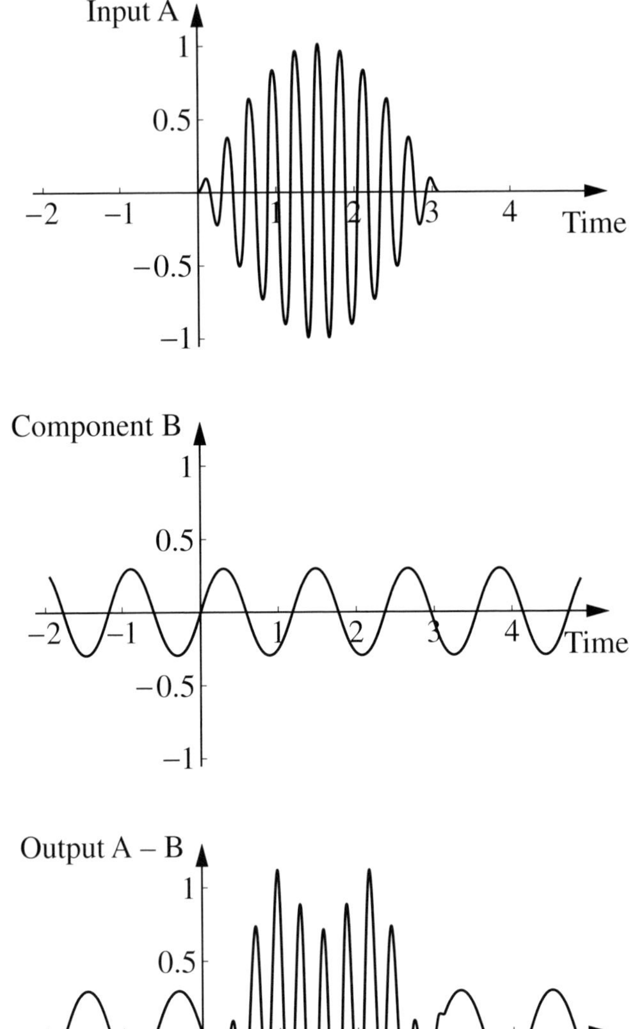

Fig. 3.2. Connection between causality and dispersion visualized by a wave package. An input $A(t)$ which is zero for times $t < 0$ is formed as a superposition of many Fourier components such as B, each of which extends from $t = -\infty$ to $t = \infty$. It is impossible to design a system which absorbs just the component $B(t)$ without affecting other components; otherwise the output would contain the complement of $B(t)$ during times before the onset of the input wave, in contradiction with causality. The lower panel shows the result of a simple subtraction of one component $A(t) - B(t)$ which is non-zero for times $t < 0$ (after [Tol56]).

causality implies that absorption of one frequency ω must be accompanied by a compensating shift in phase of other frequencies ω'; the required phase shifts are prescribed by the dispersion relation. The opposite is also true: a change in phase at one frequency is necessarily connected to an absorption at other frequencies.

Let us apply these relations to the various material parameters and optical constants. The current $\mathbf{J}$ is related to the electric field $\mathbf{E}$ by Ohm's law (2.2.11); and the complex conductivity $\hat{\sigma}(\omega) = \sigma_1(\omega) + i\sigma_2(\omega)$ is the response function describing this. The dispersion relation which connects the real and imaginary parts of the complex conductivity is then given by

$$\sigma_1(\omega) = \frac{1}{\pi}\mathcal{P}\int_{-\infty}^{\infty}\frac{\sigma_2(\omega')}{\omega' - \omega}\,\mathrm{d}\omega' \qquad (3.2.10\text{a})$$

$$\sigma_2(\omega) = -\frac{1}{\pi}\mathcal{P}\int_{-\infty}^{\infty}\frac{\sigma_1(\omega')}{\omega' - \omega}\,\mathrm{d}\omega' \quad . \qquad (3.2.10\text{b})$$

We can write these equations in a somewhat different form in order to eliminate the negative frequencies. Since $\hat{\sigma}(\omega) = \hat{\sigma}^*(-\omega)$ from Eq. (3.2.4c), the real part $\sigma_1(-\omega) = \sigma_1(\omega)$ is an even function and the imaginary part $\sigma_2(-\omega) = -\sigma_2(\omega)$ is an odd function in frequency. Thus, we can rewrite Eqs (3.2.10a) and (3.2.10b) by using the transformation for a function $f(x)$

$$\mathcal{P}\int_{-\infty}^{\infty}\frac{f(x)}{x-a}\,\mathrm{d}x = \mathcal{P}\int_{0}^{\infty}\frac{x[f(x)-f(-x)]+a[f(x)+f(-x)]}{x^2-a^2}\,\mathrm{d}x \quad ,$$

yielding for the two components of the conductivity

$$\sigma_1(\omega) = \frac{2}{\pi}\mathcal{P}\int_{0}^{\infty}\frac{\omega'\sigma_2(\omega')}{\omega'^2 - \omega^2}\,\mathrm{d}\omega' \qquad (3.2.11\text{a})$$

$$\sigma_2(\omega) = -\frac{2\omega}{\pi}\mathcal{P}\int_{0}^{\infty}\frac{\sigma_1(\omega')}{\omega'^2 - \omega^2}\,\mathrm{d}\omega' \quad . \qquad (3.2.11\text{b})$$

The dispersion relations are simple integral formulas relating a dispersive process (i.e. a change in phase of the electromagnetic wave, described by $\sigma_2(\omega)$) to an absorption process (i.e. a loss in energy, described by $\sigma_1(\omega)$), and vice versa.

Next, let us turn to the complex dielectric constant. The polarization $\mathbf{P}$ in response to an applied electric field $\mathbf{E}$ leads to a displacement $\mathbf{D}$ given by Eq. (2.2.5), and

$$4\pi\mathbf{P}(\omega) = [\hat{\epsilon}(\omega) - 1]\mathbf{E}(\omega) \quad ;$$

consequently $\hat{\epsilon}(\omega) - 1$ is the appropriate response function:

$$\epsilon_1(\omega) - 1 = \frac{2}{\pi}\mathcal{P}\int_{0}^{\infty}\frac{\omega'\epsilon_2(\omega')}{\omega'^2 - \omega^2}\,\mathrm{d}\omega' \qquad (3.2.12\text{a})$$

$$\epsilon_2(\omega) = -\frac{2}{\pi\omega}\mathcal{P}\int_0^\infty \frac{\omega'^2[\epsilon_1(\omega')-1]}{\omega'^2-\omega^2}\,d\omega' \quad . \tag{3.2.12b}$$

One immediate result of the dispersion relation is that if we have no absorption in the entire frequency range ($\epsilon_2(\omega)=0$), then there is no frequency dependence of the dielectric constant: $\epsilon_1(\omega)=1$ everywhere. Using

$$\mathcal{P}\int_0^\infty \frac{1}{\omega'^2-\omega^2}\,d\omega' = 0 \quad , \tag{3.2.13}$$

Eq. (3.2.12b) can be simplified to

$$\begin{aligned}
\int_0^\infty \frac{\omega'^2[1-\epsilon_1(\omega')]}{\omega'^2-\omega^2}\,d\omega' &= \int_0^\infty \frac{[1-\epsilon_1(\omega')]\omega'^2-\omega^2+\epsilon_1(\omega')\omega^2-\epsilon_1(\omega')\omega^2}{\omega'^2-\omega^2}\,d\omega' \\
&= \int_0^\infty \frac{[1-\epsilon_1(\omega')](\omega'^2-\omega^2)-\epsilon_1(\omega')\omega^2}{\omega'^2-\omega^2}\,d\omega' \\
&= \int_0^\infty [1-\epsilon_1(\omega')]\,d\omega' - \int_0^\infty \frac{\epsilon_1(\omega')\omega^2}{\omega'^2-\omega^2}\,d\omega' \quad .
\end{aligned}$$

The dc conductivity calculated from Eq. (3.2.11a) by setting $\omega=0$ and using $\sigma_2 = (1-\epsilon_1)\frac{\omega}{4\pi}$ from Table 2.1 is given by

$$\sigma_{\mathrm{dc}} = \sigma_1(0) = \frac{1}{2\pi^2}\int_0^\infty [1-\epsilon_1(\omega')]\,d\omega' \quad ; \tag{3.2.14}$$

we consequently find for Eq. (3.2.12b)

$$\epsilon_2(\omega) = \frac{4\pi\sigma_{\mathrm{dc}}}{\omega} - \frac{2\omega}{\pi}\mathcal{P}\int_0^\infty \frac{\epsilon_1(\omega')}{\omega'^2-\omega^2}\,d\omega' \quad . \tag{3.2.15}$$

For $\sigma_{\mathrm{dc}} \neq 0$ the imaginary part of the dielectric constant diverges for $\omega \to 0$. For insulating materials σ_{dc} vanishes, and therefore the first term of the right hand side is zero. In this case by using Eq. (3.2.13) we can add a factor 1 to Eq. (3.2.15) and obtain a form symmetric to the Kramers–Kronig relation for the dielectric constant (3.2.12a):

$$\epsilon_2(\omega) = -\frac{2\omega}{\pi}\mathcal{P}\int_0^\infty \frac{\epsilon_1(\omega')-1}{\omega'^2-\omega^2}\,d\omega' \quad ; \tag{3.2.16}$$

it should be noted that this expression is valid only for insulators.

For longitudinal fields the loss function $1/\hat{\epsilon}^{\mathrm{L}}(\omega)$ describes the response to the longitudinal displacement as written in Eq. (3.1.23). Consequently $1/\hat{\epsilon}^{\mathrm{L}}(\omega)$ can also be regarded as a response function with its components obeying the Kramers–Kronig relations

$$\mathrm{Re}\left\{\frac{1}{\hat{\epsilon}(\omega)}\right\} - 1 = \frac{1}{\pi}\mathcal{P}\int_{-\infty}^\infty \mathrm{Im}\left\{\frac{1}{\hat{\epsilon}(\omega')}\right\}\frac{d\omega'}{\omega'-\omega} \tag{3.2.17a}$$

$$\mathrm{Im}\left\{\frac{1}{\hat{\epsilon}(\omega)}\right\} = \frac{1}{\pi}P\int_{-\infty}^{\infty}\left[1-\mathrm{Re}\left\{\frac{1}{\hat{\epsilon}(\omega')}\right\}\right]\frac{d\omega'}{\omega'-\omega} \qquad . \text{ (3.2.17b)}$$

Similar arguments can be used for other optical constants which can be regarded as a response function. Thus the Kramers–Kronig relations for the two components of the complex refractive index $\hat{N}(\omega) = n(\omega) + ik(\omega)$ are as follows:

$$n(\omega) - 1 = \frac{2}{\pi}P\int_{0}^{\infty}\frac{\omega'k(\omega')}{\omega'^2-\omega^2}d\omega' \qquad (3.2.18a)$$

$$k(\omega) = -\frac{2}{\pi\omega}P\int_{0}^{\infty}\frac{(\omega')^2[n(\omega')-1]}{\omega'^2-\omega^2}d\omega' \qquad . \qquad (3.2.18b)$$

This is useful for experiments which measure only one component such as the absorption $\alpha(\omega) = 2k(\omega)\omega/c$. If this is done over a wide frequency range the refractive index $n(\omega)$ can be calculated without separate phase measurements. The square of the refractive index $\hat{N}^2 = (n^2 - k^2) + 2ink$ is also a response function.

We can write down the dispersion relation between the amplitude $\sqrt{R} = |\hat{r}|$ and the phase shift ϕ_{r} of the wave reflected off the surface of a bulk sample as in Eq. (2.4.12): $\mathrm{Ln}\,\hat{r}(\omega) = \ln|\hat{r}(\omega)| + i\phi_{\mathrm{r}}(\omega)$. The dispersion relations

$$\ln|\hat{r}(\omega)| = \frac{1}{\pi}P\int_{-\infty}^{\infty}\frac{\phi_{\mathrm{r}}(\omega')}{\omega'-\omega}d\omega' \qquad (3.2.19a)$$

$$\phi_{\mathrm{r}}(\omega) = -\frac{1}{\pi}P\int_{-\infty}^{\infty}\frac{\ln|\hat{r}(\omega')|}{\omega'-\omega}d\omega' \qquad (3.2.19b)$$

follow immediately. The response function which determines the reflected power is given by the square of the reflectivity (Eq. (2.4.15)) and therefore the appropriate dispersion relations can be utilized.

Finally, the real and imaginary parts of the surface impedance $\hat{Z}_{\mathrm{S}}$ are also related by similar dispersion relations:

$$R_{\mathrm{S}}(\omega) - Z_0 = \frac{1}{\pi}P\int_{-\infty}^{\infty}\frac{X_{\mathrm{S}}(\omega')}{\omega'-\omega}d\omega' = \frac{2}{\pi}P\int_{0}^{\infty}\frac{\omega'X_{\mathrm{S}}(\omega')}{\omega'^2-\omega^2}d\omega'$$

$$X_{\mathrm{S}}(\omega) = -\frac{1}{\pi}P\int_{-\infty}^{\infty}\frac{[R_{\mathrm{S}}(\omega')-Z_0]}{\omega'-\omega}d\omega' \qquad (3.2.20a)$$

$$= -\frac{2\omega}{\pi}P\int_{0}^{\infty}\frac{[R_{\mathrm{S}}(\omega')-Z_0]}{\omega'^2-\omega^2}d\omega' \qquad . \qquad (3.2.20b)$$

For the same reasons as discussed above, the difference from the free space impedance $R_{\mathrm{S}}(\omega) - Z_0$ has to be considered, since only this quantity approaches zero at infinite frequencies.

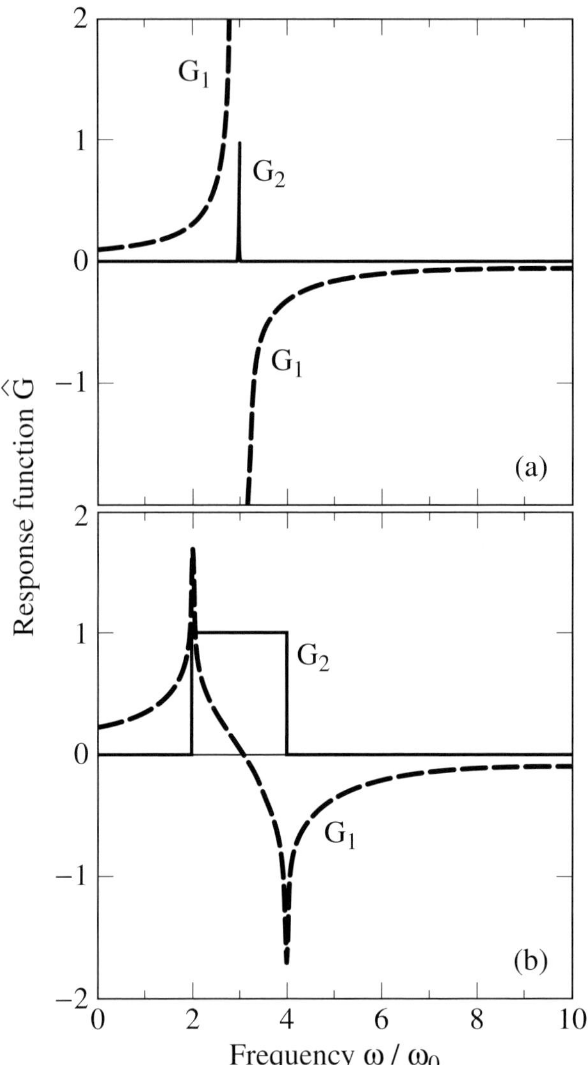

Fig. 3.3. Frequency dependence of the complex response function $\hat{G}(\omega) = G_1(\omega) + iG_2(\omega)$. (a) For $G_2(\omega) = \delta\{3\omega_0\}$ (solid line) the corresponding component $G_1(\omega)$ diverges as $1/(3\omega_0 - \omega)$ (dashed line). (b) The relationship between the real and imaginary parts of a response function if $G_2(\omega) = 1$ for $2 < \omega < 4$ and zero elsewhere.

The Kramers–Kronig relations are non-local in frequency: the real (imaginary) component of the response at a certain frequency ω is related to the behavior of the imaginary (real) part over the entire frequency range, although the influence of the contributions diminishes as $(\omega'^2 - \omega^2)^{-1}$ for larger and larger frequency differences. This global behavior leads to certain difficulties when these relations are used to analyze experimental results which cover only a finite range of frequencies. Certain qualitative statements can, however, be made, in particular when one or the other component of the response function displays a strong frequency dependence near certain frequencies.

From a δ-peak in the imaginary part of $\hat{G}(\omega)$ we obtain a divergency in $G_1(\omega)$, and vice versa, as displayed in Fig. 3.3. Also shown is the response to a pulse in $G_2(\omega)$. A quantitative discussion can be found in [Tau66, Vel61].

3.2.2 Sum rules

We can combine the Kramers–Kronig relations with physical arguments about the behavior of the real and imaginary parts of the response function to establish a set of so-called sum rules for various optical parameters.

The simplest approach is the following. For frequencies ω higher than those of the highest absorption, which is described by the imaginary part of the dielectric constant $\epsilon_2(\omega')$, the dispersion relation (3.2.12a) can be simplified to

$$\epsilon_1(\omega) \approx 1 - \frac{2}{\pi \omega^2} \int_0^\infty \omega' \epsilon_2(\omega') \, d\omega' \quad . \tag{3.2.21}$$

In order to evaluate the integral, let us consider a model for calculating the dielectric constant; in Sections 5.1 and 6.1 we will discuss such models for absorption processes in more detail. For a particular mode in a solid, at which the rearrangement of the electronic charge occurs, the equation of motion can, in general, be written as

$$m \left(\frac{d^2 \mathbf{r}}{dt^2} + \frac{1}{\tau} \frac{d\mathbf{r}}{dt} + \omega_0^2 \mathbf{r} \right) = -e\mathbf{E}(t) \quad , \tag{3.2.22}$$

where $1/\tau$ is a phenomenological damping constant, ω_0 is the characteristic frequency of the mode, and $-e$ and m are the electronic charge and mass, respectively. In response to the alternating field $\mathbf{E}(t) = \mathbf{E}_0 \exp\{-i\omega t\}$, the displacement is then given by

$$\mathbf{r} = -\frac{e\mathbf{E}}{m} \frac{1}{\omega_0^2 - \omega^2 - i\omega/\tau} \quad . \tag{3.2.23}$$

The dipole moment of the atom is proportional to the field and given by $-e\mathbf{r} = \hat{\alpha}\mathbf{E}$ if we have only one atom per unit volume; $\hat{\alpha}(\omega)$ is called the molecular polarizability. In this simple case the dielectric constant can now be calculated as

$$\hat{\epsilon}(\omega) = 1 + 4\pi\hat{\alpha} = 1 + \frac{4\pi e^2}{m} \frac{1}{\omega_0^2 - \omega^2 - i\omega/\tau} \quad . \tag{3.2.24}$$

Each electron participating in the absorption process leads to such a mode; in total we assume N modes per unit volume in the solid. At frequencies higher than the highest mode $\omega \gg \omega_0 > 1/\tau$, the excitation frequency dominates the expression for the polarizability; hence the previous equation can be simplified to

$$\lim_{\omega \gg \omega_0} \epsilon_1(\omega) = 1 - \frac{4\pi N e^2}{m\omega^2} = 1 - \frac{\omega_p^2}{\omega^2} \qquad (3.2.25)$$

for all modes, with the plasma frequency defined as

$$\omega_p = \left(\frac{4\pi N e^2}{m} \right)^{1/2} . \qquad (3.2.26)$$

From Eq. (3.2.25) we see that for high frequencies $\omega \gg \omega_0$ and $\omega > \omega_p$, the real part of the dielectric constant ϵ_1 always approaches unity from below. Comparing the real part of this expansion with the previous expression for the high frequency dielectric constant given in Eq. (3.2.21), we obtain

$$\frac{\pi}{2}\omega_p^2 = \int_0^\infty \omega \epsilon_2(\omega)\, d\omega \quad , \qquad (3.2.27)$$

which in terms of the optical conductivity $\sigma_1(\omega) = (\omega/4\pi)\epsilon_2(\omega)$ can also be written as

$$\frac{\omega_p^2}{8} = \int_0^\infty \sigma_1(\omega)\, d\omega = \frac{\pi N e^2}{2m} . \qquad (3.2.28)$$

Therefore the spectral weight $\omega_p^2/8$ defined as the area under the conductivity spectrum $\int_0^\infty \sigma_1(\omega)\, d\omega$ is proportional to the ratio of the electronic density to the mass of the electrons.

It can be shown rigorously that

$$\int_0^\infty \sigma_1(\omega)\, d\omega = \frac{\pi}{2} \sum_j \frac{(q_j)^2}{M_j} \qquad (3.2.29)$$

holds for any kind of excitation in the solid in which charge is involved; here q_j and M_j are the corresponding charge and mass, respectively. For electrons, $q = -e$ and $M = m$. For phonon excitations, for example, q may not equal e, and most importantly M is much larger than m due to the heavy ionic mass.

We still have to specify in more detail the parameter N in Eq. (3.2.26). For more than one electron per atom, N is the total number of electrons per unit volume, if the integration in Eq. (3.2.28) is carried out to infinite frequencies. This means that for high enough frequencies even the core electrons are excited, and have to be included in the sum rule. Studying the optical properties of solids, in general we are only interested in the optical response due to the conduction electrons in the case of a metal or due to the valence electrons in the case of a semiconductor. The core electron excitation energies are usually well separated in the higher frequency range $\omega > \omega_c$. If more than one mode is related to a single electron the polarizability sums up all contributions:

$$\hat{\alpha}(\omega) = \frac{e^2}{m} \sum_j \frac{f_j}{\omega_{0j}^2 - \omega^2 - i\omega/\tau_j} \quad . \tag{3.2.30}$$

Here f_j is called the oscillator strength; it measures the probability of a particular transition taking place with a characteristic frequency ω_{0j}. It obeys the so-called f sum rule $\sum_j f_j = 1$ which expresses normalization. A quantum mechanical derivation of the sum rule and a detailed discussion of the electronic transitions are given in Section 6.1.1.

Looking at the limit $\omega = 0$, the first Kramers–Kronig relation for the dielectric constant (3.2.12a) reads

$$\epsilon_1(0) - 1 = \frac{2}{\pi} \int_0^\infty \frac{\epsilon_2(\omega)}{\omega} \, d\omega \quad . \tag{3.2.31}$$

The static dielectric constant is related to the sum of all higher frequency contributions of the imaginary part; it is a measure of absorption processes and their mode strength. This relation (3.2.31) is the equivalent to Eq. (3.2.14) for the dc conductivity.

Now let us consider the second Kramers–Kronig relation (3.2.12b). We can split the integral of Eq. (3.2.14), and in the case of an insulator ($\sigma_{dc} = 0$) we obtain

$$0 = \int_0^\infty [\epsilon_1(\omega) - 1] \, d\omega = \int_0^{\omega_c} [\epsilon_1(\omega) - 1] \, d\omega + \int_{\omega_c}^\infty [\epsilon_1(\omega) - 1] \, d\omega \quad . \tag{3.2.32}$$

If we assume that above a cutoff frequency ω_c there are no excitations (i.e. the imaginary part of the dielectric constant equals zero at frequencies higher than ω_c), Eq. (3.2.25) can be applied to the second term on the right hand side of Eq. (3.2.32), yielding

$$\int_{\omega_c}^\infty [\epsilon_1(\omega) - 1] \, d\omega = \frac{\omega_p^2}{\omega_c} \quad .$$

The rest of Eq. (3.2.32) gives

$$\frac{1}{\omega_c} \int_0^{\omega_c} \epsilon_1(\omega) \, d\omega = 1 - \frac{\omega_p^2}{\omega_c^2} \quad , \tag{3.2.33}$$

which expresses that the average of the real part of the dielectric constant approaches unity in the high frequency limit $\omega_c \gg \omega_p$.

Next, we consider a response of the electron gas to an additional electron moving inside the solid; a situation which occurs when a so-called electron loss spectroscopy experiment is conducted. The moving charge produces a displacement field $\mathbf{D}$, and the rate of electronic energy density absorbed per unit volume is given by the product

$$\mathrm{Re}\left\{\mathbf{E}\cdot\frac{\partial \mathbf{D}}{\partial t}\right\} = \mathrm{Re}\left\{\frac{\mathbf{D}}{\hat{\epsilon}}\cdot\frac{\partial \mathbf{D}}{\partial t}\right\} = \frac{-\omega\epsilon_2}{|\epsilon_1^2 + \epsilon_2^2|}|D_0|^2 = \omega\,\mathrm{Im}\left\{\frac{1}{\hat{\epsilon}(\omega)}\right\}|D_0|^2 \quad,$$

and we find that $\mathrm{Im}\{1/\hat{\epsilon}\}$ describes the energy loss associated with the electrons moving in the medium. It is consequently called the loss function of solids and is the basic parameter measured by electron loss spectroscopy.[5] From the dispersion relations (3.2.17) of the inverse dielectric function, we obtain the sum rule

$$-\int_0^\infty \omega\,\mathrm{Im}\left\{\frac{1}{\hat{\epsilon}(\omega)}\right\}\,\mathrm{d}\omega = \frac{\pi}{2}\omega_{\mathrm{p}}^2 \quad, \tag{3.2.35}$$

which can also be derived using rigorous quantum mechanical arguments [Mah90]. Furthermore from Eq. (3.2.17a) we find the following relations:

$$\int_0^\infty \frac{1}{\omega}\,\mathrm{Im}\left\{\frac{1}{\hat{\epsilon}(\omega)}\right\}\,\mathrm{d}\omega = -\frac{\pi}{2} \tag{3.2.36}$$

and

$$\frac{2}{\pi}\int_0^\infty \mathrm{Im}\left\{\frac{1}{\hat{\epsilon}(\omega')}\right\}\mathcal{P}\frac{\omega'}{\omega'^2 - \omega^2}\,\mathrm{d}\omega' = \mathrm{Re}\left\{\frac{1}{\hat{\epsilon}(\omega)}\right\} - 1 \quad.$$

Using $\lim_{\omega\to 0}\mathrm{Re}\left\{1/\hat{\epsilon}(\omega)\right\} = 1 + \omega_{\mathrm{p}}^2/\omega^2 + \mathcal{O}\left\{\omega^{-4}\right\}$, the real part of the loss function obeys the following equation:

$$\int_0^\infty \left[\mathrm{Im}\left\{\frac{1}{\hat{\epsilon}(\omega)}\right\} - 1\right]\mathrm{d}\omega = 0 \quad. \tag{3.2.37}$$

As far as the surface impedance is concerned, from Eq. (3.2.20b) we obtain

$$\int_0^\infty [R(\omega) - Z_0]\,\mathrm{d}\omega = 0 \quad, \tag{3.2.38}$$

where $Z_0 = 4\pi/c$, because $\lim_{\omega\to 0}\omega X_S(\omega) = 0$ for all conducting material [Bra74].

Not surprisingly, we can also establish sum rules for other optical parameters.

---

[5] If we evaluate the energy and momentum transfer per unit time in a scattering experiment by the Born approximation, we obtain a generalized loss function $W(\omega) \approx (\omega/8\pi)\,\mathrm{Im}\left\{1/\hat{\epsilon}(\omega)\right\}$. It is closely related to the structure factor or dynamic form factor $S(\omega)$, which is the Fourier transform of the density–density correlation [Pin63, Pla73]. It allows for a direct comparison of scattering experiments with the longitudinal dielectric function:

$$S(\omega) = -\frac{\hbar q^2}{4\pi^2 e^2}\,\mathrm{Im}\left\{\frac{1}{\hat{\epsilon}(\omega)}\right\} \quad. \tag{3.2.34}$$

For example the sum rule for the components of the complex refractive index, $n(\omega)$ and $k(\omega)$, are

$$\int_0^\infty \omega n(\omega) k(\omega)\, d\omega \;=\; \frac{\pi}{2}\omega_\mathrm{p}^2 \tag{3.2.39a}$$

$$\frac{c}{2}\int_0^\infty \alpha(\omega)\, d\omega = \int_0^\infty \omega k(\omega)\, d\omega \;=\; \frac{\pi}{4}\omega_\mathrm{p}^2 \tag{3.2.39b}$$

$$\int_0^\infty [n(\omega) - 1]\, d\omega \;=\; 0 \;. \tag{3.2.39c}$$

For the reflectivity, similar arguments yield:

$$\int_0^\infty |\hat{r}(\omega)| \cos\{\phi_\mathrm{r}(\omega)\}\, d\omega \;=\; 0 \tag{3.2.40a}$$

$$\int_0^\infty \omega |\hat{r}(\omega)| \sin\{\phi_\mathrm{r}(\omega)\}\, d\omega \;=\; \frac{\pi}{8}\omega_\mathrm{p}^2 \;. \tag{3.2.40b}$$

These formulations of the sum rule do not express any new physics but may be particularly useful in certain cases.

In [Smi85] sum rules to higher powers of the optical parameters are discussed; these converge faster with frequency and thus can be used in a limited frequency range but are in general not utilized in the analysis of the experimental results. There are also dispersion relations and sum rules investigated for the cases of non-normal incidence [Ber67, Roe65], ellipsometry [Hal73, Ina79, Que74], and transmission measurements [Abe66, Neu72, Nil68, Ver68].

References

[Abe66] F. Abelès and M.L. Thèye, *Surf. Sci.* **5**, 325 (1966)

[Ber67] D.W. Berreman, *Appl. Opt.* **6**, 1519 (1967)

[Bod45] H.W. Bode, *Network Analysis and Feedback Amplifier Design* (Van Nostrand, Princeton, 1945)

[Bra74] G. Brändli, *Phys. Rev. B* **9**, 342 (1974)

[Hal73] G.M. Hale, W.E. Holland, and M.R. Querry, *Appl. Opt.* **12** 48 (1973)

[Ina79] T. Inagaki, H. Kuwata, and A. Ueda, *Phys. Rev. B* **19**, 2400 (1979); *Surf. Sci.* **96**, 54 (1980)

[Kra26] H.A. Kramers, *Nature* **117**, 775 (1926); H.A. Kramers, in: *Estratto dagli Atti del Congr. Int. Fis.*, Vol. 2 (Zanichelli, Bologna, 1927), p. 545; *Collected Scientific Papers* (North-Holland, Amsterdam, 1956).

[Kro26] R. de L. Kronig, *J. Opt. Soc. Am.* **12**, 547 (1926); *Ned. Tijdschr. Natuurk.* **9**, 402 (1942)

[Lan80] L.D. Landau and E.M. Lifshitz, *Statistical Physics*, 3rd edition (Pergamon Press, London, 1980)

[Lin54] J. Lindhard, *Dan. Mat. Fys. Medd.* **28**, no. 8 (1954)

[Mac56] J.R. Macdonald and M.K. Brachman, *Rev. Mod. Phys.* **28**, 393 (1956)

[Mah90] G.D. Mahan, *Many-Particle Physics*, 2nd edition (Plenum Press, New York,1990)

[Mar67] P.C. Martin, *Phys. Rev.* **161**, 143 (1967)

[Neu72] J.D. Neufeld and G. Andermann, *J. Opt. Soc. Am.* **62**, 1156 (1972)

[Nil68] P.-O. Nilson, *Appl. Optics* **7**, 435 (1968)

[Nye57] J.F. Nye, *Physical Properties of Crystals* (Clarendon, Oxford, 1957)

[Pin63] D. Pines, *Elementary Excitations in Solids* (Addison-Wesley, Reading, MA, 1963)

[Pin66] D. Pines and P. Noizières, *The Theory of Quantum Liquids*, Vol. 1 (Addison-Wesley, Reading, MA, 1966)

[Pla73] P.M. Platzman and P.A. Wolff, *Waves and Interactions in Solid State Plasmas* (Academic Press, New York, 1973)

[Que74] M.R. Querry and W.E. Holland, *Appl. Opt.* **13**, 595 (1974)

[Roe65] D.M. Roessler, *Brit. J. Appl. Phys.* **16**, 1359 (1965)

[Smi85] D.Y. Smith, *Dispersion Theory, Sum Rules, and Their Application to the Analysis of Optical Data*, in: *Handbook of Optical Constants of Solids*, Vol. 1, edited by E.D. Palik (Academic Press, Orlando, FL, 1985), p. 35

[Tau66] J. Tauc, *Optical Properties of Semiconductors*, in: *The Optical Properties of Solids*, edited by J. Tauc, Proceedings of the International School of Physics 'Enrico Fermi' **34** (Academic Press, New York, 1966)

[Tol56] J.S. Toll, *Phys. Rev.* **104**, 1760 (1956)

[Vel61] B. Velický, *Czech. J. Phys. B* **11**, 787 (1961)

[Ver68] H.W. Verleur, *J. Opt. Soc. Am.* **58**, 1356 (1968)

Further reading

[Agr84] V.M. Agranovich and V.L. Ginzburg, *Crystal Optics with Spatial Dispersion and Excitons*, 2nd edition (Springer-Verlag, Berlin, 1984)

[But91] P.N. Butcher and D. Cotter, *The Elements of Nonlinear Optics* (Cambridge University Press, Cambridge, 1991)

[Gei68] J. Geiger, *Elektronen und Festkörper* (Vieweg-Verlag, Braunschweig, 1968)

[Gin90] V.L. Ginzburg, *Phys. Rep.* **194**, 245 (1990)

[Kel89] L.V. Keldysh, D.A. Kirzhnitz, and A.A. Maradudin, *The Dielectric Function of Condensed Systems* (North-Holland, Amsterdam, 1989)

[Mil91] P.L. Mills, *Nonlinear Optics* (Springer-Verlag, Berlin, 1991)

[Sas70] W.M. Saslow, *Phys. Lett. A* **33**, 157 (1970)

[Smi81] D.Y. Smith and C.A. Manogue, *J. Opt. Soc. Am.* **71**, 935 (1981)

4

The medium: correlation and response functions

In the previous chapters the response of the medium to the electromagnetic waves was described in a phenomenological manner in terms of the frequency and wavevector dependent complex dielectric constant and conductivity. Our task at hand now is to relate these parameters to the changes in the electronic states of solids, brought about by the electromagnetic fields or by external potentials. Several routes can be chosen to achieve this goal. First we derive the celebrated Kubo formula: the conductivity given in terms of current–current correlation functions. The expression is general and not limited to electrical transport; it can be used in the context of different correlation functions, and has been useful in a variety of transport problems in condensed matter. We use it in the subsequent chapters to discuss the complex, frequency dependent conductivity. This is followed by the description of the response to a scalar field given in terms of the density–density correlations. Although this formalism has few limitations, in the following discussion we restrict ourselves to electronic states which have well defined momenta. In Section 4.2 formulas for the so-called semiclassical approximation are given; it is utilized in later chapters when the electrodynamics of the various broken symmetry states is discussed. Next, the response to longitudinal and transverse electromagnetic fields is treated in terms of the Bloch wavefunctions, and we derive the well known Lindhard dielectric function: the expression is used for longitudinal excitations of the electron gas; the response to transverse electromagnetic fields is accounted for in terms of the conductivity. Following Lindhard [Lin54], the method was developed in the 1960s by Pines [Pin63, Pin66] and others, [Ehr59, Noz64] and now forms an essential part of the many-body theory of solids.

Throughout the chapter, both the transverse and the longitudinal responses are discussed; and, as usual, we derive the conductivity in terms of the current–current correlation function and the dielectric constant within the framework of the Lindhard formalism. Because of the relationship $\hat{\epsilon} = 1 + 4\pi i \hat{\sigma}/\omega$ between the complex dielectric constant and the complex conductivity, it is of course a matter of choice or taste as to which optical parameter is used. In both cases, we derive the appropriate equations for the real part of the conductivity and dielectric constant; the imaginary part of these quantities is obtained either by utilizing the Kramers–Kronig relations or the adiabatic approximation.

4.1 Current–current correlation functions and conductivity

In the presence of a vector and a scalar potential, the Hamiltonian of an N-electron system in a solid is in general given by

$$\mathcal{H} = \frac{1}{2m} \sum_{i=1}^{N} \left(\mathbf{p}_i + \frac{e}{c} \mathbf{A}(\mathbf{r}_i) \right)^2 + \sum_{i=1,j=1}^{N,M} V_j^0(\mathbf{r}_i - \mathbf{R}_j)$$

$$+ \frac{1}{2} \sum_{i=1,i'=1}^{N,N} \frac{e^2}{|\mathbf{r}_i - \mathbf{r}_{i'}|} - \sum_{i=1}^{N} e\Phi(\mathbf{r}_i) \quad . \tag{4.1.1}$$

The first term refers to the coupling between the electromagnetic wave (described by the external vector potential $\mathbf{A}$) and the electrons with momenta $\mathbf{p}_i$ at their location $\mathbf{r}_i$. As usual, $-e$ is the electronic charge and m is the electron mass. The second term defines the interaction between the ions and the electrons; this interaction is given by the potential V_j^0. In a crystalline solid $V_j^0(\mathbf{r}_i - \mathbf{R}_j)$ is a periodic function, and with M ionic sites and N electrons the band filling is N/M. The summation over the ionic positions is indicated by the index j; the indices i and i' refer to the electrons. The third term describes the electron–electron interaction ($i \neq i'$); we assume that only the Coulomb repulsion is important; we have not included vibrations of the underlying lattice, and consequently electron–phonon interactions are also neglected, together with spin dependent interactions between the electrons. The last term in the Hamiltonian describes an external scalar potential Φ as produced by an external charge. Both $\mathbf{p}$ and $\mathbf{A}$ are time dependent, which for brevity will be indicated only when deemed necessary.

In general, we can split the Hamilton operator as

$$\mathcal{H} = \mathcal{H}_0 + \mathcal{H}_{\text{int}} \quad , \tag{4.1.2}$$

with the first term describing the unperturbed Hamiltonian in the absence of a vector and a scalar potential:

$$\mathcal{H}_0 = \frac{1}{2m} \sum_{i=1}^{N} \mathbf{p}_i^2 + \sum_{i=1,j=1}^{N,M} V_j^0(\mathbf{r}_i - \mathbf{R}_j) + \frac{1}{2} \sum_{i=1,i'=1}^{N,N} \frac{e^2}{|\mathbf{r}_i - \mathbf{r}_{i'}|} \quad . \tag{4.1.3}$$

The second term in Eq. (4.1.2) accounts for the interaction of the system with the electromagnetic radiation and with the electrostatic potential. This interaction with the electromagnetic field is given by

$$\mathcal{H}_{\text{int}} = \frac{e}{2mc} \sum_{i=1}^{N} [\mathbf{p}_i \cdot \mathbf{A}(\mathbf{r}_i) + \mathbf{A}(\mathbf{r}_i) \cdot \mathbf{p}_i] - e \sum_{i=1}^{N} \Phi(\mathbf{r}_i) \quad . \tag{4.1.4}$$

Here we have neglected second order terms (e.g. terms proportional to A^2). Note that in Eq. (4.1.4) $\mathbf{p}$, $\mathbf{A}$, and Φ are quantum mechanical operators which observe the commutation relations. For the moment, however, we will treat both $\mathbf{A}$ and Φ as classical fields.

As discussed in Section 3.1.1, in the Coulomb gauge the vector potential $\mathbf{A}$ is only related to the transverse response of the medium, while the scalar potential Φ determines the longitudinal response to the applied electromagnetic fields. These two cases are treated separately. As usual, magnetic effects are neglected, and throughout this and subsequent chapters we assume that $\mu_1 = 1$ and $\mu_2 = 0$.

4.1.1 Transverse conductivity: the response to the vector potential

Next we derive an expression for the complex conductivity in terms of the current–current correlation function. The operator of the electrical current density is defined as

$$\mathbf{J}^{\text{T}}(\mathbf{r}) = -\frac{e}{2} \sum_{i=1}^{N} [\mathbf{v}_i \delta\{\mathbf{r} - \mathbf{r}_i\} + \delta\{\mathbf{r} - \mathbf{r}_i\}\mathbf{v}_i] \quad , \tag{4.1.5}$$

where $\mathbf{v}_i$ is the velocity of the ith particle at position $\mathbf{r}_i$ (and we have implicitly assumed the usual commutation rules). The velocity operator of an electron in the presence of an electromagnetic field is given by $\mathbf{v} = \mathbf{p}/m + e\mathbf{A}/mc$; consequently the current density has two terms,

$$\mathbf{J}^{\text{T}}(\mathbf{r}) = \mathbf{J}_{\text{p}}(\mathbf{r}) + \mathbf{J}_{\text{d}}(\mathbf{r})$$

$$= -\frac{e}{2m} \sum_{i=1}^{N} [\mathbf{p}_i \delta\{\mathbf{r} - \mathbf{r}_i\} + \delta\{\mathbf{r} - \mathbf{r}_i\}\mathbf{p}_i] - \frac{e^2}{mc} \sum_{i=1}^{N} \mathbf{A}(\mathbf{r}_i)\delta\{\mathbf{r} - \mathbf{r}_i\} \quad . \tag{4.1.6}$$

The second term follows from the fact that $\mathbf{A}(\mathbf{r}_i)$ commutes with $\delta\{\mathbf{r} - \mathbf{r}_i\}$. The first term is called the paramagnetic and the second the diamagnetic current. As the vector potential $\mathbf{A}$ depends on the position, it will in general not commute with the momentum. However, since $\mathbf{p} = -i\hbar\nabla$ we obtain

$$\mathbf{p} \cdot \mathbf{A} - \mathbf{A} \cdot \mathbf{p} = -i\hbar\nabla \cdot \mathbf{A} \quad . \tag{4.1.7}$$

Note, if Coulomb gauge ($\nabla \cdot \mathbf{A} = 0$) is assumed, $\mathbf{A}$ and $\mathbf{p}$ commute.

Using the definition (4.1.5) of the electric current density, the interaction term

Eq. (4.1.4) for transverse fields (note that $\Phi = 0$) can be written as:

$$\mathcal{H}_{\text{int}}^{\text{T}} = -\frac{1}{c} \int \mathbf{J}^{\text{T}}(\mathbf{r}) \cdot \mathbf{A}^{\text{T}}(\mathbf{r}) \, d\mathbf{r} \quad . \tag{4.1.8}$$

We have replaced the summation over the individual positions i by an integral, and we have assumed that the current is a continuous function of the position $\mathbf{r}$. Note that the diamagnetic current term $\mathbf{J}_{\text{d}}$ leads to a term in the interaction Hamiltonian which is second order in $\mathbf{A}$; thus it is not included if we restrict ourselves to interactions proportional to $\mathbf{A}$. Our objective is to derive the wavevector and frequency dependent response, and in order to do so we define the spatial Fourier transforms (see Appendix A.1) of the current density operator $\mathbf{J}(\mathbf{q})$ and of the vector potential $\mathbf{A}(\mathbf{q})$ as

$$\begin{aligned}
\mathbf{J}(\mathbf{q}) &= \frac{1}{\Omega} \int \mathbf{J}(\mathbf{r}) \exp\{-i\mathbf{q} \cdot \mathbf{r}\} \, d\mathbf{r} \\
&= -\frac{e}{2} \frac{1}{\Omega} \sum_{i=1}^{N} [\mathbf{v}_i \exp\{-i\mathbf{q} \cdot \mathbf{r}_i\} + \exp\{-i\mathbf{q} \cdot \mathbf{r}_i\} \mathbf{v}_i] \quad ,
\end{aligned} \tag{4.1.9}$$

and

$$\mathbf{A}(\mathbf{q}) = \frac{1}{\Omega} \int d\mathbf{A}(\mathbf{r}) \exp\{-i\mathbf{q} \cdot \mathbf{r}\} \, \mathbf{r} \quad . \tag{4.1.10}$$

Here Ω denotes the volume element over which the integration is carried out; if just the unit volume is considered, Ω is often neglected. It is straightforward to show that with these definitions the $\mathbf{q}$ dependent interaction term in first order perturbation becomes

$$\mathcal{H}_{\text{int}}^{\text{T}} = -\frac{1}{c} \mathbf{J}^{\text{T}}(\mathbf{q}) \cdot \mathbf{A}^{\text{T}}(\mathbf{q}) \quad . \tag{4.1.11}$$

Next we derive the rate of absorption of electromagnetic radiation. As we have seen in Section 2.3.1, this absorption rate can also be written as $P = \sigma_1^{\text{T}}(\mathbf{E}^{\text{T}})^2$, in terms of the real part of the complex conductivity. If we equate the absorption which we obtain later in terms of the electric currents using this expression, it will then lead us to a formula for σ_1 in terms of the $\mathbf{q}$ and ω dependent current densities. The imaginary part of $\hat{\sigma}(\omega)$ is subsequently obtained by utilizing the Kramers–Kronig relation. We assume that the incident electromagnetic wave with wavevector $\mathbf{q}$ and frequency ω results in the scattering of an electron from one state to another with higher energy. Fermi's golden rule is utilized: for one electron the number of transitions per unit time and per unit volume from the initial state $|s\rangle$ to a final state $|s'\rangle$ of the system is

$$W_{s \to s'} = \frac{2\pi}{\hbar^2} \left| \langle s' | \mathcal{H}_{\text{int}}^{\text{T}} | s \rangle \right|^2 \delta\{\omega - (\omega_{s'} - \omega_s)\} \quad . \tag{4.1.12}$$

Here $\hbar\omega_s$ and $\hbar\omega_{s'}$ correspond to the energy of the initial and final states of the system, respectively. The energy difference between the final and initial states $\hbar\omega_{s'} - \hbar\omega_s$ is positive for photon absorption and negative for emission. In this notation, $|s\rangle$ and $|s'\rangle$ do not refer to the single-particle states only, but include all excitations of the electron system we consider. The average is not necessarily restricted to zero temperature, but is valid also at finite temperatures if the bracket is interpreted as a thermodynamic average. This is valid also for the expression we proceed to derive.

The matrix element for the transition is given by Eq. (4.1.11); substituting

$$\langle s'|\mathcal{H}_{\mathrm{int}}^{\mathrm{T}}|s\rangle = -\frac{1}{c}\langle s'|\mathbf{J}^{\mathrm{T}}(\mathbf{q})|s\rangle\mathbf{A}^{\mathrm{T}}(\mathbf{q})$$

into Eq. (4.1.12) leads to

$$W_{s\to s'} = \frac{2\pi}{\hbar^2 c^2}\langle s'|\mathbf{J}^{\mathrm{T}}(\mathbf{q})|s\rangle\langle s|\mathbf{J}^{\mathrm{T}*}(\mathbf{q})|s'\rangle\left|\mathbf{A}^{\mathrm{T}}(\mathbf{q})\right|^2 \delta\{\omega - \omega_{s'} + \omega_s\} \quad ,$$

where $\mathbf{J}^*(\mathbf{q}) = \mathbf{J}(-\mathbf{q})$. The summation over all occupied initial and all empty final states gives the total transfer rate per unit volume:

$$
\begin{aligned}
W &= \sum_{s,s'} W_{s\to s'} \\
&= \sum_{s,s'} \frac{2\pi}{\hbar^2 c^2}\langle s'|\mathbf{J}^{\mathrm{T}}(\mathbf{q})|s\rangle\langle s|\mathbf{J}^{\mathrm{T}*}(\mathbf{q})|s'\rangle\left|\mathbf{A}^{\mathrm{T}}(\mathbf{q})\right|^2 \delta\{\omega - \omega_{s'} + \omega_s\} \quad . \quad (4.1.13)
\end{aligned}
$$

We use the identity

$$\delta\{\omega\} = \frac{1}{2\pi}\int \exp\{-i\omega t\}\,dt \quad ;$$

substituting this into Eq. (4.1.13) yields

$$
\begin{aligned}
W &= \sum_{s,s'} \frac{1}{\hbar^2 c^2}\int dt\,\exp\{-i\omega t\}\langle s'|\mathbf{J}^{\mathrm{T}}(\mathbf{q})|s\rangle \\
&\quad \times \langle s|\exp\{i\omega_{s'}t\}\mathbf{J}^{\mathrm{T}*}(\mathbf{q})\exp\{-i\omega_s t\}|s'\rangle\left|\mathbf{A}^{\mathrm{T}}(\mathbf{q})\right|^2 \quad . \quad (4.1.14)
\end{aligned}
$$

For the complete set of states $\sum_s |s\rangle\langle s| = 1$. In the Heisenberg representation $\exp\{-i\omega_s t\}|s\rangle = \exp\{-i\mathcal{H}_0 t/\hbar\}|s\rangle$, and then the time dependence is written as

$$\mathbf{J}^{\mathrm{T}*}(\mathbf{q}, t) = \exp\{i\mathcal{H}_0 t/\hbar\}\mathbf{J}^{\mathrm{T}*}(\mathbf{q})\exp\{-i\mathcal{H}_0 t/\hbar\} = \exp\{i\omega_{s'}t\}\mathbf{J}^{\mathrm{T}*}(\mathbf{q})\exp\{-i\omega_s t\} \quad ,$$

and the absorbed energy per unit time and per unit volume then becomes

$$P = \hbar\omega W = \left|\mathbf{A}^{\mathrm{T}}(\mathbf{q})\right|^2 \sum_s \frac{\omega}{\hbar c^2}\int dt\,\langle s|\mathbf{J}^{\mathrm{T}}(\mathbf{q}, 0)\mathbf{J}^{\mathrm{T}*}(\mathbf{q}, t)|s\rangle\exp\{-i\omega t\} \quad (4.1.15)$$

where we have replaced s' by s in order to simplify the notation.

For applied ac fields the equation relating the transverse electric field and the vector potential is given by $\mathbf{E}^{\mathrm{T}} = i\omega \mathbf{A}^{\mathrm{T}}/c$, leading to our final result: the absorbed power per unit volume expressed as a function of the electric field and the current–current correlation function:

$$P = \left|\mathbf{E}(\mathbf{q})^{\mathrm{T}}\right|^2 \sum_s \frac{1}{\hbar\omega} \int dt \, \langle s|\mathbf{J}^{\mathrm{T}}(\mathbf{q}, 0)\mathbf{J}^{\mathrm{T*}}(\mathbf{q}, t)|s\rangle \exp\{-i\omega t\} \quad .$$

With $P = \sigma_1^{\mathrm{T}}(\mathbf{E}^{\mathrm{T}})^2$ the conductivity per unit volume is simply given by the current–current correlation function, averaged over the states $|s\rangle$ of our system in question

$$\sigma_1^{\mathrm{T}}(\mathbf{q}, \omega) = \sum_s \frac{1}{\hbar\omega} \int dt \, \langle s|\mathbf{J}^{\mathrm{T}}(\mathbf{q}, 0)\mathbf{J}^{\mathrm{T*}}(\mathbf{q}, t)|s\rangle \exp\{-i\omega t\} \quad , \qquad (4.1.16)$$

i.e. the Kubo formula for the $\mathbf{q}$ and ω dependent conductivity. Here the right hand side describes the fluctuations of the current in the ground state. The average value of the current is of course zero, and the conductivity depends on the time correlation between current operators, integrated over all times. The above relationship between the conductivity (i.e. the response to an external driving force) and current fluctuations is an example of the so-called fluctuation-dissipation theorem. The formula is one of the most utilized expressions in condensed matter physics, and it is used extensively, among other uses, also for the evaluation of the complex conductivity under a broad variety of circumstances, both for crystalline and non-crystalline solids. Although we have implied (by writing the absolved power as $P = \sigma_1^{\mathrm{T}}(\mathbf{E}^{\mathrm{T}})^2$) that the conductivity is a scalar quantity, it turns out that σ_1^{T} is in general a tensor for arbitrary crystal symmetry. For orthorhombic crystals the conductivity tensor reduces to a vector, with different magnitudes in the different crystallographic orientations, this difference being determined by the (anisotropic) form factor.

Now we derive a somewhat different expression for the conductivity for states for which Fermi statistics apply. At zero temperature the transition rate per unit volume between the states $|s\rangle$ and $|s'\rangle$ is given by

$$W_{s\to s'} = \frac{2\pi}{\hbar} \left|\langle s'|\mathcal{H}_{\mathrm{int}}^{\mathrm{T}}|s\rangle\right|^2 f(\mathcal{E}_s)[1 - f(\mathcal{E}_{s'})]\delta\{\hbar\omega - (\mathcal{E}_{s'} - \mathcal{E}_s)\} \quad . \qquad (4.1.17)$$

Here the interaction Hamiltonian is $\mathcal{H}_{\mathrm{int}}^{\mathrm{T}} = (e/mc)\mathbf{A}^{\mathrm{T}} \cdot \mathbf{p}$, where we do not explicitly include the $\mathbf{q}$ dependence since we are only interested in the $\mathbf{q} = 0$ limit of the conductivity at this point. (Note that in this Hamiltonian we neglected the diamagnetic current term of $\mathbf{J}^{\mathrm{T}}$.) The energies $\mathcal{E}_s = \hbar\omega_s$ and $\mathcal{E}_{s'} = \hbar\omega_{s'}$ correspond to the energy of the initial state $|s\rangle$ and of the final state $|s'\rangle$, respectively; the energy difference between these states is $\mathcal{E}_{s'} - \mathcal{E}_s = \hbar\omega_{s's}$. In the $T = 0$ limit, the Fermi function

$$f(\mathcal{E}_s) = \frac{1}{\exp\left\{\frac{\mathcal{E}_s - \mathcal{E}_F}{k_B T}\right\} + 1}$$

equals unity and $f(\mathcal{E}_{s'}) = 0$, as all states below the Fermi level are occupied while the states above are empty. Integrating over all $\mathbf{k}$ vectors we obtain

$$W(\omega) = \frac{\pi e^2}{m^2 \hbar c^2} \frac{2}{(2\pi)^3} \int |\mathbf{A}^T|^2 |\langle s'|\mathbf{p}|s\rangle|^2 f(\mathcal{E}_s)[1 - f(\mathcal{E}_{s'})]\delta\{\hbar\omega - \hbar\omega_{s's}\}\, d\mathbf{k} \quad .$$

$$(4.1.18)$$

This is the probability (per unit time and per unit volume) that the electromagnetic energy $\hbar\omega$ is absorbed by exciting an electron to a state of higher energy. The absorbed power per unit volume is $P(\omega) = \hbar\omega W(\omega)$, and by using $\mathbf{A}^T = -(ic/\omega)\mathbf{E}^T$ we obtain

$$P(\omega) = \frac{\pi e^2}{m^2 \omega} |\mathbf{E}^T|^2 \frac{2}{(2\pi)^3} \int |\langle s'|\mathbf{p}|s\rangle|^2 f(\mathcal{E}_s)[1 - f(\mathcal{E}_{s'})]\delta\{\hbar\omega - \hbar\omega_{s's}\}\, d\mathbf{k} \quad .$$

$$(4.1.19)$$

As before, the conductivity $\sigma_1(\omega)$ is related to the absorbed power through Ohm's law. In the $T = 0$ limit, $f(\mathcal{E}_s)[1 - f(\mathcal{E}_{s'})] = 1$, and we obtain

$$\sigma_1(\omega) = \frac{\pi e^2}{m^2 \omega} \frac{2}{(2\pi)^3} \int |\mathbf{p}_{s's}|^2 \delta\{\hbar\omega - \hbar\omega_{s's}\}\, d\mathbf{k} \quad , \qquad (4.1.20)$$

with the abbreviation $|\mathbf{p}_{s's}|^2 = |\langle s'|\mathbf{p}|s\rangle|^2$ used for the momentum operator, the so-called dipole matrix element.

This expression is appropriate for a transition between states $|s\rangle$ and $|s'\rangle$, and we assume that the number of transitions within an interval $d\omega$ is $N_{s's}(\omega)$. We define the joint density of states for these transitions as $D_{s's}(\omega) = dN_{s's}(\omega)/d\omega$:

$$D_{s's}(\hbar\omega) = \frac{2}{(2\pi)^3} \int \delta\{\hbar\omega - \hbar\omega_{s's}\}\, d\mathbf{k} \quad ,$$

where the factor of 2 refers to the different spin orientations. The conductivity is then given in terms of the joint density of states for both spin directions and of the transition probability as

$$\sigma_1(\omega) = \frac{\pi e^2}{m^2 \omega} |\mathbf{p}_{s's}(\omega)|^2 D_{s's}(\hbar\omega) \quad . \qquad (4.1.21)$$

This equation, often referred to as the Kubo–Greenwood formula, is most useful when interband transitions are important, with the states $|s\rangle$ and $|s'\rangle$ belonging to different bands. We utilize this expression in Section 6.2 when the optical conductivity of band semiconductors is derived and also in Section 7.3 when the absorption of the electromagnetic radiation in superconductors is discussed.

4.1.2 Longitudinal conductivity: the response to the scalar field

The above considerations led to the response of the electron gas to a transverse electromagnetic perturbation. Longitudinal currents and electric fields which occur due to the interaction of the charge density ρ and scalar external potential Φ (which we assume to be also time dependent) can be discussed along similar lines. Within the framework of linear response the interaction between this potential and the charge density is

$$\mathcal{H}_{\text{int}}^{\text{L}} = \int \rho(\mathbf{r})\Phi(\mathbf{r}, t)\,d\mathbf{r} \quad , \tag{4.1.22}$$

with

$$\rho(\mathbf{r}) = -e \sum_{i=1}^{N} \delta\{\mathbf{r} - \mathbf{r}_i\}$$

the summation over all electrons at sites $\mathbf{r}_i$. Here $\Phi(\mathbf{r}, t)$ is the selfconsistent potential[1] including the external perturbation and the induced screening potential: $\Phi = \Phi^{\text{ext}} + \Phi^{\text{ind}}$. The interaction form is similar to Eq. (4.1.8). By replacing

$$\mathbf{A}(\mathbf{r}) \rightarrow \Phi(\mathbf{r}) \qquad \text{and} \qquad -\frac{1}{c}\mathbf{J}(\mathbf{r}) \rightarrow \rho(\mathbf{r}) \quad , \tag{4.1.23}$$

the calculation of the absorbed power proceeds along the lines we performed for the vector potential and the current density. We find that

$$P = |\Phi(\mathbf{q})|^2 \sum_{s} \frac{\omega}{\hbar^2} \int dt \, \langle s|\rho(\mathbf{q}, 0)\rho^*(\mathbf{q}, t)|s\rangle \exp\{-i\omega t\} \tag{4.1.24}$$

in analogy to Eq. (4.1.15). The Fourier transform of the longitudinal electric field is given by Eq. (3.1.2):

$$-i\mathbf{q}\Phi(\mathbf{q}) = \mathbf{E}^{\text{L}}(\mathbf{q}) \quad ,$$

and on utilizing this we find that the longitudinal conductivity per unit volume is

$$\sigma_1^{\text{L}}(\mathbf{q}, \omega) = \sum_{s} \frac{\omega}{\hbar q^2} \int dt \, \langle s|\rho(\mathbf{q}, 0)\rho^*(\mathbf{q}, t)|s\rangle \exp\{-i\omega t\} \quad . \tag{4.1.25}$$

The continuity equation (2.1.9) connects the density fluctuations and the longitudinal current density. With both $\mathbf{J}$ and ρ described by the spatial and time variation as $\exp\{i(\mathbf{q} \cdot \mathbf{r} - \omega t)\}$, the continuity equation becomes

$$i\mathbf{q} \cdot \mathbf{J}(\mathbf{q})^{\text{L}} = i\omega\rho(\mathbf{q}) \quad .$$

---

[1] Note that in textbooks on solid state theory often the particle density N and the potential energy V are used, quantities which are related to the charge density and potential by $\rho = -eN$ and $V = -e\Phi$, respectively. To follow common usage, we switch to this notation in Section 4.3.

Consequently, the conductivity per unit volume can also be written in the form which includes the longitudinal current density

$$\sigma_1^{\mathrm{L}}(\mathbf{q}, \omega) = \sum_s \frac{1}{\hbar\omega} \int \mathrm{d}t \, \langle s|\mathbf{J}^{\mathrm{L}}(\mathbf{q}, 0)\mathbf{J}^{\mathrm{L}*}(\mathbf{q}, t)|s\rangle \exp\{-\mathrm{i}\omega t\} \quad . \tag{4.1.26}$$

This expression can be compared with Eq. (4.1.16) derived earlier for the transverse conductivity.

Since the longitudinal and transverse conductivities are described by identical expressions, we can formally combine the two and in general write

$$(\sigma_1)^{\mathrm{L,T}}(\mathbf{q}, \omega) = \sum_s \frac{1}{\hbar\omega} \int \mathrm{d}t \, \langle s|\mathbf{J}^{\mathrm{L,T}}(\mathbf{q}, 0)\mathbf{J}^{\mathrm{L,T}*}(\mathbf{q}, t)|s\rangle \exp\{-\mathrm{i}\omega t\} \quad .$$

As implied by the notation, the longitudinal components of the currents lead to the longitudinal conductivity and the transverse components to the transverse conductivity.

4.2 The semiclassical approach

The electric current and electronic charge density induced by external electromagnetic fields can also be discussed using second quantization formalism, with the applied potentials either described as classical fields or also quantized. The so-called semiclassical approximation, where we treat both $\mathbf{A}(\mathbf{r}, t)$ and $\Phi(\mathbf{r}, t)$ as classical fields, is valid for field strengths and frequencies for which the quantization of the electromagnetic radiation can be neglected and linear response theory applies; this is the case for all methods and optical experiments discussed in this book.

The second quantized formalism, as utilized here, is appropriate for materials where the electronic states can be described as plane waves; this is valid for free electrons or for electrons in a crystalline solid when they are described in terms of Bloch wavefunctions. While this is certainly a restriction, when compared with the applicability of the Kubo formula, the formalism has been utilized extensively to describe the response of Fermi liquids, including their broken symmetry ground states. The formalism can be used to derive the conductivity, although here we merely write down the appropriate expressions for the currents and charge densities, and these can simply replace the currents and charge densities in the expressions we have derived above. An introduction to second quantization can be found in [Hau94, Kit63, Mah90, Sch83] and similar textbooks; here we merely quote the final results.

We write the field operator $\Psi(\mathbf{r})$ which obeys the equation of motion

$$\mathrm{i}\hbar\frac{\partial}{\partial t}\Psi(\mathbf{r}) = -[\mathcal{H}, \Psi(\mathbf{r})] \tag{4.2.1}$$

in the so-called second quantized form

$$\Psi(\mathbf{r}) = \sum_{\mathbf{k},\sigma} a_{\mathbf{k},\sigma}\exp\{i\mathbf{k}\cdot\mathbf{r}\}|\sigma\rangle \quad , \tag{4.2.2}$$

where $|\sigma\rangle$ denotes the spin part of the one-electron wavefunction. In terms of the coefficient $a_{\mathbf{k},\sigma}$, the electronic density yields

$$\rho(\mathbf{r}) = -e\sum_{j}\delta\left\{\mathbf{r}-\mathbf{r}_{j}\right\} = -e\Psi^{*}(\mathbf{r})\Psi(\mathbf{r}) = -e\sum_{\mathbf{k},\mathbf{k}',\sigma} a_{\mathbf{k},\sigma}^{+}a_{\mathbf{k}',\sigma}\exp\{i\left(\mathbf{k}'-\mathbf{k}\right)\cdot\mathbf{r}\} \tag{4.2.3}$$

and its Fourier component

$$\rho(\mathbf{q}) = \int \rho(\mathbf{r})\exp\{-i\mathbf{q}\cdot\mathbf{r}\}\,d\mathbf{r} = -e\sum_{\mathbf{k},\sigma} a_{\mathbf{k},\sigma}^{+}a_{\mathbf{k}+\mathbf{q},\sigma} \quad , \tag{4.2.4}$$

where $\mathbf{k}' = \mathbf{k} + \mathbf{q}$. The electric current from Eq. (4.1.6) then yields

$$\begin{aligned}
\mathbf{J}(\mathbf{r}) &= \frac{ie\hbar}{2m}\left(\Psi^{*}\nabla\Psi + \Psi\nabla\Psi^{*}\right) - \frac{e^{2}}{mc}\Psi^{*}\mathbf{A}\Psi \\
&= -\frac{e\hbar}{2m}\sum_{\mathbf{k},\mathbf{q},\sigma} a_{\mathbf{k}+\mathbf{q},\sigma}^{+}a_{\mathbf{k},\sigma}\exp\{i\mathbf{q}\cdot\mathbf{r}\}(2\mathbf{k}+\mathbf{q}) \\
&\quad -\frac{e^{2}}{mc}\sum_{\mathbf{k},\mathbf{q},\sigma} a_{\mathbf{k}+\mathbf{q},\sigma}^{+}a_{\mathbf{k},\sigma}\exp\{-i\mathbf{q}\cdot\mathbf{r}\}\mathbf{A}(\mathbf{r}) \\
&= \mathbf{J}_{\mathrm{p}}(\mathbf{r}) + \mathbf{J}_{\mathrm{d}}(\mathbf{r}) \quad , \tag{4.2.5}
\end{aligned}$$

consisting of two contributions, the paramagnetic and diamagnetic current.

As we have discussed in the previous section, the interaction Hamiltonian $\mathcal{H}_{\mathrm{int}}$ can be split into two contributions: the transverse response related to the vector potential and the longitudinal response related to the scalar potential. The interaction with the vector potential $\mathbf{A}(\mathbf{r})$ becomes

$$\begin{aligned}
\mathcal{H}_{\mathrm{int}}^{\mathrm{T}} &= -\frac{1}{c}\int \mathbf{J}^{\mathrm{T}}(\mathbf{r})\cdot\mathbf{A}^{\mathrm{T}}(\mathbf{r})\,d\mathbf{r} = \frac{e}{mc}\int d\mathbf{r}\,\Psi^{*}(\mathbf{r})\,\mathbf{p}\cdot\mathbf{A}^{\mathrm{T}}(\mathbf{r})\,\Psi(\mathbf{r}) \\
&= \frac{e\hbar}{mc}\int d\mathbf{r}\sum_{\mathbf{k},\mathbf{q},\sigma} a_{\mathbf{k}+\mathbf{q},\sigma}^{+}a_{\mathbf{k},\sigma}\,\mathbf{k}\cdot\mathbf{A}^{\mathrm{T}}(\mathbf{r})\exp\{-i\mathbf{q}\cdot\mathbf{r}\} \\
&= \frac{e\hbar}{mc}\sum_{\mathbf{k},\mathbf{q},\sigma} a_{\mathbf{k}+\mathbf{q},\sigma}^{+}a_{\mathbf{k},\sigma}\,\mathbf{k}\cdot\mathbf{A}^{\mathrm{T}}(\mathbf{q}) \quad , \tag{4.2.6}
\end{aligned}$$

where $\mathbf{A}(\mathbf{q})$ is given by Eq. (4.1.10). The interaction with the scalar field $\Phi(\mathbf{r})$ is

$$\mathcal{H}_{\mathrm{int}}^{\mathrm{L}} = \int \rho(\mathbf{r})\Phi(\mathbf{r})\,d\mathbf{r} = -e\int d\mathbf{r}\,\Psi^{*}(\mathbf{r})\,\Phi(\mathbf{r})\,\Psi(\mathbf{r}) = -e\sum_{\mathbf{k},\mathbf{q},\sigma} a_{\mathbf{k}+\mathbf{q},\sigma}^{*}a_{\mathbf{k},\sigma}\Phi(\mathbf{q}) \quad . \tag{4.2.7}$$

The appropriate expressions of the charge density and current given above can be inserted into the Kubo formula (4.1.16) and into the expression which gives the conductivity in terms of the density fluctuations; with these substitutions these are cast into a form using the second quantization formalisms.

4.3 Response function formalism and conductivity

Next we relate the dielectric constant and the conductivity to the changes in the electronic states brought about by the electromagnetic fields; for simplicity we restrict ourselves to cubic crystals for which the conductivity and dielectric constant are scalar. The formalism is applied for a collection of electrons which are described by Bloch wavefunctions, and as such is applied to crystalline solids. We will make use of what is usually called the selfconsistent field approximation as follows. Starting from the considerations of Section 3.1, we first write the total potential as

$$\Phi(\mathbf{q}, t) = \Phi_{\text{ext}}(\mathbf{q}, t) + \Phi_{\text{ind}}(\mathbf{q}, t) \qquad (4.3.1)$$

and define the dielectric constant $\hat{\epsilon}(\mathbf{q}, t)$ as

$$\Phi(\mathbf{q}, t) = \frac{\Phi_{\text{ext}}(\mathbf{q}, t)}{\hat{\epsilon}(\mathbf{q}, t)} \quad . \qquad (4.3.2)$$

The potential $\Phi(\mathbf{q}, t)$ leads to changes in the electronic density of states; these changes in turn can be obtained using the Heisenberg picture by treating $\Phi(\mathbf{q}, t)$ as a perturbation. The variations in the density of states cause an induced potential $\Phi_{\text{ind}}(\mathbf{q}, t)$ through Poisson's equation, and, through the selfconsistent equations above, this finally yields the expression of the dielectric constant in terms of the changes in the electronic states.

The procedure we describe eventually leads to expressions for the longitudinal response; the transverse response can be arrived at by similar arguments, but calculating the modification of the electronic states brought about by the vector potential $\mathbf{A}(\mathbf{q}, t)$.

4.3.1 Longitudinal response: the Lindhard function

The derivation is somewhat tedious but the end result is transparent and easily understood. As usual, we derive the expressions for the complex frequency and wavevector dependent dielectric constant; the conductivity can be subsequently written in a straightforward fashion. For the derivation of the response functions we begin with the Heisenberg equation, and write the time evolution of the electron number density operator N as

$$i\hbar\frac{\partial}{\partial t}N = [\mathcal{H}, N] \quad . \tag{4.3.3}$$

Let us introduce the unperturbed Hamiltonian $\mathcal{H}_0$

$$\mathcal{H}_0|\mathbf{k}l\rangle = \mathcal{E}_{\mathbf{k}l}|\mathbf{k}l\rangle \quad ,$$

and describe the interaction with the local electromagnetic potential by the interaction energy $V(\mathbf{r}, t)$ with $\mathcal{H}^{\mathrm{L}} = \mathcal{H}_0 + V$. The particle density is decomposed into the unperturbed density N_0 and the density fluctuation induced by the interaction $N(\mathbf{r}, t) = N_0 + \delta N(\mathbf{r}, t)$. Now we obtain

$$i\hbar\frac{\partial}{\partial t}\delta N = [\mathcal{H}_0, \delta N] + [V, N_0] \quad , \tag{4.3.4}$$

where the second order terms in the perturbation (e.g. $V\,\delta N$) have been neglected. In what follows we use Bloch wavefunctions to describe the electron states

$$|\mathbf{k}l\rangle = \Omega^{-1/2}\exp\{i\mathbf{k}\cdot\mathbf{r}\}u_{\mathbf{k}l}(\mathbf{r}) \quad , \tag{4.3.5}$$

where l is a band index, Ω is the volume of the system, and $u_{\mathbf{k}l}(\mathbf{r} + \mathbf{R}) = u_{\mathbf{k}l}(\mathbf{r})$, with $\mathbf{R}$ a Bravais lattice vector. These states are eigenstates of both H_0 and N_0, with eigenvalues $\mathcal{E}_{\mathbf{k}l}$ and $f^0(\mathcal{E}_{\mathbf{k}l})$, respectively.

$$f^0(\mathcal{E}_{\mathbf{k}l}) = [\exp\{(\mathcal{E}_{\mathbf{k}l} - \mathcal{E}_{\mathrm{F}})/k_{\mathrm{B}}T\} + 1]^{-1}$$

is the Fermi–Dirac distribution function, $\mathcal{E}_{\mathrm{F}}$ is the Fermi energy, and k_{B} is the Boltzmann constant. $\mathcal{E}_{\mathbf{k}l}$ indicates the energy of the $|\mathbf{k}l\rangle$ state; note that this energy depends on the wavevector: $\mathcal{E}_{\mathbf{k}} = \mathcal{E}(\mathbf{k})$.

First, we relate the induced charge density to the total potential energy $V(\mathbf{r}, t)$ acting on our electron system. This potential includes both the external and the screening potential; we will evaluate the latter in a selfconsistent fashion. Taking the matrix elements of the Heisenberg equation between states $|\mathbf{k}l\rangle$ and $|\mathbf{k} + \mathbf{q}, l'\rangle$ gives

$$i\hbar\frac{\partial}{\partial t}\langle\mathbf{k} + \mathbf{q}, l'|\delta N|\mathbf{k}l\rangle = \langle\mathbf{k} + \mathbf{q}, l'|[\mathcal{H}_0, \delta N]|\mathbf{k}l\rangle + \langle\mathbf{k} + \mathbf{q}, l'|[V(\mathbf{r}, t), N_0]|\mathbf{k}l\rangle \quad .$$
$$\tag{4.3.6}$$

Because these states are eigenstates of $\mathcal{H}_0$, the first term on the right hand side can be written as

$$\langle\mathbf{k} + \mathbf{q}, l'|[\mathcal{H}_0, \delta N]|\mathbf{k}l\rangle = \left(\mathcal{E}_{\mathbf{k}+\mathbf{q},l'} - \mathcal{E}_{\mathbf{k}l}\right)\langle\mathbf{k} + \mathbf{q}, l'|\delta N|\mathbf{k}l\rangle \quad ,$$

and the second term can be expressed as

$$\langle \mathbf{k} + \mathbf{q}, l' | [V(\mathbf{r}, t), N_0] | \mathbf{k}l \rangle = [f^0(\mathcal{E}_{\mathbf{k}l}) - f^0(\mathcal{E}_{\mathbf{k}+\mathbf{q},l'})]$$
$$\times \langle \mathbf{k} + \mathbf{q}, l' | \sum_{\mathbf{q}'} V(\mathbf{q}', t) \exp\{i\mathbf{q}' \cdot \mathbf{r}\} | \mathbf{k}l \rangle \quad , \quad (4.3.7)$$

where the perturbing potential energy $V(\mathbf{r}, t)$ has been expanded in its Fourier components. Note that $V(\mathbf{r}, t)$ includes both the external and the induced potential.

Now we substitute the explicit form of the Bloch states given above into the matrix element of Eq. (4.3.7), yielding

$$\langle \mathbf{k} + \mathbf{q}, l' | \sum_{\mathbf{q}'} V(\mathbf{q}', t) \exp\{i\mathbf{q}' \cdot \mathbf{r}\} | \mathbf{k}l \rangle = \Omega^{-1} \sum_{\mathbf{q}'} V(\mathbf{q}', t) \int_{\Omega} d\mathbf{r}\, u^*_{\mathbf{k}+\mathbf{q},l'} u_{\mathbf{k}l}$$
$$\times \exp\{i(\mathbf{q}' - \mathbf{q}) \cdot \mathbf{r}\} \quad , \quad (4.3.8)$$

where the volume integral involves the entire medium. As a consequence of the periodicity of $u_{\mathbf{k}l}$, it is possible to rewrite the above integral as an integral over the unit cell with an additional summation over all the unit cells. A change of variables $\mathbf{r} = \mathbf{R}_n + \mathbf{r}'$, where $\mathbf{R}_n$ is the position of the nth unit cell, and $\mathbf{r}'$ is the position within the unit cell, yields

$$\langle \mathbf{k} + \mathbf{q}, l' | \sum_{\mathbf{q}'} V(\mathbf{q}', t) \exp\{i\mathbf{q}' \cdot \mathbf{r}\} | \mathbf{k}l \rangle = \Omega^{-1} \sum_{\mathbf{q}'} \sum_n \exp\{i(\mathbf{q}' - \mathbf{q}) \cdot \mathbf{R}_n\} V(\mathbf{q}', t)$$
$$\times \int_{\Delta} d\mathbf{r}' u^*_{\mathbf{k}+\mathbf{q},l'} u_{\mathbf{k}l} \exp\{i(\mathbf{q}' - \mathbf{q}) \cdot \mathbf{r}'\} \quad ,$$
$$(4.3.9)$$

where the integral extends over the volume of a single unit cell Δ, and the index n represents the summation over all unit cells. We note that $\sum_n \exp\{i(\mathbf{q}' - \mathbf{q}) \cdot \mathbf{R}_n\}$ is negligible unless $\mathbf{q}' - \mathbf{q} = \mathbf{K}$, where $\mathbf{K} = 2\pi/\mathbf{R}$ is a reciprocal lattice vector. In this case, the summation is equal to N_c, the total number of unit cells. In the reduced zone scheme, we take $\mathbf{K} = 0$ and therefore

$$\sum_n \exp\{i(\mathbf{q}' - \mathbf{q}) \cdot \mathbf{R}_n\} = N_c \delta_{\mathbf{q}'\mathbf{q}} = (\Omega/\Delta)\delta_{\mathbf{q}'\mathbf{q}} \quad , \quad (4.3.10)$$

where we have rewritten N_c in terms of the volume of the unit cell Δ. Substituting this into Eq. (4.3.9) and using the δ-function to perform the summation over $\mathbf{q}'$ leads to

$$\langle \mathbf{k} + \mathbf{q}, l' | \sum_{\mathbf{q}'} V(\mathbf{q}', t) \exp\{i\mathbf{q}' \cdot \mathbf{r}\} | \mathbf{k}l \rangle = V(\mathbf{q}, t) \langle \mathbf{k} + \mathbf{q}, l' | \exp\{i\mathbf{q} \cdot \mathbf{r}\} | \mathbf{k}l \rangle_* \quad ,$$
$$(4.3.11)$$

where we have made use of

$$\langle \mathbf{k} + \mathbf{q}, l' | \exp\{i\mathbf{q} \cdot \mathbf{r}\} | \mathbf{k}l \rangle_* = \Delta^{-1} \int_\Delta u^*_{\mathbf{k}+\mathbf{q}l'}(\mathbf{r}) u_{\mathbf{k}l}(\mathbf{r})\, d\mathbf{r} \quad , \tag{4.3.12}$$

as the definition of $\langle \ \rangle_*$. With these expressions, Eq. (4.3.6) is cast into the following form:

$$
\begin{aligned}
i\hbar \frac{\partial}{\partial t} \langle \mathbf{k} + \mathbf{q}, l' | \delta N | \mathbf{k}l \rangle \ = \ & (\mathcal{E}_{\mathbf{k}+\mathbf{q},l'} - \mathcal{E}_{\mathbf{k}l}) \langle \mathbf{k} + \mathbf{q}, l' | \delta N | \mathbf{k}, l \rangle \\
& + [f^0(\mathcal{E}_{\mathbf{k}l}) - f^0(\mathcal{E}_{\mathbf{k}+\mathbf{q},l'})] \\
& \times V(\mathbf{q}, t) \langle \mathbf{k} + \mathbf{q}, l' | \exp\{i\mathbf{q} \cdot \mathbf{r}\} | \mathbf{k}l \rangle_* \quad .
\end{aligned}
\tag{4.3.13}
$$

This is a time dependent equation for δN, the contribution to the density operator resulting from the induced particle density.

Next we use Eq. (4.3.13) to derive an expression for the complex dielectric function $\hat{\epsilon}(\mathbf{q}, \omega)$. In doing so, we use the so-called adiabatic approximation, where we assume that the perturbation is turned on gradually starting at $t = -\infty$ with a time dependence $\exp\{\eta t\}$, and we will take the limit $\eta \to 0$ after appropriate expressions for the response have been derived. Consequently we assume that the time dependence of the external scalar potential has the form

$$\Phi_{\text{ext}}(\mathbf{r}, t) = \lim_{\eta \to 0} \Phi_{\text{ext}}(\mathbf{r}, 0) \exp\{-i\omega t + \eta t\} \quad . \tag{4.3.14}$$

Since the Fourier components are independent of each other, it is sufficient to consider only one component. We also assume that the induced screening potential energy, the total potential energy, and the density fluctuations all have the same $\exp\{-i\omega t + \eta t\}$ time dependence. With this we can rewrite Eq. (4.3.13) as

$$
\begin{aligned}
\lim_{\eta \to 0}(\hbar\omega - i\hbar\eta) \langle \mathbf{k} + \mathbf{q}, l' | \delta N | \mathbf{k}l \rangle \ = \ & (\mathcal{E}_{\mathbf{k}+\mathbf{q},l'} - \mathcal{E}_{\mathbf{k}l}) \langle \mathbf{k} + \mathbf{q}, l' | \delta N | \mathbf{k}l \rangle \\
& + [f^0(\mathcal{E}_{\mathbf{k}l}) - f^0(\mathcal{E}_{\mathbf{k}+\mathbf{q},l'})] \\
& \times V(\mathbf{q}, t) \langle \mathbf{k} + \mathbf{q}, l' | \exp\{i\mathbf{q} \cdot \mathbf{r}\} | \mathbf{k}l \rangle_*
\end{aligned}
$$

or, after some rearrangements,

$$\langle \mathbf{k}+\mathbf{q}, l' | \delta N | \mathbf{k}l \rangle = \lim_{\eta \to 0} \frac{f^0(\mathcal{E}_{\mathbf{k}+\mathbf{q},l'}) - f^0(\mathcal{E}_{\mathbf{k}l})}{\mathcal{E}_{\mathbf{k}+\mathbf{q},l'} - \mathcal{E}_{\mathbf{k}l} - \hbar\omega - i\hbar\eta} V(\mathbf{q}, t) \langle \mathbf{k}+\mathbf{q}, l' | \exp\{i\mathbf{q}\cdot\mathbf{r}\} | \mathbf{k}l \rangle_* \ , \tag{4.3.15}$$

connecting δN, the induced density, to $V(\mathbf{q}, t)$, the total selfconsistent perturbing potential energy.

What we apply to the system is the external potential Φ_{ext}; however, Eq. (4.3.15) is given in terms of $\Phi(\mathbf{q}, t)$, the total potential, which includes also the screening potential Φ_{ind}. Rewriting Eq. (4.3.2) yields

$$\Phi(\mathbf{q}, t) = \frac{\Phi_{\text{ind}}(\mathbf{q}, t)}{[1 - \hat{\epsilon}(\mathbf{q}, t)]} \quad .$$

Our task now is to establish a relationship between the induced potential and the changes in the electronic density. The energy $V_{\text{ind}}(\mathbf{r}, t) = -e\Phi_{\text{ind}}(\mathbf{r}, t)$ of the induced screening potential is related by Poisson's equation:

$$\nabla^2 V_{\text{ind}}(\mathbf{r}, t) = -4\pi e^2 \langle \delta N(\mathbf{r}, t) \rangle \quad .$$

The electronic charge density ρ can be written in terms of the particle density operator N as

$$\langle \rho \rangle = -e\langle N \rangle = -e \,\text{Tr}\,\{N \,\delta\{\mathbf{r} - \mathbf{r}_0\}\} \quad ,$$

where Tr indicates the trace. Using the identity $\sum_{\mathbf{k},l} |\mathbf{k}l\rangle\langle \mathbf{k}l| = 1$ we obtain

$$\langle \delta N \rangle = \text{Tr}\,\{\delta N \,\delta\{\mathbf{r} - \mathbf{r}_0\}\} = \sum_{\mathbf{k},\mathbf{q}} \sum_{l,l'} \langle \mathbf{k} + \mathbf{q}, l'|\delta N|\mathbf{k}l\rangle \langle \mathbf{k}l|\delta\{\mathbf{r} - \mathbf{r}_0\}|\mathbf{k} + \mathbf{q}, l'\rangle \quad ,$$

$$(4.3.16)$$

where $\mathbf{r}_0$ indicates the electron positions and l, l' are the band indices. Now we replace the states in the second matrix element with Bloch functions and use $\delta\{\mathbf{r} - \mathbf{r}_0\}$ to perform the integration over $\mathbf{r}$. Replacing $\mathbf{r}_0$ by $\mathbf{r}$, this yields

$$\langle \delta N \rangle = \Omega^{-1} \sum_{\mathbf{k},\mathbf{q}} \sum_{l,l'} u^*_{\mathbf{k}l} u_{\mathbf{k}+\mathbf{q},l'} \exp\{i\mathbf{q} \cdot \mathbf{r}\} \langle \mathbf{k} + \mathbf{q}, l'|\delta N|\mathbf{k}l\rangle \quad .$$

Putting this expression for the change in particle density into the Poisson equation gives the response to the change in the potential

$$\nabla^2 V_{\text{ind}}(\mathbf{r}, t) = -\frac{4\pi e^2}{\Omega} \sum_{\mathbf{k},\mathbf{q}} \sum_{l,l'} u^*_{\mathbf{k}l} u_{\mathbf{k}+\mathbf{q},l'} \exp\{i\mathbf{q} \cdot \mathbf{r}\} \langle \mathbf{k} + \mathbf{q}, l'|\delta N|\mathbf{k}l\rangle \quad ;$$

and taking the Fourier transform of $\nabla^2 V_{\text{ind}}(\mathbf{r}, t)$ leads to

$$-(q')^2 V_{\text{ind}}(\mathbf{q}', t) = -\frac{4\pi e^2}{\Omega} \sum_{\mathbf{k},\mathbf{q}} \sum_{l,l'} \langle \mathbf{k} + \mathbf{q}, l'|\delta N|\mathbf{k}l\rangle$$
$$\times \int d\mathbf{r}\, u^*_{\mathbf{k}l} u_{\mathbf{k}+\mathbf{q},l'} \exp\{i(\mathbf{q} - \mathbf{q}') \cdot \mathbf{r}\} \quad . \quad (4.3.17)$$

We can utilize the periodicity of the function $u(\mathbf{r})$ in the same way as before, and hence convert the integral over the entire system to one over a single unit cell and a sum over the unit cells. Following the same procedure as we employed in deriving Eqs (4.3.9) and (4.3.11), we arrive at the following expression for the induced potential energy:

$$V_{\text{ind}}(\mathbf{q}, t) = \frac{4\pi e^2}{q^2 \Omega} \sum_{\mathbf{k}} \sum_{l,l'} \langle \mathbf{k}+\mathbf{q}, l' | \delta N | \mathbf{k}l \rangle \langle \mathbf{k}l | \exp\{-i\mathbf{q}\cdot\mathbf{r}\} | \mathbf{k}+\mathbf{q}, l' \rangle_* \quad , \quad (4.3.18)$$

where we have used the definition of Eq. (4.3.12). Substituting Eq. (4.3.15) into Eq. (4.3.18) gives a relationship between the induced and total potential energy:

$$\begin{aligned} V_{\text{ind}}(\mathbf{q}, t) \;=\; & \lim_{\eta \to 0} \frac{4\pi e^2}{q^2 \Omega} V(\mathbf{q}, t) \sum_{\mathbf{k}} \sum_{l,l'} \frac{f^0(\mathcal{E}_{\mathbf{k}+\mathbf{q},l'}) - f^0(\mathcal{E}_{\mathbf{k}l})}{\mathcal{E}_{\mathbf{k}+\mathbf{q},l'} - \mathcal{E}_{\mathbf{k}l} - \hbar\omega - i\hbar\eta} \\ & \times \left| \langle \mathbf{k}+\mathbf{q} l' | \exp\{i\mathbf{q}\cdot\mathbf{r}\} | \mathbf{k}l \rangle_* \right|^2 \quad . \end{aligned} \quad (4.3.19)$$

Now we have prepared all the ingredients necessary to evaluate the dielectric constant or other material parameters as a function of wavevector and frequency. Rearranging Eq. (4.3.2), we have

$$\hat{\epsilon}(\mathbf{q}, t) = 1 - \frac{\Phi_{\text{ind}}(\mathbf{q}, t)}{\Phi(\mathbf{q}, t)} = 1 - \frac{V_{\text{ind}}(\mathbf{q}, t)}{V(\mathbf{q}, t)} \quad ;$$

and the previous two expressions allow us to write the Fourier component of the dielectric constant in terms of Bloch functions as follows:

$$\begin{aligned} \hat{\epsilon}(\mathbf{q}, \omega) \;=\; & 1 - \lim_{\eta \to 0} \frac{4\pi e^2}{q^2 \Omega} \sum_{\mathbf{k}} \sum_{l,l'} \frac{f^0(\mathcal{E}_{\mathbf{k}+\mathbf{q},l'}) - f^0(\mathcal{E}_{\mathbf{k}l})}{\mathcal{E}_{\mathbf{k}+\mathbf{q},l'} - \mathcal{E}_{\mathbf{k}l} - \hbar\omega - i\hbar\eta} \\ & \times \left| \langle \mathbf{k}+\mathbf{q}, l' | \exp\{i\mathbf{q}\cdot\mathbf{r}\} | \mathbf{k}l \rangle_* \right|^2 \quad , \end{aligned} \quad (4.3.20)$$

an expression usually referred to as the Lindhard form for the dielectric constant; $\hat{\epsilon}(\mathbf{q}, \omega)$ refers to the longitudinal dielectric constant, the component related to the scalar potential. The same information about the longitudinal response is contained in $\hat{\chi}(\mathbf{q}, \omega)$, the so-called Lindhard function defined in Eq. (3.1.32)

$$\hat{\chi}(\mathbf{q}, \omega) = \lim_{\eta \to 0} \frac{e^2}{\Omega} \sum_{\mathbf{k}} \sum_{l,l'} \frac{f^0(\mathcal{E}_{\mathbf{k}+\mathbf{q},l'}) - f^0(\mathcal{E}_{\mathbf{k}l})}{\mathcal{E}_{\mathbf{k}+\mathbf{q},l'} - \mathcal{E}_{\mathbf{k}l} - \hbar\omega - i\hbar\eta} |\langle \mathbf{k}+\mathbf{q}, l' | \exp\{i\mathbf{q}\cdot\mathbf{r}\} | \mathbf{k}l \rangle_*|^2 \quad ,$$

$$(4.3.21)$$

the function which describes the change of particle density δN due to an external scalar potential energy V.

With the general relation introduced in Eq. (3.2.7), the expression involving the differences in the occupation of the states and their energies, the so-called polarization function found in Eqs (4.3.20) and (4.3.21), becomes

$$\begin{aligned} \lim_{\eta \to 0} \frac{f_0(\mathcal{E}_{\mathbf{k}+\mathbf{q},l'}) - f_0(\mathcal{E}_{\mathbf{k}l})}{\mathcal{E}_{\mathbf{k}+\mathbf{q},l'} - \mathcal{E}_{\mathbf{k}l} - \hbar\omega - i\hbar\eta} \;=\; & \mathcal{P}\left[\frac{f_0(\mathcal{E}_{\mathbf{k}+\mathbf{q},l'}) - f_0(\mathcal{E}_{\mathbf{k}l})}{\mathcal{E}_{\mathbf{k}+\mathbf{q},l'} - \mathcal{E}_{\mathbf{k}l} - \hbar\omega} \right] \\ & + i\pi \left[f_0(\mathcal{E}_{\mathbf{k}+\mathbf{q},l'}) - f_0(\mathcal{E}_{\mathbf{k}l}) \right] \\ & \times \delta\{\mathcal{E}_{\mathbf{k}+\mathbf{q},l'} - \mathcal{E}_{\mathbf{k}l} - \hbar\omega\} \quad . \end{aligned} \quad (4.3.22)$$

In addition, the summation over all $\mathbf{k}$ values in the first term on the right hand side can be written as

$$\sum_{\mathbf{k}} \frac{f_0(\mathcal{E}_{\mathbf{k+q},l'}) - f_0(\mathcal{E}_{\mathbf{k}l})}{\mathcal{E}_{\mathbf{k+q},l'} - \mathcal{E}_{\mathbf{k}l} - \hbar\omega} = \sum_{\mathbf{k}} \frac{f_0(\mathcal{E}_{\mathbf{k+q},l'})}{\mathcal{E}_{\mathbf{k+q},l'} - \mathcal{E}_{\mathbf{k}l} - \hbar\omega} - \sum_{\mathbf{k}} \frac{f_0(\mathcal{E}_{\mathbf{k}l})}{\mathcal{E}_{\mathbf{k+q},l'} - \mathcal{E}_{\mathbf{k}l} - \hbar\omega} \quad .$$

This separation allows us to make the substitution $\mathbf{k} + \mathbf{q} \to \mathbf{k}$ in the first term on the right hand side; hence, this change does not affect the summation.

Next we calculate the real part of $\hat{\epsilon} = \epsilon_1 + i\epsilon_2$ by substituting the previous displayed expression into Eq. (4.3.20); we find

$$\epsilon_1(\mathbf{q}, \omega) = 1 - \frac{4\pi e^2}{q^2 \Omega} \sum_{\mathbf{k}} \sum_{l,l'} f_0(\mathcal{E}_{\mathbf{k}l}) \left(\frac{1}{\mathcal{E}_{\mathbf{k}l} - \mathcal{E}_{\mathbf{k-q},l'} - \hbar\omega} - \frac{1}{\mathcal{E}_{\mathbf{k+q},l'} - \mathcal{E}_{\mathbf{k}l} - \hbar\omega} \right)$$

$$\times |\langle \mathbf{k}+\mathbf{q}, l' | \exp\{i\mathbf{q}\cdot\mathbf{r}\}|\mathbf{k}l\rangle_*|^2 \tag{4.3.23a}$$

$$= 1 - \frac{4\pi e^2}{q^2 \Omega} \sum_{\mathbf{k}} \sum_{l,l'} f_0(\mathcal{E}_{\mathbf{k}l}) \left(\frac{\mathcal{E}_{\mathbf{k}l} - \mathcal{E}_{\mathbf{k-q},l'} + \hbar\omega}{(\mathcal{E}_{\mathbf{k}l} - \mathcal{E}_{\mathbf{k-q},l'})^2 - (\hbar\omega)^2} \right.$$

$$\left. - \frac{\mathcal{E}_{\mathbf{k+q},l'} - \mathcal{E}_{\mathbf{k}l} + \hbar\omega}{(\mathcal{E}_{\mathbf{k+q},l'} - \mathcal{E}_{\mathbf{k}l})^2 - (\hbar\omega)^2} \right) |\langle \mathbf{k}+\mathbf{q}, l' | \exp\{i\mathbf{q}\cdot\mathbf{r}\}|\mathbf{k}l\rangle_*|^2 \quad .$$

The imaginary part has the form

$$\epsilon_2(\mathbf{q}, \omega) = \frac{4\pi^2 e^2}{\Omega q^2} \sum_{\mathbf{k}} \sum_{l,l'} f^0(\mathcal{E}_{\mathbf{k}l}) \left[\delta\left\{\mathcal{E}_{\mathbf{k+q},l'} - \mathcal{E}_{\mathbf{k}l} - \hbar\omega\right\} \right.$$

$$\left. - \delta\left\{\mathcal{E}_{\mathbf{k}l} - \mathcal{E}_{\mathbf{k} \; \mathbf{q},l'} - \hbar\omega\right\} \right] |\langle \mathbf{k}+\mathbf{q}, l' | \exp\{i\mathbf{q}\cdot\mathbf{r}\}|\mathbf{k}l\rangle_*|^2 \quad . \tag{4.3.23b}$$

These expressions for the real and imaginary parts of the dielectric constant include both transitions between bands corresponding to different l and l' indices, and also transitions between states within one band. They will be used later for both metals and semiconductors, with intraband excitations more important for the former and interband transitions more important for the latter. As shown in Appendix C, the matrix element

$$|\langle \mathbf{k}+\mathbf{q}, l' | \exp\{i\mathbf{q}\cdot\mathbf{r}\}|\mathbf{k}l\rangle_*|^2$$

can be evaluated using the so-called $\mathbf{k}\cdot\mathbf{p}$ perturbation theory for small $\mathbf{q}$ values.

4.3.2 Response function for the transverse conductivity

The considerations which lead to the longitudinal response can also be applied to the response to a transverse electromagnetic field defined by $\mathbf{A} = (-\mathrm{i}c/\omega)\mathbf{E}$. As before we start from Heisenberg's equation (4.3.3) with the interaction term

$$\mathcal{H}_{\text{int}}^{\text{T}} = \frac{e}{mc}\,\mathbf{p}(\mathbf{r}) \cdot \mathbf{A}(\mathbf{r}, t) \quad , \tag{4.3.24}$$

where we have assumed Coulomb gauge $\Delta \cdot \mathbf{A} = 0$. Using Bloch states, we can write

$$
\begin{aligned}
i\hbar\frac{\partial}{\partial t}\langle \mathbf{k} + \mathbf{q}, l' | \delta N | \mathbf{k}l\rangle &= \langle \mathbf{k} + \mathbf{q}, l' | [H_0, \delta N] | \mathbf{k}l\rangle \\
&\quad + \langle \mathbf{k} + \mathbf{q}, l' | [(e/mc)\,\mathbf{p} \cdot \mathbf{A}(\mathbf{r}, t), N_0] | \mathbf{k}l\rangle \\
&= (\mathcal{E}_{\mathbf{k}+\mathbf{q},l'} - \mathcal{E}_{\mathbf{k}l})\langle \mathbf{k} + \mathbf{q}, l' | \delta N | \mathbf{k}l\rangle \\
&\quad + [f^0(\mathcal{E}_{\mathbf{k}l}) - f^0(\mathcal{E}_{\mathbf{k}+\mathbf{q},l'})] \\
&\quad \times \langle \mathbf{k} + \mathbf{q}, l' | \frac{e}{mc}\sum_{\mathbf{q}'}\mathbf{p} \cdot \mathbf{A}(\mathbf{q}', t)\exp\{i(\mathbf{q} \cdot \mathbf{r}')\}\,| \mathbf{k}l\rangle \quad .
\end{aligned}
$$

$$\tag{4.3.25}$$

With the explicit form of the Bloch wavefunctions from before, the matrix element of the second term becomes

$$
\langle \mathbf{k} + \mathbf{q}, l' | \frac{e}{mc}\sum_{\mathbf{q}'}\mathbf{p} \cdot \mathbf{A}(\mathbf{q}', t)\exp\{i(\mathbf{q}' \cdot \mathbf{r})\}\,| \mathbf{k}l\rangle = \Omega^{-1}\sum_{\mathbf{q}'}\sum_{n}\exp\{i(\mathbf{q}' - \mathbf{q}) \cdot \mathbf{R}_n\}
$$

$$
\times \int_{\Delta} d\mathbf{r}'\,\mathbf{p} \cdot \mathbf{A}(\mathbf{q}, t)\,u_{\mathbf{k}+\mathbf{q},l'}^{*}u_{\mathbf{k}l}\exp\{i(\mathbf{q}' - \mathbf{q}) \cdot \mathbf{r}'\} \quad . \tag{4.3.26}
$$

We again exploit the periodicity of the lattice ($\mathbf{r} = \mathbf{R}_n + \mathbf{r}'$) and use expression (4.3.10). Then it is sufficient to examine the sum over states within one unit cell

$$
\langle \mathbf{k}+\mathbf{q}, l' | \frac{e}{mc}\sum_{\mathbf{q}'}\mathbf{p}\cdot\mathbf{A}(\mathbf{q}', t)\exp\{i(\mathbf{q}'\cdot\mathbf{r})\}| \mathbf{k}l\rangle = \langle \mathbf{k}+\mathbf{q}, l' | \frac{e}{mc}\mathbf{p}\cdot\mathbf{A}(\mathbf{q}, t)\exp\{i\mathbf{q}\cdot\mathbf{r}\}| \mathbf{k}l\rangle_*.
$$

$$\tag{4.3.27}$$

As before we describe the adiabatic switch-on by the factor $\exp\{\eta t\}$, and the left hand side of Eq. (4.3.25) becomes

$$
i\hbar\frac{\partial}{\partial t}\langle \mathbf{k} + \mathbf{q}, l' | \delta N | \mathbf{k}l\rangle = (\hbar\omega - i\hbar\eta)\langle \mathbf{k} + \mathbf{q}, l' | \delta N | \mathbf{k}l\rangle \quad ,
$$

and thus we finally obtain with $\eta \to 0$

$$
\langle \mathbf{k} + \mathbf{q}, l' | \delta N | \mathbf{k}l\rangle = \lim_{\eta \to 0}\frac{f^0(\mathcal{E}_{\mathbf{k}l}) - f^0(\mathcal{E}_{\mathbf{k}+\mathbf{q},l'})}{\mathcal{E}_{\mathbf{k}+\mathbf{q},l'} - \mathcal{E}_{\mathbf{k}l} - \hbar\omega - i\hbar\eta}
$$

$$
\times \langle \mathbf{k} + \mathbf{q}, l' | \frac{e}{mc}\mathbf{p} \cdot \mathbf{A}(\mathbf{q}, t)\exp\{i\mathbf{q} \cdot \mathbf{r}\}\,| \mathbf{k}l\rangle_* \quad . \tag{4.3.28}
$$

In analogy to Eq. (4.3.15), we have now related the induced density matrix δN to the perturbation applied.

Our aim is to evaluate the conductivity or dielectric constant as a function of frequency and wavevector. Starting from the velocity operator of an electron in an electromagnetic field

$$\mathbf{v} = \frac{e\mathbf{A}}{mc} + \frac{\mathbf{p}}{m},$$

in first order of $\mathbf{A}$ we can write for the current density operator

$$
\begin{aligned}
\mathbf{J}(\mathbf{q}) \;=\; & -\operatorname{Tr}\left\{\exp\{-\mathrm{i}\mathbf{q}\cdot\mathbf{r}\}\frac{e^2}{mc}N_0\mathbf{A}(\mathbf{q})\right\} - \operatorname{Tr}\left\{\exp\{-\mathrm{i}\mathbf{q}\cdot\mathbf{r}\}\frac{e}{m}\delta N\,\mathbf{p}\right\} \\[2mm]
\;=\; & \frac{-e^2}{mc}\operatorname{Tr}\{N_0\mathbf{A}(\mathbf{q})\} - \frac{e}{m}\operatorname{Tr}\{\exp\{-\mathrm{i}\mathbf{q}\cdot\mathbf{r}\}\delta N\,\mathbf{p}\} \quad.
\end{aligned}
\tag{4.3.29}
$$

From comparison with Eq. (4.1.6) the term containing $\mathbf{A}(\mathbf{q})$ is the diamagnetic current and is given as

$$
\mathbf{J}_{\mathrm{d}}(\mathbf{q}) = -\frac{e^2 N_0}{mc}\mathbf{A}(\mathbf{q}) \quad,
\tag{4.3.30}
$$

where N_0 is the total electron density. The second term in Eq. (4.3.29) is known as the paramagnetic current; it can be evaluated by taking $|\mathbf{k}', l'\rangle = |\mathbf{k} + \mathbf{q}, l'\rangle$ and by substituting the Bloch wavefunctions. Following the procedure as in Eqs (4.3.16)–(4.3.19) finally leads to

$$
\begin{aligned}
\mathbf{J}(\mathbf{q}) = & -\frac{e^2 N_0}{mc}\mathbf{A}(\mathbf{q}) + \frac{1}{\Omega}\frac{e^2}{m^2 c}\mathbf{A}(\mathbf{q}) \\[2mm]
& \times \sum_{\mathbf{k}}\sum_{l,l'}\frac{f^0(\mathcal{E}_{\mathbf{k}l}) - f^0(\mathcal{E}_{\mathbf{k}+\mathbf{q},l'})}{\mathcal{E}_{\mathbf{k}+\mathbf{q},l'} - \mathcal{E}_{\mathbf{k}l} - \hbar\omega - \mathrm{i}\hbar\eta}\left|\langle\mathbf{k}+\mathbf{q},l'|\mathbf{p}|\mathbf{k}l\rangle_*\right|^2 \quad.
\end{aligned}
\tag{4.3.31}
$$

Applying Ohm's law we immediately obtain the transverse conductivity and dielectric constant

$$
\begin{aligned}
\hat{\sigma}(\mathbf{q},\omega) \;=\; & \mathrm{i}\frac{N_0 e^2}{\omega m} \\[2mm]
& + \lim_{\eta\to 0}\frac{\mathrm{i}}{\Omega}\frac{e^2}{\omega m^2}\sum_{\mathbf{k}}\sum_{l,l'}\frac{f^0(\mathcal{E}_{\mathbf{k}+\mathbf{q},l'}) - f^0(\mathcal{E}_{\mathbf{k}l})}{\mathcal{E}_{\mathbf{k}+\mathbf{q},l'} - \mathcal{E}_{\mathbf{k}l} - \hbar\omega - \mathrm{i}\hbar\eta}\left|\langle\mathbf{k}+\mathbf{q},l'|\mathbf{p}|\mathbf{k}l\rangle_*\right|^2,
\end{aligned}
\tag{4.3.32}
$$

$$
\begin{aligned}
\hat{\epsilon}(\mathbf{q},\omega) = & 1 + \frac{4\pi\mathrm{i}}{\omega}\hat{\sigma}(\mathbf{q},\omega) \\[2mm]
= & 1 - \frac{4\pi N_0 e^2}{\omega^2 m} - \lim_{\eta\to 0}\frac{4\pi}{\Omega}\frac{e^2}{\omega^2 m^2}\sum_{\mathbf{k}}\sum_{l,l'}\frac{f^0(\mathcal{E}_{\mathbf{k}+\mathbf{q},l'}) - f^0(\mathcal{E}_{\mathbf{k}l})}{\mathcal{E}_{\mathbf{k}+\mathbf{q},l'} - \mathcal{E}_{\mathbf{k}l} - \hbar\omega - \mathrm{i}\hbar\eta} \\[2mm]
& \times \left|\langle\mathbf{k}+\mathbf{q},l'|\mathbf{p}|\mathbf{k}l\rangle_*\right|^2 \quad,
\end{aligned}
\tag{4.3.33}
$$

an expression similar to the Lindhard equation for the longitudinal response. There are several differences between these two expressions. First we find an additional contribution (the second term on the right hand side), the so-called diamagnetic current. In the interaction Hamiltonian it shows up as a second order term in $\mathbf{A}$ and thus is often neglected. Second the matrix element contains the momentum $\mathbf{p} = -i\hbar\nabla$ instead of the dipole matrix element. Because of this difference, the pre-factor to the summation in Eq. (4.3.20) is proportional to $1/q^2$, whereas in Eq. (4.3.33) a $1/\omega^2$ factor appears.

Using a similar procedure to before, we can write down the expression for the real and imaginary parts; however, due to the fact that we have to take the limit and finally the integral over $\mathbf{k}$, this procedure is correct only for the imaginary part:

$$
\epsilon_2(\mathbf{q}, \omega) = \frac{4\pi^2}{\Omega}\frac{e^2}{\omega^2 m^2}\sum_{\mathbf{k}}\sum_{l,l'} f^0(\mathcal{E}_{\mathbf{k}l})\left[\delta\left\{\mathcal{E}_{\mathbf{k}l} - \mathcal{E}_{\mathbf{k}-\mathbf{q},l'} - \hbar\omega\right\}\right.
$$
$$
\left. - \delta\left\{\mathcal{E}_{\mathbf{k}+\mathbf{q},l'} - \mathcal{E}_{\mathbf{k}l} - \hbar\omega\right\}\right]\left|\langle\mathbf{k}+\mathbf{q}, l'|\mathbf{p}|\mathbf{k}l\rangle_*\right|^2 \quad . \tag{4.3.34a}
$$

The real part can then be obtained from $\epsilon(\mathbf{q}, \omega)$ as given above by applying the Kramers–Kronig relation, and we find:

$$
\epsilon_1(\mathbf{q}, \omega) = 1 - \frac{4\pi N_0 e^2}{\omega^2 m} - \frac{4\pi}{\Omega}\frac{e^2}{m^2}\sum_{\mathbf{k}}\sum_{l,l'}\left[\frac{1}{\mathcal{E}_{\mathbf{k}l} - \mathcal{E}_{\mathbf{k}-\mathbf{q},l'}}\frac{f^0(\mathcal{E}_{\mathbf{k}l})}{(\mathcal{E}_{\mathbf{k}l} - \mathcal{E}_{\mathbf{k}-\mathbf{q},l'})^2 - \hbar^2\omega^2}\right.
$$
$$
\left. - \frac{1}{\mathcal{E}_{\mathbf{k}+\mathbf{q},l'} - \mathcal{E}_{\mathbf{k}l}}\frac{f^0(\mathcal{E}_{\mathbf{k}l})}{(\mathcal{E}_{\mathbf{k}+\mathbf{q},l'} - \mathcal{E}_{\mathbf{k}l})^2 - \hbar^2\omega^2}\right]\left|\langle\mathbf{k}+\mathbf{q}, l'|\mathbf{p}|\mathbf{k}l\rangle_*\right|^2. \tag{4.3.34b}
$$

A general expression of the response function, valid when both scalar and vector potentials are present, can be found in [Adl62].

Here the transition matrix element is different from the matrix element which enters into the expression of the longitudinal response Eq. (4.3.12), and is given, for Bloch functions by

$$
\langle\mathbf{k}+\mathbf{q}, l'\,|\mathbf{p}|\,\mathbf{k}, l\rangle_* = -\frac{i\hbar}{\Delta}\int_\Delta u_{\mathbf{k}+\mathbf{q},l'}\,\nabla u_{\mathbf{k},l}\,d\mathbf{r} \quad . \tag{4.3.35}
$$

After these preliminaries, we are ready to describe the electrodynamic response of metals and semiconductors, together with the response of various ground states which arise as the consequence of electron–electron and electron–phonon interactions. Of course, this can be done using the current–current (or charge–charge) correlation functions, or using the response function formalism as outlined above, and in certain cases transition rate arguments. Which route one follows is a matter of choice. All the methods discussed here will be utilized in the subsequent chapters, and to a large extent the choice is determined by the usual procedures and conventions adopted in the literature.

References

[Adl62] S.L. Adler, *Phys. Rev.* **126**, 413 (1962)

[Ehr59] H. Ehrenreich and M.H. Cohen, *Phys. Rev.* **115**, 786 (1959)

[Hau94] H. Haug and S.W. Koch, *Quantum Theory of the Optical and Electronic Properties of Semiconductors*, 3rd edition (World Scientific, Singapore, 1994)

[Kit63] C. Kittel, *Quantum Theory of Solids* (John Wiley & Sons, New York, 1963)

[Lin54] J. Lindhard, *Dan. Mat. Fys. Medd.* **28**, no. 8 (1954)

[Mah90] G.D. Mahan, *Many-Particle Physics*, 2nd edition (Plenum Press, New York, 1990)

[Noz64] P. Nozières, *The Theory of Interacting Fermi Systems* (Benjamin, New York, 1964)

[Pin63] D. Pines, *Elementary Excitations in Solids* (Addison-Wesley, Reading, MA, 1963)

[Pin66] D. Pines and P. Nozières, *The Theory of Quantum Liquids*, Vol. 1 (Addison-Wesley, Reading, MA, 1966)

[Sch83] J.R. Schrieffer, *Theory of Superconductivity*, 3rd edition (Benjamin, New York, 1983)

Further reading

[Ehr66] H. Ehrenreich, *Electromagnetic Transport in Solids: Optical Properties and Plasma Effects*, in: *The Optical Properties of Solids*, edited by J. Tauc, Proceedings of the International School of Physics 'Enrico Fermi' **34** (Academic Press, New York, 1966)

[Hak76] H. Haken, *Quantum Field Theory of Solids* (North-Holland, Amsterdam, 1976)

[Jon73] W. Jones and N.H. March, *Theoretical Solid State Physics* (John Wiley & Sons, New York, 1973)

[Kub57] R. Kubo, *J. Phys. Soc. Japan* **12**, 570 (1957); *Lectures in Theoretical Physics*, Vol. 1 (John Wiley & Sons, New York, 1959)

[Pla73] P.M. Platzman and P.A. Wolff, *Waves and Interactions in Solid State Plasmas* (Academic Press, New York, 1973)

[Wal86] R.F. Wallis and M. Balkanski, *Many-Body Aspects of Solid State Spectroscopy* (North-Holland, Amsterdam, 1986)

5

Metals

In this chapter we apply the formalisms developed in the previous chapter to the electrodynamics of metals, i.e. materials with a partially filled electron band. Optical transitions between electron states in the partially filled band – the so-called intraband transitions – together with transitions between different bands – the interband transitions – are responsible for the electrodynamics. Here the focus will be on intraband excitations, and interband transitions will be dealt with in the next chapter. We first describe the frequency dependent optical properties of the phenomenological Drude–Sommerfeld model. Next we derive the Boltzmann equation, which along with the Kubo formula will be used in the zero wavevector limit to derive the Drude response of metals. The response to small $\mathbf{q}$ values (i.e. for wavevectors $q \ll k_F$, the Fermi wavevector) is discussed within the framework of the Boltzmann theory, both in the $\omega \gg q v_F$ (homogeneous) limit and $\omega \ll q v_F$ (static or quasi-static) limit; here v_F is the Fermi velocity. The treatment of the non-local conductivity due to Chambers is also valid in this limit and consequently will be discussed here. An example where the non-local response is important is the so-called anomalous skin effect; this will be discussed using heuristic arguments, and the full discussion is in Appendix E.1. The response for arbitrary $\mathbf{q}$ values is described using the selfconsistent field approximation developed in Section 4.3, where we derived expressions for $\hat{\sigma}(\mathbf{q}, \omega)$, the wavevector and frequency dependent complex conductivity for metals.

The longitudinal response is presented along similar lines. The Thomas–Fermi approximation leads to the static response in the small $\mathbf{q}$ limit, but the same expressions are also recovered using the Boltzmann equation. The Lindhard formalism is employed extensively to evaluate the dielectric response function $\hat{\chi}(\mathbf{q}, \omega)$ and also $\hat{\epsilon}(\mathbf{q}, \omega)$ and $\hat{\sigma}(\mathbf{q}, \omega)$, the various expressions bringing forth the different important aspects of the electrodynamics of the metallic state.

For sake of simplicity, we again assume cubic symmetry, and avoid complications which arise from the tensor character of the conductivity.

5.1 The Drude and the Sommerfeld models

5.1.1 The relaxation time approximation

The model due to Drude regards metals as a classical gas of electrons executing a diffusive motion. The central assumption of the model is the existence of an average relaxation time τ which governs the relaxation of the system to equilibrium, i.e. the state with zero average momentum $\langle \mathbf{p} \rangle = 0$, after an external field $\mathbf{E}$ is removed. The rate equation is

$$\frac{d\langle \mathbf{p} \rangle}{dt} = -\frac{\langle \mathbf{p} \rangle}{\tau} \quad . \tag{5.1.1}$$

In the presence of an external electric field $\mathbf{E}$, the equation of motion becomes

$$\frac{d}{dt}\langle \mathbf{p} \rangle = -\frac{\langle \mathbf{p} \rangle}{\tau} - e\mathbf{E} \quad .$$

The current density is given by $\mathbf{J} = -Ne\mathbf{p}/m$, with N the density of charge carriers; m is the carrier mass, and $-e$ is the electronic charge. For dc fields, the condition $d\langle \mathbf{p} \rangle/dt = 0$ leads to a dc conductivity

$$\sigma_{dc} = \frac{\mathbf{J}}{\mathbf{E}} = \frac{Ne^2\tau}{m} \quad . \tag{5.1.2}$$

Upon the application of an ac field of the form $\mathbf{E}(t) = \mathbf{E}_0 \exp\{-i\omega t\}$, the solution of the equation of motion

$$m\frac{d^2\mathbf{r}}{dt^2} + \frac{m}{\tau}\frac{d\mathbf{r}}{dt} = -e\mathbf{E}(t) \tag{5.1.3}$$

gives a complex, frequency dependent conductivity

$$\hat{\sigma}(\omega) = \frac{Ne^2\tau}{m}\frac{1}{1 - i\omega\tau} = \sigma_1(\omega) + i\sigma_2(\omega) = \frac{Ne^2\tau}{m}\frac{1 + i\omega\tau}{1 + \omega^2\tau^2} \quad . \tag{5.1.4}$$

There is an average distance traveled by the electrons between collisions, called the mean free path ℓ. Within the framework of the Drude model, $\ell = \langle v \rangle_{th}\tau$, where $\langle v \rangle_{th}$ is the average thermal velocity of classical particles, and the kinetic energy $\frac{1}{2}m\langle v^2 \rangle_{th} = \frac{3}{2}k_B T$ at temperature T.

The picture is fundamentally different for electrons obeying quantum statistics, and the consequences of this have been developed by Sommerfeld. Within the framework of this model, the concept of the Fermi surface plays a central role. In the absence of an electric field, the Fermi sphere is centered around zero momentum, and $\langle \mathbf{p} \rangle = \hbar\langle \mathbf{k} \rangle = 0$, as shown in Fig. 5.1. The Fermi sphere is displaced in the presence of an applied field $\mathbf{E}$, with the magnitude of the displacement given by $-e\mathbf{E}\tau/\hbar$; electrons are added to the region A and removed from region B. The equation of motion for the average momentum is the same as given above. Again

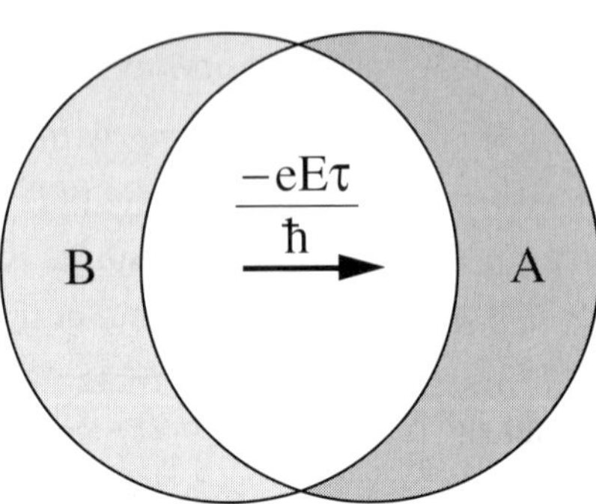

Fig. 5.1. Displaced Fermi sphere in the presence of an applied electric field $\mathbf{E}$. Electrons are removed from region B by the momentum $\mathbf{p} = -e\mathbf{E}\tau$, and electrons are added in region A. The figure applies to a metal with a spherical Fermi surface of radius k_{F} – or a circular Fermi surface in two dimensions.

with $\langle \mathbf{J} \rangle = -Ne\hbar \langle \mathbf{k} \rangle / m$ and wavevector $\mathbf{k} = \mathbf{p}/\hbar$ expression (5.1.4) is recovered. However, the scattering processes which establish the equilibrium in the presence of the electric field involve only electrons near to the Fermi surface; states deep within the Fermi sea are not influenced by the electric field. Consequently the expression for the mean free path is $\ell = v_{\mathrm{F}}\tau$, with v_{F} the Fermi velocity, and differs dramatically from the mean free path given by the original Drude model. The difference has important consequences for the temperature dependences, and also for non-linear response to large electric fields, a subject beyond the scope of this book. The underlying interaction with the lattice, together with electron–electron and electron–phonon interactions, also are of importance and lead to corrections to the above description. Broadly speaking these effects can be summarized assuming an effective mass which is different from the free-electron mass [Pin66] and also frequency dependent; this issue will be dealt with later in Section 12.2.2.

The sum rules which have been derived in Section 3.2 are of course obeyed, and can be easily proven by direct integration. The f sum rule (3.2.28) for example follows as

$$
\int_0^\infty \sigma_1(\omega)\, \mathrm{d}\omega = \frac{Ne^2}{m} \int_0^\infty \frac{\tau\, \mathrm{d}\omega}{1+\omega^2\tau^2} = \int_0^\infty \frac{\omega_{\mathrm{p}}^2}{4\pi} \frac{\mathrm{d}(\omega\tau)}{1+(\omega\tau)^2}
$$

$$
= \frac{\omega_{\mathrm{p}}^2}{4\pi} \arctan\{\omega\tau\}\Big|_0^\infty = \frac{\omega_{\mathrm{p}}^2}{4\pi}\frac{\pi}{2} = \frac{\omega_{\mathrm{p}}^2}{8} \quad ,
$$

where we have defined

$$
\omega_{\mathrm{p}} = \left(\frac{4\pi Ne^2}{m} \right)^{1/2} \quad . \tag{5.1.5}
$$

This is called the plasma frequency, and its significance will be discussed in subsequent sections.

Before embarking on the discussion of the properties of the Drude model, let us first discuss the response in the limit when the relaxation time $\tau \to 0$. In this limit

$$\sigma_1(\omega) = \frac{\pi}{2}\frac{Ne^2}{m}\delta\{\omega = 0\} \qquad \text{and} \qquad \sigma_2(\omega) = \frac{Ne^2}{m\omega} \quad .$$

Here $\sigma_2(\omega)$ reflects the inertial response, and this component does not lead to absorption but only to a phase lag. $\sigma_1(\omega)$ is zero everywhere except at $\omega = 0$; a degenerate collisionless free gas of electrons cannot absorb photons at finite frequencies. This can be proven directly by showing that the Hamiltonian $\mathcal{H}$ which describes the electron gas commutes with the momentum operator, and thus $\mathbf{p}$ has no time dependence. This is valid even when electron–electron interactions are present, as such interactions do not change the total momentum of the electron system.

5.1.2 Optical properties of the Drude model

Using the optical conductivity, as obtained by the Drude–Sommerfeld model, the various optical parameters can be evaluated in a straightforward manner. The dc limit of the conductivity is

$$\sigma_1(\omega = 0) = \sigma_{\mathrm{dc}} = \frac{Ne^2\tau}{m} = \frac{1}{4\pi}\omega_{\mathrm{p}}^2\tau \quad ; \tag{5.1.6}$$

its frequency dependence can be written as

$$\hat{\sigma}(\omega) = \frac{\sigma_{\mathrm{dc}}}{1 - i\omega\tau} = \frac{Ne^2}{m}\frac{1}{1/\tau - i\omega} = \frac{\omega_{\mathrm{p}}^2}{4\pi}\frac{1}{1/\tau - i\omega} \tag{5.1.7}$$

with the components

$$\sigma_1(\omega) = \frac{\omega_{\mathrm{p}}^2\tau}{4\pi}\frac{1}{1 + \omega^2\tau^2} \qquad \text{and} \qquad \sigma_2(\omega) = \frac{\omega_{\mathrm{p}}^2\tau}{4\pi}\frac{\omega\tau}{1 + \omega^2\tau^2} \quad . \tag{5.1.8}$$

Thus, within the framework of the Drude model the complex conductivity and consequently all the various optical parameters are fully characterized by two frequencies: the plasma frequency ω_{p} and the relaxation rate $1/\tau$; in general $1/\tau \ll \omega_{\mathrm{p}}$. These lead to three regimes with distinctively different frequency dependences of the various quantities.

Using the general relation (2.2.12), the frequency dependence of the dielectric constant is

$$\hat{\epsilon}(\omega) = \epsilon_1(\omega) + i\epsilon_2 = 1 - \frac{\omega_{\mathrm{p}}^2}{\omega^2 - i\omega/\tau} \tag{5.1.9}$$

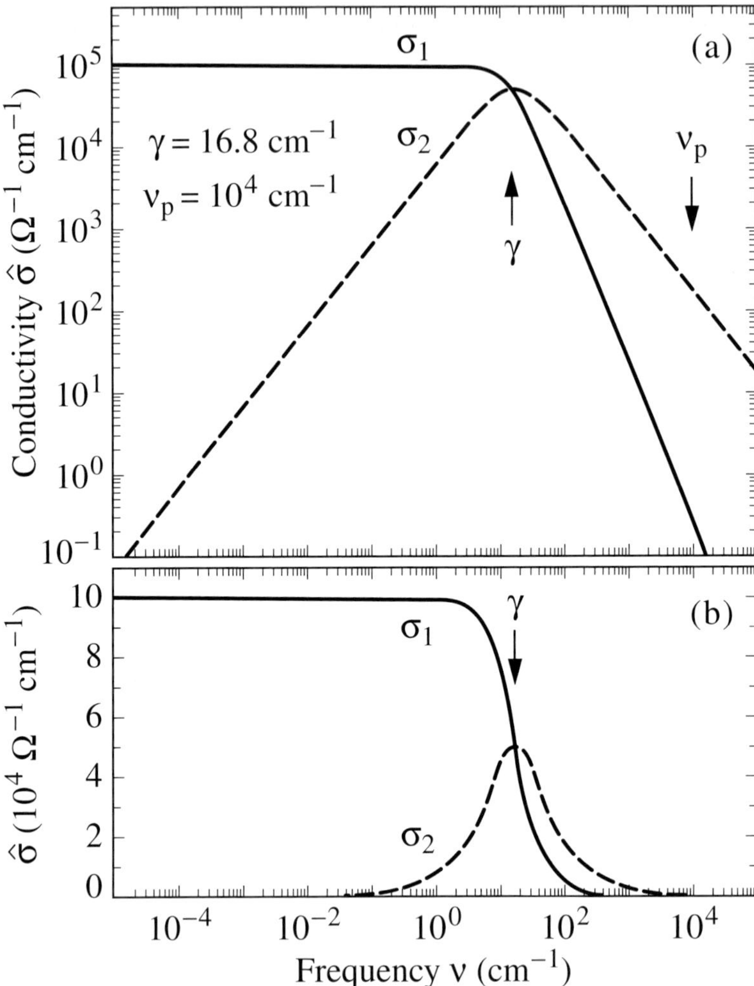

Fig. 5.2. Frequency dependent conductivity $\hat{\sigma}(\omega)$ calculated after the Drude model (5.1.7) for the plasma frequency $\omega_p/(2\pi c) = \nu_p = 10^4$ cm^{-1} and the scattering rate $1/(2\pi c\tau) = \gamma = 16.8$ cm^{-1} in (a) a logarithmic and (b) a linear conductivity scale. Well below the scattering rate γ, the real part of the conductivity σ_1 is frequency independent with a dc value $\sigma_{dc} = 10^5$ Ω^{-1} cm^{-1}, above γ it falls off with ω^{-2}. The imaginary part $\sigma_2(\omega)$ peaks at γ where $\sigma_1 = \sigma_2 = \sigma_{dc}/2$; for low frequencies $\sigma_2(\omega) \propto \omega$, for high frequencies $\sigma_2(\omega) \propto \omega^{-1}$.

with the real and imaginary parts

$$\epsilon_1(\omega) = 1 - \frac{\omega_p^2}{\omega^2 + \tau^{-2}} \quad \text{and} \quad \epsilon_2(\omega) = \frac{1}{\omega\tau}\frac{\omega_p^2}{\omega^2 + \tau^{-2}} \quad . \tag{5.1.10}$$

The components of the complex conductivity $\hat{\sigma}(\omega)$ as a function of frequency

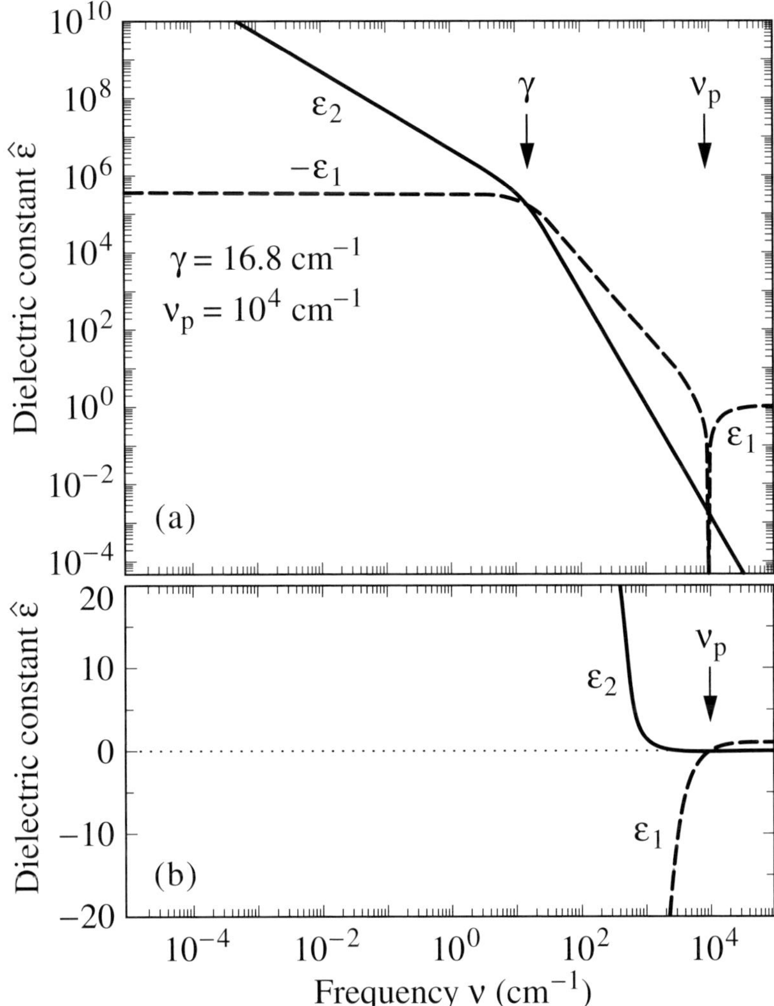

Fig. 5.3. Dielectric constant $\hat{\epsilon}(\omega)$ of the Drude model (5.1.9) plotted as a function of frequency in (a) a logarithmic and (b) a linear scale for $\nu_p = 10^4 \ \text{cm}^{-1}$ and $\gamma = 16.8 \ \text{cm}^{-1}$. For frequencies up to the scattering rate γ, the real part of the dielectric constant $\epsilon_1(\omega)$ is negative and independent of frequency; for $\nu > \gamma$ it increases with ω^{-2} and finally changes sign at the plasma frequency ν_p before ϵ_1 approaches 1. The imaginary part $\epsilon_2(\omega)$ stays always positive, decreases monotonically with increasing frequencies, but changes slope from $\epsilon_2(\omega) \propto \omega$ to ω^{-3} at $\nu = \gamma$.

are shown in Fig. 5.2 for the parameters $\nu_p = \omega_p/(2\pi c) = 10^4 \ \text{cm}^{-1}$ and $\gamma = 1/(2\pi c\tau) = 16.8 \ \text{cm}^{-1}$; these are typical for good metals at low temperatures. The frequency dependence of both components of the complex dielectric constant $\hat{\epsilon}(\omega)$ is displayed in Fig. 5.3 for the same parameters.

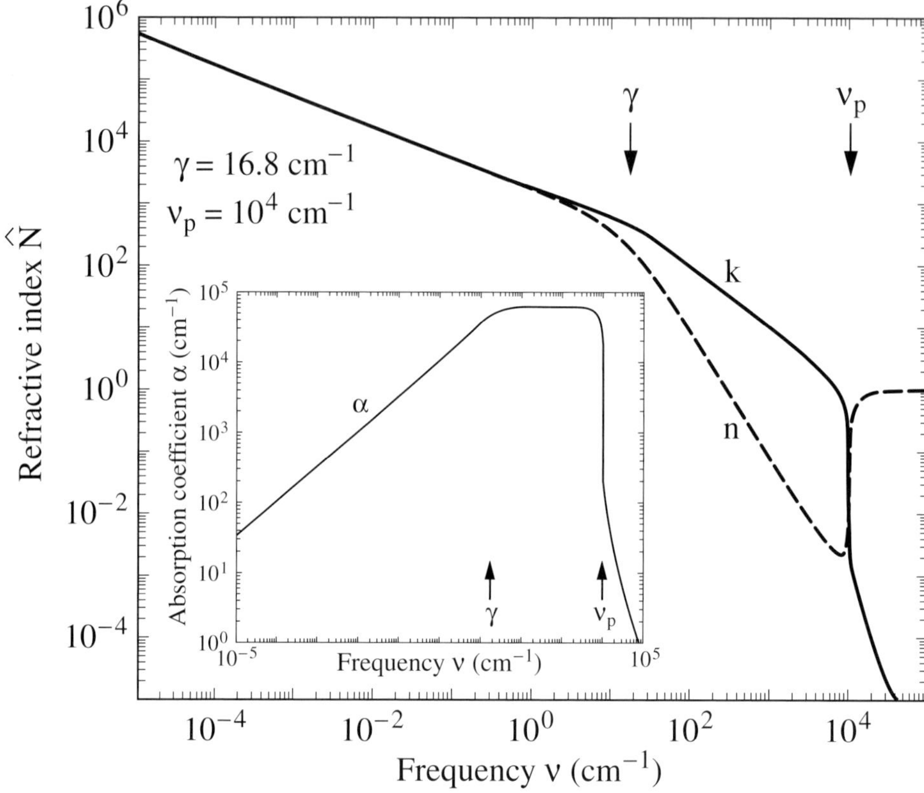

Fig. 5.4. Real and imaginary parts of the refractive index $\hat{N}(\omega) = n + ik$ of the Drude model as a function of frequency for $\nu_\mathrm{p} = 10^4$ cm^{-1} and $\gamma = 16.8$ cm^{-1}. Well below the scattering rate γ (i.e. in the Hagen–Rubens regime) $n = k \propto \omega^{-1/2}$. In the relaxation regime, we have $k > n$; also $k(\omega) \propto \omega^{-1}$ in this region, and $n(\omega) \propto \omega^{-2}$. Above the plasma frequency ν_p, the metal becomes transparent, indicated by the drop in k; the real part of the refractive index goes to 1. The inset shows the absorption coefficient α, which describes the attenuation of the electromagnetic wave in the material. At low frequencies $\alpha(\omega) \propto \omega^{1/2}$, whereas in the range $\gamma < \nu < \nu_\mathrm{p}$ α is basically constant before it falls off at ν_p.

The complex refractive index $\hat{N} = n + ik$, the components of which (n and k) are displayed in Fig. 5.4; the inset shows the absorption coefficient α. The reflectivity R and the absorptivity $A = 1 - R$ of a bulk material can be calculated by utilizing Eq. (2.4.15), while the phase shift of the reflected wave is given by Eq. (2.4.14). The frequency dependent reflectivity R is plotted in Fig. 5.5, with the inset showing the low frequency reflectivity on a linear frequency scale. $R(\omega)$ is close to 1 up to the plasma frequency where it drops rapidly to almost zero; this is called the plasma edge. In Fig. 5.6 the absorptivity A and the phase shift ϕ_r corresponding to

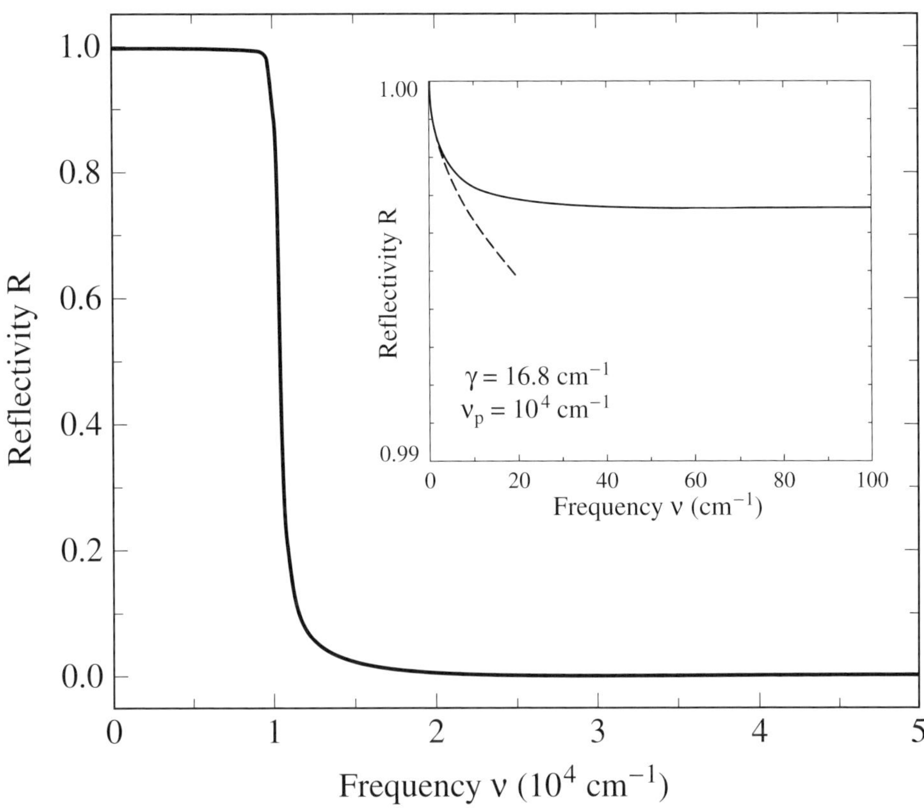

Fig. 5.5. Frequency dependent reflectivity $R(\omega)$ of the Drude model for $\nu_p = 10^4$ cm^{-1} and $\gamma = 16.8$ cm^{-1} on a linear scale. At the plasma frequency ν_p the reflectivity drops drastically ($R \rightarrow 0$) and the material becomes transparent. The inset shows the low frequency reflectivity $R(\omega)$. The square root frequency dependence $R(\omega) = 1 - (2\omega/\pi\sigma_{\mathrm{dc}})^{1/2}$ of the Hagen–Rubens equation (5.1.17) is indicated by the dashed line, which deviates considerably from the results of the Drude model (solid line) for frequencies below the scattering frequency γ.

the reflection of the waves are displayed as a function of frequency. The skin depth

$$\delta_0(\omega) = \frac{2}{\alpha(\omega)} = \frac{2}{\omega\, k(\omega)}$$

was given in Eq. (2.3.15a). Finally Fig. 5.7 shows the components of the surface impedance $\hat{Z}_S$ as a function of frequency calculated for the same parameters of the metal. For completeness we also give the expression of the dielectric loss function

$$\frac{1}{\hat{\epsilon}(\omega)} = \frac{\epsilon_1(\omega) - i\epsilon_2(\omega)}{[\epsilon_1(\omega)]^2 + [\epsilon_2(\omega)]^2} \tag{5.1.11}$$

introduced in Section 3.2.2. Within the framework of the Drude model the two

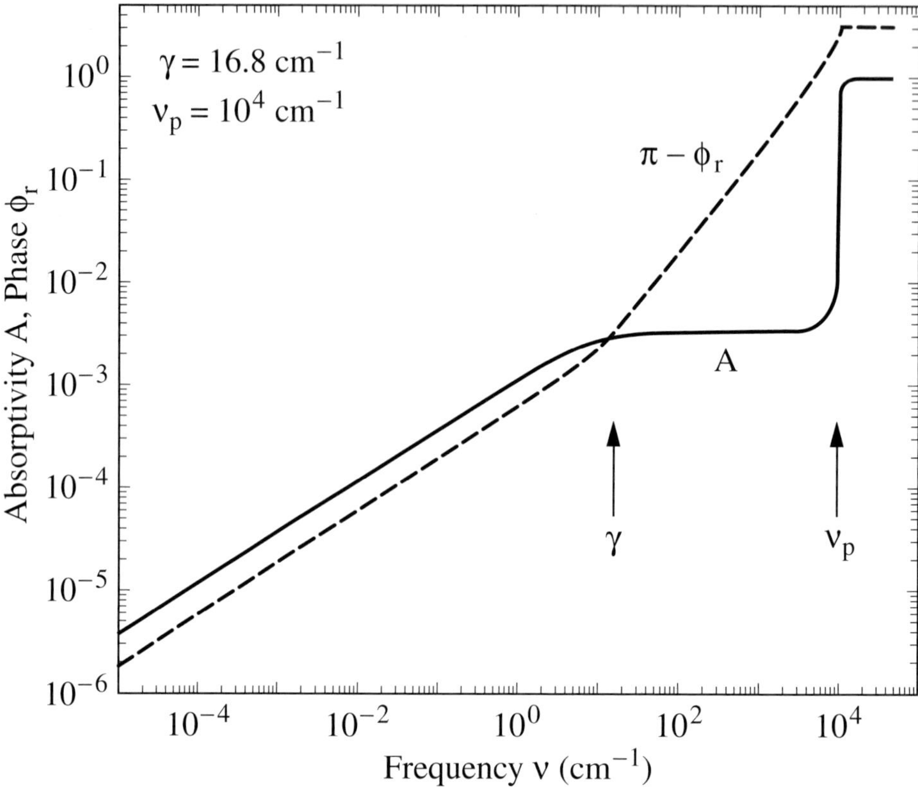

Fig. 5.6. Frequency dependent absorptivity $A = 1 - R$ and phase shift ϕ_r of the Drude model for $\nu_p = 10^4$ cm^{-1} and $\gamma = 16.8$ cm^{-1}. In the Hagen–Rubens regime, A and ϕ_r increase with $\omega^{1/2}$, whereas in the relaxation regime the absorptivity stays constant. The plasma frequency is indicated by a sudden change in phase and the dramatic increase in absorptivity.

components are

$$\mathrm{Re}\left\{\frac{1}{\hat{\epsilon}(\omega)}\right\} = 1 + \frac{(\omega^2 - \omega_p^2)\omega_p^2}{(\omega^2 - \omega_p^2)^2 + \omega^2\tau^{-2}} \qquad (5.1.12a)$$

$$-\mathrm{Im}\left\{\frac{1}{\hat{\epsilon}(\omega)}\right\} = \frac{\omega_p^2\omega/\tau}{(\omega^2 - \omega_p^2)^2 + \omega^2\tau^{-2}} \quad , \qquad (5.1.12b)$$

and these relations are displayed in Fig. 5.8 for the same $1/\tau$ and ω_p values as used before.

With $1/\tau$ and ω_p the fundamental parameters which enter the frequency dependent conductivity, three different regimes can be distinguished in the spectra, and they will be discussed separately: the so-called Hagen–Rubens regime for $\omega \ll 1/\tau$, the relaxation regime for $1/\tau \ll \omega \ll \omega_p$, and the transparent regime

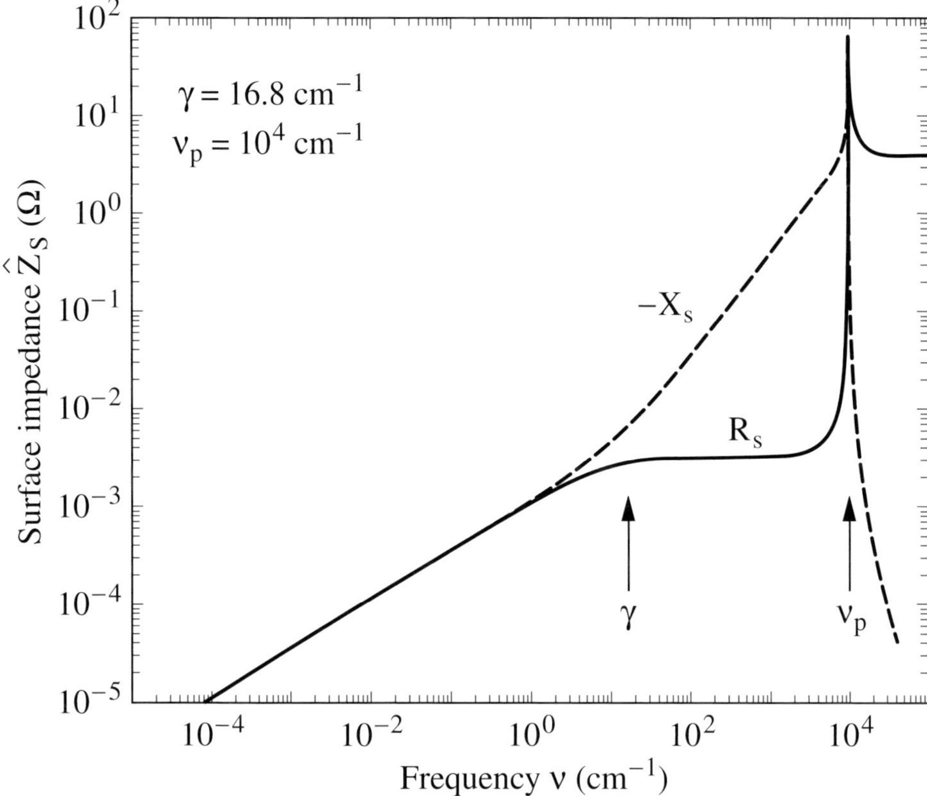

Fig. 5.7. Real and imaginary parts of the frequency dependent surface impedance, R_{S} and X_{S}, respectively, of the Drude model for $\nu_{\mathrm{p}} = 10^4$ cm^{-1} and $\gamma = 16.8$ cm^{-1}. For frequencies below the scattering rate γ, the surface resistance equals the surface reactance, $R_{\mathrm{S}} = -X_{\mathrm{S}}$, and both are proportional to $\omega^{1/2}$.

for $\omega \gg \omega_{\mathrm{p}}$. All three regimes have a characteristic optical response which is fundamentally different from the response found in other regimes.

Hagen–Rubens regime

The low frequency or Hagen–Rubens regime is defined by the condition $\omega\tau \ll 1$. In this regime the optical properties are mainly determined by the dc conductivity σ_{dc}: the real part of the conductivity σ_1 is frequency independent in this range (Fig. 5.2):

$$\sigma_{\mathrm{dc}} \approx \sigma_1(\omega) \gg \sigma_2(\omega) \quad ; \tag{5.1.13a}$$

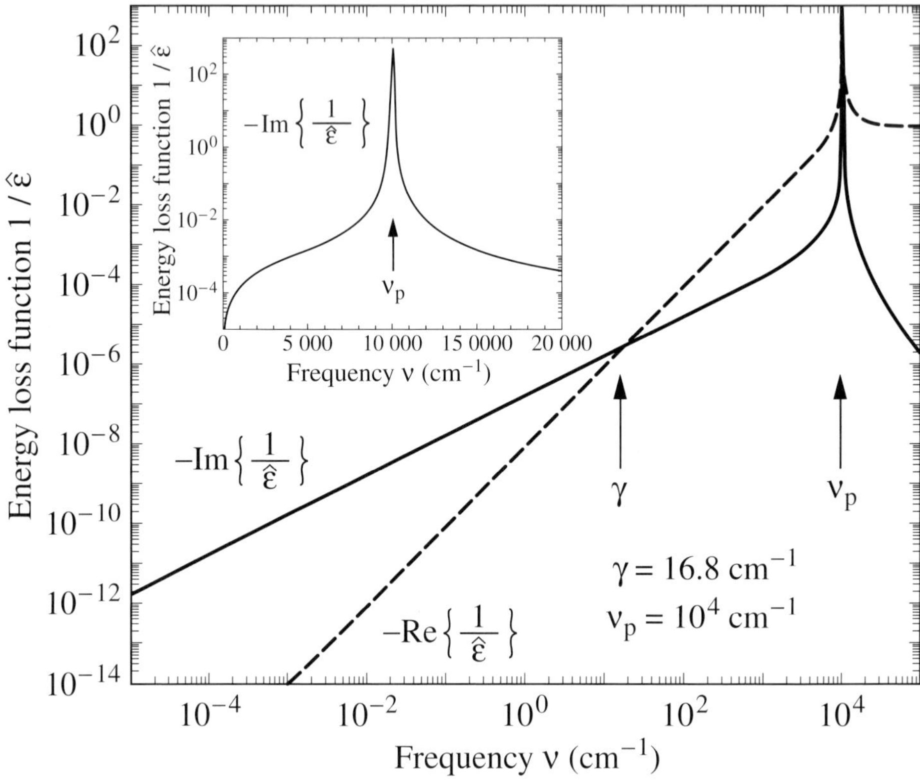

Fig. 5.8. Frequency dependence of the energy loss function $1/\hat{\epsilon}$ of the Drude model for $\nu_\mathrm{p} = 10^4$ cm^{-1} and $\gamma = 16.8$ cm^{-1}. The two components are equal at γ, and both peak at the plasma frequency ν_p. The inset shows $-\mathrm{Im}\{1/\hat{\epsilon}\}$ on a linear frequency scale to stress the extremely narrow peak of width γ.

while the imaginary part increases linearly with frequency:

$$\sigma_2(\omega) \approx \sigma_{\mathrm{dc}}\, \omega\tau = \frac{\omega_\mathrm{p}^2\tau^2}{4\pi}\omega \ll \sigma_1(\omega) \quad . \tag{5.1.13b}$$

The real part of the dielectric constant ϵ_1 is negative and large; for $\omega < 1/\tau$ Eq. (5.1.10) can be simplified to yield the constant value

$$\epsilon_1(\omega) = \epsilon_1(0) \approx 1 - \omega_\mathrm{p}^2\tau^2 \quad , \tag{5.1.14a}$$

while

$$\epsilon_2(\omega) \approx \frac{\omega_\mathrm{p}^2\tau}{\omega} \quad . \tag{5.1.14b}$$

From Eq. (2.3.1) we obtain

$$\hat{N} = n + \mathrm{i}k \approx (\mathrm{i}\epsilon_2)^{1/2} \approx (\epsilon_2/2)^{1/2}(1 + \mathrm{i}) \tag{5.1.15}$$

for the refractive index, and both components are equal,

$$n(\omega) = k(\omega) = \left[\frac{\epsilon_2(\omega)}{2}\right]^{1/2} = \left(\frac{2\pi\sigma_{dc}}{\omega}\right)^{1/2} \gg 1 \quad , \tag{5.1.16}$$

in this regime. The reflectivity R can now be written as

$$\begin{aligned} R(\omega) &\approx \frac{k(\omega)-1}{k(\omega)+1} \approx 1 - \frac{2}{k(\omega)} + \frac{2}{[k(\omega)]^2} \\ &\approx 1 - \left(\frac{2\omega}{\pi\sigma_{dc}}\right)^{1/2} = 1 - \left(\frac{8\omega}{\omega_p^2\tau}\right)^{1/2} = 1 - A(\omega) \quad ; \end{aligned} \tag{5.1.17}$$

the absorptivity $A(\omega)$ increases as the square root of the frequency. This relation was first found by Hagen and Rubens in emission experiments. In this frequency range, the skin depth (Eq. (2.3.15b)) is determined by the dc conductivity:

$$\delta_0 = \left(\frac{c^2}{2\pi\omega\sigma_{dc}}\right)^{1/2} \quad ;$$

the components of the surface impedance (Fig. 5.7) are equal

$$R_S(\omega) = -X_S(\omega) = \left(\frac{2\pi\omega}{c^2\sigma_{dc}}\right)^{1/2} = \frac{1}{\delta_0(\omega)\sigma_{dc}} \tag{5.1.18}$$

and display a characteristic $\omega^{1/2}$ dependence on the frequency.

Relaxation regime

For frequencies which lie in the intermediate spectral range between the scattering rate $1/\tau$ and the plasma frequency ω_p – the so-called relaxation regime – the term $(\omega\tau)^2$ in Eq. (5.1.8) cannot be neglected. As can be seen from Figs 5.2 and 5.3, the scattering rate defines the crossover frequency at which $\sigma_1 = \sigma_2$ and $|\epsilon_1| = \epsilon_2$. For $\omega \gg 1/\tau$ to first approximation we have

$$\sigma_1(\omega) \approx \frac{\sigma_{dc}}{(\omega\tau)^2} \quad \text{and} \quad \sigma_2(\omega) \approx \frac{\sigma_{dc}}{\omega\tau} \quad ; \tag{5.1.19}$$

for high frequencies this implies $\sigma_1(\omega) \ll \sigma_2(\omega)$. In the same frequency range the real and imaginary parts of the dielectric constant decrease with increasing frequency as

$$\epsilon_1(\omega) \approx 1 - \frac{\omega_p^2}{\omega^2} \quad \text{and} \quad \epsilon_2(\omega) \approx \frac{\omega_p^2}{\omega^3\tau} \quad , \tag{5.1.20}$$

and also n and k are not identical and show a different frequency dependence. Substituting the above expression into Eqs (2.3.3) and (2.3.4), we find that

$$n(\omega) \approx \frac{\omega_p}{2\tau\omega^2} \quad \text{and} \quad k(\omega) \approx \frac{\omega_p}{\omega} \quad . \tag{5.1.21}$$

In this regime the absorptivity A becomes frequency independent and

$$A(\omega) = 1 - R(\omega) \approx \frac{2}{\omega_p \tau} = \frac{1}{(\pi \sigma_{dc} \tau)^{1/2}} \quad ; \qquad (5.1.22)$$

and because $A \approx 4R_S/Z_0$ the surface resistance R_S is frequency independent too. Since $\alpha(\omega) = 2k\omega/c$, the absorption rate α is also constant in the relaxation regime.

Transparent regime

The reflectivity of a metal drops significantly at the plasma frequency, and above ω_p the material becomes transparent. The spectral region $\omega > \omega_p$ is hence called the transparent regime. However, the behavior of the complex conductivity shows no clear indication of the plasma frequency, and both components continue to fall off monotonically with increasing frequency: $\sigma_1(\omega) \propto \omega^{-2}$ and $\sigma_2(\omega) \propto \omega^{-1}$. From Eq. (3.2.25) we find that the dielectric constant

$$\epsilon_1(\omega) \approx 1 - \left(\frac{\omega_p}{\omega}\right)^2 \qquad (5.1.23)$$

approaches unity in the high frequency limit ($\omega\tau \gg 1$), and the Drude model leads to a zero-crossing of $\epsilon_1(\omega)$ at

$$\omega = \left[\omega_p^2 - (1/\tau)^2\right]^{1/2} \quad . \qquad (5.1.24)$$

For $1/\tau \ll \omega_p$ (the usual situation encountered in a metal), the dielectric constant $\epsilon_1(\omega)$ becomes positive for ω exceeding the plasma frequency (Fig. 5.3). This sign change of the dielectric constant has a certain significance. The attenuation of the electromagnetic wave in the medium is given by the extinction coefficient (Eq. (2.3.4))

$$k = \frac{1}{2} \left\{ \left[\epsilon_1^2 + \left(\frac{4\pi\sigma_1}{\omega}\right)^2\right]^{1/2} - \epsilon_1 \right\}^{1/2} \quad ,$$

and $4\pi\sigma_1/\omega \ll |\epsilon_1|$ in the region well above $1/\tau$. For $\epsilon_1 < 0$, i.e. for $\omega \leq \omega_p$, the extinction $k = (|\epsilon_1|/2)^{1/2}$ is approximately $\frac{1}{2}|\epsilon_1|$, while for frequencies $\omega \geq \omega_p$, it vanishes. Consequently the power absorption coefficient $\alpha = 2\omega k/c$ undergoes a sudden decrease near ω_p, and hence the name transparent regime for the spectral range $\omega \geq \omega_p$. Both components of the energy loss function $1/\hat{\epsilon}(\omega)$ peak at the plasma frequency; such peaks when detected by experiments can also be used to identify ω_p.

5.1.3 Derivation of the Drude expression from the Kubo formula

Next, we derive the Drude response in a somewhat more rigorous fashion. In Section 4.1.1 we arrived at an expression of the optical conductivity in terms of the current–current correlation function, the so-called Kubo formula (4.1.16), which we reproduce here:

$$\hat{\sigma}(\mathbf{q}, \omega) = \frac{1}{\hbar\omega} \sum_s \int dt \, \langle s|\mathbf{J}(\mathbf{q}, 0)\mathbf{J}^*(\mathbf{q}, t)|s\rangle \exp\{-i\omega t\} \quad . \tag{5.1.25}$$

We now assume that the current–current correlation time τ is the same for all states, and

$$\mathbf{J}(\mathbf{q}, t) = \mathbf{J}(\mathbf{q}, 0) \exp\{-t/\tau\} \tag{5.1.26}$$

describes the correlation of the current density as the time elapses; broadly speaking it is the measure of how long the current remains at a particular value $\mathbf{J}$. Inserting this into Kubo's expression, we find

$$\hat{\sigma}(\mathbf{q}, \omega) = \frac{1}{\hbar\omega} \sum_s \int dt \, \exp\{-i\omega t - |t|/\tau\}\langle s|\mathbf{J}^2(\mathbf{q})|s\rangle \quad . \tag{5.1.27}$$

Next we have to find an expression for the fluctuations of the current. The average current density $\langle\mathbf{J}\rangle$ is of course zero, but fluctuations lead to finite $\langle\mathbf{J}^2\rangle$. By inserting a complete set of states s' for which $\sum_{s'} |s'\rangle\langle s'| = 1$ we find that

$$\langle s|\mathbf{J}^2|s\rangle = \sum_{s'} \left|\langle s'|\mathbf{J}|s\rangle\right|^2 \quad .$$

Using this identity, we rewrite the expression for the conductivity as

$$\hat{\sigma}(\mathbf{q}, \omega) = \frac{1}{\hbar\omega} \int dt \, \exp\{-i\omega t - |t|/\tau\} \sum_{s,s'} \left|\langle s'|\mathbf{J}(\mathbf{q})|s\rangle\right|^2 \quad . \tag{5.1.28}$$

In Eq. (4.1.5) we described the current density operator $\mathbf{J}$ in terms of the velocity $\mathbf{v}$, respectively the momenta $\mathbf{p}$, of the individual particles j

$$\mathbf{J}(\mathbf{r}) = -\frac{e}{2m} \sum_j \left[\mathbf{p}_j\delta\{\mathbf{r} - \mathbf{r}_j\} + \delta\{\mathbf{r} - \mathbf{r}_j\}\mathbf{p}_j\right] \quad . \tag{5.1.29}$$

The Fourier component of the operator becomes

$$\mathbf{J}(\mathbf{q}) = \int \mathbf{J}(\mathbf{r}) \exp\{-i\mathbf{q} \cdot \mathbf{r}\} \, d\mathbf{r} = -\frac{e}{m} \sum_j \mathbf{p}_j \tag{5.1.30}$$

by assuming that the momentum $\mathbf{q}$ is small and $\exp\{i\mathbf{q}\cdot r\} \approx 1$, the so-called dipole approximation. With this approximation

$$\hat{\sigma}(\omega) = \frac{e^2}{m^2\hbar\omega} \int dt \, \exp\{-i\omega t - |t|/\tau\} \sum_{s,s',j} \left|\langle s'|\mathbf{p}_j|s\rangle\right|^2 \quad , \tag{5.1.31}$$

where the frequency $\hbar\omega = \hbar\omega_{s's} = \hbar\omega_{s'} - \hbar\omega_s$ corresponds to the energy difference between the two states s and s', involved in the optical transitions.

In Appendix D we introduce the so-called oscillator strength[1]

$$2 \sum_{s,s',j} \frac{\left|\langle s'|\mathbf{p}_j|s\rangle\right|^2}{m\hbar\omega_{s's}} = f_{s's} \tag{5.1.32}$$

where $f_{ss'} = -f_{s's}$ (see Eq. (D.2)). With this notation we arrive at an expression for the complex conductivity

$$\hat{\sigma}(\omega) = \frac{e^2\tau}{m}\frac{f_{s's}}{1+i\omega\tau} \quad . \tag{5.1.33}$$

The oscillator strength can easily be calculated for the simple case of free electrons, for which the energy and the average momentum are given by

$$\hbar\omega = \frac{\hbar^2 k^2}{2m} \qquad \text{and} \qquad \left|\langle s'|\mathbf{p}_j|s\rangle\right|^2 = \langle \mathbf{p}_j^2\rangle = \left(\frac{m\mathbf{v}}{2}\right)^2 = \frac{\hbar^2 k^2}{4} \quad .$$

We find that $f_{ss'} = N$, the number of electrons per unit volume in the conduction band, and consequently the frequency dependent conductivity is

$$\hat{\sigma}(\omega) = \frac{Ne^2\tau}{m}\frac{1}{1+i\omega\tau} \quad .$$

To arrive at $\hat{\sigma}(\omega)$, some approximations were made: first we assumed that the current–current correlation function exponentially decays in time (the relaxation time approximation); this means that only a single, frequency independent relaxation time exists. Second we restricted our considerations to zero wavevector (the so-called local limit). As discussed later, electron–lattice and electron–electron interactions all may lead to frequency dependent parameters such as τ and m. Even in the absence of these parameters, complications may arise. If the relaxation time is extremely long, the phase shift between two collisions becomes appreciable. Thus we cannot expect the dipole approximation to be valid. In addition, non-local effects also have to be included; this may occur at low temperatures in extremely pure metals, as will be discussed later. We have also assumed that the momenta $\mathbf{p}_j$ are well defined, by no means an obvious assumption, in particular for the short relaxation times, and thus for short mean free paths.

5.2 Boltzmann's transport theory

In order to describe a variety of transport phenomena Boltzmann developed a useful approximation, assuming classical particles with well defined momenta. We will,

---

[1] For a more detailed discussion of the oscillator strength see Section 6.1.1 and Appendix D.

however, go one step beyond and use quantum statistics for the electrons; the equations are not modified by this change. The relation derived is widely used to account for thermo- and galvano-electrical effects but also helps in understanding the electrical transport due to an external (frequency and – with certain limitations – wavevector dependent) electric field. There is a considerable literature on this transport equation [Bla57, Cha90, Mad78, Zim72], and subtle issues, such as the difference between the steady state and equilibrium, have been addressed. None of these issues will be mentioned here; we merely outline the steps which lead to the final result: Eq. (5.2.7).

5.2.1 Liouville's theorem and the Boltzmann equation

Owing to the large number of particles we cannot consider the motion of each particular electron, and thus we define a distribution function $f(\mathbf{r}, \mathbf{k}, t)$ which describes the carrier density at a point $\mathbf{r}$, with a particular momentum $\hbar\mathbf{k}$, and at a certain time t. Although the total number of carriers $\int d\mathbf{r} \int d\mathbf{k} \, f(\mathbf{r}, \mathbf{k}, t)$ is constant, at a particular point $(\mathbf{r}, \mathbf{k})$ in phase space (Liouville's theorem), it may change due to various effects. First, the distribution function can be time dependent, which we write as

$$\left. \frac{\partial f}{\partial t} \right|_{\text{time}} . \tag{5.2.1}$$

This time dependence can arise, for instance, because of a time dependent external perturbation. The location $\mathbf{r}$ of the carrier may change by diffusion; assuming that the carriers move with a velocity $\mathbf{v_k} = \partial\mathbf{r}/\partial t$ and using

$$f_\mathbf{k}(\mathbf{r}, t) = f_\mathbf{k}(\mathbf{r} + \mathbf{v_k}t, 0) \quad , \tag{5.2.2}$$

we can write

$$\left. \frac{\partial f}{\partial t} \right|_{\text{diff}} = \frac{\partial f}{\partial\mathbf{r}} \cdot \frac{\partial\mathbf{r}}{\partial t} = \mathbf{v_k} \cdot \frac{\partial f}{\partial\mathbf{r}} \quad . \tag{5.2.3}$$

Here we assume that the temperature is constant throughout the sample, $\nabla T = 0$, and thus there is no diffusion due to the temperature gradient. If external forces are present, such as the Coulomb force and the Lorentz force, the wavevector of the electrons is affected,

$$\frac{\partial\mathbf{k}}{\partial t} = -\frac{e}{\hbar}\left(\mathbf{E} + \frac{1}{c}\mathbf{v_k} \times \mathbf{H} \right) \quad , \tag{5.2.4}$$

leading to a change in the distribution function:

$$\left. \frac{\partial f}{\partial t} \right|_{\text{field}} = \frac{\partial f}{\partial\mathbf{k}} \cdot \frac{\partial\mathbf{k}}{\partial t} = -\frac{e}{\hbar}\left(\mathbf{E} + \frac{1}{c}\mathbf{v_k} \times \mathbf{H} \right) \cdot \frac{\partial f}{\partial\mathbf{k}} \quad . \tag{5.2.5}$$

Here $\mathbf{E}$ and $\mathbf{H}$ are uniform, frequency dependent electric and magnetic fields. Finally, collisions may cause a transition from one state $|\mathbf{k}\rangle$ to another $|\mathbf{k}'\rangle$:

$$\left.\frac{\partial f}{\partial t}\right|_{\text{scatter}} = \int [f_{\mathbf{k}'}(1-f_{\mathbf{k}}) - f_{\mathbf{k}}(1-f_{\mathbf{k}'})] W(\mathbf{k}, \mathbf{k}')\,\mathrm{d}\mathbf{k}' = \int [f_{\mathbf{k}'} - f_{\mathbf{k}}] W(\mathbf{k}, \mathbf{k}')\,\mathrm{d}\mathbf{k}' \quad .$$

$$(5.2.6)$$

In this equation $W(\mathbf{k}, \mathbf{k}')$ represents the transition probability between the initial state $|\mathbf{k}\rangle$ and the final state $|\mathbf{k}'\rangle$ with the corresponding wavevectors; we assume microscopic reversibility $W(\mathbf{k}, \mathbf{k}') = W(\mathbf{k}', \mathbf{k})$. Here $f_{\mathbf{k}}$ denotes the number of carriers in the state $|\mathbf{k}\rangle$, and $1 - f_{\mathbf{k}}$ refers to the number of unoccupied states $|\mathbf{k}\rangle$. Restrictions like energy, momentum, and spin conservations are taken into account by the particular choice of $W(\mathbf{k}, \mathbf{k}')$. With all terms included, the continuity equation in the phase space reads as

$$\left.\frac{\partial f}{\partial t}\right|_{\text{time}} + \left.\frac{\partial f}{\partial t}\right|_{\text{diff}} + \left.\frac{\partial f}{\partial t}\right|_{\text{field}} = \left.\frac{\partial f}{\partial t}\right|_{\text{scatter}} \quad .$$

We can now write the linearized Boltzmann equation as

$$0 = \frac{\partial f}{\partial t} + \mathbf{v}_{\mathbf{k}} \cdot \nabla_{\mathbf{r}} f - \frac{e}{\hbar}\left(\mathbf{E} + \frac{1}{c}\mathbf{v}_{\mathbf{k}} \times \mathbf{H}\right) \cdot \nabla_{\mathbf{k}} f - \left.\frac{\partial f}{\partial t}\right|_{\text{scatter}} \quad . \qquad (5.2.7)$$

This form of Boltzmann's equation is valid within the framework of linear response theory. Because we have retained only the gradient term $\nabla_{\mathbf{r}} f$, the theory is appropriate in the small $\mathbf{q}$ limit, but has no further limitations.

Next we solve the Boltzmann equation for a free gas of electrons in the absence of an external magnetic field ($\mathbf{H} = 0$), with the electrons obeying quantum statistics and subjected only to scattering which leads to changes in their momenta. Let us assume that the distribution function varies only slightly from its equilibrium state $f_{\mathbf{k}}^{0}$, and $f_{\mathbf{k}} = f_{\mathbf{k}}^{0} + f_{\mathbf{k}}^{1}$, where

$$f_{\mathbf{k}}^{0} = f^{0}(\mathcal{E}_{\mathbf{k}}) = \left(\exp\left\{\frac{\mathcal{E}_{\mathbf{k}} - \mathcal{E}_{\mathrm{F}}}{k_{\mathrm{B}}T}\right\} + 1\right)^{-1} \qquad (5.2.8)$$

is the Fermi–Dirac distribution function. In this limit of small perturbation the scattering term reduces to

$$\left.\frac{\partial f_{\mathbf{k}}}{\partial t}\right|_{\text{scatter}} = \int \left(f_{\mathbf{k}'}^{1} - f_{\mathbf{k}}^{1}\right) W(\mathbf{k}, \mathbf{k}')\,\mathrm{d}\mathbf{k}' \quad .$$

If we make the assumption that the system relaxes exponentially to its equilibrium state after the perturbation is switched off (the relaxation time approximation), the distribution function has the time dependence $f_{\mathbf{k}}^{1}(t) = f_{\mathbf{k}}^{1}(0)\exp\{-t/\tau\}$; therefore we find that the (**k** independent) scattering term is given by

$$\left.\frac{\partial f_{\mathbf{k}}}{\partial t}\right|_{\text{scatter}} = -\frac{1}{\tau} f_{\mathbf{k}}^{1} \quad . \tag{5.2.9}$$

If the momentum change is small compared with the Fermi momentum ($k \ll k_{\mathrm{F}}$), scattering across the Fermi surface can be neglected. Since $\frac{\partial f}{\partial k} = \frac{\partial f}{\partial \mathcal{E}}\frac{\partial \mathcal{E}}{\partial k}$ and (again assuming $\mathbf{H} = 0$) the carrier velocity $\mathbf{v_k} = \nabla_{\mathbf{k}}\omega(\mathbf{k}) = \hbar^{-1}\nabla_{\mathbf{k}}\mathcal{E}(\mathbf{k})$ (which is the group velocity of the electrons), we can write Eq. (5.2.7) as

$$-\frac{\partial f_{\mathbf{k}}^{1}}{\partial t} = \mathbf{v_k} \cdot \nabla_{\mathbf{r}} f_{\mathbf{k}}^{1} - e\mathbf{E} \cdot \mathbf{v_k}\frac{\partial f_{\mathbf{k}}^{0}}{\partial \mathcal{E}} + \frac{f_{\mathbf{k}}^{1}}{\tau} \quad . \tag{5.2.10}$$

We can solve this equation by assuming that the distribution function follows the spatial and time dependent perturbation of $\exp\{i(\mathbf{q}\cdot\mathbf{r} - \omega t)\}$ and has itself the form $f_{\mathbf{k}}^{1}(t) \propto \exp\{i(\mathbf{q}\cdot\mathbf{r} - \omega t)\}$. Taking the Fourier components in $\mathbf{r}$ and t, we solve Eq. (5.2.10) for f^{1}, obtaining

$$f^{1}(\mathbf{q}, \mathbf{k}, \omega) = \frac{-e\mathbf{E}(\mathbf{q}, \omega) \cdot \mathbf{v_k}(-\frac{\partial f^{0}}{\partial \mathcal{E}})\tau}{1 - i\omega\tau + i\mathbf{v_k}\cdot\mathbf{q}\tau} \quad . \tag{5.2.11}$$

The linearized distribution function yields a finite relaxation time τ and, consequently a finite mean free path $\ell = v_{\mathrm{F}}\tau$. The current density can in general be written as

$$\mathbf{J}(\mathbf{q}, \omega) = -\frac{2e}{(2\pi)^{3}} \int f^{1}(\mathbf{q}, \mathbf{k}, \omega)\, \mathbf{v_k}\, d\mathbf{k} \tag{5.2.12}$$

in terms of the distribution function $f^{1}(\mathbf{q}, \mathbf{k}, \omega)$, where $\mathbf{v_k}$ is the velocity of the electrons for a given wavevector $\mathbf{k}$ and the factor of 2 takes both spin directions into account. Using Eq. (5.2.11) for $f^{1}(\mathbf{q}, \mathbf{k}, \omega)$ we obtain

$$\mathbf{J}(\mathbf{q}, \omega) = \frac{2e^{2}}{(2\pi)^{3}} \int d\mathbf{k}\, \frac{\tau \mathbf{E}(\mathbf{q}, \omega) \cdot \mathbf{v_k}(-\frac{\partial f^{0}}{\partial \mathcal{E}})}{1 - i\omega\tau + i\mathbf{v_k}\cdot\mathbf{q}\tau}\mathbf{v_k} \quad . \tag{5.2.13}$$

As $\mathbf{J}(\mathbf{q}, \omega) = \hat{\sigma}(\mathbf{q}, \omega)\mathbf{E}(\mathbf{q}, \omega)$, the wavevector and frequency dependent conductivity can now be calculated. Let us simplify the expression first; for cubic symmetry

$$\hat{\sigma}(\mathbf{q}, \omega) = \frac{2e^{2}}{(2\pi)^{3}} \int d\mathbf{k}\, \frac{\tau(\mathbf{n_E} \cdot \mathbf{v_k})v_{\mathbf{k}}(-\frac{\partial f^{0}}{\partial \mathcal{E}})}{1 - i\omega\tau + i\mathbf{v_k}\cdot\mathbf{q}\tau} \quad , \tag{5.2.14}$$

where $\mathbf{n_E}$ represents the unit vector along the direction of the electric field $\mathbf{E}$. In the spirit of the Sommerfeld theory the derivative $(-\partial f^{0}/\partial \mathcal{E})$ is taken at $\mathcal{E} = \mathcal{E}_{\mathrm{F}}$, as only the states near the Fermi energy contribute to the conductivity. Second, we consider this equation for metals in the limit $T = 0$, then the Fermi–Dirac distribution (5.2.8) is a step function at $\mathcal{E}_F$. With

$$\lim_{T \to 0} \left(-\frac{\partial f^0}{\partial \mathcal{E}} \right) = \delta\{\mathcal{E} - \mathcal{E}_F\} \tag{5.2.15}$$

the integral has contributions only at the Fermi surface S_F. Under such circumstances we can reduce Eq. (5.2.14) to

$$
\begin{aligned}
\hat{\sigma}(\mathbf{q}, \omega) &= \frac{2e^2}{(2\pi)^3} \int\int \frac{\tau(\mathbf{n_E} \cdot \mathbf{v_k})v_k}{1 - i\omega\tau + i\mathbf{v_k} \cdot \mathbf{q}\tau} \left(-\frac{\partial f^0}{\partial \mathcal{E}} \right) \frac{dS}{\hbar v_k} d\mathcal{E} \\
&= \frac{2e^2}{(2\pi)^3} \int_{\mathcal{E}=\mathcal{E}_F} \frac{\tau(\mathbf{n_E} \cdot \mathbf{v_k})v_k}{1 - i\omega\tau + i\mathbf{v_k} \cdot \mathbf{q}\tau} \frac{dS_F}{\hbar v_k} ,
\end{aligned}
\tag{5.2.16}
$$

where we have also converted the volume integral $d\mathbf{k}$ into one over surfaces dS of constant energy as

$$d\mathbf{k} = dS \, dk_\perp = dS \frac{d\mathcal{E}}{|\nabla_\mathbf{k} \mathcal{E}|} = dS \frac{d\mathcal{E}}{\hbar v_k} , \tag{5.2.17}$$

and we have assumed a parabolic energy dispersion $\mathcal{E}(k) = \frac{\hbar^2}{2m} k^2$.

5.2.2 The $\mathbf{q} = 0$ limit

In the so-called local limit $\mathbf{q} \to 0$, the conductivity simplifies to

$$\hat{\sigma}(0, \omega) = \frac{e^2}{4\pi^3 \hbar} \int \frac{\tau(\mathbf{n_E} \cdot \mathbf{v_k})v_k}{v_k} \frac{1}{1 - i\omega\tau} dS_F = \sigma_{dc} \frac{1}{1 - i\omega\tau} \quad ; \tag{5.2.18}$$

and we recover the familiar equation of the Drude model for the frequency dependent conductivity. The velocity term $(\mathbf{n_E} \cdot \mathbf{v_k})v_k/v_k$ averaged over the Fermi surface is simply $\frac{1}{3} v_F$ and therefore

$$\sigma_{dc} = \frac{e^2}{8\pi^3 \hbar} \int dS_F \frac{\tau v_F^2}{v_F} = \frac{e^2 \tau}{4\pi^3 \hbar} \frac{8\pi \hbar k_F^3}{3m} = \frac{Ne^2}{m} \tau$$

by considering the spherical Fermi surface for which $\int_{\mathcal{E}_F} dS = 2(4\pi k_F^2)$ and the density of charge carriers $N = k_F^3/(3\pi^2)$. With our prior assumption of quantum statistics, only electrons near the Fermi energy $\mathcal{E}_F$ are important, and therefore the mean free path $\ell = v_F \tau$.

5.2.3 Small q limit

Next we explore the response to long wavelength excitations and zero temperature starting from Eq. (5.2.16). With the dc conductivity $\sigma_{dc} = Ne^2\tau/m$, straightforward integration yields[2]

$$\hat{\sigma}(\mathbf{q}, \omega) = \frac{3\sigma_{dc}}{4} \frac{i}{\tau} \left[2\frac{\omega + i/\tau}{q^2 v_F^2} - \left(\frac{1 - (\omega + i/\tau)^2/(q v_F)^2}{q v_F} \right) \mathrm{Ln} \left\{ \frac{\omega - q v_F + i/\tau}{\omega + q v_F + i/\tau} \right\} \right];$$

$$(5.2.19)$$

where we recall the definition of the logarithm (Ln) of a complex value $\hat{x} = |\hat{x}| \exp\{i\phi\}$ as the principal value

$$\mathrm{Ln}\{\hat{x}\} = \ln\{|\hat{x}|\} + i\phi \qquad (5.2.20)$$

with $-\pi < \phi \leq \pi$. The first terms of the expansion

$$\mathrm{Ln}\{(\hat{z} + 1)/(\hat{z} - 1)\} = 2(1/\hat{z} + 1/3\hat{z}^2 + 1/5\hat{z}^5 + \cdots)$$

give

$$\hat{\sigma}(\mathbf{q}, \omega) \approx \frac{Ne^2\tau}{m} \frac{1}{1 - i\omega\tau} \left[1 - \frac{1}{5}\left(\frac{q v_F \tau}{1 - i\omega\tau} \right)^2 + \cdots \right] \qquad (5.2.21)$$

for $q v_F < |\omega + i/\tau|$. This is called the homogeneous limit, the name referring to the case when the variation of the wavevector $\mathbf{q}$ is small. The second term in the square brackets is neglected for $\mathbf{q} = 0$ and we recover the Drude form. An expansion in terms of $(\omega + i/\tau)/(q v_F)$ leads to

$$\hat{\sigma}(\mathbf{q}, \omega) \approx \frac{3\pi Ne^2}{4q v_F \tau m} \left[1 - \frac{\omega^2}{q^2 v_F^2} + i\frac{4\omega}{\pi q v_F} \right] \qquad (5.2.22)$$

for $|\omega + i/\tau| < q v_F$, in the so-called quasi-static limit. The name refers to the fact that in this limit the phase velocity of the electromagnetic field ω/q is small compared with the velocity of the particles v_F. If the relaxation rate goes to zero ($\tau \rightarrow \infty$), or more generally for $\omega\tau \gg 1$, the real and imaginary parts of the conductivity can be derived without having to limit ourselves to the long wavelength fluctuations. By utilizing the general transformation

$$\lim_{1/\tau \rightarrow 0} \mathrm{Ln}\{-|x| + i/\tau\} = \ln|x| + i\pi \quad , \qquad (5.2.23)$$

---

[2] The integration (or summation) is first performed over the angle between $\mathbf{k}$ and $\mathbf{q}$, leading to the logarithmic term

$$\int_0^\pi \frac{x}{a + b\cos x}\,dx = \frac{1}{b}\ln\frac{a + b}{a - b} \quad ,$$

and then from 0 to v_F by using the relation

$$\int x \ln\{ax + b\}\,dx = \frac{b}{2a}x - \frac{1}{4}x^2 + \frac{1}{2}\left(x^2 - \frac{b^2}{a^2} \right)\ln\{ax + b\}.$$

after some algebra we find that

$$\lim_{\tau \to \infty} \sigma_1(\mathbf{q}, \omega) = \begin{cases} \frac{3\pi}{4} \frac{Ne^2}{mq v_F} \left[1 - \frac{\omega^2}{q^2 v_F^2} \right] & \text{for} \quad \omega < q v_F \\ 0 & \text{for} \quad \omega > q v_F \end{cases} \qquad (5.2.24a)$$

$$\lim_{\tau \to \infty} \sigma_2(\mathbf{q}, \omega) = \frac{3}{4} \frac{Ne^2}{mq v_F} \left[\frac{2\omega}{q v_F} + \left(\frac{\omega^2}{q^2 v_F^2} - 1 \right) \ln \left| \frac{q v_F - \omega}{q v_F + \omega} \right| \right] \qquad (5.2.24b)$$

for both $\omega < q v_F$ and $\omega > q v_F$. The conductivity σ_1 is finite only for $\omega < q v_F$, there are no losses in the opposite, homogeneous, limit. The reason for this is clear: $\omega > q v_F$ would imply a wave with velocity ω/q greater than the Fermi velocity v_F; the electron gas clearly cannot respond to a perturbation traveling with this velocity.

5.2.4 The Chambers formula

The $\mathbf{q}$ dependent response is intimately related to the non-local conduction where the current density $\mathbf{J}$ at the position $\mathbf{r}'$ is determined also by fields at other locations $\mathbf{r} \neq \mathbf{r}'$. Here we develop an approximate expression for the current which depends on the spatial distribution of the applied electric field. Such a situation may occur in the case of clean metals at low temperatures when the mean free path ℓ is large. Let us consider an electron moving from a point $\mathbf{r}$ to another position taken to be the origin of the coordinate system. At the initial location, the electron is subjected to an electric field $\mathbf{E}(\mathbf{r})$ which is different from that at the origin. However, because of collisions with the lattice or impurities, the momentum acquired by the electron from the field at $\mathbf{r}$ decays exponentially as the origin is approached. The characteristic decay length defines the mean free path ℓ, and the currents at the origin are the result of the fields $\mathbf{E}(\mathbf{r})$ within the radius of $\ell = v_F \tau$.

The argument which accounts for such a non-local response is as follows. When an electron moves from a position $(\mathbf{r} - d\mathbf{r})$ to $\mathbf{r}$, it is influenced by an effective field $\mathbf{E}(\mathbf{r}) \exp\{-r/\ell\}$ for a time dr/v_F. The momentum gained in the direction of motion is

$$d\mathbf{p}(0) = -\frac{e \, dr}{v_F} \frac{\mathbf{r}}{r} \cdot \mathbf{E}(\mathbf{r}) \exp\{-r/\ell\} \quad ; \qquad (5.2.25)$$

in the following we drop the indication of the position (0). By integrating the above equation from the origin to infinity, the total change in momentum for an electron at the origin is found. Performing this calculation for all directions allows us to map out the momentum surface in a non-uniform field, and the deviations from a sphere centered at the origin constitute a current $\mathbf{J}$. The following arguments lead to the expression of the current. Only electrons residing in regions of momentum space not normally occupied when the applied field is zero contribute to the current.

The density of electrons ΔN moving in a solid angle $d\Omega$ and occupying the net displaced volume in momentum space ΔP is

$$\Delta N = \frac{\Delta P}{P} N = \frac{(mv_{\mathrm{F}})^2 \, d\Omega \, dp}{\frac{4}{3}\pi (mv_{\mathrm{F}})^3} N = \frac{3N \, d\Omega \, dp}{4\pi v_{\mathrm{F}} m} \quad , \tag{5.2.26}$$

where P is the total momentum space volume. The contribution to the current density from these electrons is

$$d\mathbf{J} = -\Delta N e v_{\mathrm{F}} \frac{\mathbf{r}}{r} = -\frac{3Ne}{4\pi m} \frac{\mathbf{r}}{r} \, d\Omega \, dp \quad .$$

Substituting Eq. (5.2.25) into this equation and integrating over the currents given above yields

$$\mathbf{J}(\mathbf{r}=0) = \frac{3\sigma_{\mathrm{dc}}}{4\pi \ell} \int \frac{\mathbf{r}[\mathbf{r} \cdot \mathbf{E}(r) \exp\{-r/\ell\}]}{r^4} \, d\mathbf{r} \quad , \tag{5.2.27}$$

since a volume element in real space is $r^2 \, dr \, d\Omega$ and $\sigma_{\mathrm{dc}} = Ne^2\tau/m = Ne^2\ell/(mv_{\mathrm{F}})$. Equation (5.2.27) represents the non-local generalization of Ohm's law for free electrons, and reduces to $\mathbf{J} = \sigma_{\mathrm{dc}}\mathbf{E}$ for the special case where $\ell \to 0$, as expected. The Chambers formula [Cha90, Pip54b] is valid for finite momentum, but as the Fermi momentum is not explicitly included its use is restricted to $q < k_{\mathrm{F}}$, and in general to the small $\mathbf{q}$ limit. A quantum mechanical derivation of Chambers' result was given by Mattis and Dresselhaus [Mat58]. In Appendix E the non-local response is discussed in more detail.

5.2.5 Anomalous skin effect

In Section 2.2 we introduced the skin effect of a metal and found that the skin depth is given by

$$\delta_0(\omega) = \frac{c}{(2\pi \omega \sigma_{\mathrm{dc}})^{1/2}}$$

in the low frequency, so-called Hagen–Rubens regime. In this regime we also obtained the following equation for the surface impedance

$$\hat{Z}_{\mathrm{S}} = R_{\mathrm{S}} + iX_{\mathrm{S}} = \frac{(2\pi)^2}{c} \frac{\delta_0}{\lambda_0}(1 - i) \quad , \tag{5.2.28}$$

where the frequency dependence enters through the skin depth δ_0 and through the free space wavelength λ_0 of the electromagnetic field. The condition for the above results is that the mean free path ℓ is smaller than δ_0 and hence local electrodynamics apply. At low temperatures in clean metals this condition is not met because ℓ becomes large, and the consequences of the non-local electrodynamics have to be

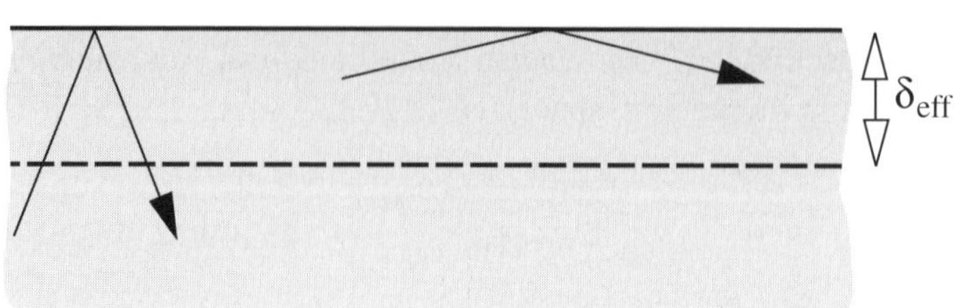

Fig. 5.9. Electron trajectories near the surface of a material (solid line) in the case of the anomalous regime, where the mean free path ℓ is much larger than the effective skin depth δ_{eff} (dashed line). According to the ineffectiveness concept, only electrons traveling nearly parallel to the surface are effective in absorbing and screening electromagnetic radiation when $\ell \gg \delta_{\text{eff}}$, since the carriers moving perpendicular to the surface leave the skin layer before they scatter.

explored in detail. This can be done by using the Chambers formula (5.2.27), or alternatively the solution of the non-local electrodynamics, as in Appendix E.1.

An elegant argument, the so-called ineffectiveness concept due to Pippard [Pip47, Pip54a, Pip62] reproduces all the essential results. If the mean free path is larger than the skin depth, the effect on the electrons which move perpendicular to the surface is very different from those traveling parallel to the surface. Due to the long mean free path, electrons in the first case leave the skin depth layer without being scattered; the situation is illustrated in Fig. 5.9. Thus, only those electrons which are moving approximately parallel to the surface (i.e. their direction of motion falls within an angle $\pm \gamma \delta_{\text{eff}}/\ell$ parallel to the surface of the metal) contribute to the absorption [Pip54b, Reu48]. The (reduced) number of effective electrons $N_{\text{eff}} = \gamma N \delta_{\text{eff}}/\ell$ also changes the conductivity, which we write as:

$$\sigma_{\text{eff}} = \gamma \sigma_{\text{dc}} \frac{\delta_{\text{eff}}}{\ell} \quad , \tag{5.2.29}$$

where γ is a numerical factor of the order of unity, and can be evaluated by using Chamber's expression [Abr72, Pip54b, Reu48]. The factor depends on the nature of the scattering of the electrons at the surface; for specular reflection $\gamma = 8/9$ and for diffuse reflection $\gamma = 1$. In turn the effective conductivity leads to a modified skin depth $\delta_{\text{eff}} = c/(2\pi\omega\sigma_{\text{eff}})^{1/2}$, and by substituting this into Eq. (5.2.29) we obtain through selfconsistence for the effective skin depth

$$\delta_{\text{eff}} = \left(\frac{c^2}{2\pi\omega} \frac{\ell}{\gamma \sigma_{\text{dc}}} \right)^{1/3} = \left(\frac{c^2}{2\pi\omega} \frac{m v_{\text{F}}}{\gamma N e^2} \right)^{1/3} \quad , \tag{5.2.30}$$

and for the effective conductivity

$$\sigma_{\text{eff}} = \left(\frac{\gamma c}{\sqrt{2\pi\omega}} \frac{\sigma_{\text{dc}}}{\ell} \right)^{2/3} = \left(\frac{\gamma c}{\sqrt{2\pi\omega}} \frac{N e^2}{m v_{\text{F}}} \right)^{2/3} \quad . \tag{5.2.31}$$

We can then use this σ_{eff} for evaluating the optical constants $n + ik = (4\pi i \sigma_{\text{eff}} \omega)^{1/2} / c$ and the surface impedance $\hat{Z}$, utilizing Eq. (2.3.28). The end result for the surface impedance is

$$\hat{Z}_S = R_S + iX_S = \frac{1 - i\sqrt{3}}{\delta_{\text{eff}} \sigma_{\text{eff}}} = \left[\left(\frac{2\pi\omega}{c^2} \right)^2 \frac{\ell}{\gamma \sigma_{\text{dc}}} \right]^{1/3} (1 - i\sqrt{3})$$

$$= \left[\left(\frac{2\pi\omega}{c^2} \right)^2 \frac{m v_F}{\gamma N e^2} \right]^{1/3} (1 - i\sqrt{3}) \quad . \tag{5.2.32}$$

Both R_S and X_S increase with the $\omega^{2/3}$ power of frequency, and the ratio $X_S/R_S = -\sqrt{3}$; both features are dramatically different from those obtained in the case of the normal skin effect. There is also another significant difference. The effective conductivity and surface resistance are independent of the mean free path or any other temperature dependent parameter for the anomalous skin effect. This can be used to explore the characteristics of the Fermi surface, as will be discussed in Chapter 12.

5.3 Transverse response for arbitrary q values

By employing Boltzmann's equation we have limited ourselves to long wavelength fluctuations with the Drude–Sommerfeld model representing the $\mathbf{q} = 0$ limit. This limit is set by performing an expansion in terms of $\mathbf{q}$; however, more importantly, we have neglected the limits in the transition probability $W_{\mathbf{k}\to\mathbf{k}'}$ which are set by the existence of the Fermi surface separating occupied and unoccupied states. These have been included in the selfconsistent field approximation, which was discussed in Section 4.3. We now utilize this formalism when we evaluate $\hat{\sigma}(\mathbf{q}, \omega)$ for arbitrary $\mathbf{q}$ and ω parameters.

Let us first look at the condition for electron–hole excitations to occur; these will determine the absorption of electromagnetic radiation. In the ground state, such pair excitations with momentum $\mathbf{q}$ are allowed only if the state $\mathbf{k}$ is occupied and the state $\mathbf{k}' = \mathbf{k} + \mathbf{q}$ is empty, as indicated in Fig. 5.10. The energy for creating such electron–hole pairs is

$$\Delta\mathcal{E} = \mathcal{E}(\mathbf{k} + \mathbf{q}) - \mathcal{E}(\mathbf{k}) = \frac{\hbar^2}{2m} \left[(\mathbf{k} + \mathbf{q})^2 - k^2 \right] = \frac{\hbar^2}{2m} (q^2 + 2\mathbf{k} \cdot \mathbf{q}) \quad . \tag{5.3.1}$$

For any $\mathbf{q}$ the energy difference $\Delta\mathcal{E}$ has a maximum value when $\mathbf{q} \parallel \mathbf{k}$ and $|\mathbf{k}| = k_F$, and

$$\Delta\mathcal{E}_{\text{max}}(q) = \frac{\hbar^2}{2m} (q^2 + 2q k_F) \quad . \tag{5.3.2}$$

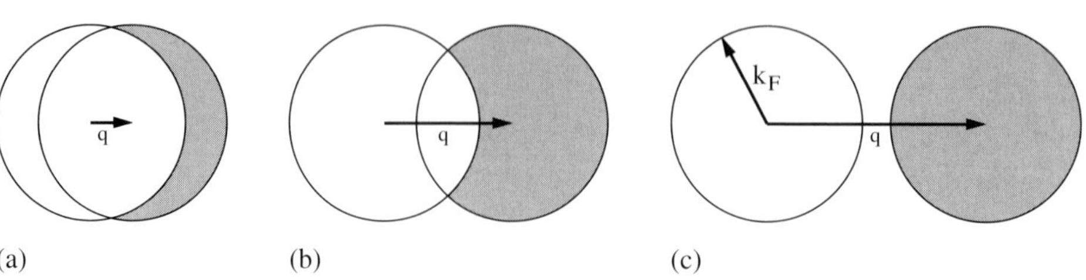

(a)   (b)   (c)

Fig. 5.10. Fermi sphere with radius k_F corresponding to electronic states with momentum $\hbar\mathbf{k}$ and $\hbar(\mathbf{k} + \mathbf{q})$. The shaded region represents the momentum space to which single-particle–hole pairs can be excited with momentum $\hbar\mathbf{q}$ so that $|\mathbf{k}| < |\mathbf{k}_F|$ and $|\mathbf{k}+\mathbf{q}| > |\mathbf{k}_F|$ (only values of $\mathbf{k}$ where $\mathbf{k}$ is occupied and $\mathbf{k} + \mathbf{q}$ is empty, and vice versa, contribute): (a) small $\mathbf{q}$ values, (b) intermediate $\mathbf{q}$ values, (c) large momentum $|\mathbf{q}| > |2\mathbf{k}_F|$.

The minimum excitation energy is zero for $|\mathbf{q}| \leq 2k_F$. However, for $|\mathbf{q}| > 2k_F$ the minimum energy occurs for $-\mathbf{q} \parallel \mathbf{k}$ and $|\mathbf{k}| = k_F$, and

$$\Delta\mathcal{E}_{\min}(q) = \begin{cases} \frac{\hbar^2}{2m}(q^2 - 2qk_F) & \text{for } |\mathbf{q}| > 2k_F \\ 0 & \text{otherwise} \end{cases} \quad (5.3.3)$$

finally leading to the following condition for electron–hole excitations to occur:

$$\left| \frac{\omega}{q\,v_F} - \frac{q}{2k_F} \right| < 1 < \left(\frac{\omega}{q\,v_F} + \frac{q}{2k_F} \right) \quad .$$

This region is indicated in Fig. 5.11 by the hatched area.

Next we want to evaluate the conductivity for intraband transitions ($l = l'$) starting from Eq. (4.3.32). In this case,

$$\hat{\sigma}(\mathbf{q}, \omega) = i\frac{Ne^2}{\omega m} - \lim_{\eta \to 0} \frac{i}{\Omega} \frac{e^2}{\omega m^2} \sum_{\mathbf{k}} \left[\frac{f^0(\mathcal{E}_{\mathbf{k}})}{\mathcal{E}_{\mathbf{k}} - \mathcal{E}_{\mathbf{k}-\mathbf{q}} - \hbar\omega - i\hbar\eta} \right.$$

$$\left. - \frac{f^0(\mathcal{E}_{\mathbf{k}})}{\mathcal{E}_{\mathbf{k}+\mathbf{q}} - \mathcal{E}_{\mathbf{k}} - \hbar\omega - i\hbar\eta} \right] |\langle \mathbf{k} + \mathbf{q}|\mathbf{p}|\mathbf{k} \rangle_*|^2 \quad , \quad (5.3.4)$$

if we consider one single band. Here we have to split up the summation as

$$\sum_{\mathbf{k}} \frac{f(\mathcal{E}_{\mathbf{k}+\mathbf{q}}) - f(\mathcal{E}_{\mathbf{k}})}{\mathcal{E}_{\mathbf{k}+\mathbf{q}} - \mathcal{E}_{\mathbf{k}} - \hbar\omega - i\hbar\eta} = \sum_{\mathbf{k}} \frac{f(\mathcal{E}_{\mathbf{k}})}{\mathcal{E}_{\mathbf{k}} - \mathcal{E}_{\mathbf{k}-\mathbf{q}} - \hbar\omega - i\hbar\eta}$$

$$- \sum_{\mathbf{k}} \frac{f(\mathcal{E}_{\mathbf{k}})}{\mathcal{E}_{\mathbf{k}+\mathbf{q}} - \mathcal{E}_{\mathbf{k}} - \hbar\omega - i\hbar\eta} \quad .$$

We have replaced $\mathbf{k} + \mathbf{q}$ by $\mathbf{k}$ in the first term, which is allowed as the summation over $\mathbf{k}$ involves all the states, just as the summation over $\mathbf{k}+\mathbf{q}$ does. We now replace the summation $\Omega^{-1} \sum_{\mathbf{k}}$ by the integration over the $\mathbf{k}$ space $2/(2\pi)^3 \int d\mathbf{k}$ and, in the spirit of the relaxation time approximation, η is substituted by the relaxation

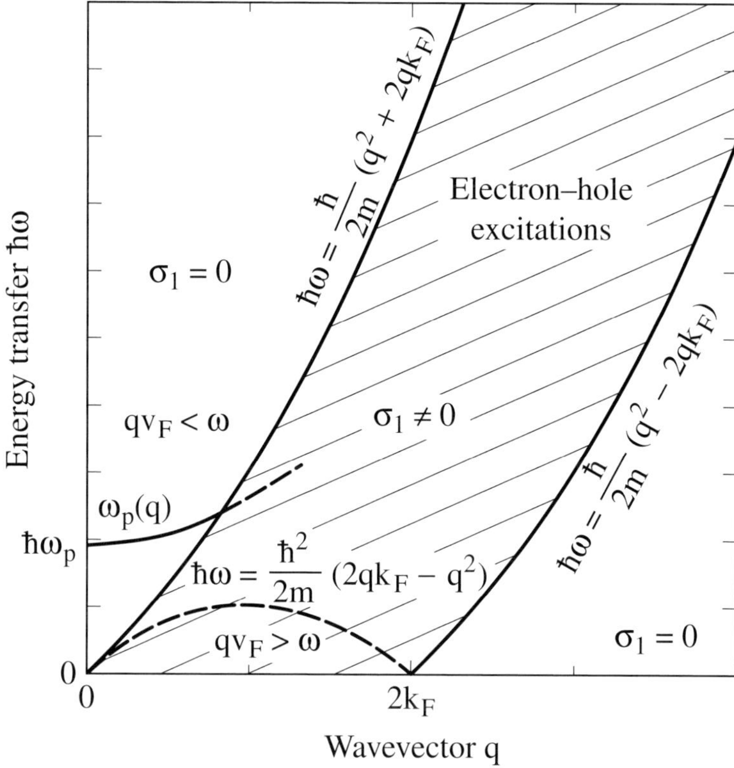

Fig. 5.11. Excitation spectrum of a three-dimensional free-electron gas; the transferred energy is plotted as a function of transferred momentum. The pair excitations fall within the shaded area. In the region $\hbar\omega > (h^2/2m)(q + 2k_F)q$, the absorption vanishes since $\hbar\omega$ is larger than energies possible for pair creation. For $\hbar\omega < (\hbar^2/2m)(q - 2k_F)q$ we find $\sigma_1 = 0$. Also shown is the dispersion of the plasma frequency $\omega_p(q)$ calculated from Eq. (5.4.28) using typical values for sodium: $\omega_p = 5.9$ eV and $v_F = 1.1 \times 10^8$ cm s^{-1}.

rate $1/\tau$. Evaluating the matrix element for the conduction band in the case of a free-electron gas $\mathcal{E}_\mathbf{k} = \frac{\hbar^2 k^2}{2m}$, we can rewrite the diagonal elements of Eq. (5.3.4) as [Lin54]

$$\hat{\sigma}(\mathbf{q}, \omega) = i\frac{e^2}{m\omega} \sum_\mathbf{k} 2f^0(\mathcal{E}_\mathbf{k}) \left\{ \left[k^2 - \left(\frac{\mathbf{k} \cdot \mathbf{q}}{q} \right)^2 \right] \left[\frac{1}{q^2 + 2\mathbf{k} \cdot \mathbf{q} - \frac{2m}{\hbar}(\omega + i/\tau)} \right. \right.$$

$$\left. \left. + \frac{1}{q^2 - 2\mathbf{k} \cdot \mathbf{q} + \frac{2m}{\hbar}(\omega + i/\tau)} \right] + \frac{1}{2} \right\} \quad . \tag{5.3.5}$$

The first square bracket is the matrix element $|\langle \mathbf{k} + \mathbf{q}|\mathbf{p}|\mathbf{k}\rangle_*|^2$ evaluated using Eq. (3.1.5) to consider the transverse components. For transverse coupling between the vector potential **A** and momentum **p**, we keep only those components of **p** which are perpendicular to $\mathbf{q}$. The matrix element is indeed zero if $\mathbf{k}$ and $\mathbf{q}$ are parallel. Combining all terms yields for the complex conductivity

$$
\hat{\sigma}(\mathbf{q}, \omega) = \frac{\mathrm{i}Ne^2}{\omega m} \left(\frac{3}{8} \left[\left(\frac{q}{2k_{\mathrm{F}}} \right)^2 + 3 \left(\frac{\omega + \mathrm{i}/\tau}{q v_{\mathrm{F}}} \right)^2 + 1 \right] \right.
$$
$$
- \frac{3k_{\mathrm{F}}}{16q} \left[1 - \left(\frac{q}{2k_{\mathrm{F}}} - \frac{\omega + \mathrm{i}/\tau}{q v_{\mathrm{F}}} \right)^2 \right]^2 \mathrm{Ln} \left\{ \frac{\frac{q}{2k_{\mathrm{F}}} - \frac{\omega + \mathrm{i}/\tau}{q v_{\mathrm{F}}} + 1}{\frac{q}{2k_{\mathrm{F}}} - \frac{\omega + \mathrm{i}/\tau}{q v_{\mathrm{F}}} - 1} \right\}
$$
$$
\left. - \frac{3k_{\mathrm{F}}}{16q} \left[1 - \left(\frac{q}{2k_{\mathrm{F}}} + \frac{\omega + \mathrm{i}/\tau}{q v_{\mathrm{F}}} \right)^2 \right]^2 \mathrm{Ln} \left\{ \frac{\frac{q}{2k_{\mathrm{F}}} + \frac{\omega + \mathrm{i}/\tau}{q v_{\mathrm{F}}} + 1}{\frac{q}{2k_{\mathrm{F}}} + \frac{\omega + \mathrm{i}/\tau}{q v_{\mathrm{F}}} - 1} \right\} \right) . \tag{5.3.6}
$$

In the limit of small relaxation rate ($\tau \to \infty$) and using Eq. (5.2.23) for the expression for the logarithm Ln, these equations reduce to

$$
\sigma_1(\mathbf{q}, \omega) = \begin{cases} \frac{3\pi}{4} \frac{Ne^2}{mq v_{\mathrm{F}}} \left[1 - \left(\frac{\omega}{q v_{\mathrm{F}}} \right)^2 - \left(\frac{q}{2k_{\mathrm{F}}} \right)^2 \right]^2 & \frac{\omega}{q v_{\mathrm{F}}} + \frac{q}{2k_{\mathrm{F}}} < 1 \\[2ex] \frac{Ne^2}{\omega m} \frac{3\pi k_{\mathrm{F}}}{16q} \left[1 - \left(\frac{\omega}{q v_{\mathrm{F}}} - \frac{q}{2k_{\mathrm{F}}} \right)^2 \right]^2 & \left| \frac{\omega}{q v_{\mathrm{F}}} - \frac{q}{2k_{\mathrm{F}}} \right| < 1 < \frac{\omega}{q v_{\mathrm{F}}} + \frac{q}{2k_{\mathrm{F}}} \\[2ex] 0 & \left| \frac{\omega}{q v_{\mathrm{F}}} - \frac{q}{2k_{\mathrm{F}}} \right| > 1 \end{cases}
$$
$$
\tag{5.3.7a}
$$

$$
\sigma_2(\mathbf{q}, \omega) = \frac{Ne^2}{\omega m} \left\{ \frac{3}{8} \left[\left(\frac{q}{2k_{\mathrm{F}}} \right)^2 + 3 \left(\frac{\omega}{q v_{\mathrm{F}}} \right)^2 + 1 \right] \right.
$$
$$
- \frac{3k_{\mathrm{F}}}{16q} \left[1 - \left(\frac{q}{2k_{\mathrm{F}}} - \frac{\omega}{q v_{\mathrm{F}}} \right)^2 \right]^2 \ln \left| \frac{\frac{q}{2k_{\mathrm{F}}} - \frac{\omega}{q v_{\mathrm{F}}} + 1}{\frac{q}{2k_{\mathrm{F}}} - \frac{\omega}{q v_{\mathrm{F}}} - 1} \right|
$$
$$
\left. - \frac{3k_{\mathrm{F}}}{16q} \left[1 - \left(\frac{q}{2k_{\mathrm{F}}} + \frac{\omega}{q v_{\mathrm{F}}} \right)^2 \right]^2 \ln \left| \frac{\frac{q}{2k_{\mathrm{F}}} + \frac{\omega}{q v_{\mathrm{F}}} + 1}{\frac{q}{2k_{\mathrm{F}}} + \frac{\omega}{q v_{\mathrm{F}}} - 1} \right| \right\} . \tag{5.3.7b}
$$

Both the real and imaginary parts of the conductivity are plotted in Fig. 5.12 as functions of frequency ω and wavevector q. Of course, it is straightforward to calculate the dielectric constant $\hat{\epsilon}(\mathbf{q}, \omega)$ and Lindhard response function $\hat{\chi}(\mathbf{q}, \omega)$.

The overall qualitative behavior of the conductivity components can easily be inferred from Fig. 5.11. First, absorption can occur only when electron–hole excitations are possible; outside this region $\sigma_1(\mathbf{q}, \omega) = 0$. On the left hand side of Fig. 5.11, in this dissipationless limit, for small $\mathbf{q}$ a power expansion gives

$$
\sigma_2(\mathbf{q}, \omega) = \frac{Ne^2}{\omega m} \left[1 + \frac{q^2 v_{\mathrm{F}}^2}{5\omega^2} + \frac{3q^4 v_{\mathrm{F}}^2}{16k_{\mathrm{F}}^2 \omega^2} + \cdots \right] . \tag{5.3.8}
$$

(a)

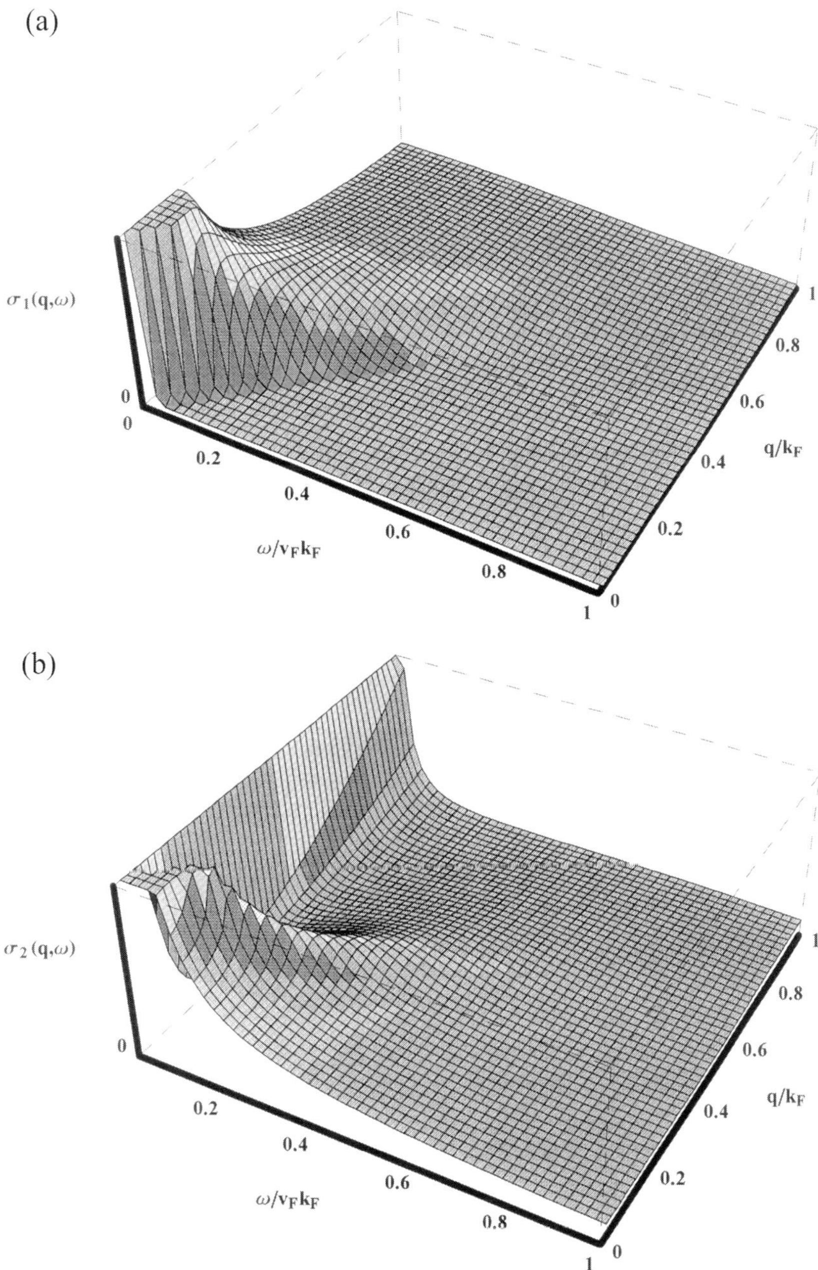

(b)

Fig. 5.12. (a) Real and (b) imaginary parts of the frequency and wavevector dependent conductivity $\hat{\sigma}(\mathbf{q},\omega)$ calculated after Eqs (5.3.7) in the limit $\tau \to \infty$. The conductivity is plotted in arbitrary units, the frequency is normalized to the Fermi frequency $\omega_F = v_F k_F/2$, and the wavevector is normalized to the Fermi wavevector k_F. There are no excitations for $\omega < q v_F - 2q^2 v_F/k_F$.

The region $\frac{\omega}{q v_{\mathrm{F}}} + \frac{q}{2k_{\mathrm{F}}} < 1$, also indicated in Fig. 5.11, is called the quasi-static limit; for small $\mathbf{q}$ values this indicates the region where, due to the low frequency of the fluctuations, screening is important. For small $\mathbf{q}$ values, series expansion gives in the $\tau \to \infty$ limit

$$\sigma_1(\mathbf{q}, \omega) = \frac{3}{4} \frac{N e^2 \pi}{q v_{\mathrm{F}} m} \left[1 - \frac{\omega^2}{q^2 v_{\mathrm{F}}^2} - \frac{q^2}{4 k_{\mathrm{F}}^2} \right] \quad , \tag{5.3.9}$$

where we have retained an extra term compared with the Boltzmann equation (5.2.22).

5.4 Longitudinal response

So far the discussion in this chapter has focused on the transverse response, i.e. the response of the medium to a vector potential $\mathbf{A}(\mathbf{q}, \omega)$. Now we discuss the response to static (but later also time dependent) charges. First, the low frequency limit ($\omega \to 0$) is analyzed; this leads to the well known Thomas–Fermi screening of the electron gas. Next, the response as obtained from the selfconsistent field approximation (Section 4.3.1) is discussed in detail; this gives the response functions appropriate for arbitrary $\mathbf{q}$ values. Finally, we examine the collective and single-particle excitations of an electron gas at zero temperature. Although we confine ourselves to the framework of the classical electrodynamics, and first quantization is used, the problem can also be treated in the corresponding second quantized formalism [Hau94, Mah90].

5.4.1 Thomas–Fermi approximation: the static limit for $q < k_{\mathrm{F}}$

If we place an additional charge $-e$ into a metal at position $\mathbf{r} = 0$, the conduction electrons will be rearranged due to Coulomb repulsion; this effect, called screening, can be simply estimated by the so-called Thomas–Fermi method.

Starting from Poisson's equation (2.1.7) in Coulomb gauge

$$\nabla^2 \Phi_{\mathrm{ind}}(\mathbf{r}) = -4\pi \,\delta\rho(\mathbf{r}) = -4\pi \left[\rho(\mathbf{r}) - \rho_0 \right] = 4\pi e [N(\mathbf{r}) - N_0] \quad , \tag{5.4.1}$$

where $\delta\rho(\mathbf{r}) = \rho(\mathbf{r}) - \rho_0$ is the deviation from the uniform charge density, $\delta N(\mathbf{r}) = N(\mathbf{r}) - N_0$ is the deviation from the uniform particle density, and $\Phi(\mathbf{r})$ is the electrostatic potential at position $\mathbf{r}$. We assume that $\Phi(\mathbf{r})$ is a slowly varying function of distance r, and that the potential energy of the field $V(\mathbf{r}) = -e\Phi(\mathbf{r})$ is simply added to the kinetic energy of all the electrons (the so-called rigid band approximation); i.e.

$$\mathcal{E}(\mathbf{r}) \;=\; \mathcal{E}_{\mathrm{F}} - e\Phi_{\mathrm{ind}}(\mathbf{r}) = \frac{\hbar^2}{2m} \left[3\pi^2 N(\mathbf{r}) \right]^{2/3}$$

$$\mathcal{E}_{\mathrm{F}} = \frac{\hbar^2}{2m}\left[3\pi^2 N_0\right]^{2/3} = \frac{\hbar^2 k_{\mathrm{F}}^2}{2m} \quad . \tag{5.4.2}$$

The carrier density can be expanded around the Fermi energy $\mathcal{E}_{\mathrm{F}}$ in terms of $\Phi(\mathbf{r})$

$$\begin{aligned}
N(\mathbf{r}) &= \frac{1}{3\pi^2}\left(\frac{2m}{\hbar^2}\right)^{3/2}\left[\mathcal{E}_{\mathrm{F}} - e\Phi_{\mathrm{ind}}(\mathbf{r})\right]^{3/2} \\
&\approx \frac{1}{3\pi^2}\left(\frac{2m}{\hbar^2}\right)^{3/2}\mathcal{E}_{\mathrm{F}}^{3/2}\left[1 - \frac{3e\Phi_{\mathrm{ind}}(\mathbf{r})}{2\mathcal{E}_{\mathrm{F}}} + \cdots\right] \\
&= N_0 - \frac{3}{2}N_0\frac{e\Phi_{\mathrm{ind}}(\mathbf{r})}{\mathcal{E}_{\mathrm{F}}} \quad ,
\end{aligned} \tag{5.4.3}$$

where the unperturbed carrier density is (see Eq. (5.4.2))

$$N_0 = \frac{1}{3\pi^2}\left(2m\mathcal{E}_{\mathrm{F}}/\hbar^2\right)^{3/2} \quad . \tag{5.4.4}$$

The Poisson equation is then written as

$$\nabla^2\Phi_{\mathrm{ind}}(\mathbf{r}) = 4\pi e\,\delta N(\mathbf{r}) = -\frac{6\pi N_0 e^2}{\mathcal{E}_{\mathrm{F}}}\Phi_{\mathrm{ind}}(\mathbf{r}) = -\lambda^2\Phi_{\mathrm{ind}}(\mathbf{r}) \quad ; \tag{5.4.5}$$

where the Thomas–Fermi screening parameter λ

$$\lambda = \left(\frac{6\pi N_0 e^2}{\mathcal{E}_{\mathrm{F}}}\right)^{1/2} = \left[4\pi D(\mathcal{E}_{\mathrm{F}})e^2\right]^{1/2} \tag{5.4.6}$$

is closely related to the density of states at the Fermi level

$$D(\mathcal{E}_{\mathrm{F}}) = \frac{3N}{2\mathcal{E}_{\mathrm{F}}} = \frac{mk_{\mathrm{F}}}{\pi^2\hbar^2} \quad . \tag{5.4.7}$$

Interestingly, the screening does not depend on the mass of the particles or other transport properties because it is calculated in the static limit. Note that without assuming a degenerate free-electron gas we can write more generally

$$\lambda_{\mathrm{class}} = \left(4\pi e^2\frac{\partial N_0}{\partial\mathcal{E}}\right)^{1/2} \quad ,$$

which in the classical (non-degenerate) limit leads to the Debye–Hückel screening

$$\lambda_{\mathrm{DH}} = \left(4\pi N_0 e^2/k_{\mathrm{B}}T\right)^{1/2}$$

since $\partial\mathcal{E}/\partial N_0 = 3k_{\mathrm{B}}T/2N_0$. For an isotropic metal the solution

$$\Phi_{\mathrm{ind}}(r) \propto \frac{\exp\{-\lambda r\}}{r} \tag{5.4.8}$$

is called the screened Coulomb or Yukava potential. It is appropriate only close to the impurity; at distances larger than $1/2k_{\mathrm{F}}$ the functional form is fundamentally different [Kit63]. The screening length $1/\lambda$ is about 1 Å for a typical metal, i.e. slightly smaller than the lattice constant, suggesting that the extra charge is screened almost entirely within one unit cell.

The Fourier transform of Eq. (5.4.8) can be inverted to give the screened potential

$$\Phi(q,0) = \frac{4\pi e}{q^2 + \lambda^2} \quad ; \tag{5.4.9}$$

and, using Eq. (3.1.31),

$$\epsilon_1(q,0) = 1 - \frac{4\pi}{q^2}\frac{\delta\rho(q,0)}{\Phi(q,0)} = 1 + \frac{\lambda^2}{q^2} = 1 + \frac{6\pi N_0 e^2}{\mathcal{E}_F q^2} \quad . \tag{5.4.10}$$

For $\mathbf{q} \rightarrow 0$, the dielectric constant diverges; at distances far from the origin the electron states are not perturbed.

5.4.2 *Solution of the Boltzmann equation: the small* **q** *limit*

The Thomas–Fermi approximation is appropriate for static screening ($\omega = 0$). If we are interested in the frequency dependence, we have to utilize the Boltzmann equation (5.2.7) for the case of a longitudinal field; from there we obtain the longitudinal current, in a fashion similar to Eq. (5.2.13). Assuming quasi-free electrons or a spherical Fermi surface, after some algebra we arrive at the following expression for the longitudinal dielectric constant:

$$\hat{\epsilon}(\mathbf{q},\omega) = 1 + \frac{3\omega_p^2}{q^2 v_F^2}\left[1 + \frac{\omega + i/\tau}{2q v_F}\,\mathrm{Ln}\left\{\frac{q v_F - \omega - i/\tau}{-q v_F - \omega - i/\tau}\right\}\right] \quad , \tag{5.4.11}$$

in analogy to what is derived for the transverse dielectric constant using Eq. (5.2.19). If $\mathbf{q}$ is small, utilizing the Taylor expansion in $\mathbf{q}$ yields the expression

$$\hat{\epsilon}(\mathbf{q},\omega) = 1 - \frac{\omega_p^2}{(\omega + i/\tau)^2}\left[1 + \frac{3q^2 v_F^2}{5(\omega + i/\tau)^2} + \cdots\right] \tag{5.4.12}$$

for $q v_F < |\omega + i/\tau|$. In the limit $1/\tau \ll \omega$, this expression is identical to Eq. (5.2.21) obtained for the transverse response and resembles the Drude formula (5.1.9) as the $\mathbf{q} = 0$ case. This is not surprising, for at $\mathbf{q} = 0$ the distinction between longitudinal and transverse response becomes obsolete. For finite $\mathbf{q}$, the expression tends to the static limit

$$\hat{\epsilon}(\mathbf{q},\omega) = 1 + \frac{3\omega_p^2}{q^2 v_F^2}\left[1 - \frac{\omega^2}{q^2 v_F^2} + i\frac{\pi\omega}{2q v_F}\right] \quad . \tag{5.4.13}$$

As the plasma frequency $\omega_p^2 = 4\pi N e^2/m$, the first term is identical to that derived within the framework of the Thomas–Fermi approximation.

5.4.3 *Response functions for arbitrary* q *values*

The semiclassical calculation of the $\mathbf{q}$ dependent conductivity is limited to small $\mathbf{q}$ values since it does not include the scattering across the Fermi surface; the appearance of v_F in appropriate expressions merely reflects through the quantum statistics the existence of the Fermi surface. If we intend to take into account these processes, we must consider the time evolution of the density matrix as in Section 4.3.1. A more detailed discussion of the problems can be found in various textbooks [Cal91, Mah90, Pin66, Zim72].

The response functions to be evaluated are

$$\hat{\epsilon}(\mathbf{q}, \omega) = 1 - \frac{4\pi}{q^2}\hat{\chi}(\mathbf{q}, \omega) \qquad \text{and} \qquad \hat{\sigma}(\mathbf{q}, \omega) = \frac{i\omega}{q^2}\hat{\chi}(\mathbf{q}, \omega) \qquad (5.4.14)$$

with the Lindhard function

$$\hat{\chi}(\mathbf{q}, \omega) = \frac{2e^2}{(2\pi)^3}\int d\mathbf{k}\,\frac{f^0_{\mathbf{k}+\mathbf{q}} - f^0_{\mathbf{k}}}{\mathcal{E}(\mathbf{k}+\mathbf{q}) - \mathcal{E}(\mathbf{k}) - \hbar(\omega + i/\tau)} \qquad (5.4.15)$$

as derived in Eq. (4.3.21). Following a traditional treatment of the problem we evaluate the dielectric constant, but we also give the appropriate expressions for the conductivity. We will evaluate these expressions in three dimensions: the discussion of the one-dimensional and two-dimensional response functions can be found in Appendix F.

For $T = 0$, the Fermi–Dirac distribution $f^0_{\mathbf{k}} = 1$ for $\mathbf{k} < \mathbf{k}_F$ and 0 for $\mathbf{k} > \mathbf{k}_F$, which allows the integral to be solved analytically,[3] and we find:

$$\hat{\chi}(\mathbf{q}, \omega) = -\frac{e^2 D(\mathcal{E}_F)}{2}\left(1 + \frac{k_F}{2q}\left[1 - \left(\frac{q}{2k_F} - \frac{\omega + 1/\tau}{qv_F}\right)^2\right]\mathrm{Ln}\left\{\frac{\frac{q}{2k_F} - \frac{\omega + i/\tau}{qv_F} + 1}{\frac{q}{2k_F} - \frac{\omega + i/\tau}{qv_F} - 1}\right\}\right.$$

$$\left. + \frac{k_F}{2q}\left[1 - \left(\frac{q}{2k_F} + \frac{\omega + 1/\tau}{qv_F}\right)^2\right]\mathrm{Ln}\left\{\frac{\frac{q}{2k_F} + \frac{\omega + i/\tau}{qv_F} + 1}{\frac{q}{2k_F} + \frac{\omega + i/\tau}{qv_F} - 1}\right\}\right) \quad , (5.4.16)$$

where $D(\mathcal{E}_F)$ is the density of states of both spin directions at the Fermi level,

---

[3] Following the method given in footnote [2] on p. 111, the steps are as follows:

$$\int d\mathbf{k}\,\frac{\Theta(\mathbf{k} - \mathbf{k}_F)}{\omega - q^2 v_F/2k_F - (v_F/k_F)(\mathbf{k}\cdot\mathbf{q}) + i/\tau}$$

$$= \pi\int_0^{k_F} k\,dk\,\mathrm{Ln}\left\{\frac{\omega - q^2 v_F/2k_F - qv_F k/k_F + i/\tau}{\omega - q^2 v_F/2k_F + qv_F k/k_F + i/\tau}\right\}$$

$$= \frac{\pi}{2q}\left[\frac{(\omega - q^2 v_F/2k_F)}{qv_F/k_F^2} + \frac{q^2 v_F^2 - (\omega + q^2 v_F/2k_F)^2}{2q^2 v_F^2/k_F^2}\mathrm{Ln}\left\{\frac{\omega - q^2 v_F/2k_F - qv_F + i/\tau}{\omega - q^2 v_F/2k_F + qv_F + i/\tau}\right\}\right]$$

$$= \frac{\pi k_F^2}{2q}\left[\left(\frac{\omega}{qv_F} - \frac{q}{2k_F}\right) + \frac{1}{2}\left[1 - \left(\frac{\omega}{qv_F} + \frac{q}{2k_F}\right)^2\right]\mathrm{Ln}\left\{\frac{q^2 v_F/2k_F - (\omega + i/\tau) + qv_F}{q^2 v_F/2k_F - (\omega + i/\tau) - qv_F}\right\}\right] \quad .$$

The first part is solved in a similar way by using $\mathbf{k} + \mathbf{q}$ instead of $\mathbf{k}$.

the well known Lindhard response function. Again, Ln denotes the principal part
of the complex logarithm. The frequency and wavevector dependence of the real
and imaginary parts of the dielectric response function $\hat{\chi}(\mathbf{q}, \omega)$ are calculated by
utilizing Eq. (3.2.7). In the $1/\tau \to 0$ limit,

$$
\begin{aligned}
\chi_1(\mathbf{q}, \omega) = -\frac{e^2 D(\mathcal{E}_F)}{2} &\left\{ 1 + \frac{k_F}{2q}\left[1 - \left(\frac{q}{2k_F} - \frac{\omega}{q v_F}\right)^2\right] \ln\left|\frac{\frac{q}{2k_F} - \frac{\omega}{q v_F} + 1}{\frac{q}{2k_F} - \frac{\omega}{q v_F} - 1}\right| \right. \\
&\left. + \frac{k_F}{2q}\left[1 - \left(\frac{q}{2k_F} + \frac{\omega}{q v_F}\right)^2\right] \ln\left|\frac{\frac{q}{2k_F} + \frac{\omega}{q v_F} + 1}{\frac{q}{2k_F} + \frac{\omega}{q v_F} - 1}\right| \right\}
\end{aligned}
\tag{5.4.17a}
$$

$$
\chi_2(\mathbf{q}, \omega) = -\frac{e^2 D(\mathcal{E}_F)}{2}
\begin{cases}
\dfrac{2\pi\omega}{q v_F} & \dfrac{q}{2k_F} + \dfrac{\omega}{q v_F} < 1 \\[2ex]
\dfrac{\pi k_F}{q}\left[1 - \left(\dfrac{q}{2k_F} - \dfrac{\omega}{q v_F}\right)^2\right] & \left|\dfrac{q}{2k_F} - \dfrac{\omega}{q v_F}\right| < 1 < \dfrac{q}{2k_F} + \dfrac{\omega}{q v_F} \\[2ex]
0 & \left|\dfrac{q}{2k_F} - \dfrac{\omega}{q v_F}\right| > 1
\end{cases}
$$

$$
\tag{5.4.17b}
$$

Both functions are displayed in Fig. 5.13, where the frequency axis is normalized to
$v_F k_F$ and the wavevector axis to k_F. It is obvious that both for $\mathbf{q} \to \infty$ and $\omega \to \infty$
the response function $\chi_1(\mathbf{q}, \omega) \to 0$, expressing the fact that the variations of the
charge distribution cannot follow a fast temporal or rapid spatial variation of the
potential.

In the static limit (Thomas–Fermi approximation, $\omega \to 0$), the real part of the
Lindhard dielectric response function of Eq. (5.4.17a) is reduced to

$$
\chi_1(\mathbf{q}, 0) = -\frac{e^2 D(\mathcal{E}_F)}{2}\left[1 + \left(\frac{k_F}{q} - \frac{1}{4}\frac{q}{k_F}\right)\ln\left|\frac{q + 2k_F}{q - 2k_F}\right|\right]
\tag{5.4.18}
$$

for a spherical Fermi surface. For small $\mathbf{q}$ values

$$
\chi_1(\mathbf{q}, 0) = \frac{\delta\rho(\mathbf{q}, 0)}{\Phi(\mathbf{q}, 0)} = \frac{\lambda^2}{4\pi} = \frac{3}{2}\frac{N_0 e^2}{\mathcal{E}_F} \quad ,
\tag{5.4.19}
$$

which is independent of $\mathbf{q}$. For finite q values, $\chi_1(\mathbf{q})$ decreases with increasing $\mathbf{q}$
and the derivative has a logarithmic singularity at $\mathbf{q} = 2\mathbf{k}_F$. The functional form
is displayed in Fig. 5.14 along with the results found for one and two dimensions.
The singularity is well understood and can be discussed by referring to Figs 5.10
and 5.11. By increasing $\mathbf{q}$ towards $\mathbf{k}_F$, the number of states which contribute to the
integral (5.4.15), $f^0_{\mathbf{k}+\mathbf{q}} - f^0_{\mathbf{k}}$, progressively increases, until $\mathbf{q}$ is equal to $2\mathbf{k}_F$, and
the sum does not change when $\mathbf{q}$ increases beyond this value. When $\mathbf{q} \approx 2\mathbf{k}_F$ there
is a vanishingly small number of states, for which $\mathcal{E}(\mathbf{k} + \mathbf{q}) \approx \mathcal{E}(\mathbf{k})$ (these states
giving, however, a large contribution to the integral), and a combination of these

(a)

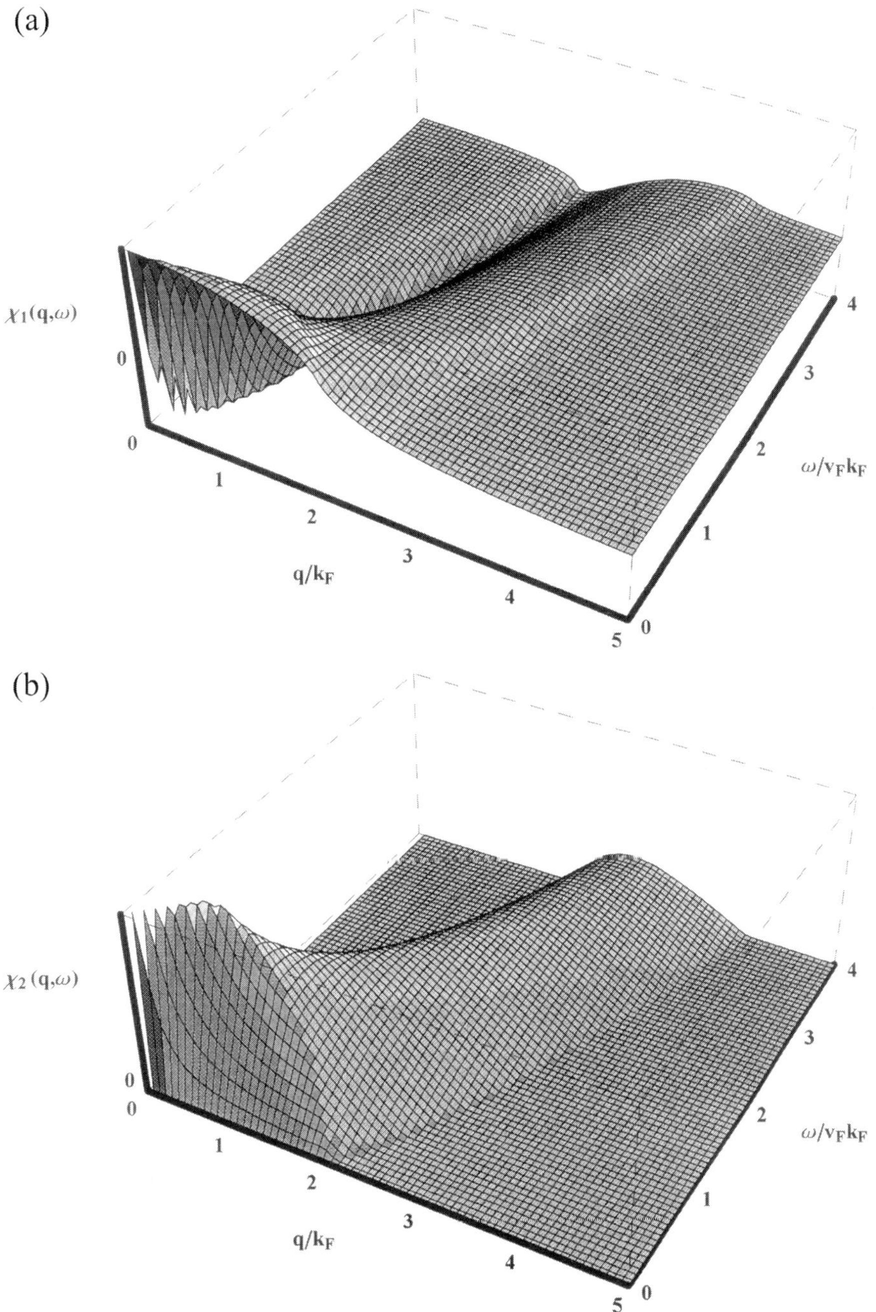

Fig. 5.13. Wavevector and frequency dependence of the Lindhard response function $\hat{\chi}(\mathbf{q}, \omega)$ of a free-electron gas at $T = 0$ in three dimensions calculated by Eqs (5.4.17). (a) The real part $\chi_1(\mathbf{q}, \omega)$ and (b) the imaginary part $\chi_2(\mathbf{q}, \omega)$ are both in arbitrary units.

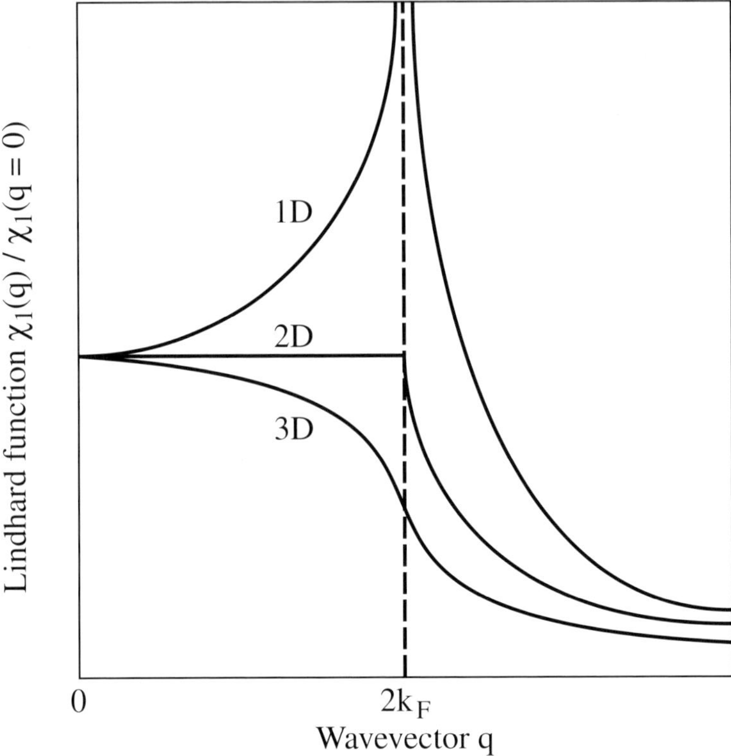

Fig. 5.14. The static Lindhard dielectric response function χ_1 as a function of wavevector q at $T = 0$ in one, two, and three dimensions (1D, 2D, and 3D, respectively). Here $\chi_1(\mathbf{q})/\chi_1(0)$ is plotted for normalization purposes. At the wavevector $2k_\mathrm{F}$ the Lindhard function $\chi_1(\mathbf{q})$ diverges in one dimension, has a cusp in two dimensions, and shows a singularity in the derivative in three dimensions.

two effects leads to a decreasing $\chi_1(\mathbf{q}, \omega)$ near $\mathbf{q} \approx 2\mathbf{k}_\mathrm{F}$ and to a logarithmical singularity in the derivative. For one dimension, near $q = 2k_\mathrm{F}$ there is a large number of states with $\mathcal{E}(\mathbf{k} + \mathbf{q}) \approx \mathcal{E}(\mathbf{k})$, and this in turn leads to a singularity in the response function in one dimension also shown in Fig. 5.14. As expected, the two-dimensional case lies between these two situations, and there is a cusp at $\mathbf{q} = 2\mathbf{k}_\mathrm{F}$.

All this is related to what is in general referred to as the Kohn anomaly [Koh59], the strong modification of the phonon spectrum, brought about (through screening) by the logarithmic singularity at $2k_\mathrm{F}$. The effect is enhanced in low dimensions due to the more dramatic changes at $2k_\mathrm{F}$. In addition to these effects, various broken symmetry states, with a wavevector dependent oscillating charge or spin density may also develop for electron–electron interactions of sufficient strength; these are discussed in Chapter 7.

The singularity of the response function at $\mathbf{q} = 2\mathbf{k}_F$ leads also to a characteristic spatial dependence of the screening charge around a charged impurity. While in the Thomas–Fermi approximation the screening charge decays exponentially with distance far from the impurity, the selfconsistent field approximations yield the well known Friedel oscillations with a period $2k_F$, which in three dimensions reads

$$\Delta\rho = \frac{A}{r^3} \cos\{2k_F + \phi\} \quad ; \qquad (5.4.20)$$

here both A and ϕ depend on the scattering potential [Gru77].

After having evaluated $\hat{\chi}(\mathbf{q}, \omega)$ it is straightforward to write down the following expression for the dielectric constant:

$$\hat{\epsilon}(\mathbf{q}, \omega) = 1 + \frac{3\omega_p^2}{q^2 v_F^2} \left\{ \frac{1}{2} + \frac{k_F}{4q} \left[1 - \left(\frac{q}{2k_F} - \frac{\omega + 1/\tau}{q v_F} \right)^2 \right] \mathrm{Ln} \frac{\frac{q}{2k_F} - \frac{\omega + i/\tau}{q v_F} + 1}{\frac{q}{2k_F} - \frac{\omega + i/\tau}{q v_F} - 1} \right.$$

$$\left. + \frac{k_F}{4q} \left[1 - \left(\frac{q}{2k_F} + \frac{\omega + 1/\tau}{q v_F} \right)^2 \right] \mathrm{Ln} \frac{\frac{q}{2k_F} + \frac{\omega + i/\tau}{q v_F} + 1}{\frac{q}{2k_F} + \frac{\omega + i/\tau}{q v_F} - 1} \right\} \quad , \quad (5.4.21)$$

which for negligible damping, $1/\tau \to 0$, reduces to

$$\epsilon_1(\mathbf{q}, \omega) = 1 + \frac{\lambda^2}{q^2} \left\{ \frac{1}{2} + \frac{k_F}{4q} \left[1 - \left(\frac{q}{2k_F} - \frac{\omega}{q v_F} \right)^2 \right] \ln \left| \frac{\frac{q}{2k_F} - \frac{\omega}{q v_F} + 1}{\frac{q}{2k_F} - \frac{\omega}{q v_F} - 1} \right| \right.$$

$$\left. + \frac{k_F}{4q} \left[1 - \left(\frac{q}{2k_F} + \frac{\omega}{q v_F} \right)^2 \right] \ln \left| \frac{\frac{q}{2k_F} + \frac{\omega}{q v_F} + 1}{\frac{q}{2k_F} + \frac{\omega}{q v_F} - 1} \right| \right\} \quad , \quad (5.4.22a)$$

where $\lambda^2 = 3\omega_p^2/v_F^2$ is the Thomas–Fermi screening parameter introduced above. In the same limit the imaginary part becomes

$$\epsilon_2(\mathbf{q}, \omega) = \begin{cases} \frac{3\pi \omega_p^2 \omega}{2q^3 v_F^3} & \frac{q}{2k_F} + \frac{\omega}{q v_F} < 1 \\[2mm] \frac{3\pi \omega_p^2 k_F}{4q^3 v_F^2} \left[1 - \left(\frac{q}{2k_F} - \frac{\omega}{q v_F} \right)^2 \right] & \left| \frac{q}{2k_F} - \frac{\omega}{q v_F} \right| < 1 < \frac{q}{2k_F} + \frac{\omega}{q v_F} \\[2mm] 0 & \left| \frac{q}{2k_F} - \frac{\omega}{q v_F} \right| > 1 \end{cases} \quad . \quad (5.4.22b)$$

In the static limit $\omega \to 0$, the imaginary part disappears and the expression for $\epsilon(\mathbf{q}, \omega)$ from Eq. (5.4.10) is recovered, as expected. In Fig. 5.15 these components are shown as a function of frequency and wavevector. For small $\mathbf{q}$ values we can expand the energy as

$$\mathcal{E}_{\mathbf{k}+\mathbf{q}} = \mathcal{E}_{\mathbf{k}} + \frac{\partial \mathcal{E}_{\mathbf{k}}}{\partial k} q + \frac{1}{2} \frac{\partial^2 \mathcal{E}_{\mathbf{k}}}{\partial k^2} q^2 \quad , \qquad (5.4.23)$$

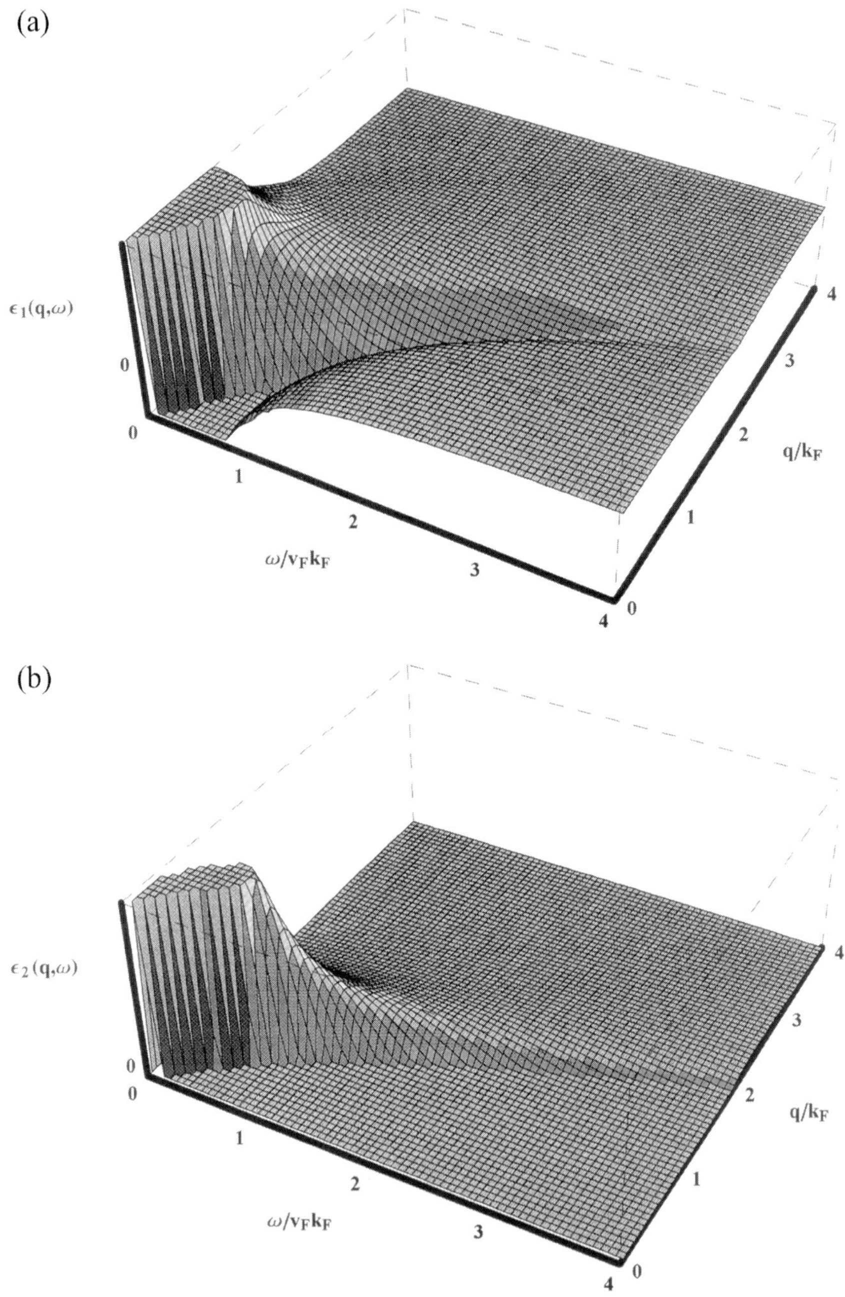

Fig. 5.15. (a) Real part and (b) imaginary part of the dielectric constant $\hat{\epsilon}(\mathbf{q}, \omega)$ of a free-electron gas as a function of frequency and wavevector at $T = 0$ in three dimensions calculated using Eq. (5.4.21). The frequency axis is normalized to the Fermi frequency $\omega_F = v_F k_F/2$, and the wavevector axis is normalized to the Fermi wavevector k_F.

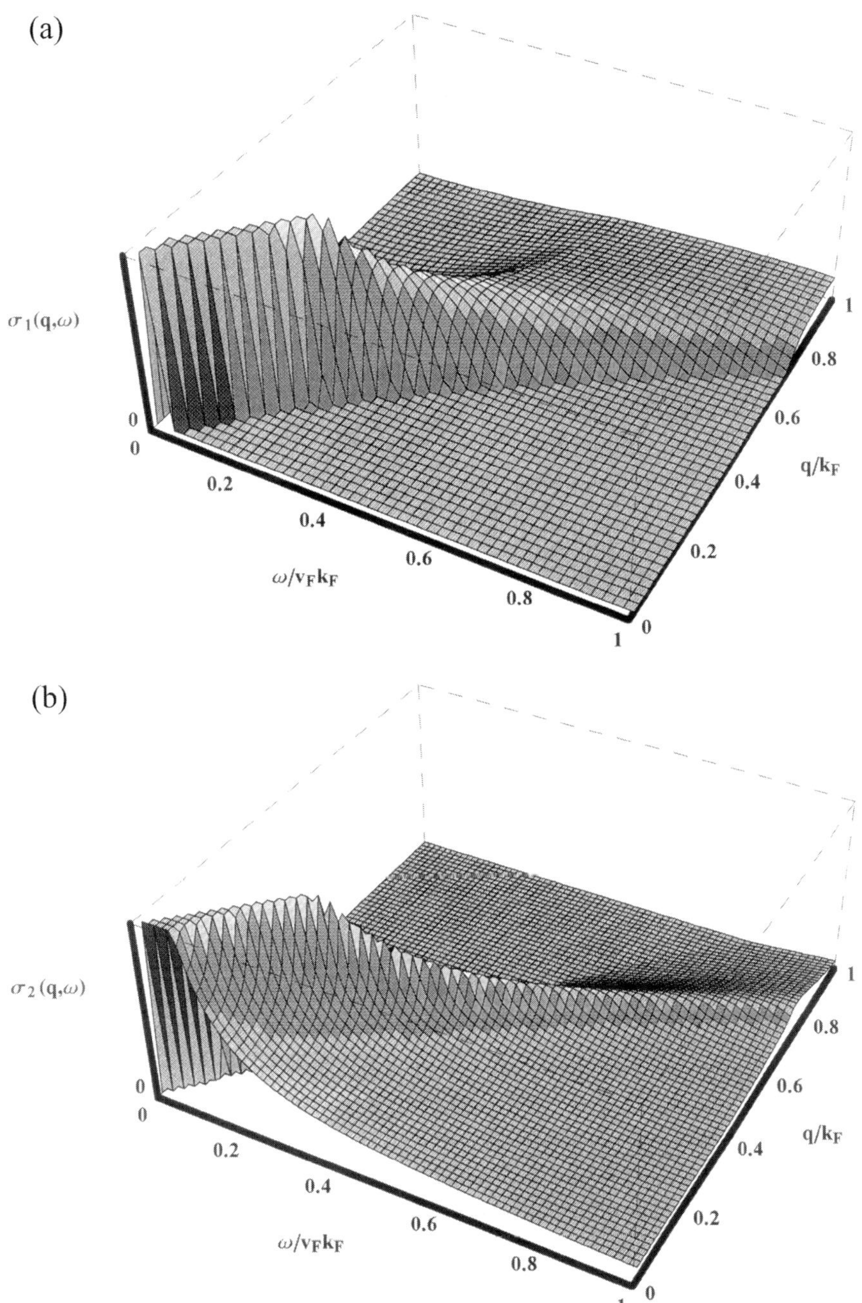

Fig. 5.16. Frequency and wavevector dependence of (a) the real and (b) the imaginary part of the conductivity $\hat{\sigma}(\mathbf{q}, \omega)$ of a free-electron gas at $T = 0$ in three dimensions after Eq. (5.4.25). The frequency axis is normalized to the Fermi frequency $\omega_{\mathrm{F}} = k_{\mathrm{F}} v_{\mathrm{F}}/2 = \mathcal{E}_{\mathrm{F}}/\hbar$, and the wavevector axis is normalized to the Fermi wavevector k_{F}.

and on substituting this into Eq. (5.4.15) we obtain the following expression for the real part of the dielectric constant:

$$\epsilon_1(\mathbf{q}, \omega) = 1 - \frac{e^2}{\pi^2 \hbar^2 \omega^2} \int d f^0(\mathcal{E}_{\mathbf{k}}) \frac{\partial^2 \mathcal{E}_{\mathbf{k}}}{\partial k^2} \mathbf{k} \quad .$$

If we also expand the Fermi function in the numerator of Eq. (5.4.15),

$$f^0_{\mathbf{k}+\mathbf{q}} = f^0_{\mathbf{k}} + \frac{\partial f^0_{\mathbf{k}}}{\partial \mathcal{E}} \mathbf{q} \cdot \nabla_{\mathbf{k}} \mathcal{E}(\mathbf{k}) + \mathcal{O}(q^2) = f^0_{\mathbf{k}} + \frac{\partial f^0_{\mathbf{k}}}{\partial \mathcal{E}} \mathbf{q} \cdot \frac{\hbar^2}{m} \mathbf{k} + \mathcal{O}(q^2) \quad ,$$

we find that for $\omega \to 0$ the term linear in q gives the Thomas–Fermi result (5.4.19). In this limit the real part of the dielectric constant has the simple form

$$\epsilon_1(\mathbf{q}, 0) = 1 - \frac{4\pi e^2}{q^2} \int \frac{\partial f^0_{\mathbf{k}}}{\partial \mathcal{E}} D(\mathcal{E}) \, d\mathcal{E} \quad .$$

The integral yields at $T = 0$

$$\lim_{q \to 0} \epsilon_1(\mathbf{q}, 0) = 1 + \frac{4\pi e^2}{q^2} D(\mathcal{E}_{\mathrm{F}}) = 1 + \frac{4\pi e^2}{q^2} \frac{3 N_0 m^2}{\hbar^2 k_{\mathrm{F}}^2} = 1 + \frac{4\pi e^2}{q^2} \frac{3 N_0}{2 \mathcal{E}_{\mathrm{F}}} = 1 + \frac{\lambda^2}{q^2} \quad ;$$

$$(5.4.24)$$

the result we arrived at in Eq. (5.4.10).

We can also use the relationship between $\hat{\sigma}(\mathbf{q}, \omega)$ and $\hat{\chi}(\mathbf{q}, \omega)$ to obtain

$$\hat{\sigma}(\mathbf{q}, \omega) = \frac{3\omega}{4\pi \mathrm{i}} \frac{\omega_{\mathrm{p}}^2}{q^2 v_{\mathrm{F}}^2} \left(\frac{1}{2} + \frac{k_{\mathrm{F}}}{4q} \left[1 - \left(\frac{q}{2k_{\mathrm{F}}} - \frac{\omega + \mathrm{i}/\tau}{q v_{\mathrm{F}}} \right)^2 \right] \mathrm{Ln} \left\{ \frac{\frac{q}{2k_{\mathrm{F}}} - \frac{\omega + \mathrm{i}/\tau}{q v_{\mathrm{F}}} + 1}{\frac{q}{2k_{\mathrm{F}}} - \frac{\omega + \mathrm{i}/\tau}{q v_{\mathrm{F}}} - 1} \right\} \right.$$

$$\left. + \frac{k_{\mathrm{F}}}{4q} \left[1 - \left(\frac{q}{2k_{\mathrm{F}}} + \frac{\omega + \mathrm{i}/\tau}{q v_{\mathrm{F}}} \right)^2 \right] \mathrm{Ln} \left\{ \frac{\frac{q}{2k_{\mathrm{F}}} + \frac{\omega + \mathrm{i}/\tau}{q v_{\mathrm{F}}} + 1}{\frac{q}{2k_{\mathrm{F}}} + \frac{\omega + \mathrm{i}/\tau}{q v_{\mathrm{F}}} - 1} \right\} \right) \quad . \quad (5.4.25)$$

The real and imaginary parts of the complex, wavevector dependent conductivity are displayed in Fig. 5.16 as a function of frequency ω and wavevector $\mathbf{q}$. As discussed in Section 3.1.2, for small $\mathbf{q}$ values this expression is the same as $\hat{\sigma}(\mathbf{q}, \omega)$ derived for the transverse conductivity, and when $\mathbf{q} \to 0$ the Drude model is recovered. For large $\mathbf{q}$ values – or short wavelength fluctuations – the results are significantly different: there is an appreciable conductivity in the case of the low frequency transverse response. However, in the longitudinal case $\sigma_1(\mathbf{q}, \omega)$ rapidly drops to zero with increasing $\mathbf{q}$; this is the consequence of the screening for longitudinal excitations.

5.4.4 Single-particle and collective excitations

In the previous section we have addressed the issue of screening, the rearrangement of the electronic charge in response to transverse and longitudinal electric fields.

The formalism leads to the dielectric constant or conductivity in terms of the excitations of electron–hole pairs. Besides those excitations, a collective mode also occurs, and it involves coherent motion of the system as a whole.

Collective excitations of the electron gas are longitudinal plasma oscillations. As we saw in the derivation leading to Eq. (4.3.2) for vanishing $\epsilon_1(\omega)$, an infinitesimally small external perturbation Φ^{ext} produces a strong internal field Φ; and in the absence of damping the electron gas oscillates collectively. For the uniform $\mathbf{q} = 0$ mode simple considerations lead to the frequency of these oscillations. Let us consider a situation in which a region of the electron cloud is displaced by a distance x without affecting the rest of the system. The result is a layer of net positive charge on one side of this region and an identical negative layer on the opposite side, both of thickness x. The polarization is simply given by $\mathbf{P} = -Ne\mathbf{r}$, and this leads to an electric field $\mathbf{E} = -4\pi\mathbf{P}$. In the absence of damping $(1/\tau \to 0)$ the equation of motion (3.2.22) reduces to

$$
m\frac{d^2\mathbf{r}}{dt^2} = -e\mathbf{E}(t) = 4\pi Ne^2\mathbf{r} \quad . \tag{5.4.26}
$$

The solution is an oscillation of the entire charged electron gas, with an oscillation frequency

$$
\omega_{\text{p}} = \left(\frac{4\pi Ne^2}{m}\right)^{1/2}
$$

called the plasma frequency. Of course such collective oscillations of the electron plasma could be sustained only in the absence of a dissipative mechanism; in reality the oscillations are damped, albeit weakly in typical materials. Note that ω_{p} here is the frequency of longitudinal oscillations of the electron gas. There is – as we have discussed in Section 5.1 – also a significant change in the transverse response at this frequency. The dielectric constant $\epsilon_1(\mathbf{q} = 0, \omega)$ switches from negative to positive sign at ω_{p}, and this leads to a sudden drop of the reflectivity.

In order to explore the finite frequency and long wavelength (small $\mathbf{q}$) limit, we expand ϵ_1 in terms of $\mathbf{q}$; we obtain

$$
\epsilon_1(\mathbf{q}, \omega) = 1 - \frac{\omega_{\text{p}}^2}{\omega^2}\left(1 + \frac{3}{5}\frac{q^2 v_{\text{F}}^2}{\omega_{\text{p}}^2} + \cdots\right) \tag{5.4.27}
$$

at frequencies $\omega \approx \omega_{\text{p}}$. Setting $\epsilon_1(q, \omega) = 0$ yields for the dispersion of the plasma frequency

$$
\omega_{\text{p}}^2(q) \approx \omega_{\text{p}}^2\left(1 + \frac{3}{5}\frac{q^2 v_{\text{F}}^2}{\omega_{\text{p}}^2}\right)
$$

or

$$\omega_{\rm p}(q) \approx \omega_{\rm p}\left[1 + \frac{3}{10}\frac{v_{\rm F}^2}{\omega_{\rm p}^2}q^2 + \left(\frac{\hbar}{8m^2\omega^2} - \frac{3}{280}\frac{v_{\rm F}^4}{\omega_{\rm p}^4}\right)q^4\right] \quad , \tag{5.4.28}$$

where the term $\frac{3}{10}\frac{v_{\rm F}^2}{\omega_{\rm p}^2}q^2 \approx q^2/\lambda^2$, with λ the Thomas–Fermi screening parameter as defined earlier in Eq. (5.4.6). The collective oscillations can occur at various wavelengths, and we can refer to these quanta of oscillations as plasmons. In Eq. (5.4.28) we have added the last term in $\mathcal{O}(q^4)$ without having it explicitly carried through the calculation [Gei68, Mah90]. Plasma oscillations are well defined for small $\mathbf{q}$ since there is no damping due to the lack of single-particle excitations in the region of plasma oscillations. For larger $\mathbf{q}$, the plasmon dispersion curve shown in Fig. 5.11 merges into the continuum of single-electron excitations at a wavevector $q_{\rm c}$ and for $q > q_{\rm c}$ the oscillations will be damped and decay into the single-particle continuum. Using the dispersion relation $\omega_{\rm p}(q)$ and the spectrum of electron–hole excitations, the onset of the damping of the plasmon called the Landau damping can be derived for particular values of the parameters involved. For small dispersion, we obtain by equating Eq. (5.3.2) to $\omega_{\rm p}(q = 0)$ the simple approximate relation

$$q_{\rm c} \approx \frac{\omega_{\rm p}}{v_{\rm F}} \tag{5.4.29}$$

for the onset of damping of plasma oscillations by the creation of electron–hole pairs.

5.5 Summary of the ω dependent and q dependent response

Let us finally recall the results we have obtained for the complex $\mathbf{q}$ and ω dependent conductivity (or, equivalently, dielectric constant). Electron–hole excitations – and the absorption of electromagnetic radiation – are possible only for certain $\mathbf{q}$ and ω values, indicated by the shaded area on Fig. 5.11. This area is defined by the boundaries $\hbar\omega = \frac{\hbar^2}{2m}\left(2qk_{\rm F} - q^2\right)$ and $\hbar\omega = \frac{\hbar^2}{2m}\left(2qk_{\rm F} + q^2\right)$.

Outside this region $\sigma_1(\mathbf{q}, \omega) = 0$ and no absorption occurs. In the so-called homogeneous limit, for $qv_{\rm F} \ll \omega$, we find that the transverse and longitudinal responses are equivalent (we cannot distinguish between transverse and longitudinal fields in the $\mathbf{q} \to 0$ limit), and to leading order we find

$$\hat{\sigma}^{\rm L}(\mathbf{q}, \omega) = \hat{\sigma}^{\rm T}(\mathbf{q}, \omega) = \frac{iNe^2}{m\omega}\left(1 + \frac{q^2v_{\rm F}^2}{5\omega^2} + \cdots\right) \tag{5.5.1}$$

with, of course, the associated $\delta\{\omega = 0\}$ response for the real part. The dielectric constant has the form

$$\epsilon_1^{\mathrm{L}}(\mathbf{q}, \omega) = \epsilon_1^{\mathrm{T}}(\mathbf{q}, \omega) = 1 - \left(\frac{\omega_{\mathrm{p}}^2}{\omega}\right)^2 + \cdots \quad ; \tag{5.5.2}$$

the zeros of which lead to the ultraviolet transparency of metals and the longitudinal plasma oscillations with the dispersion relation $\omega_{\mathrm{p}}(\mathbf{q})$ given above.

In the opposite, quasi-static, limit for $q v_{\mathrm{F}} \ll \omega$ screening becomes important, and indeed in this limit the transverse and longitudinal responses are fundamentally different. Here we find

$$\hat{\sigma}^{\mathrm{T}}(\mathbf{q}, \omega) = \frac{3\pi N e^2}{4 q v_{\mathrm{F}} m} \left[1 - \frac{\omega^2}{q^2 v_{\mathrm{F}}^2} + \cdots\right] \tag{5.5.3}$$

and the response is primarily real; we recover dissipation in the small $\mathbf{q}$ (but still $q v_{\mathrm{F}} \ll \omega$) limit. In contrast,

$$\hat{\sigma}^{\mathrm{L}}(\mathbf{q}, \omega) = \frac{3 N e^2 \omega}{q^2 v_{\mathrm{F}}^2} \, \mathrm{i} \left[1 - \frac{k_{\mathrm{F}} \omega^2}{2 q^3 v_{\mathrm{F}}^2} + \cdots\right] \tag{5.5.4}$$

is primarily imaginary; there is no dissipation, and we also recover the Thomas–Fermi screening in this limit.

All this is valid for a three-dimensional degenerate electron gas. Both the single-particle and collective excitations are somewhat different in two- and one-dimensional electron gases for various reasons. First, phase space arguments, which determine the spectrum of single-particle excitations, are somewhat different for a Fermi sphere in three dimensions, a Fermi circle in two dimensions and two points (at $+\mathbf{k}_{\mathrm{F}}$ and $-\mathbf{k}_{\mathrm{F}}$) in one dimension. Second, polarization effects are different in reduced dimensions, with the consequence that the frequency of the plasmon excitations approach zero as $\mathbf{q} = 0$. These features are discussed in Appendix F.

A final word on temperature dependent effects is in order. Until now, we have considered $T = 0$, and this leads to a sharp, well defined sudden change in the occupation numbers at $\mathbf{q} = \mathbf{k}_{\mathrm{F}}$. At finite temperatures, the thermal screening of the Fermi distribution function leads to similar smearing for the single-particle excitations. With increasing temperature, we progressively cross over to a classical gas of particles, and in this limit classical statistics prevails; the appropriate expressions can be obtained by replacing $m v_{\mathrm{F}}^2$ by $3 k_{\mathrm{B}} T$.

References

[Abr72]   A.A. Abrikosov, *Introduction to the Theory of Normal Metals* (Academic Press, New York and London, 1972)

[Bla57]   F.J. Blatt, *Theory of Mobility of Electrons in Solids*, in: *Solid State Physics*, **4**, edited by F. Seitz and D. Turnbull (Academic Press, New York, 1957), p. 199

[Cal91]   J. Callaway, *Quantum Theory of Solids*, 2nd edition (Academic Press, New York, 1991)

[Cha90]   R.G. Chambers, *Electrons in Metals and Semiconductors* (Chapman and Hall, London, 1990)

[Gei68]   J. Geiger, *Elektronen und Festkörper* (Vieweg Verlag, Braunschweig, 1968)

[Gru77]   G. Grüner and M. Minier, *Adv. Phys.* **26**, 231 (1977)

[Hau94]   H. Haug and S.W. Koch, *Quantum Theory of the Optical and Electronic Properties of Semiconductors*, 3rd edition (World Scientific, Singapore, 1994)

[Kit63]   C. Kittel, *Quantum Theory of Solids* (John Wiley & Sons, New York, 1963)

[Koh59]   W. Kohn, *Phys. Rev. Lett.* **2**, 395 (1959)

[Lin54]   J. Lindhard, *Dan. Mat. Fys. Medd.* **28**, no. 8 (1954)

[Mad78]   O. Madelung, *Introduction to Solid State Theory* (Springer-Verlag, Berlin, 1978)

[Mah90]   G.D. Mahan, *Many-Particle Physics*, 2nd edition (Plenum Press, New York, 1990)

[Mat58]   D.C. Mattis and G. Dresselhaus, *Phys. Rev.* **111**, 403 (1958)

[Pin66]   D. Pines and P. Nozières, *The Theory of Quantum Liquids*, Vol. 1 (Addison-Wesley, Reading, MA, 1966)

[Pip47]   A.B. Pippard, *Proc. Roy. Soc. A* **191**, 385 (1947)

[Pip54a] A.B. Pippard, *Proc. Roy. Soc. A* **224**, 273 (1954)

[Pip54b] A.B. Pippard, *Metallic Conduction at High Frequencies and Low Temperatures*, in: *Advances in Electronics and Electron Physics* **6**, edited by L. Marton (Academic Press, New York, 1954), p. 1

[Pip62]   A.B. Pippard, *The Dynamics of Conduction Electrons*, in: *Low Temperature Physics*, edited by C. DeWitt, B. Dreyfus, and P.G. deGennes (Gordon and Breach, New York, 1962)

[Reu48]   G.E.H. Reuter and E.H. Sondheimer, *Proc. Roy. Soc. A* **195**, 336 (1948)

[Zim72]   J.M. Ziman, *Principles of the Theory of Solids*, 2nd edition (Cambridge University Press, London, 1972)

Further reading

[Car68]   M. Cardona, *Electronic Optical Properties of Solids*, in: *Solid State Physics, Nuclear Physics, and Particle Physics*, edited by I. Saavedra (Benjamin, New York, 1968), p. 737

[Dru00a] P. Drude, *Phys. Z.* **1**, 161 (1900)

[Dru00b] P. Drude, *Ann. Physik* **1**, 566 (1900); **3**, 369 (1900)

[Gol89]   A.I. Golovashkin, ed., *Metal Optics and Superconductivity* (Nova Science Publishers, New York, 1989)

[Gro79]   P. Grosse, *Freie Elektronen in Festkörpern* (Springer-Verlag, Berlin, 1979)

[Huf96]   S. Hüfner, *Photoelectron Spectroscopy*, 2nd edition, *Springer Series in Solid-State Sciences* **82** (Springer-Verlag, Berlin, 1996)

[Hum71] R.E. Hummel, *Optische Eigenschaften von Metallen und Legierungen* (Springer-Verlag, Berlin, 1971)

[Jon73] W. Jones and N.H. March, *Theoretical Solid State Physics* (John Wiley & Sons, New York, 1973)

[Lif73] I.M. Lifshitz, M.Ya. Azbel, and M.I. Kaganov, *Electron Theory of Metals* (Consultants Bureau, New York, 1973)

[Pin63] D. Pines, *Elementary Excitations in Solids* (Addison-Wesley, Reading, MA, 1963)

[Pla73] P.M. Platzman and P.A. Wolff, *Waves and Interactions in Solid State Plasmas* (Academic Press, New York, 1973)

[Sin77] K.P. Sinha, *Electromagnetic Properties of Metals and Superconductors*, in: *Interaction of Radiation with Condensed Matter*, Vol. 2 IAEA-SMR-20/20 (International Atomic Energy Agency, Vienna, 1977), p. 3

[Sok67] A.V. Sokolov, *Optical Properties of Metals* (American Elsevier, New York, 1967)

[Som33] A. Sommerfeld and H. Bethe, *Handbuch der Physik*, **24/2**, edited by H. Geiger and K. Scheel (Springer-Verlag, Berlin, 1933)

6

Semiconductors

The focus of this chapter is on the optical properties of band semiconductors and insulators. The central feature of these materials is the appearance of a single-particle gap, separating the valence band from the conduction band. The former is full and the latter is empty at zero temperature. The Fermi energy lies between these bands, leading to zero dc conduction at $T = 0$, and to a finite static dielectric constant. In contrast to metals, interband transitions from the valence band to the conduction band are of superior importance, and these excitations are responsible for the main features of the electrodynamic properties. Many of the phenomena discussed in this chapter also become relevant for higher energy excitations in metals when the transition between bands becomes appreciable for the optical absorption.

Following the outline of Chapter 5, we first introduce the Lorentz model, a phenomenological description which, while obviously not the appropriate description of the state of affairs, reproduces many of the optical characteristics of semiconductors. The transverse conductivity of a semiconductor is then described, utilizing the formalisms which we have developed in Chapter 4, and the absorption near the bandgap is discussed in detail, followed by a summary of band structure effects. After discussing longitudinal excitations and the $\mathbf{q}$ dependent optical response, we briefly mention indirect transitions and finite temperature and impurity effects; some of the discussion of these phenomena, however, is relegated to Chapter 13.

Optical properties of semiconductors are among the best studied phenomena in solid state physics, mainly because of their technological relevance. Both experimental data and theoretical description are quite advanced and are the subject of many excellent textbooks and monographs, e.g. [Coh88, Hau94, Yu96].

6.1 The Lorentz model

6.1.1 Electronic transitions

Before discussing optical transitions induced by the electromagnetic fields between different bands in a solid, let us first examine a simple situation: the transitions between a ground state and excited states of N identical atoms, the wavefunctions of which are $\Psi_0(\mathbf{r})$ and $\Psi_l(\mathbf{r})$. The electrodynamic field is assumed to be of the form

$$\mathbf{E}(t) = \mathbf{E}_0 \left(\exp\{-i\omega t\} + \exp\{i\omega t\}\right) \tag{6.1.1}$$

and leads to the admixture of the excited states to the ground state. This can be treated by using the time dependent Schrödinger equation with the perturbation given for an electric field by

$$\mathcal{H}' = e\mathbf{E} \cdot \mathbf{r} \quad .$$

Because of this admixture, the resulting time dependent wavefunction is

$$\Psi(\mathbf{r}, t) = \Psi_0 \exp\left\{-\frac{i\mathcal{E}_0 t}{\hbar}\right\} + \sum_l a_l(t) \Psi_l \exp\left\{-\frac{i\mathcal{E}_l t}{\hbar}\right\} \tag{6.1.2}$$

with $\mathcal{E}_l$ denoting the various energy levels. The coefficients $a_l(t)$ are obtained from the solution of the Schrödinger equation

$$i\hbar \frac{\partial}{\partial t} \Psi(\mathbf{r}, t) = - \left(\mathcal{H}_0 + \mathcal{H}'\right) \Psi(\mathbf{r}, t) \quad , \tag{6.1.3}$$

where $\mathcal{H}_0 \Psi_l = \mathcal{E}_l \Psi_l$ describes the energy levels in the absence of the applied electromagnetic field. Inserting the wavefunction Ψ into Eq. (6.1.3) we find that the coefficients are

$$a_l(t) = \frac{\mathbf{E}}{2\hbar} \left(\frac{1 - \exp\left\{\frac{i(\hbar\omega + \mathcal{E}_l - \mathcal{E}_0)t}{\hbar}\right\}}{\hbar\omega + (\mathcal{E}_l - \mathcal{E}_0)} - \frac{1 - \exp\left\{\frac{i(-\hbar\omega + \mathcal{E}_l - \mathcal{E}_0)t}{\hbar}\right\}}{\hbar\omega - (\mathcal{E}_l - \mathcal{E}_0)} \right) \int \Psi_l^*(-e\mathbf{r})\Psi_0 \, d\mathbf{r}. \tag{6.1.4}$$

Here

$$\int \Psi_l^* \, e\mathbf{r} \, \Psi_0 \, d\mathbf{r} = e\mathbf{r}_{l0}$$

is the matrix element of the dipole moment $\mathbf{P} = -e\mathbf{r}$. The induced dipole moment

$$\mathbf{P}(t) = \langle -e\mathbf{r}(t)\rangle = \int \Psi^*(\mathbf{r}, t)(-e\,\mathbf{r})\Psi(\mathbf{r}, t) \, d\mathbf{r} = \frac{\mathbf{E}}{2\hbar} \sum_l e^2 \, |\mathbf{r}_{l0}|^2 \, \frac{2\omega_l}{\omega_l^2 - \omega^2} = \alpha\mathbf{E} \tag{6.1.5}$$

results after some calculations [Woo72], and we define a polarizability

$$\hat{\alpha}(t) = \frac{\mathbf{P}(t)}{\mathbf{E}(t)} = \sum_l \frac{e^2 |r_{l0}|^2}{\hbar} \frac{2\omega_{l0}}{\omega_{l0}^2 - \omega^2} \quad ,$$

which is purely real as no absorption has been considered. We use $\hbar\omega_{l0} = \mathcal{E}_l - \mathcal{E}_0$ for the energy difference between the two states. The dielectric constant, for N atoms, is then

$$\epsilon_1(\omega) = 1 + 4\pi N\alpha(\omega) = 1 + \frac{4\pi Ne^2}{m} \sum_l \frac{2m}{\hbar^2} \frac{|r_{l0}|^2 \hbar\omega_{l0}}{\omega_{l0}^2 - \omega^2} \quad . \tag{6.1.6}$$

The polarizability can be regarded as a parameter accounting for the effectiveness of the transition, and the fraction of energy absorbed can be expressed in a dimensionless unit as

$$f_{l0} = \frac{2m}{\hbar^2} \hbar\omega_{l0} |r_{l0}|^2 \quad . \tag{6.1.7}$$

This is called the oscillator strength, which was introduced in Eq. (5.1.32) in terms of the momentum matrix element $|\mathbf{p}_{l0}|^2$:

$$f_{l0} = \frac{2|\mathbf{p}_{l0}|^2}{m\hbar\omega_{l0}} \quad ; \tag{6.1.8}$$

both can readily be converted by using the commutation relation $[p, x] = -i\hbar$. The oscillator strength is related to the power absorbed by the transmission, which reads as $P = W_{l0}\hbar\omega_{l0}$, with the number of transitions per second and per volume

$$W_{l0} = \frac{\mathrm{d}}{\mathrm{d}t} \left(a_l^* a_l \right) = \frac{\pi e^2 E^2}{2\hbar^2} |r_{l0}|^2 \quad . \tag{6.1.9}$$

As discussed in Appendix D in more detail, the sum of all transitions should add up to unity; this is the so-called oscillator strength sum rule

$$\sum_l f_{l0} = 1 \quad . \tag{6.1.10}$$

With this notation

$$\epsilon_1(\omega) = 1 + \frac{4\pi Ne^2}{m} \sum_l \frac{f_{l0}}{\omega_{l0}^2 - \omega^2} \quad ; \tag{6.1.11}$$

this refers to the real part of the dielectric constant. By virtue of the Kramers–Kronig relation we obtain also the imaginary part, and the complex dielectric constant reads

$$\hat{\epsilon}(\omega) = 1 + \frac{4\pi Ne^2}{m} \sum_l f_{l0} \left[\frac{1}{\omega_{l0}^2 - \omega^2} + \frac{i\pi}{2\omega}\delta\left\{\omega^2 - \omega_{l0}^2\right\} \right] \quad . \tag{6.1.12}$$

For N atoms, all with one excited state, at energy $\hbar\omega_{l0}$, the dielectric constant is shown in Fig. 6.1. $\epsilon_1(\omega = 0)$ is a positive quantity, and $\epsilon_1(\omega)$ increases as the frequency is raised and eventually diverges at $\omega = \omega_1$. It changes sign at ω_{l0} and decreases as the frequency increases. Finally we find a second zero-crossing (with positive slope) from $\epsilon_1 < 0$ to $\epsilon_1 > 0$ at the so-called plasma frequency

$$\omega_{\mathrm{p}} = \left(\frac{4\pi N e^2}{m}\right)^{1/2} \quad ,$$

in analogy to the free-electron case of metals. Here we have assumed that there is no absorption of energy by the collection of N atoms, the exception being the absorption process associated with the transitions between the various energy levels. When the system is coupled to a bath, the energy levels assume a lifetime broadening; this is described by a factor $\exp\{-t/2\tau\}$. With such broadening included, the transitions spread out to have a finite width, and the complex dielectric constant becomes

$$\hat{\epsilon}(\omega) = 1 + \frac{4\pi N e^2}{m} \sum_l \frac{f_{l0}}{\left(\omega_{l0}^2 - \omega^2\right) - i\omega/\tau} \quad . \tag{6.1.13}$$

This will remove the singularity at the frequency ω_{l0}, and $\epsilon_1(\omega)$ for broadened energy levels is also displayed in Fig. 6.1.

Some of the above results can be recovered by assuming the response of a classical harmonic oscillator. This of course is a highly misleading representation of the actual state of affairs in semiconductors, where the absence of dc conduction is due to full bands and not to the localization of electron states. Nevertheless, to examine the consequences of this description, a comparison with the Drude model is a useful exercise. We keep the same terms that we have included for the Drude description, the inertial and relaxational response, and supplement these with a restoring force K. This so-called Lorentz model is that of a harmonic oscillator

$$\frac{d^2\mathbf{r}}{dt^2} + \frac{1}{\tau}\frac{d\mathbf{r}}{dt} + \omega_0^2\mathbf{r} = -\frac{e}{m}\mathbf{E}(t) \quad . \tag{6.1.14}$$

Here $\omega_0 = (K/m)^{1/2}$, where m is the mass of the electron, and τ takes into account damping effects. A model such as that given by the above equation obviously does not lead to a well defined energy gap, but it is nevertheless in accord with the major features of the electrodynamics of a non-conducting state and is widely used to describe the optical properties of non-conducting materials. When the local electric field $\mathbf{E}$ has an $\exp\{-i\omega t\}$ time dependence, we find

$$\hat{\mathbf{r}}(\omega) = \frac{-e\mathbf{E}/m}{(\omega_0^2 - \omega^2) - i\omega/\tau} \quad ,$$

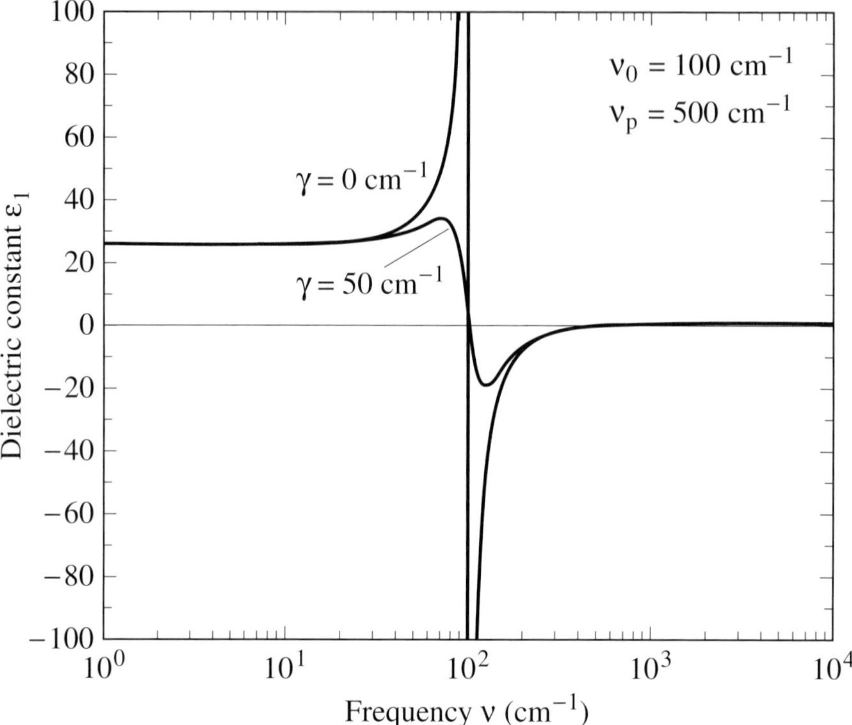

Fig. 6.1. Frequency dependence of the dielectric constant $\hat{\epsilon}(\omega)$ for a transition between well defined energy levels with energy separation $\omega_0/(2\pi c) = v_0 = 100$ cm^{-1}, together with the dielectric constant for a transition between energy levels of width $1/(2\pi c\tau) = \gamma = 50$ cm^{-1}; the spectral weight is given by the plasma frequency $\omega_p/(2\pi c) = v_p = 500$ cm^{-1}.

and for the induced dipole moment $\hat{\mathbf{p}}(\omega) = -e\hat{\mathbf{r}}(\omega)$ we obtain

$$\hat{\mathbf{p}}(\omega) = \frac{e^2}{m}\mathbf{E}\,\frac{1}{(\omega_0^2 - \omega^2) - i\omega/\tau} = \hat{\alpha}_a(\omega)\mathbf{E} \quad ,$$

where $\hat{\alpha}_a(\omega)$ is the atomic polarizability. The macroscopic polarizability is the sum over all the atoms N per unit volume involved in this excitation: $\mathbf{P} = N\langle\hat{\mathbf{p}}\rangle = N\hat{\alpha}_a\mathbf{E} = \hat{\chi}_e\mathbf{E}$. Since the dielectric constant is related to the dielectric susceptibility via $\hat{\epsilon}(\omega) = 1 + 4\pi\hat{\chi}_e(\omega)$, we obtain for the frequency dependent complex dielectric constant

$$\hat{\epsilon}(\omega) = 1 + \frac{4\pi N e^2}{m}\,\frac{1}{(\omega_0^2 - \omega^2) - i\omega/\tau} = 1 + \frac{\omega_p^2}{(\omega_0^2 - \omega^2) - i\omega/\tau} \quad . \tag{6.1.15}$$

We assume here that each atom contributes one electron to the absorption process.

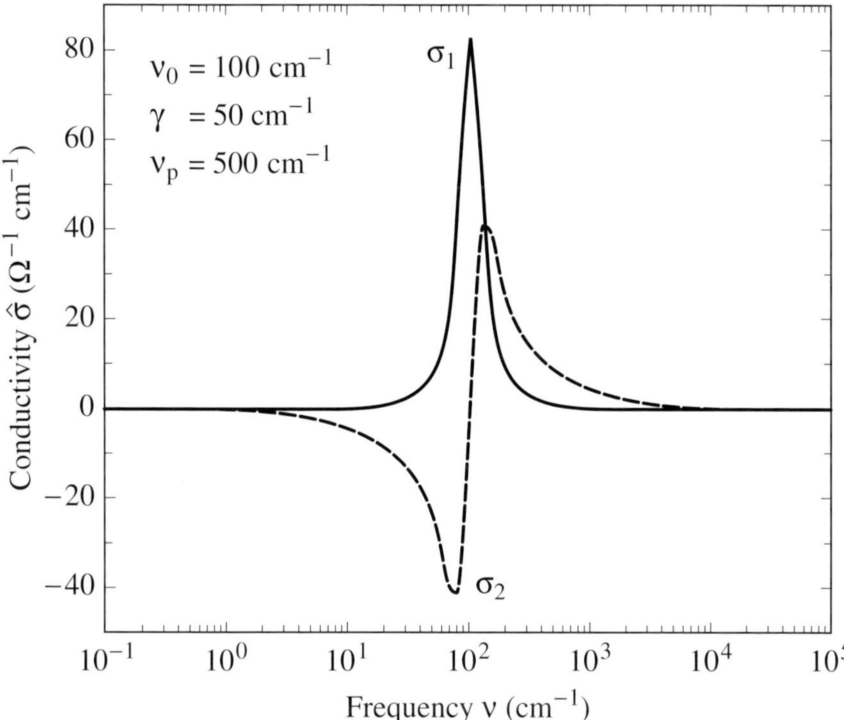

Fig. 6.2. Real and imaginary parts of the conductivity $\hat{\sigma}(\omega)$ versus frequency calculated after the Lorentz model (6.1.16) for center frequency $\nu_0 = \omega_0/(2\pi c) = 100\ \mathrm{cm}^{-1}$, the width $\gamma = 1/(2\pi c\tau) = 50\ \mathrm{cm}^{-1}$, and the plasma frequency $\nu_\mathrm{p} = \omega_\mathrm{p}/(2\pi c) = 500\ \mathrm{cm}^{-1}$.

For the complex conductivity we can write

$$\hat{\sigma}(\omega) = \frac{Ne^2}{m}\frac{\omega}{\mathrm{i}(\omega_0^2 - \omega^2) + \omega/\tau}\quad, \tag{6.1.16}$$

where ω_0 is the center frequency, often called the oscillator frequency, $1/\tau$ denotes the broadening of the oscillator due to damping, and $\omega_\mathrm{p} = \left(4\pi Ne^2/m\right)^{1/2}$ describes the oscillator strength, and is referred to as the plasma frequency. It is clear by comparing this expression with Eq. (5.1.4) that the Drude model can be obtained from the Lorentz model by setting $\omega_0 = 0$. This is not surprising, as we have retained only the inertial and damping terms to describe the properties of the metallic state.

6.1.2 Optical properties of the Lorentz model

The frequency dependence of the various optical constants can be evaluated in a straightforward manner. Fig. 6.2 displays the real and imaginary parts of the

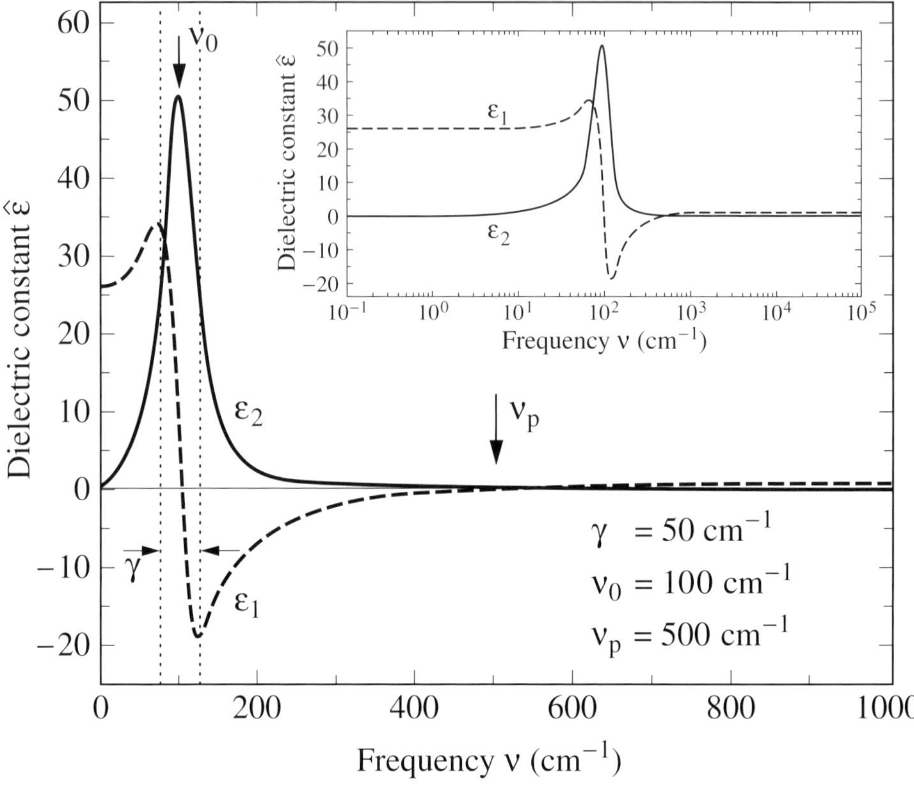

Fig. 6.3. Frequency dependent dielectric constant $\hat{\epsilon}(\omega)$ of a Lorentz oscillator (Eq. (6.1.15)) with $\nu_0 = 100$ cm^{-1}, $\gamma = 50$ cm^{-1}, and $\nu_p = 500$ cm^{-1}. The extrema of ϵ_1 are at $\pm\gamma/2$ around the oscillator frequency ν_0; $\epsilon_1(\omega)$ crosses zero at approximately ν_0 and ν_p. The inset shows the real and imaginary parts of the dielectric constant on a logarithmic frequency scale.

complex conductivity $\hat{\sigma}(\omega)$,

$$\sigma_1(\omega) = \frac{\omega_p^2}{4\pi} \frac{\omega^2/\tau}{(\omega_0^2 - \omega^2)^2 + \omega^2/\tau^2} \quad \text{and} \quad \sigma_2(\omega) = -\frac{\omega_p^2}{4\pi} \frac{\omega(\omega_0^2 - \omega^2)}{(\omega_0^2 - \omega^2)^2 + \omega^2/\tau^2}$$

$$(6.1.17)$$

as a function of frequency. We have chosen the center frequency $\nu_0 = \omega_0/(2\pi c) = 100$ cm^{-1}, the width $\gamma = 1/(2\pi c\tau) = 50$ cm^{-1}, and the plasma frequency $\nu_p = \omega_p/(2\pi c) = 500$ cm^{-1}, typical of a narrow gap semiconductor. The real and imaginary parts of the dielectric constant,

$$\epsilon_1(\omega) = 1 + \frac{\omega_p^2(\omega_0^2 - \omega^2)}{(\omega_0^2 - \omega^2)^2 + \omega^2/\tau^2} \qquad (6.1.18a)$$

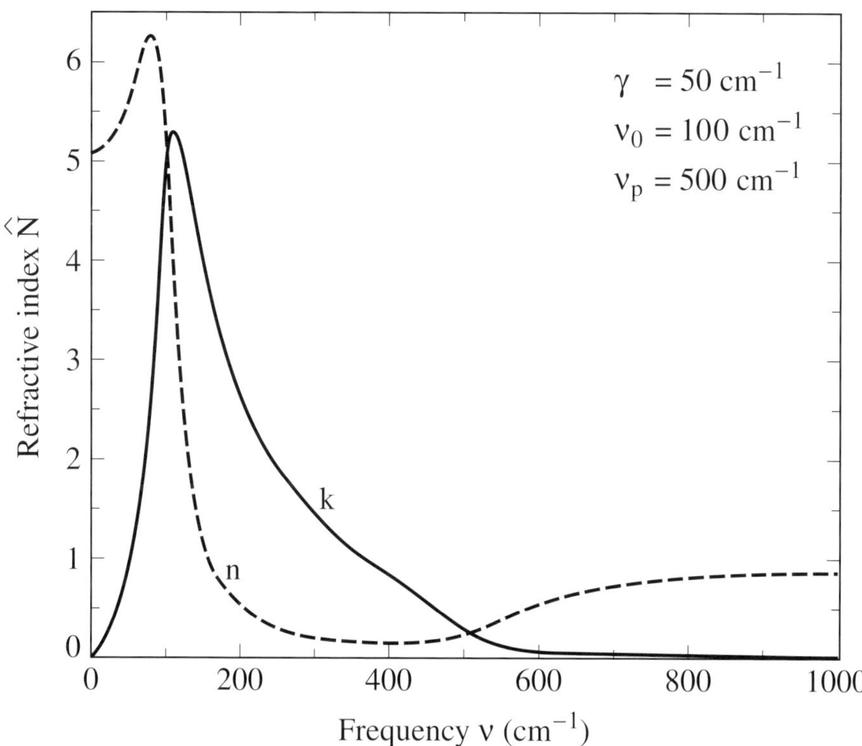

Fig. 6.4. Refractive index n and extinction coefficient k versus frequency calculated using the Lorentz model for $\nu_0 = 100$ cm^{-1}, $\gamma = 50$ cm^{-1}, and $\nu_p = 500$ cm^{-1}. Absorption (corresponding to large k) occurs mainly in the range $\pm\gamma$ around ν_0.

and

$$\epsilon_2(\omega) = \frac{\omega_p^2 \omega/\tau}{(\omega_0^2 - \omega^2)^2 + \omega^2/\tau^2} \tag{6.1.18b}$$

are displayed in Fig. 6.3 on both linear and logarithmic frequency scales. The refraction coefficient $\hat{N}$ can be calculated using Eqs (2.3.3) and (2.3.4); the frequency dependence of the real and imaginary parts of $\hat{N}$ are shown in Fig. 6.4. Using these parameters, we can, by employing Eqs (2.4.15) and (2.4.14), calculate the reflectivity $R(\omega)$ and the phase shift $\phi_r(\omega)$, and the results are displayed in Fig. 6.5. With the help of Eqs (2.3.34) the real and imaginary parts of the complex surface impedance, $R_S(\omega)$ and $X_S(\omega)$, are calculated and displayed in Fig. 6.6. The electronic loss function $1/\hat{\epsilon}$ of a Lorentz oscillator is shown in Fig. 6.7.

Three spectral ranges, the Hagen–Rubens, the relaxation, and the transparent regime, could be distinguished in the case of the Drude model; for the Lorentz model here four regimes with distinctively different spectral characteristics are of importance: a low frequency range at which the material does not absorb; the

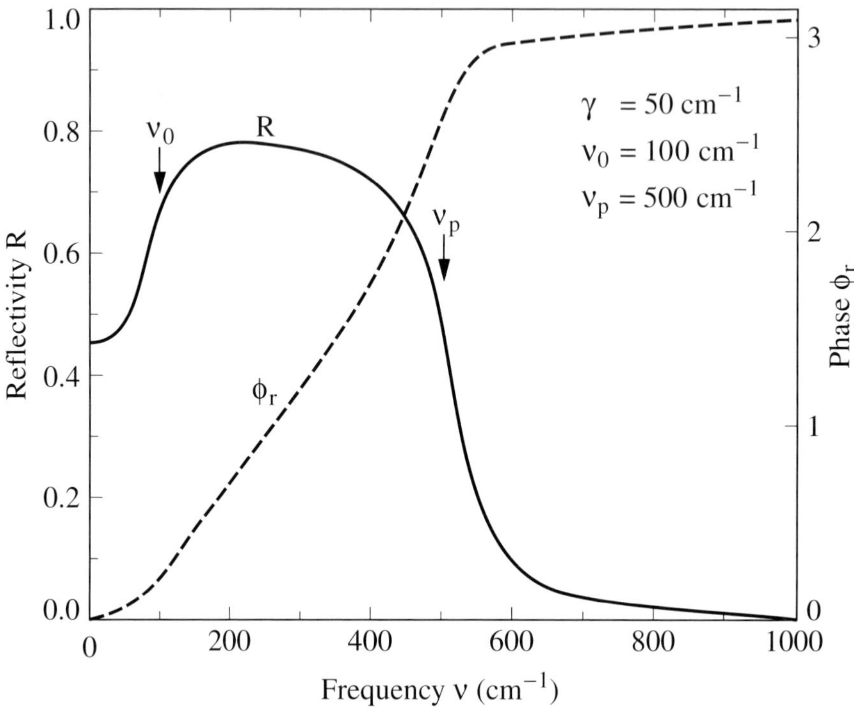

Fig. 6.5. Reflectivity R and the phase angle ϕ_r as a function of frequency in the Lorentz model. The reflectivity shows a strong increase at the center frequency of the oscillator $\nu_0 = 100$ cm^{-1}, has a plateau between ν_0 and ν_p, and drops at $\nu_p = 500$ cm^{-1}.

spectral range close to the center frequency ω_0 at which electrons are excited and thus absorption dominates; a range of high reflectivity for $\omega_0 < \omega < \omega_p$; and the transparent regime for large frequencies $\omega > \omega_p$.

Low frequency range

At low frequencies $\omega < (\omega_0 - 1/\tau)$ the real part of the conductivity σ_1 is small and there is little absorption. The real part of the dielectric constant saturates at a constant value $\epsilon_1(\omega \to 0)$ as the frequency is reduced. It can be evaluated by the Kramers–Kronig relation (3.2.31) with $\epsilon_1(\omega \to \infty) = 1$:

$$\epsilon_1(0) = 1 + \frac{2}{\pi} \int_0^\infty \frac{\epsilon_2(\omega)}{\omega}\, d\omega \approx 1 + \frac{2}{\pi \omega_0^2} \int_0^\infty \omega \epsilon_2(\omega)\, d\omega = 1 + \frac{\omega_p^2}{\omega_0^2} \quad . \quad (6.1.19)$$

In this region, the behavior of the refractive index $n(\omega)$ follows the dielectric constant. Depending on its value in the low frequency range the reflectivity can

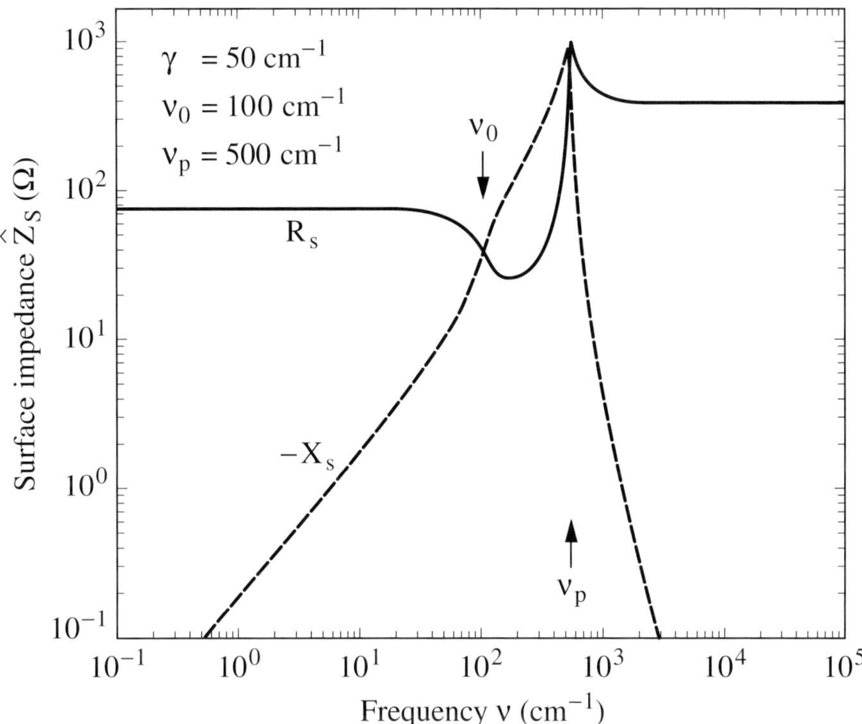

Fig. 6.6. Frequency dependence of the complex surface impedance $\hat{Z}_S(\omega) = R_S(\omega) + iX_S(\omega)$ calculated after the Lorentz model. There is a peak in both the real and imaginary parts at the plasma frequency $\nu_p = 500 \text{ cm}^{-1}$.

be large

$$R = \left(\frac{1 - \sqrt{\epsilon_1}}{1 + \sqrt{\epsilon_1}} \right)^2 = \left(\frac{1 - n}{1 + n} \right)^2$$

according to Eq. (2.4.16), as illustrated in Fig. 6.5. However, the light which is not reflected due to the impedance mismatch – caused by the step in the dielectric constant or refractive index – is transmitted with almost no attenuation: $\alpha \rightarrow 0$ and $k \rightarrow 0$ for $\omega \rightarrow 0$. In contrast to $X_S(\omega)$, which increases linearly with frequency up to ω_0, for $\omega \rightarrow 0$ the surface resistance $R_S(\omega)$ approaches the constant value Z_0/n according to Eq. (2.3.34a).

Absorption range

As soon as the frequency approaches the center frequency, electrons can be excited across the bandgap. Thus the vicinity of the oscillator frequency ω_0 is characterized by strong absorption due to the large conductivity σ_1 (although the loss function $-\,\text{Im}\{1/\hat{\epsilon}\}$ does not indicate any change by entering this spectral range). The width

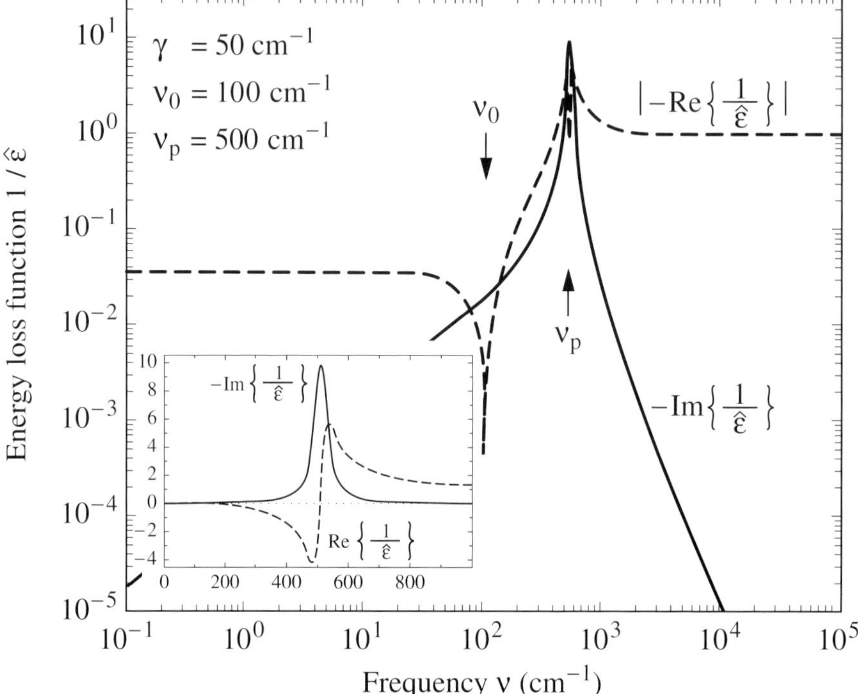

Fig. 6.7. Real and imaginary parts, $\mathrm{Re}\{1/\hat{\epsilon}\}$ and $-\mathrm{Im}\{1/\hat{\epsilon}\}$, of the frequency dependent loss function. The imaginary part peaks at ν_p and $\mathrm{Re}\{1/\hat{\epsilon}\}$ shows an anti-resonance-like shape. The inset shows the behavior on linear scales.

of the absorption range is characterized by the damping rate: $(\omega_0 - 1/2\tau) < \omega < (\omega_0 + 1/2\tau)$; it is the full width at half the conductivity maximum. As seen in Fig. 6.3 the real part of the dielectric constant crosses zero at ω_0 with negative slope, and it comes back to positive values at ω_p. Between the two extrema of the dielectric constant at $\omega_{\mathrm{extr}} = \omega_0 \pm 1/2\tau$ (in the case of small damping) the dispersion is negative. As displayed in the figures, most optical quantities, such as the refractive index n, the reflectivity $R(\omega)$, or $\mathrm{Re}\{1/\hat{\epsilon}(\omega)\}$, exhibit strong changes in this range.

Reflection range

The third frequency range, between $(\omega_0 + 1/\tau)$ and ω_p, is characterized by a high and almost frequency independent reflectivity; the phase angle, however, varies strongly, as displayed in Fig. 6.5. At the oscillator frequency ω_0 and at the plasma frequency ω_p the curvature of the reflectivity changes. We have already mentioned that the Drude model is a special case of the Lorentz model with $\omega_0 = 0$; hence the reflection range corresponds to the relaxation regime in the former limit. The conductivity drops as

$$\sigma_1(\omega) \propto (\omega\tau)^{-2} \qquad \text{and} \qquad \sigma_2(\omega) \propto (\omega\tau)^{-1} \quad . \tag{6.1.20}$$

Similar considerations hold for the dielectric constant:

$$\epsilon_1(\omega) \approx 1 - \frac{\omega_p^2}{\omega^2} \qquad \text{and} \qquad \epsilon_2(\omega) \approx \frac{\omega_p^2}{\omega^3 \tau} \quad ; \tag{6.1.21}$$

both decrease with increasing frequency; of course ϵ_1 is still negative.

It is only in this spectral range that the extinction coefficient k which describes the losses of the system is larger than the refractive index n (Fig. 6.4). Similarly, for the surface impedance, the absolute value of the surface reactance X_S becomes larger than the surface resistance R_S in the range above the center frequency of the oscillator ω_0 but still below the plasma frequency ω_p, as displayed in Fig. 6.6. The surface resistance exhibits a minimum in the range of high reflectivity because of the phase change between the electric and magnetic fields.

Transparent regime

Finally, at frequencies above ω_p, transmission is again important, for the same reasons as discussed in the case of the Drude model. Since k is small, the optical properties such as reflectivity or surface impedance are dominated by the behavior of $n(\omega)$. The high frequency dielectric constant $\epsilon_1(\omega \to \infty) = \epsilon_\infty$ approaches unity from below, thus the reflectivity drops to zero above the plasma frequency, and the material becomes transparent. The imaginary part of the energy loss function $1/\hat{\epsilon}(\omega)$ plotted in Fig. 6.7 is only sensitive to the plasma frequency, where it peaks. The real part

$$\text{Re}\left\{\frac{1}{\hat{\epsilon}(\omega)}\right\} = \frac{\epsilon_1(\omega)}{\epsilon_1^2(\omega) + \epsilon_2^2(\omega)}$$

shows a zero-crossing at ω_p and at ω_0.

As expected, all the sum rules derived in Section 3.2.2 also apply to the Lorentz model. If $1/\tau$ is small compared to ω_0, the spectral weight is obtained by substituting the expression (6.1.18b) of the imaginary part of the dielectric constant into Eq. (3.2.27):

$$\begin{aligned}
\int_0^\infty \omega\epsilon_2(\omega)\, d\omega &= \frac{\omega_p^2}{\tau} \int_0^\infty \frac{\omega^2}{(\omega_0 - \omega)^2(\omega_0 + \omega)^2 + \omega^2/\tau^2}\, d\omega \\
&\approx \frac{\omega_p^2}{\tau} \int_0^\infty \frac{d\omega}{4(\omega_0 - \omega)^2 + 1/\tau^2} \\
&= \frac{\omega_p^2 \tau}{\tau} \arctan\{2(\omega_0 - \omega)\tau\}\,\big|_0^\infty = \omega_p^2 \frac{\pi}{2} = \frac{2\pi^2 N e^2}{m} \quad .
\end{aligned}$$

Here N is the density of electrons in the valence band and m is the mass of the charge carriers, which may be replaced by the bandmass. The verification of the sum rule is somewhat more involved for $1/\tau$ comparable to ω_0, and in the limit $1/\tau \gg \omega_0$ we recover the sum rule as derived for the Drude model.

6.2 Direct transitions

6.2.1 General considerations on energy bands

Within the framework of the band theory of solids, insulators and semiconductors have a full band at zero temperature, called the valence band, separated by a single-particle gap $\mathcal{E}_g$ from the conduction band, which at $T = 0$ is empty. Because there is a full band and a gap, the static electrical conductivity $\sigma_{dc}(T = 0) = 0$, and optical absorption occurs only at finite frequencies. The valence and conduction bands are sketched in Fig. 6.8a in the so-called reduced Brillouin zone representation with two different dispersion relations displayed in the upper and lower parts of the figure. In general, excitation of an electron from the valence band to the conduction band (indicated by subscripts c and v, respectively) leads to an extra electron with momentum $\mathbf{k}'$ in the conduction band, leaving a hole with momentum $\mathbf{k}$ behind in the valence band. In a crystal, momentum conservation requires that

$$\mathbf{k}' = \mathbf{k} + \mathbf{K} \tag{6.2.1}$$

where $\mathbf{K}$ is the reciprocal lattice vector. Neglecting umklapp processes, we consider only vertical direct transitions, for which $\mathbf{k}' = \mathbf{k}$. Such transitions obey

$$\hbar\omega = \mathcal{E}_c(\mathbf{k}') - \mathcal{E}_v(\mathbf{k}) \tag{6.2.2}$$

due to the energy conservation, and are indicated in Fig. 6.8 by d. Of course these transitions appear to be vertical in this reduced zone representation where the dispersions at the higher Brillouin zone are folded back to the first zone; in the extended zone representation the electron gains the momentum $\mathbf{K}$ from the crystal lattice, in a fashion similar to that which leads to the band structure itself.

Processes where momentum is absorbed by the lattice vibrations during the creation of electrons and holes, and

$$\mathbf{k}' = \mathbf{k} + \mathbf{P} \quad , \tag{6.2.3}$$

with $\mathbf{P}$ the wavevector of the phonon involved, are also possible, and we refer to these processes as indirect transitions. With both direct and indirect transitions of importance, Fig. 6.8 suggests different scenarios for the onset of optical absorptions. In Fig. 6.8a, the lowest energy transition is a direct transition, and will dominate the onset of optical absorption. In contrast, the situation shown in

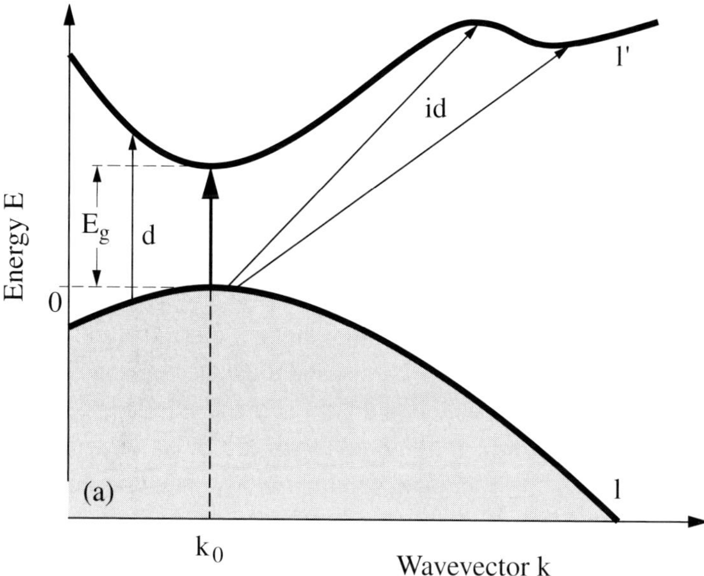

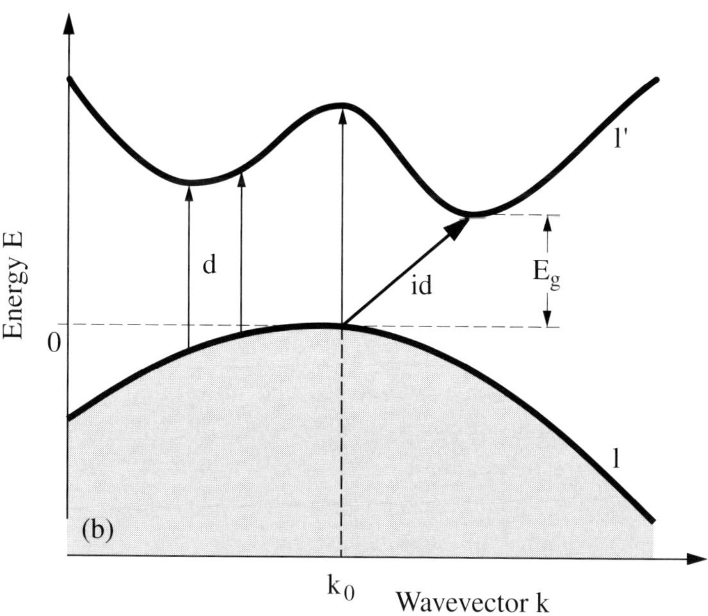

Fig. 6.8. Transitions between occupied bands l and unoccupied bands l' in a semiconductor, i.e. between the valence band and the conduction band. (a) Absorption edge corresponding to direct transition, d. If the minimum of the conduction band and the maximum of the valence band have the same **k** value, direct transitions are the lowest energy transitions. (b) Indirect transitions, id, when the extrema of the two bands are at different **k** values. For higher frequencies direct transitions also become possible.

Fig. 6.8b is an indirect transition for which

$$\hbar\omega = \mathcal{E}_{\mathrm{c}}(\mathbf{k}') - \mathcal{E}_{\mathrm{v}}(\mathbf{k}) \pm \hbar\omega_{\mathbf{P}} \quad , \tag{6.2.4}$$

where $\omega_{\mathbf{P}}$ is the frequency of the relevant phonon which determines the onset of optical absorption.

First let us consider direct transitions from the top of the valence band to the bottom of the conduction band as shown in Fig. 6.8a. In the simplest case these transitions determine the optical properties near to the bandgap. As expected, the solution of the problem involves two ingredients: the transition matrix element between the conduction and valence bands, and the density of states of these bands. As far as the transition rate is concerned, knowing the solution of the one-electron Schrödinger equation at these extreme points in the Brillouin zone, it is possible to obtain solutions in their immediate neighborhood by treating the scalar product $\mathbf{k} \cdot \mathbf{p}$ as a perturbation (see Appendix C) [Jon73, Woo72].

Some assumptions will be made concerning the density of states. We assume the energy band to be parabolic near the absorption edge, and the dispersion relations are given by

$$\mathcal{E}_{\mathrm{v}} = -\frac{\hbar^2 k^2}{2m_{\mathrm{v}}} \qquad \text{and} \qquad \mathcal{E}_{\mathrm{c}} = \mathcal{E}_{\mathrm{g}} + \frac{\hbar^2 k^2}{2m_{\mathrm{c}}} \quad , \tag{6.2.5}$$

where $\mathcal{E}_{\mathrm{g}}$ is the single-particle gap, and m_{v} and m_{c} refer to the bandmass in the valence and conduction bands, respectively. $\hbar\mathbf{k}$ is measured from the momentum which corresponds to the smallest gap value $\mathbf{k}_0$ in Fig. 6.8, and zero energy corresponds to the top of the valence band.

6.2.2 Transition rate and energy absorption for direct transitions

We first discuss the transition probabilities and absorption rates; subsequently the absorption rate is related to the complex conductivity and complex dielectric constant. We follow the procedure used in the previous treatment of the electrodynamics of metals.

The states in the valence and conduction bands are described by Bloch wavefunctions

$$\Psi_l \;=\; \Omega^{-1/2} \exp\{\mathrm{i}\mathbf{k} \cdot \mathbf{r}\} u_l(\mathbf{k}) \tag{6.2.6a}$$

$$\Psi_{l'} \;=\; \Omega^{-1/2} \exp\{\mathrm{i}\mathbf{k}' \cdot \mathbf{r}\} u_{l'}(\mathbf{k}') \quad . \tag{6.2.6b}$$

We use the indices l and l' to indicate that this applies generally to any interband transition; Ω refers to the volume element over which the integration is carried out. The Hamiltonian which describes the interaction of the electromagnetic field with the electronic states is given by Eq. (4.3.24), which, if $\mathbf{p} = -i\hbar\nabla$, can be rewritten as

$$\mathcal{H} = -\frac{ie\hbar}{2mc}\,(\mathbf{A}\cdot\nabla + \nabla\cdot\mathbf{A}) \quad . \tag{6.2.7}$$

We assume, for the sake of generality, that the vector has the following momentum dependence

$$\mathbf{A}(\mathbf{q}) = \mathbf{A}\exp\{i\mathbf{q}\cdot\mathbf{r}\} \quad ,$$

and we neglect the frequency dependence for the moment. The matrix element of the transition is

$$\mathcal{H}_{ll'}(\mathbf{r}) = \int \Psi_l^* \left[-\frac{ie\hbar}{2mc}\,(\mathbf{A}\exp\{i\mathbf{q}\cdot\mathbf{r}\}\cdot\nabla + \nabla\cdot\mathbf{A}\exp\{i\mathbf{q}\cdot\mathbf{r}\}) \right] \Psi_l\,d\mathbf{r} \quad . \tag{6.2.8}$$

By substituting the Bloch functions for the valence and conduction bands, the transition matrix element becomes

$$\mathcal{H}_{ll'}^{\text{int}}(\mathbf{r}) = -\frac{ie\hbar}{2mc}\frac{1}{\Omega}\int u_l^* \left[\mathbf{A}\cdot\nabla u_{l'} + iu_{l'}\mathbf{A}\cdot\mathbf{k}'\right]\exp\{i(\mathbf{k}'+\mathbf{q}-\mathbf{k})\cdot\mathbf{r}\}\,d\mathbf{r} \tag{6.2.9}$$

$$= -\frac{ie\hbar}{2mc}\frac{1}{\Delta}\sum_n \exp\{i(\mathbf{k}'+\mathbf{q}-\mathbf{k})\cdot\mathbf{R}_n\}\int_\Delta u_l^* \left[\mathbf{A}\cdot\nabla u_{l'} + iu_{l'}\mathbf{A}\cdot\mathbf{k}'\right]d\mathbf{r}$$

with m the electron mass and Δ the unit cell volume. For a periodic crystal structure

$$\exp\{i(\mathbf{k}'+\mathbf{q}-\mathbf{k})\cdot\mathbf{r}\} = \exp\{i(\mathbf{k}'+\mathbf{q}-\mathbf{k})\cdot(\mathbf{R}_n+\mathbf{r}')\} \quad ,$$

where $\mathbf{R}_n$ is the position of the nth unit cell. Here

$$\sum_n \exp\{i(\mathbf{k}'+\mathbf{q}-\mathbf{k})\cdot\mathbf{R}_n\} \approx 0$$

unless $(\mathbf{k}'+\mathbf{q}-\mathbf{k}) = \mathbf{K}$, where $\mathbf{K}$ is a reciprocal lattice vector. In the reduced zone scheme, we can take $\mathbf{K} = 0$, so that $\mathbf{k} = \mathbf{k}' + \mathbf{q}$. For direct or vertical transitions $\mathbf{k} = \mathbf{k}'$, and therefore $\mathbf{q} = 0$. Thus we obtain

$$\mathcal{H}_{ll'}^{\text{int}}(\mathbf{r}) = -\frac{ie\hbar}{2mc}\frac{1}{\Delta}\int_\Delta u_l^*(\mathbf{k})[\mathbf{A}\cdot\nabla u_{l'}(\mathbf{k}) + i\mathbf{A}\cdot\mathbf{k}u_{l'}(\mathbf{k})]\,d\mathbf{r} \quad . \tag{6.2.10}$$

For a pure semiconductor the second term is zero due to the orthogonality of the Bloch functions, and transitions associated with this matrix element are forbidden. Of course this restriction can be lifted by scattering processes (due to impurity scattering for example), and the optical properties when this occurs will be discussed later. In the majority of cases, the first term in the brackets dominates the absorption process, and we call the situation when this term is finite allowed transition [Coh88, Rid93].

By substituting the momentum operator

$$\mathbf{p}_{l'l} = \langle \mathbf{k}l'|\mathbf{p}|\mathbf{k}l\rangle_* = -\frac{i\hbar}{\Delta}\int_\Delta u_{l'}^*\nabla u_l\,\mathrm{d}\mathbf{r} \quad , \tag{6.2.11}$$

which is sometimes called the electric dipole transition matrix element, we obtain

$$H_{l'l}^{\mathrm{int}} = \frac{e}{mc}\mathbf{A}\cdot\mathbf{p}_{l'l} \quad ,$$

neglecting the forbidden transitions for the moment. The transition rate of an electron from state $|\mathbf{k}l\rangle$ to state $|\mathbf{k}l'\rangle$ can now be calculated by using Fermi's golden rule:

$$W_{l\to l'} = W_{ll'} = (2\pi/\hbar)|H_{ll'}^{\mathrm{int}}|^2\delta\{\mathcal{E}_{l'l} - \hbar\omega\} \quad . \tag{6.2.12}$$

Here we consider the number of transitions per unit time. These lead to absorption, and are therefore related to the real part of the conductivity $\sigma_1(\omega)$, which will be evaluated shortly. Subsequently we have to perform a Kramers–Kronig transformation to evaluate the imaginary part. To obtain the transition rate from the initial band l to band l', we have to integrate over all allowed values of $\mathbf{k}$, where we have $1/(2\pi)^3$ states per unit volume in each spin direction:

$$W_{ll'}(\omega) = \frac{\pi e^2}{2m^2\hbar c^2}\frac{2}{(2\pi)^3}\int_{\mathrm{BZ}}|\mathbf{A}\cdot\mathbf{p}_{l'l}|^2\delta\{\hbar\omega - \mathcal{E}_{l'l}\}\,\mathrm{d}\mathbf{k} \quad , \tag{6.2.13}$$

where the energy difference $\mathcal{E}_{l'} - \mathcal{E}_l = \mathcal{E}_{l'l} = \hbar\omega_{l'l}$.

Using the relation we derived in Section 2.3.1 for the absorbed power,

$$P = \sigma_1\langle\mathbf{E}^2\rangle = \hbar\omega W_{l'l} \quad ,$$

the absorption coefficient, which according to Eq. (2.3.26) is defined as the power absorbed in the unit volume divided by the energy density times the energy velocity, becomes

$$\alpha_{l'l}(\omega) = \frac{4\pi\hbar\omega_{l'l}W_{l'l}(\omega)}{nc\langle\mathbf{E}^2\rangle} \quad . \tag{6.2.14}$$

Substituting Eq. (6.2.13) into Eq. (6.2.14) we obtain our final result: the absorption coefficient corresponding to the transition from state l to state l',

$$\alpha_{l'l}(\omega) = \frac{e^2}{\pi ncm^2\omega}\int_{\mathrm{BZ}}|\mathbf{p}_{l'l}|^2\delta\{\hbar\omega - \hbar\omega_{l'l}\}\,\mathrm{d}\mathbf{k} \quad . \tag{6.2.15}$$

As discussed earlier, $\alpha_{l'l}(\omega)$ is related to the imaginary part of the refractive index. Note that in general $n = n(\omega)$, but with Eq. (2.3.26) (the expression which relates the refractive index to the conductivity) we can immediately write down the contribution of the transition rate between the l and l' bands to the conductivity as

$$\sigma_1(\omega) = \frac{c}{4\pi}\alpha_{l'l}(\omega)n(\omega) \quad . \tag{6.2.16}$$

The conductivity $\sigma_1(\omega)$ is proportional to the transition rate times the real part of the refractive index – which itself has to be evaluated in a selfconsistent way. Hence, the procedure is not trivial as $n(\omega)$ depends on both $\sigma_1(\omega)$ and $\sigma_2(\omega)$.

Next we consider the electrodynamics of semiconductors by starting from a somewhat different point. In Eq. (4.3.33) we arrived at a general formalism for the transverse dielectric constant which includes both intraband and interband transitions. The total complex dielectric constant is the sum of both processes:

$$\hat{\epsilon}(\mathbf{q}, \omega) = \hat{\epsilon}^{\text{inter}}(\mathbf{q}, \omega) + \hat{\epsilon}^{\text{intra}}(\mathbf{q}, \omega) \quad . \tag{6.2.17}$$

In Chapter 5, the discussion focused on the dielectric response due to intraband excitations; these are particularly important for metals. Now we discuss interband transitions due to direct excitations across the bandgap $\mathcal{E}_g$ of the semiconductor. We can split the dielectric constant into its real and imaginary parts following Eqs (4.3.34b) and (4.3.34a), and in the $\mathbf{q} = 0$ limit we arrive at

$$\epsilon_1(\omega) = 1 - \frac{4\pi}{\Omega}\frac{e^2}{m^2}\sum_{\mathbf{k}}\sum_{l,l'}\left[\frac{f^0(\mathcal{E}_{\mathbf{k}l})}{\mathcal{E}_{\mathbf{k}l} - \mathcal{E}_{\mathbf{k},l'}}\frac{1}{(\mathcal{E}_{\mathbf{k}l} - \mathcal{E}_{\mathbf{k},l'})^2 - \hbar^2\omega^2}\right]|\mathbf{p}_{l'l}|^2 \tag{6.2.18a}$$

$$\epsilon_2(\omega) = \frac{4\pi^2}{\Omega}\frac{e^2}{\omega^2 m^2}\sum_{\mathbf{k}}\sum_{l\neq l'}f^0(\mathcal{E}_{\mathbf{k}l})\delta\left\{\mathcal{E}_{\mathbf{k}l'} - \mathcal{E}_{\mathbf{k}l} - \hbar\omega\right\}|\mathbf{p}_{l'l}|^2 \quad , \tag{6.2.18b}$$

where we have used Eq. (6.2.11). At $T = 0$ the Fermi distribution becomes a step function; after replacing the sum over $\mathbf{k}$ by an integral over a surface of constant energy, the imaginary part of the dielectric constant has the form

$$\epsilon_2(\omega) = \frac{4\pi^2 e^2}{\omega^2 m^2}\sum_{l\neq l'}\frac{2}{(2\pi)^3}\int_{\hbar\omega=\mathcal{E}_{l'l}}\delta\left\{\mathcal{E}_{\mathbf{k}l'} - \mathcal{E}_{\mathbf{k}l} - \hbar\omega\right\}|\mathbf{p}_{l'l}|^2\,d\mathbf{k} \quad . \tag{6.2.19}$$

This expression can be cast into a form identical to Eq. (6.2.15) by noting that the energy difference $\mathcal{E}_{\mathbf{k}l'} - \mathcal{E}_{\mathbf{k}l} = \hbar\omega_{ll'}$, using the relationships $\epsilon_2 = 1 + 4\pi\sigma_1/\omega$ and $\alpha = 4\pi\sigma_1/(nc)$ between the dielectric constant, the conductivity, and the absorption coefficient.

6.3 Band structure effects and van Hove singularities

Besides the matrix element of the transition, the dielectric properties of semiconductors are determined by the electronic density in the valence and in the conduction bands. For a band l with the dispersion relation $\nabla_{\mathbf{k}}\mathcal{E}_l(\mathbf{k})$ the so-called density of states (DOS) is given by

$$D_l(\mathcal{E})\,d\mathcal{E} = \frac{2}{(2\pi)^3}\left(\int_{\mathcal{E}_l(\mathbf{k})=\text{const.}} \frac{dS_{\mathcal{E}}}{|\nabla_{\mathbf{k}}\mathcal{E}_l(\mathbf{k})|}\right)d\mathcal{E} \qquad (6.3.1)$$

with the factor of 2 referring to the two spin directions. The expression gets most contributions from states where the dispersion is flat, and $\nabla_{\mathbf{k}}\mathcal{E}_l(\mathbf{k})$ is small. For direct optical transitions between two bands with an energy difference $\mathcal{E}_{l'l} = \mathcal{E}_{l'} - \mathcal{E}_l$, however, a different density is relevant; this is the so-called combined or joint density of states, which is defined as

$$\begin{aligned} D_{l'l}(\hbar\omega) &= \frac{2}{(2\pi)^3}\int_{\text{BZ}} \delta\{\mathcal{E}_l(\mathbf{k}) - \mathcal{E}_{l'}(\mathbf{k}) - \hbar\omega\}\,d\mathbf{k} \\[2mm] &= \frac{2}{(2\pi)^3}\int_{\hbar\omega=\mathcal{E}_{l'l}} \frac{dS_{\mathcal{E}}}{|\nabla_{\mathbf{k}}[\mathcal{E}_{l'}(\mathbf{k}) - \mathcal{E}_l(\mathbf{k})]|} \quad . \end{aligned} \qquad (6.3.2)$$

The critical points, where the denominator in the expression becomes zero,

$$\nabla_{\mathbf{k}}[\mathcal{E}_{l'}(\mathbf{k}) - \mathcal{E}_l(\mathbf{k})] = \nabla_{\mathbf{k}}\mathcal{E}_{l'}(\mathbf{k}) - \nabla_{\mathbf{k}}\mathcal{E}_l(\mathbf{k}) = 0 \quad , \qquad (6.3.3)$$

are called van Hove singularities in the joint density of states. This is the case for photon energies $\hbar\omega_{l'l}$ for which the two energy bands separated by $\mathcal{E}_{l'l}$ are parallel at a particular $\mathbf{k}$ value.

Rewriting the integral in Eq. (6.2.19), we immediately see that the imaginary part of the dielectric constant is directly proportional to the joint density of states, and

$$\begin{aligned} \epsilon_2(\omega) &= \frac{4\pi e^2}{m^2\omega^2}\frac{2}{(2\pi)^3}\sum_{l,l'}\int_{\mathcal{E}_{l'l}=\text{const.}} \frac{dS_{\mathcal{E}}}{|\nabla_{\mathbf{k}}[\mathcal{E}_{l'}(\mathbf{k}) - \mathcal{E}_l(\mathbf{k})]|}|\mathbf{p}_{l'l}|^2 \\[2mm] &= \left(\frac{2\pi e}{m\omega}\right)^2\sum_{l,l'}D_{l'l}(\hbar\omega)|\mathbf{p}_{l'l}|^2 \quad . \end{aligned} \qquad (6.3.4)$$

The message of this expression is clear: the critical points – where the joint density of states peaks – determine the prominent structures in the imaginary part of the dielectric constant $\epsilon_2(\omega)$ and thus of the absorption coefficient.

Next we evaluate the optical parameters of semiconductors using previously derived expressions.

6.3.1 The dielectric constant below the bandgap

For a semiconductor in the absence of interactions, and also in the absence of lattice imperfections, at $T = 0$, there is no absorption for energies smaller than the bandgap. The conductivity σ_1, and therefore the absorption coefficient, are zero, but the dielectric constant ϵ_1 is finite, positive, and it is solely responsible for the optical properties in this regime. In the static limit, $\omega \to 0$, ϵ_1 can be calculated if the details of the dispersion relations of the valence and conduction bands are known. Let us first assume that we have two, infinitely narrow, bands with the maximum of the valence band at the same $\mathbf{k}$ position as the minimum of the conduction band, separated by the energy $\mathcal{E}_g = \hbar\omega_g$. In this case, from Eq. (6.1.11) we find

$$\epsilon_1(\omega = 0) = 1 + \frac{4\pi N e^2}{m\omega_g^2} = 1 + \left(\frac{\omega_p}{\omega_g}\right)^2 \qquad (6.3.5)$$

by virtue of the sum rule which is $\sum_l f_l = 1$. An identical result can be derived by starting from Eq. (6.2.19) and utilizing the Kramers–Kronig relation (3.2.31)

$$\epsilon_1(\omega = 0) = 1 + \frac{2}{\pi} \int_0^\infty \frac{\epsilon_2(\omega)}{\omega} \quad .$$

This also indicates that $\epsilon_1(\omega = 0)$ can, in general, be cast into the form of

$$\epsilon_1(\omega = 0) = 1 + C\left(\frac{\omega_p}{\omega_g}\right)^2 \quad , \qquad (6.3.6)$$

where C is a numerical constant less than unity which depends on the details of the joint density of states.

6.3.2 Absorption near to the band edge

If the valence band maximum and the conduction band minimum in a semiconductor have the same wavevector $\mathbf{k}$, as shown in Fig. 6.8a, the lowest energy absorption corresponds to vertical, so-called direct, transitions. Near to the band edge, for a photon energy $\mathcal{E} = \hbar\omega \geq \mathcal{E}_g$, the dispersion relations of the valence and conduction bands are given by Eqs (6.2.5), and we obtain

$$\mathcal{E}_{l'}(\mathbf{k}) - \mathcal{E}_l(\mathbf{k}) = \mathcal{E}_c(\mathbf{k}) - \mathcal{E}_v(\mathbf{k}) = \mathcal{E}_g + \frac{\hbar^2}{2}\left(\frac{1}{m_c} + \frac{1}{m_v}\right)k^2 = \mathcal{E}_g + \frac{\hbar^2 k^2}{2\mu} \quad , \quad (6.3.7)$$

where

$$\mu = \frac{m_c m_v}{m_c + m_v} \qquad (6.3.8)$$

is the reduced electron–hole mass. The analytical behavior of the joint density of states $D_{cv}(\hbar\omega)$ near a singularity may be found by expanding $\mathcal{E}_c(\mathbf{k}) - \mathcal{E}_v(\mathbf{k})$ in a Taylor series around the critical point of energy difference $\mathcal{E}_k(\mathbf{k}_0)$.

First let us consider the three-dimensional case. For simplicity we assume isotropic effective masses m_v and m_c in the valence and conduction bands. Consequently

$$\mathcal{E}_c(\mathbf{k}) - \mathcal{E}_v(\mathbf{k}) = \mathcal{E}_g + (\hbar^2/2\mu)|\mathbf{k} - \mathbf{k}_0|^2 \quad . \tag{6.3.9}$$

Such an expression should be appropriate close to $\mathbf{k}_0$, the momentum for which the band extrema occur. The joint density of states at the absorption edge can now be calculated by Eq. (6.3.2), and we obtain

$$D_{cv}(\hbar\omega) = \frac{1}{2\pi^2} \left(\frac{2\mu}{\hbar^2}\right)^{3/2} (\hbar\omega - \mathcal{E}_g)^{1/2} \tag{6.3.10}$$

for $\hbar\omega > \mathcal{E}_g$. This term determines the form of the absorption, and also $\epsilon_2(\omega)$ (or accordingly $\sigma_1(\omega)$) in the vicinity of the bandgap. For photon energies below the gap, $\epsilon_2(\omega) = 0$, i.e. the material is transparent. Substituting Eq. (6.3.10) into Eq. (6.3.4), we obtain for energies larger than the gap $\hbar\omega > \mathcal{E}_g$

$$\epsilon_2(\omega) = \frac{2e^2}{m^2\omega^2} \left(\frac{2\mu}{\hbar^2}\right)^{3/2} |\mathbf{p}_{cv}|^2 (\hbar\omega - \mathcal{E}_g)^{1/2} \quad . \tag{6.3.11}$$

Note that m is the mass of the charge carriers which are excited; this mass, but also the reduced electron–hole mass μ, enters into the expression of the dielectric constant. The relation for the real part of the dielectric constant $\epsilon_1(\omega)$ can be obtained by applying the Kramers–Kronig relation (3.2.12a); and we find

$$\epsilon_1(\omega) = 1 + \frac{2e^2}{m^2\omega^2} \left(\frac{2\mu}{\hbar^2}\right)^{3/2} |\mathbf{p}_{cv}|^2$$
$$\times \left[2\mathcal{E}_g - \mathcal{E}_g^{1/2}(\mathcal{E}_g + \hbar\omega)^{1/2} - \mathcal{E}_g^{1/2}(\mathcal{E}_g - \hbar\omega)^{1/2}\Theta\{\mathcal{E}_g - \hbar\omega\}\right] \tag{6.3.12}$$

with the Heaviside step function

$$\Theta\{x - x'\} = \begin{cases} 1 & \text{if } x > x' \\ 0 & \text{if } x < x' \end{cases} \quad .$$

Equation (6.3.12) holds in the vicinity of the bandgap. Due to the square root dependence of $\epsilon_2(\omega)$ above the energy gap, the real part of the dielectric constant $\epsilon_1(\omega)$ displays a maximum at $\omega_g = \mathcal{E}_g/\hbar$ and falls off as $(\mathcal{E}_g - \hbar\omega)^{1/2}$ at frequencies right below the gap. Both $\epsilon_1(\omega)$ and $\epsilon_2(\omega)$ are displayed in Fig. 6.9. The frequency dependence of the dielectric constant leads to a characteristic behavior of the reflectivity $R(\omega)$ in the frequency range of the bandgap as displayed in Fig. 6.10; most remarkable is the peak in the reflectivity at the gap frequency ω_g.

For crystals, in which the energy depends only on two components of the wavevector $\mathbf{k}$, say k_y and k_z, the expression of the two-dimensional joint density of

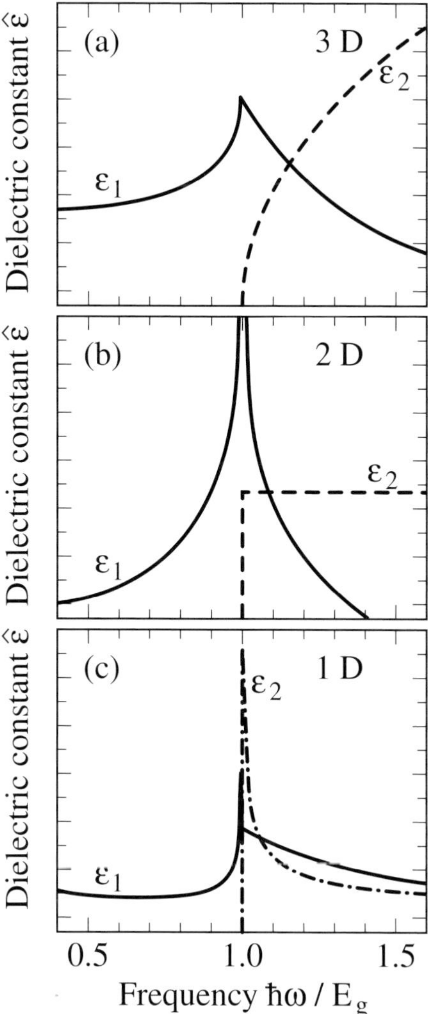

Fig. 6.9. Frequency dependence of the real and imaginary parts of the dielectric constant of a semiconductor, $\epsilon_1(\omega)$ and $\epsilon_2(\omega)$, near the bandgap with direct transition for three (3D), two (2D), and one (1D) dimensions.

states becomes

$$D_{cv}(\hbar\omega) = \frac{1}{2a\pi^2} \int_{BZ} \delta\{\mathcal{E}_c(k_y, k_z) - \mathcal{E}_v(k_y, k_z) - \hbar\omega\}\, d\mathbf{k} \quad , \qquad (6.3.13)$$

where BZ now indicates the two-dimensional Brillouin zone and $2\pi/a$ appears because of the integration involving k_x. At the band edge, D_{cv} is a step function of

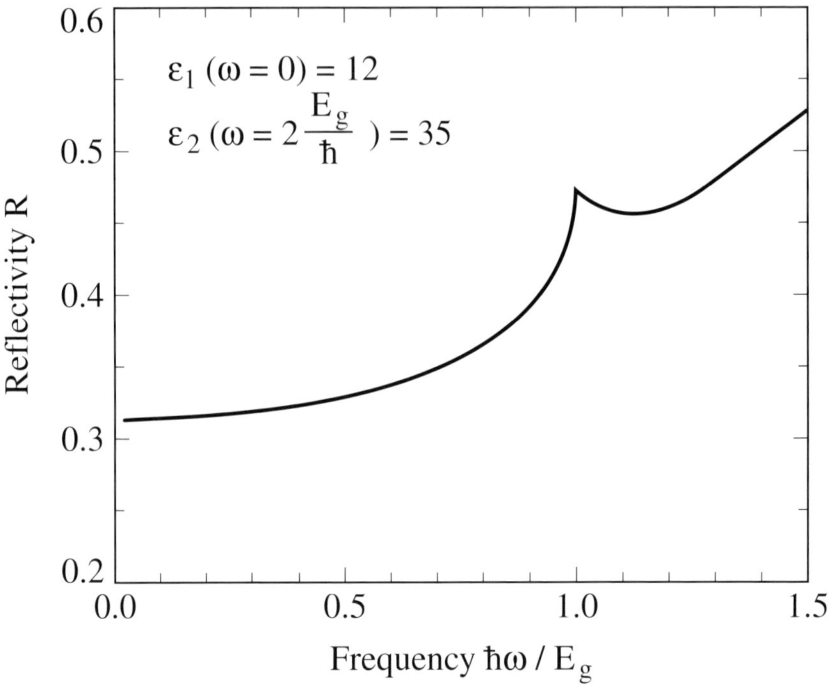

Fig. 6.10. Frequency dependent reflectivity $R(\omega)$ in the frequency range of the bandgap $\mathcal{E}_g$ calculated by using Eqs (6.3.11) and (6.3.12) with values typical for a semiconductor.

height $\mu/\pi\hbar^2$, and we find that, for $\hbar\omega > \mathcal{E}_g$,

$$\epsilon_2(\omega) = \frac{2\pi e^2}{m^2\omega^2}\left(\frac{2\mu}{\hbar^2}\right)|\mathbf{p}_{\mathrm{cv}}|^2 \quad . \tag{6.3.14}$$

In the vicinity of the bandgap, $1/\omega^2$ can be assumed constant, leading to a frequency independent absorption in two dimensions. The real part of the dielectric constant is

$$\epsilon_1(\omega) = 1 + \frac{2e^2}{m\omega^2}\left(\frac{2\mu}{\hbar^2}\right)|\mathbf{p}_{\mathrm{cv}}|^2\left|1 - \frac{\hbar^2\omega^2}{\mathcal{E}_g^2}\right| \tag{6.3.15}$$

for frequencies ω close to the gap ω_g.

If the energy depends only on one component of $\mathbf{k}$, the one-dimensional case, the arguments given above lead to a singularity of the density of states at the gap of the type $(\hbar\omega - \mathcal{E}_g)^{-1/2}$. For the imaginary part of the dielectric constant we find

$$\epsilon_2(\omega) = \frac{4\pi e^2}{m^2\omega^2}\left(\frac{2\mu}{\hbar^2}\right)^{1/2}|\mathbf{p}_{\mathrm{cv}}|^2(\hbar\omega - \mathcal{E}_g)^{-1/2} \tag{6.3.16}$$

for $\hbar\omega > \mathcal{E}_{\mathrm{g}}$, and for the real part of the dielectric constant we obtain

$$\epsilon_1(\omega) = 1 + \frac{e^2}{m^2\omega^2}\left(\frac{2\mu}{\hbar^2}\right)^{1/2}|\mathbf{p}_{\mathrm{cv}}|^2$$
$$\times \left[2\mathcal{E}_{\mathrm{g}}^{-1/2} - (\mathcal{E}_{\mathrm{g}} + \hbar\omega)^{-1/2} - (\mathcal{E}_{\mathrm{g}} - \hbar\omega)^{-1/2}\Theta\{\mathcal{E}_{\mathrm{g}} - \hbar\omega\}\right] \qquad (6.3.17)$$

for frequencies near the gap energy; these frequency dependences are displayed in Fig. 6.9. In plotting these frequency dependences, it can be assumed that the matrix element $|\mathbf{p}_{\mathrm{cv}}|$ is constant, i.e. independent of frequency – by no means an obvious assumption.

6.4 Indirect and forbidden transitions

The results obtained above are the consequence of particular features of the dispersion relations and of the transition rate – the two factors which determine the optical transition. We have assumed that the maximum in the valence band coincides – in the reduced zone scheme – with the minimum in the conduction band, and we have also assumed that the first term in Eq. (6.2.10), the term $u_l^*(\mathbf{k})\mathbf{A} \cdot \nabla u_{l'}(\mathbf{k})$, is finite. In fact, either of these assumptions may break down, leading to interband transitions with features different from those derived above.

6.4.1 Indirect transitions

In a large number of semiconductors, the energy maxima in the valence and minima in the conduction bands do not occur for the same momentum $\mathbf{k}$ and $\mathbf{k}'$, but for different momenta; a situation sketched in Fig. 6.8b. Optical direct transitions between these states cannot take place due to momentum conservation; however, such transitions become possible when the excitation of phonons is involved. The two ways in which such so-called indirect transitions can occur are indicated in Fig. 6.11. One scenario involves the creation of a photon for wavevector $\mathbf{k}$ and the subsequent phonon emission, which absorbs the energy and momentum, necessary to reach the conduction band at momentum $\mathbf{k}'$. An opposite process, in which a phonon-involved transition leads to a valence state at $\mathbf{k}'$, followed by a photon-involved vertical transition also contributes to the absorption process. In the two cases energy and momentum conservation requires that

$$\mathcal{E}_{\mathrm{g}} \pm \mathcal{E}_{\mathbf{P}} = \hbar\omega$$
$$\mathbf{k}_{l'} - \mathbf{k}_l = \pm\mathbf{P} \quad ,$$

where $\mathbf{P}$ is the phonon momentum, and $\mathcal{E}_{\mathbf{P}}$ is the energy of the phonon involved. As before, it is understood that we use the reduced zone scheme, with all of its

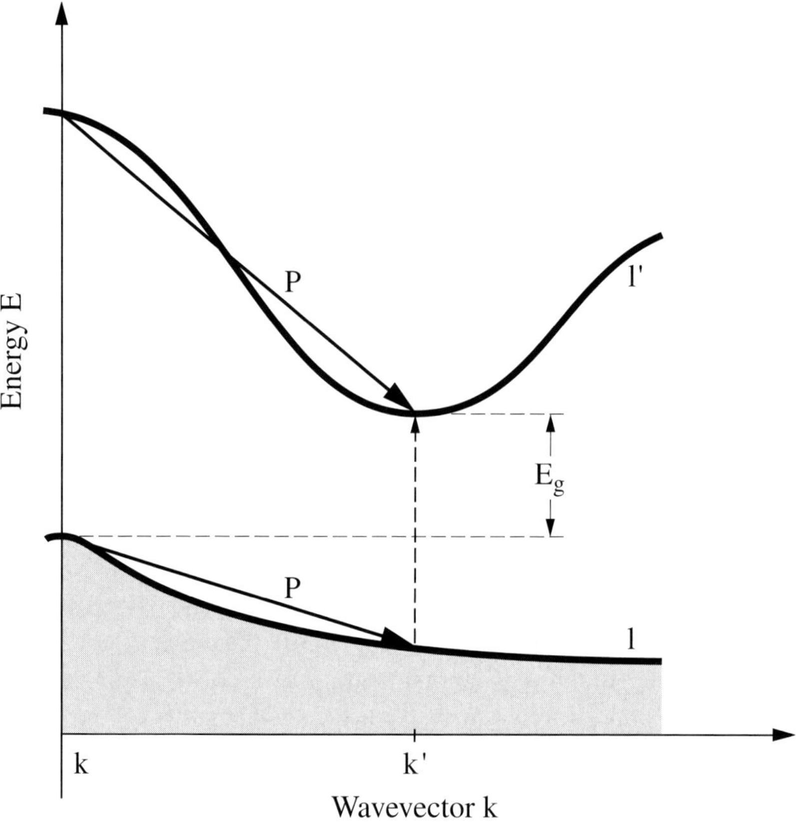

Fig. 6.11. Indirect transitions between the valence and conduction band states with different wavevectors. The dashed lines indicate the (vertical) transitions due to the interaction with photons, and the full straight lines indicate the transitions involving phonons with momenta $\mathbf{P} = \mathbf{k}' - \mathbf{k}$ with energy $\mathcal{E}_\mathbf{P}$. For both transitions $\mathcal{E}_\mathrm{g} \pm \mathcal{E}_\mathbf{P} = \hbar\omega$, the energy of the vertical transition.

implications. We can use second order perturbation theory for the transition rate to obtain

$$W_{(\mathbf{k},l) \to (\mathbf{k}+\mathbf{P},l')} = \frac{\pi e^2}{2m^2\hbar c^2} \left| \frac{\langle \mathbf{k}+\mathbf{q}, l' | V_q(\mathbf{P}, \mathbf{r}) | \mathbf{k}l'' \rangle N_\mathbf{P}^{1/2} \langle \mathbf{k}l'' | \mathbf{A} \cdot P_{l''l} | \mathbf{k}l' \rangle}{\mathcal{E}_{\mathbf{k}l''} - \mathcal{E}_{\mathbf{k}l} - \hbar\omega} \right|^2$$
$$\times \, \delta\{\mathcal{E}_{\mathbf{k}+\mathbf{P},l'} - \mathcal{E}_{\mathbf{k}l} - \hbar\omega - \hbar\omega_\mathbf{P}\} \quad , \tag{6.4.1}$$

where $N_\mathbf{P}$ denotes the phonon occupation number, and $\omega_\mathbf{P}$ is the frequency of the phonons involved. This in thermal equilibrium is given by the Bose–Einstein expression

$$N_{\mathbf{P}} = \frac{1}{\exp\left\{\frac{\hbar\omega_{\mathbf{P}}}{k_{\mathrm{B}}T}\right\} - 1} \quad . \tag{6.4.2}$$

In Eq. (6.4.1) the first angled bracket involves the transition associated with the phonons, and the second involves the vertical transition. Similar expressions hold for other transitions indicated in Fig. 6.11.

The matrix elements are usually assumed to be independent of the $\mathbf{k}$ states involved, and the transitions may occur from valence states with different momenta to the corresponding conduction states, again with different momentum states. Therefore we can simply sum up the contributions over the momenta $\mathbf{k}'$ and $\mathbf{k}$ in the delta function. Converting the summation to an integral as before, we obtain

$$W_{ll'}^{\mathrm{ind}}(\omega) \propto N_{\mathbf{P}} \int \int D_l(\mathcal{E}_l) D_{l'}(\mathcal{E}_{l'}) \delta\{\mathcal{E}_{l'} - \mathcal{E}_l - \hbar\omega \pm \hbar\omega_{\mathbf{P}}\} \, d\mathcal{E}_{l'} \, d\mathcal{E}_l \quad . \tag{6.4.3}$$

The two transitions indicated in Fig. 6.11 involve different phonons, and also different electron states in the valence and conduction bands. Unlike the case of direct vertical transitions, there is no restriction to the momenta $\mathbf{k}$ and $\mathbf{k}'$, as different phonons can absorb the momentum and energy required to make the transition possible. Consequently, the product $D_l(\mathcal{E}_l) \cdot D_{l'}(\mathcal{E}_{l'})$ – instead of the joint density of states $D_{l'l}(\mathcal{E})$ – appears in the integral describing the transition rate. This then leads to an energy dependence for the indirect optical transition which is different from the square root behavior we have found for the vertical transition (Eq. (6.3.11)), and calculations along these lines performed for the direct transition lead to

$$\epsilon_2(\omega) \propto N_{\mathbf{P}}(\mathcal{E}_{\mathrm{g}} - \mathcal{E}_{\mathbf{P}} \pm \hbar\omega)^2 \tag{6.4.4}$$

for $\hbar\omega \geq \mathcal{E}_{\mathrm{g}} \pm \mathcal{E}_{\mathbf{P}}$.

Such indirect transitions have several distinct characteristics when compared with direct transitions. First, the absorption coefficient increases with frequency as the square of the energy difference, in contrast to the square root dependence found for direct transitions. Second, the absorption is strongly temperature dependent, reflecting the phonon population factor $N_{\mathbf{P}}$; and third, because of the two processes for which the phonon energy appears in different combinations with the electronic energies, two onsets for absorption appearing at different frequencies are expected. All these make the distinction between direct and indirect optical transitions relatively straightforward.

6.4.2 *Forbidden transitions*

Let us now return to Eq. (6.2.10), which describes the transition between two bands. The first integral on the right hand side

$$\frac{ie\hbar}{2mc}\frac{1}{\Delta}\int_{\Delta} u_l^*(\mathbf{k})[\mathbf{A}\cdot\nabla u_{l'}(\mathbf{k})]\,\mathrm{d}\mathbf{r}$$

is involved in the direct transitions we discussed in Section 6.2.2. For a perfect semiconductor, for which all states in the valence and conduction bands can be described by Bloch functions, the second part of the Hamiltonian, i.e. the matrix element

$$\frac{ie\hbar}{2mc}\frac{1}{\Delta}\int_{\Delta} u_l^*(\mathbf{k})\,[\mathrm{i}\mathbf{A}\cdot\mathbf{k}\,u_{l'}(\mathbf{k})]\,\mathrm{d}\mathbf{r}\quad,$$

would be identical to zero simply because the wavefunctions are orthogonal in the different bands. Small changes in $\mathbf{k}$ (due to phonons or other scatterers) may, however, violate this strict rule and lead to a finite integral – and thus contribute to the optical transitions between the valence and conduction bands. Of course, the integral remains small compared to the integral which describes the direct transitions, and can be of importance only when the direct allowed transitions do not occur – due to the vanishingly small matrix element involved in these transitions. Let us assume that due to these scattering processes transitions between (slightly) different $\mathbf{k}$ states are possible, and expand the product for small wavevector differences $\mathbf{k}'' = \mathbf{k} - \mathbf{k}'$:

$$\int u_l^*(\mathbf{k}')u_{l'}(\mathbf{k})\,\mathrm{d}\mathbf{r} \approx \int u_l(\mathbf{k})\left[1 + \mathbf{k}''\cdot\nabla_{\mathbf{k}} + \frac{1}{2}(\mathbf{k}''\cdot\nabla_{\mathbf{k}})^2 + \cdots\right]u_{l'}^*(\mathbf{k})\,\mathrm{d}\mathbf{r}\quad.$$

$$(6.4.5)$$

The first term in the integral is zero; the second term is also zero if the allowed transition has a zero matrix element (see the first term in Eq. (6.2.10)) – as should be the case when forbidden transitions make the important contribution to the absorption; and the third term is proportional to $(\mathbf{k}'')^2$ and is therefore small. Thus, as a rule, forbidden transitions have significantly smaller matrix elements than direct transitions.

Calculation of the optical absorption for such higher order transitions proceeds along the lines we have followed before. The term introduces extra k^2 factors, and thus additional energy dependences; we find upon integration that

$$\epsilon_2(\omega) \propto \left(\hbar\omega - \mathcal{E}_{\mathrm{g}}\right)^{3/2}\quad,\qquad\qquad(6.4.6)$$

a functional form distinctively different from that which governs the direct absorption.

6.5 Excitons and impurity states

Finally we discuss optical transitions in which states brought about by Coulomb interactions are of importance. Electron–hole interactions lead to mobile states, called excitons, whereas Coulomb interactions between an impurity potential and electrons lead to bound states localized to the impurity sites. In both cases, these states can be described simply, by borrowing concepts developed for the energy levels of single atoms.

6.5.1 Excitons

When discussing the various optical transitions, we used the simple one-electron model, and have neglected the interaction between the electron in the conduction band and the hole remaining in the valence band. Coulomb interaction between these oppositely charged entities may lead to a bound state in a semiconductor, and such bound electron–hole pairs are called excitons. Such states will have an energy below the bottom of the conduction band, in the gap region of the semiconductor or insulator. As the net charge of the pair is zero, excitons obviously do not contribute to the dc conduction, but can be excited by an applied electromagnetic field, and will thus contribute to the optical absorption process.

For an isotropic semiconductor, the problem is much like the hydrogen atom problem, with the Coulomb interaction between the hole and the electron screened by the background dielectric constant ϵ_1 of the semiconductor. Both hole and electron can be represented by spherical wavefunctions, and the energy of the bound state is given by the usual Rydberg series,

$$\mathcal{E}_j^{\mathrm{exc}} = \mathcal{E}_{\mathrm{g}} - \frac{e^4 \mu}{2\hbar^2 \epsilon_1^2 j^2} \quad , \tag{6.5.1}$$

where $j = 1, 2, 3, \ldots$ and $\mu^{-1} = m_{\mathrm{v}}^{-1} + m_{\mathrm{c}}^{-1}$ denotes the effective mass we have encountered before. The low frequency dielectric constant is $\epsilon_1 = 1 + C \left(\hbar\omega_{\mathrm{p}}/\mathcal{E}_{\mathrm{g}}\right)^2$ with C a numerical constant of the order of one, as in Eq. (6.3.6). The spatial extension of this state can be estimated by the analogy to the hydrogen atom, and we find that

$$r_j^{\mathrm{exc}} = \frac{\epsilon_1 \hbar^2 j^2}{e^2 \mu} \quad .$$

For small bandgap semiconductors, the static dielectric constant ϵ_1 is large, and consequently the spatial extension of the excitons is also large. At the same time, according to Eq. (6.5.1), the energy of the excitons is small. This type of electron–hole bound state is called the Mott exciton and is distinguished from the strongly

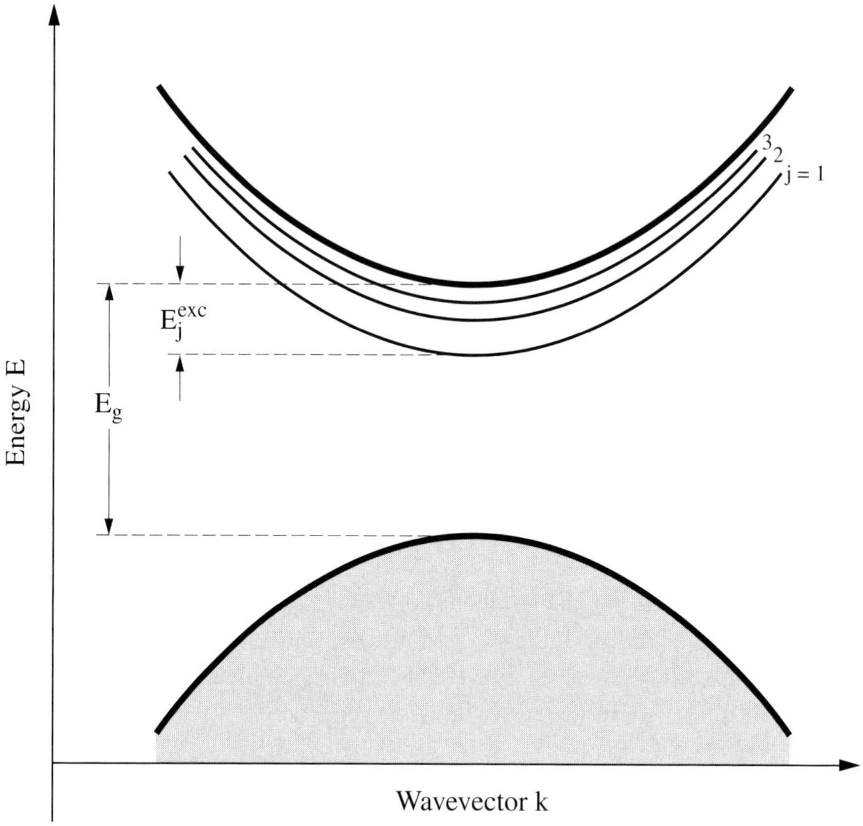

Fig. 6.12. Exciton energy levels for a direct gap semiconductor. In the absence of Coulomb interactions, transitions take place from the top of the valence band to the conduction band with the energy $\mathcal{E}_g$. With excitons transitions occur from the top of the valence band to the excitonic energy level $\mathcal{E}_j^{exc}$.

bound exciton which occurs in insulators with large energy gaps (and consequently small ϵ_1), the so-called Frenkel exciton.

The previous equation is for an exciton at rest. This is not necessarily the case, and the motion of the bound electron–hole pair can be decomposed into the relative motion of the particles about their center of mass, and the motion of the center of mass by itself. The Rydberg series describes the relative motion of the two particles. The motion of the center of mass is that of a free particle with a total mass of $m_c + m_v$, and in the effective mass approximation the kinetic energy associated with this translational motion is

$$\mathcal{E}_{kin}^{exc} = \frac{\hbar^2 \mathbf{K}^2}{2(m_c + m_v)} = \frac{\hbar^2 (\mathbf{k}_c + \mathbf{k}_v)^2}{2(m_c + m_v)} \ . \tag{6.5.2}$$

The energy levels of such exciton states are displayed in Fig. 6.12.

Here we are interested in the optical signature of these exciton states. Instead of free electrons and holes – the energy of creating these is the gap energy $\mathcal{E}_g$ or larger – the final state now corresponds to the (ideally) well defined, sharp, atomic like energy levels. Transitions to these levels then may lead to a series of well defined absorption peaks, each corresponding to the different j values in the Rydberg series. The energy separation between these levels is appreciable only if screening is weak, and this happens for semiconductors with large energy gaps. For small bandgap semiconductors, the different energy levels merge into a continuum, especially with increasing quantum numbers j. Transitions to this continuum of states – with energy close to $\mathcal{E}_g$ – lead to absorption near the gap, in addition to the absorption due to the creation of a hole in the valence band and an electron in the conduction band. The calculation of the transition probabilities is a complicated problem, beyond the scope of this book.

6.5.2 Impurity states in semiconductors

Impurity states, created by acceptor or donor impurities, and the extrinsic conductivity associated with these states (together with the properties of semiconductor junctions) have found an enormous number of applications in the semiconductor industry. Of interest to us are the different types of energy levels – often broadened into a band – which these impurity states lead to in the region of the energy gap. Various optical transitions between these energy levels and the conduction or valence bands are found, together with transitions between these impurity levels.

The impurity states are usually described in the following way. One atom in the crystal is replaced by an atom with a different charge – say a boron atom in a silicon crystal. It is regarded as an excess negative charge at the site of the doping and a positive charge loosely bound to this position (loosely because of the screening by the electrons in the underlying semiconductor). The model which can account for this, at least for an isotropic medium, is Bohr's model of an atom, and the energy levels are

$$\mathcal{E}_j^{\mathrm{imp}} = \frac{e^4 m}{2\hbar^2 \epsilon_1^2 j^2} \tag{6.5.3}$$

where $j = 1, 2, 3, \ldots$, m is the effective mass of electrons or holes, and ϵ_1 is the background dielectric constant; here the energy is counted from the top of the valence band. There is a mirror image of this in the case of doping with a donor; the energy levels as given by the above equation are then negative in sign, and are measured from the bottom of the conduction band. These states are shown in a schematic way in Fig. 6.13. The energy scales associated with impurity states are

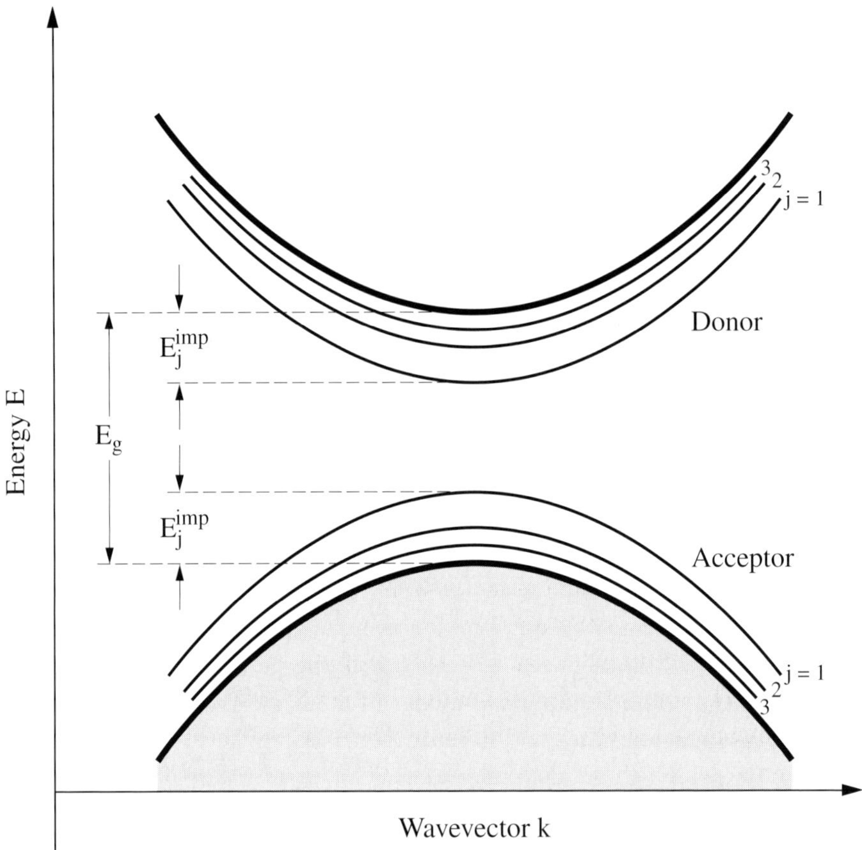

Fig. 6.13. Energy levels $\mathcal{E}_j^{imp}$ of impurity states in the vicinity of the valence and conduction bands. Donors lead to energy levels below the bottom of the conduction band; the impurity levels of acceptors are above the valence band.

small because of the small effective mass and also because of the large background dielectric constant. Typically for semiconductors such as silicon or germanium, they are of the order of 10 meV, significantly smaller than the single-particle gap (of the order of 1 eV). These states have also a spatial extension, significantly larger than the interatomic spacing, given by an expression similar to that for excitons:

$$r^{imp} = \frac{\epsilon_1 \hbar^2}{e^2 m}$$

for $j = 1$. At zero temperature the impurity states comprise an electron (or hole) weakly bound to the (oppositely charged) donor, and there is no dc conduction. Thermally excited hopping between these states, and also – because of the small energy scales involved – thermal ionization, leads to finite dc conduction, which increases strongly with increasing temperature. The various aspects of this so-called extrinsic conduction mechanism are well known and understood.

Electron states can also be delocalized by increasing the impurity concentration, and – because of the large spatial extension of the impurity states – the overlap of impurity wavefunctions, and thus delocalization occurs at low impurity concentrations. Such delocalization leads to a metallic impurity band, and this is a prototype example of (zero temperature) insulator-to-metal transition.

All this has important consequences on the optical properties of doped semiconductors. For small impurity concentrations, transitions from the valence band to the impurity levels, or from these levels to the continuum of states, are possible, leading to sharp absorption lines, corresponding to expression (6.5.3), together with transitions between states, corresponding to the different quantum numbers j. Again, the transition states between the continuum and impurity states are difficult to evaluate; but transitions between the impurity levels can be treated along the lines which have been developed to treat transitions between atomic energy levels. Interactions between impurities at finite temperatures all lead to broadening of the (initially sharp) optical absorption lines.

The situation is different when impurity states are from a delocalized, albeit narrow, band. Then what is used to account for the optical properties is the familiar Drude model, and we write

$$\sigma_1(\omega) = \frac{N_j e^2 \tau}{\epsilon_1 m_j} \frac{1}{1 + \omega^2 \tau^2} \tag{6.5.4}$$

where N_j is the number of impurity states, m_j is the mass of the impurity state, and τ is the relaxation time; ϵ_1 is again the background dielectric constant. This then leads to the so-called free-electron absorption together with a plasma edge at

$$\omega_p^* = \left(\frac{4\pi N_j e^2}{\epsilon_1 m_j} \right)^{1/2} \approx \left(\frac{N_j m}{N m_j} \right)^{1/2} \mathcal{E}_g \quad . \tag{6.5.5}$$

As $m \approx m_j$ and $N_j \ll N$, the plasma edge lies well in the gap region and is also dependent on the dopant concentration. The optical conductivity and reflectivity expected for a doped semiconductor (in the regime where the impurities form a band, and the impurity conductivity is finite at zero frequency) are displayed in Fig. 6.14; the parameters chosen are appropriate for a dc conductivity $\sigma_{dc} = 2 \times 10^4 \ \Omega^{-1} \ \mathrm{cm}^{-1}$ and a scattering rate $1/(2\pi \tau c) = 0.2 \ \mathrm{cm}^{-1}$; the energy gap between the conduction and valence bands is $v_g = \mathcal{E}_g / hc = 1000 \ \mathrm{cm}^{-1}$.

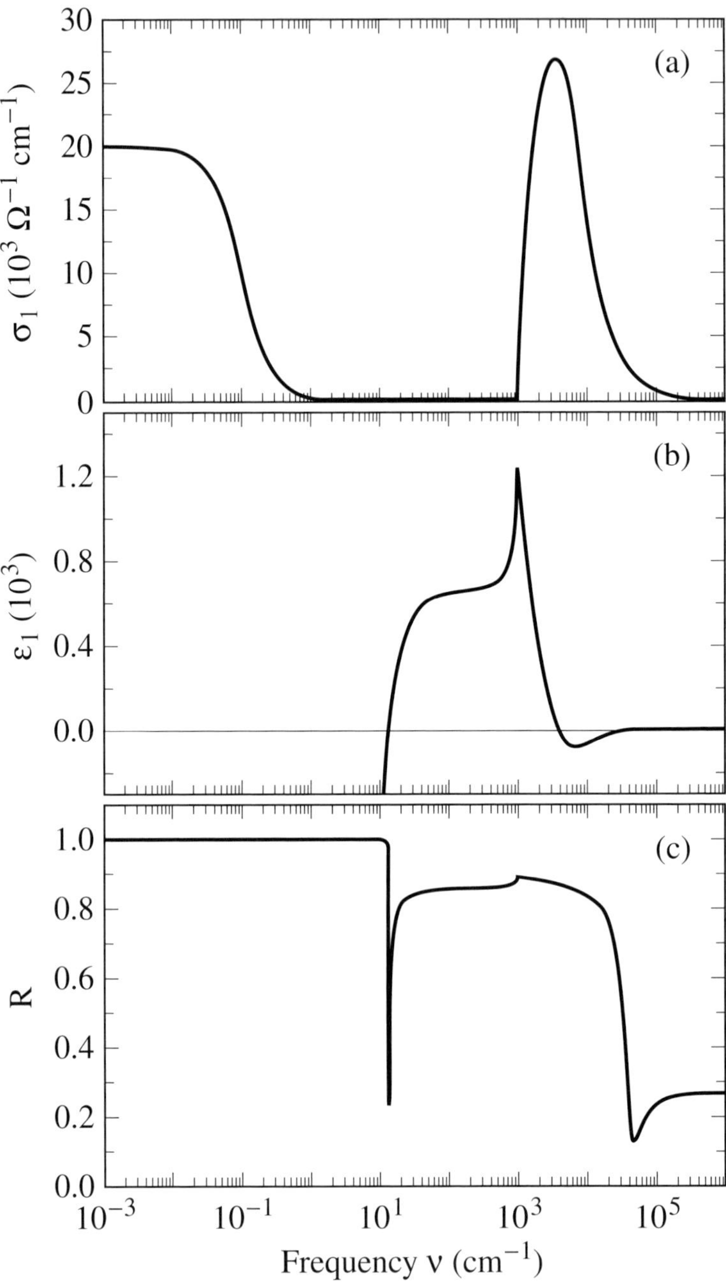

Fig. 6.14. (a) Optical conductivity $\sigma_1(\omega)$, (b) dielectric constant $\epsilon_1(\omega)$, and (c) reflectivity $R(\omega)$ for a doped semiconductor with an impurity band having finite dc conductivity $\sigma_{dc} = 2 \times 10^4 \ \Omega^{-1} \ cm^{-1}$ and scattering rate $1/(2\pi c\tau) = \gamma = 0.2 \ cm^{-1}$. The excitations across the energy gap $\mathcal{E}_g/hc = \omega_g/(2\pi c) = 1000 \ cm^{-1}$ are modeled according to Eq. (6.3.11) by a square root onset $\sigma_1(\omega) \propto (\omega - \omega_g)^{1/2}(\omega^2 + \omega_p^2)^{-1}$ with $\omega_p/(2\pi c) = 50\,000 \ cm^{-1}$.

6.6 The response for large ω and large q

Much of what has been said before refers to electronic transitions at energies near to the single-particle gap. Also, with one notable exception, namely indirect transitions, we have considered only transitions in the $\mathbf{q} = 0$ limit; a reasonable assumption for optical processes. The emergence of the gap, and in general band structure effects, lead to an optical response fundamentally different from optical properties of metals where, at least for simple metals, the band structure can be incorporated into parameters such as the bandmass m_{b} – leaving the overall qualitative picture unchanged.

Obviously, opening up a gap at the Fermi level has fundamental consequences as far as the low energy excitations are concerned; there is no dc conduction, but a finite, positive dielectric constant at zero frequency – and also at zero temperature – for example. It is expected, however, that such a (small) gap has little influence on excitations at large frequencies, and also with large momenta; and we should in this limit recover much of what has been said about the high ω and large $\mathbf{q}$ response of the metallic state. This is more than academic interest, as these excitations are readily accessible by optical and electron energy loss experiments – among other spectroscopic tools.

The evaluation of the full ω and $\mathbf{q}$ dependent response is complicated, and is discussed in several publications [Bas75, Cal59, Coh88]. Here we recall some results, first only for $\omega = 0$ and only for three-dimensional, isotropic semiconductors. Starting from Eq. (4.3.20) derived for the longitudinal response in the static limit ($\omega \to 0$), the complex dielectric constant reads

$$\hat{\epsilon}(\mathbf{q}) = 1 - \frac{4\pi e^2}{q^2 \Omega} \sum_{\mathbf{k}} \sum_{l,l'} \frac{f^0(\mathcal{E}_{\mathbf{k}+\mathbf{q},l'}) - f^0(\mathcal{E}_{\mathbf{k}l})}{\mathcal{E}_{\mathbf{k}+\mathbf{q},l'} - \mathcal{E}_{\mathbf{k}l}} |\langle \mathbf{k} + \mathbf{q}, l' | \exp\{i\mathbf{q} \cdot \mathbf{r}\} | \mathbf{k}l \rangle_* |^2 \quad .$$

$$(6.6.1)$$

Let us consider large $\mathbf{q}$ values first. Here the wavevector dependence of the dielectric constant can, in a first approximation, be described by a free-electron gas, with the bandgap and band structure effects, in general, of no importance. This approach leads to Eq. (5.4.18), and we find for the real part of the dielectric constant

$$\epsilon_1(\mathbf{q}) = 1 + \frac{3}{8} \left(\frac{\hbar \omega_{\mathrm{p}} k_{\mathrm{F}}}{q \mathcal{E}_{\mathrm{F}}} \right)^2 \left[1 + \left(\frac{k_{\mathrm{F}}}{q} - \frac{q}{4 k_{\mathrm{F}}} \right) \ln \left| \frac{q + 2 k_{\mathrm{F}}}{q - 2 k_{\mathrm{F}}} \right| \right] \quad . \qquad (6.6.2)$$

For large $\mathbf{q}$

$$\epsilon_1(\mathbf{q}) = 1 + \left(\frac{\hbar \omega_{\mathrm{p}}}{\mathcal{E}_{\mathrm{g}}} \right)^2 \left(\frac{k_{\mathrm{F}}}{q} \right)^2 \qquad (6.6.3)$$

approaches the free-electron behavior that we recovered for simple metals in Chapter 5.

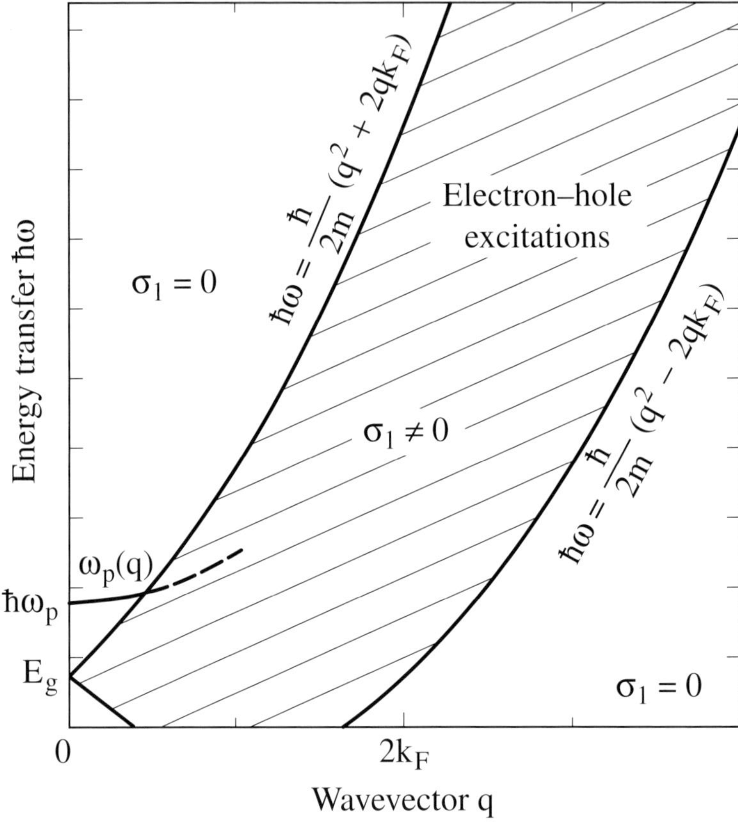

Fig. 6.15. Excitation spectrum of a three-dimensional semiconductor. The pair excitations fall within the shaded area. In the region $\hbar\omega > \mathcal{E}_\mathrm{g} + (\hbar^2/2m)(2k_\mathrm{F} + q)q$, the absorption vanishes since $\hbar\omega$ is larger than the energy difference possible. Also for $\hbar\omega < \mathcal{E}_\mathrm{g} + (\hbar^2/2m)(q - 2k_\mathrm{F})q$ we find $\sigma_1 = 0$.

Next, let us examine what happens at high frequencies, in the $\mathbf{q} = 0$ limit; i.e. at frequencies larger than the spectral range where band structure effects are of importance. In this limit, the effects related to the density of states can be neglected, and the inertial response of the electron gas is responsible for the optical response. The Lorentz model in the limit $\omega \gg \omega_0$ is an appropriate guide to what happens. First, we recover a high frequency roll-off for the conductivity $\sigma_1(\omega)$, and

$$\sigma_1(\omega) \approx \frac{\omega_\mathrm{p}^2}{4\pi} \frac{\tau}{\omega^2 \tau^2} \quad . \tag{6.6.4}$$

Second, there is a zero-crossing of the dielectric constant at frequency $\omega = \omega_\mathrm{p}$; this has two important consequences that we have already encountered for metals, namely that ω_p is the measure of the onset of transparency (here also just as in the metallic state), and $\epsilon_1 = 0$ at this frequency leads to plasma oscillations. The arguments which lead to these consequences are identical to those we have advanced in Chapter 5.

Finally let us sketch and discuss the excitation spectrum for finite ω and $\mathbf{q}$ – assuming a hypothetical small and isotropic gap. As pointed out, for small bandgaps $\mathcal{E}_\mathrm{g}$, the existence of the single-particle gap modifies the low energy part of the excitation spectrum; for $\mathcal{E}_\mathrm{g} \ll \hbar\omega_\mathrm{p}$ – the situation we usually encounter in semiconductors – the high energy single-particle and collective excitations are only slightly modified. Due to the single-particle gap, electron–hole excitations require a minimum energy $\mathcal{E}_\mathrm{g}$ for zero momentum. This leads to an upward displacement of the excitation spectrum of the electron gas, the situation appropriate for a small bandgap semiconductor (see Fig. 6.15). As for metals, we recover the region of single-particle electron–hole excitations, together with the finite plasma frequency at $\omega_\mathrm{p} = \left(4\pi N e^2/m_\mathrm{b}\right)^{1/2}$, which is in general significantly larger than the bandgap $\mathcal{E}_\mathrm{g}$.

Of course, the situation becomes dramatically different when the gap becomes comparable to the plasma frequency; the situation which may then occur is wide bandgap insulation. The situation can, at least qualitatively, be examined by utilizing the Lorentz model in the $\omega_0 \gg \omega_\mathrm{p}$ limit; this will give some insight into the $\mathbf{q} = 0$ limit.

References

[Bas75] G.F. Bassani and G. Pastori Parravicini, *Electronic States and Optical Transitions in Solids* (Pergamon Press, Oxford, 1975)

[Cal59] J. Callaway, *Phys. Rev.* **116**, 1638 (1959)

[Coh88] M.L. Cohen and J.R. Chelikowsky, *Electronic Structure and Optical Properties of Semiconductors* (Springer-Verlag, Berlin, 1988)

[Hau94] H. Haug and S.W. Koch, *Quantum Theory of the Optical and Electronic Properties of Semiconductors*, 3rd edition (World Scientific, Singapore, 1994)

[Jon73] W. Jones and N.H. March, *Theoretical Solid State Physics* (John Wiley & Sons, New York, 1973)

[Rid93] B.K. Ridley, *Quantum Processes in Semiconductors*, 3rd edition (Clarendon Press, Oxford, 1993)

[Woo72] F. Wooten, *Optical Properties of Solids* (Adademic Press, San Diego, CA, 1972)

[Yu96] P.Y. Yu and M. Cardona, *Fundamentals of Semiconductors* (Springer-Verlag, Berlin, 1996)

Further reading

[Car68] M. Cardona, *Electronic Optical Properties of Solids*, in: *Solid State Physics, Nuclear Physics, and Particle Physics*, edited by I. Saavedra (Benjamin, New York, 1968), p. 737

[Cha83] G.W. Chantry, *Properties of Dielectric Materials*, in: *Infrared and Millimeter Waves*, Vol. 8, edited by K.J. Button (Academic Press, New York, 1983)

[Coh88] M.L. Cohen and J.R. Chelikowsky, *Electronic Structure and Optical Properties of Semiconductors* (Springer-Verlag, Berlin, 1988)

[Gre68] D.L. Greenaway and G. Harbeke, *Optical Properties and Band Structure of Semiconductors* (Pergamon Press, Oxford, 1968)

[Har72] G. Harbeke, in: *Optical Properties of Solids*, edited by F. Abelès (North-Holland, Amsterdam, 1972)

[Kli95] C.F. Klingshirn, *Semiconductor Optics* (Springer-Verlag, Berlin, 1995)

[Pan71] J.I. Pankove, *Optical Processes in Semiconductors* (Prentice-Hall, Englewood Cliffs, NJ, 1971)

[Wal86] R.F. Wallis and M. Balkanski, *Many-Body Aspects of Solid State Spectroscopy* (North-Holland, Amsterdam, 1986)

7

Broken symmetry states of metals

The role of electron–electron and electron–phonon interactions in renormalizing the Fermi-liquid state has been mentioned earlier. These interactions may also lead to a variety of so-called broken symmetry ground states, of which the superconducting ground state is the best known and most studied. The ground states are superpositions of electron–electron or electron–hole pairs all in the same quantum state with total momenta of zero or $2\mathbf{k}_F$; these are the Cooper pairs for the superconducting case. There is an energy gap Δ, the well known BCS gap, introduced by Bardeen, Cooper, and Schrieffer [Bar57], which separates the ground state from the single-particle excitations. The states develop with decreasing temperature as the consequence of a second order phase transition.

After a short review of the various ground states, the collective modes and their response will be discussed. The order parameter is complex and can be written as $\Delta \exp\{i\phi\}$; the phase plays an important role in the electrodynamics of the ground state. Many aspects of the various broken symmetry states are common, but the distinct symmetries also lead to important differences in the optical properties. The absorption induced by an external probe will then be considered; it is usually discussed in terms of the so-called coherence effects, which played an important role in the early confirmation of the BCS theory. Although these effects are in general discussed in relation to the nuclear magnetic relaxation rate and ultrasonic attenuation, the electromagnetic absorption also reflects these coherence features, which are different for the various broken symmetry ground states. As usual, second quantized formalism, as introduced in Section 4.2, is used to describe these effects, and we review what is called the weak coupling theory of these ground states.

7.1 Superconducting and density wave states

The various ground states of the electron gas are, as a rule, discussed using second quantized formalism, and this route is followed here. The kinetic energy of the

173

Table 7.1. *Various broken symmetry ground states of one-dimensional metals.*

	Pairing	Total spin	Total momentum	Broken symmetry
Singlet superconductor	electron–electron $e_+, \sigma; e_-, -\sigma$	$S = 0$	$q = 0$	gauge
Triplet superconductor	electron–electron $e_+, \sigma; e_-, \sigma$	$S = 1$	$q = 0$	gauge
Spin density wave	electron–hole $e_+, \sigma; h_-, -\sigma$	$S = 1$	$q = 2k_\mathrm{F}$	translational
Charge density wave	electron–hole $e_+, \sigma; h_-, \sigma$	$S = 0$	$q = 2k_\mathrm{F}$	translational

electron gas is

$$\mathcal{H}_{\mathrm{kin}} = \int d\mathbf{r}\, \Psi^*(\mathbf{r}) \frac{\mathbf{p}^2}{2m} \Psi(\mathbf{r}) = \sum_{\mathbf{k},\sigma} \frac{\hbar^2 \mathbf{k}^2}{2m} a^+_{\mathbf{k},\sigma} a_{\mathbf{k},\sigma} = \sum_{\mathbf{k},\sigma} \mathcal{E}_\mathbf{k} a^+_{\mathbf{k},\sigma} a_{\mathbf{k},\sigma} \qquad (7.1.1)$$

in terms of the creation and elimination operators defined in Section 4.2. Next we describe the interaction between the electrons; this, in its general form, is given by

$$\begin{aligned}
\mathcal{H}_{\mathrm{ee}} &= \int d\mathbf{r}\, d\mathbf{r}'\, N(\mathbf{r}) V(\mathbf{r}, \mathbf{r}') N(\mathbf{r}') = \int d\mathbf{r}\, d\mathbf{r}'\, \Psi^*(\mathbf{r}) \Psi(\mathbf{r}) V(\mathbf{r}, \mathbf{r}') \Psi^*(\mathbf{r}') \Psi(\mathbf{r}') \\
&= \sum_{\mathbf{k},\mathbf{k}',l,l',\sigma,\sigma'} V_{\mathbf{k},\mathbf{k}',l,l'} a^+_{\mathbf{k},\sigma} a^+_{\mathbf{k}',\sigma'} a_{l,\sigma} a_{l',\sigma'} \quad ;
\end{aligned}$$

where N is the particle density, and V denotes the potential energy due to the interaction. With these – the kinetic and the interaction – terms, the pairing Hamiltonian is cast into the form

$$\mathcal{H} = \sum_{\mathbf{k},\sigma} \mathcal{E}_k c^+_{\mathbf{k},\sigma} c_{\mathbf{k},\sigma} + \sum_{\mathbf{k},\mathbf{k}',l,l',\sigma,\sigma'} V_{\mathbf{k},\mathbf{k}',l,l'} a^+_{\mathbf{k},\sigma} a^+_{\mathbf{k}',\sigma'} a_{l,\sigma} a_{l',\sigma'} \quad . \qquad (7.1.2)$$

In order to see its consequences, the interaction term has to be specified to include only terms which lead to – in the spirit of the BCS theory – formation of electron (or hole) pairs. For the term

$$\sum_{\mathbf{k},l,\sigma} V_{\mathbf{k},l} a^+_{\mathbf{k},\sigma} a^+_{-\mathbf{k},\sigma} a_{-l,-\sigma} a_{l,\sigma} \quad , \qquad (7.1.3)$$

for example, electron pairs are formed with total momentum $\mathbf{q} = 0$ and total spin $\mathbf{S} = 0$, the well known Cooper pairs. This, however, is not the only possibility. The situation is simple in the case of a one-dimensional metal where the Fermi surface consists of two points at $\mathbf{k}_F$ and $-\mathbf{k}_F$. Then the following pair formations can occur:

$$
\begin{array}{llll}
\mathbf{k}' = -\mathbf{k} & \mathbf{l}' = -\mathbf{l} & \sigma' = -\sigma & \text{singlet superconductor} \\
\mathbf{k}' = -\mathbf{k} & \mathbf{l}' = -\mathbf{l} & \sigma' = \sigma & \text{triplet superconductor} \\
\mathbf{k}' = \mathbf{k} - 2\mathbf{k}_F & \mathbf{l}' = \mathbf{l} - 2\mathbf{k}_F & \sigma' = -\sigma & \text{spin density wave} \\
\mathbf{k}' = \mathbf{k} - 2\mathbf{k}_F & \mathbf{l}' = \mathbf{l} - 2\mathbf{k}_F & \sigma' = \sigma & \text{charge density wave}
\end{array}
$$

The first two of these states develop in response to the interaction $V_{\mathbf{k},l} = V_{\mathbf{q}}$ for which $\mathbf{q} = 0$; this is called the particle–particle or Cooper channel. The resulting ground states are the well known (singlet or triplet) superconducting states of metals and will be discussed shortly. The last two states, with a finite total momentum for the pairs, develop as a consequence of the divergence of the fluctuations at $\mathbf{q} = 2\mathbf{k}_F$ (Table 7.1); this is the particle–hole channel, usually called the Peierls channel. For these states one finds a periodic variation of the charge density or spin density, and consequently they are called the charge density wave (CDW) and spin density wave (SDW) ground states. In the charge density wave ground state

$$
\Delta\rho = \rho_1 \cos\{2\mathbf{k}_F \cdot \mathbf{r} + \phi\} \tag{7.1.4}
$$

where ρ_1 is the amplitude of the charge density. In the spin density wave case, the spin density has a periodic spatial variation

$$
\Delta\mathbf{S} = \mathbf{S}_1 \cos\{2\mathbf{k}_F \cdot \mathbf{r} + \phi\} \quad . \tag{7.1.5}
$$

Throughout this chapter we are concerned with density waves where the period $\lambda_{\mathrm{DW}} = \pi/|\mathbf{Q}|$ is not a simple multiple of the lattice translation vector $\mathbf{R}$, and therefore the density wave is incommensurate with the underlying lattice. Commensurate density waves do not display many of the interesting phenomena discussed here, as the condensate is tied to the lattice and the phase of the ground state wavefunction does not play a role.

A few words about dimensionality effects are in order. The pairing, which leads to the density wave states with electron and hole states differing by $2\mathbf{k}_F$, is the consequence of the Fermi surface being two parallel sheets in one dimension. The result of this nesting is the divergence of the response function $\hat{\chi}(\mathbf{q}, T)$ at $\mathbf{Q} = 2\mathbf{k}_F$ at zero temperature as displayed in Fig. 5.14. Parallel sheets of the Fermi surface may occur also in higher dimensions, and this could lead to density wave formation, with a wavevector $\mathbf{Q}$ related to the Fermi-surface topology. In this case the charge or spin density has a spatial variation $\cos\{\mathbf{Q} \cdot \mathbf{r} + \phi\}$.

We will outline the solution which is obtained for Cooper pairs, i.e. pairs with total momentum zero and with spin zero. The Hamiltonian in this case is

$$\mathcal{H} = \sum_{\mathbf{k},\sigma} \mathcal{E}_{\mathbf{k}} a^{+}_{\mathbf{k},\sigma} a_{\mathbf{k},\sigma} + \sum_{\mathbf{k},\mathbf{l},\sigma} V_{\mathbf{k},\mathbf{l}} \, a^{+}_{\mathbf{k},\sigma} a^{+}_{-\mathbf{k},-\sigma} a_{-\mathbf{l},-\sigma} a_{\mathbf{l},\sigma} \quad . \tag{7.1.6}$$

It is modified by seeking a mean field solution; this is done by assuming that $a_{-\mathbf{k},-\sigma} a_{\mathbf{k},\sigma}$ have non-zero expectation values, but that fluctuations away from average are small. We write formally

$$a_{-\mathbf{k},-\sigma} a_{\mathbf{k},\sigma} = b_{\mathbf{k}} + (a_{-\mathbf{k},-\sigma} a_{\mathbf{k},\sigma} - b_{\mathbf{k}})$$

and neglect bilinear terms. Inserting these into the original Hamiltonian results in

$$\mathcal{H} = \sum_{\mathbf{k},\sigma} \mathcal{E}_{\mathbf{k}} a^{+}_{\mathbf{k},\sigma} a_{\mathbf{k},\sigma} + \sum_{\mathbf{k},\mathbf{l}} V_{\mathbf{k},\mathbf{l}} (a^{+}_{\mathbf{k},\sigma} a^{+}_{-\mathbf{k},-\sigma} b_{\mathbf{l}} + b^{+}_{\mathbf{k}} a_{-\mathbf{l},-\sigma}, a_{\mathbf{l},\sigma} - b^{+}_{\mathbf{k}} b_{\mathbf{l}}) \quad , \tag{7.1.7}$$

with

$$b_{\mathbf{k}} = \langle a_{-\mathbf{k},-\sigma}, a_{\mathbf{k},\sigma} \rangle_{\text{th}}$$

where $\langle \ \rangle_{\text{th}}$ denotes the thermal average. We also introduce the notation

$$\Delta_{\mathbf{k}} = - \sum_{\mathbf{l}} V_{\mathbf{k},\mathbf{l}} \langle a_{-\mathbf{l},-\sigma}, a_{\mathbf{l},\sigma} \rangle_{\text{th}} \quad , \tag{7.1.8}$$

a complex gap, as we will see later. With this we have

$$\mathcal{H} = \sum_{\mathbf{k},\sigma} \mathcal{E}_{\mathbf{k}} a^{+}_{\mathbf{k},\sigma} a_{\mathbf{k},\sigma} - \sum_{\mathbf{k}} \left(\Delta_{\mathbf{k}} a^{+}_{\mathbf{k},\sigma} a^{+}_{-\mathbf{k},-\sigma} + \Delta^{*}_{\mathbf{k}} a_{-\mathbf{k},-\sigma} a_{\mathbf{k},\sigma} - \Delta_{\mathbf{k}} b^{+}_{\mathbf{k}} \right) \quad . \tag{7.1.9}$$

We intend to find a solution by diagonalization: we introduce new states – which will be the quasi-particle states – which are related to our original states by a linear transformation, and require that the Hamiltonian is diagonal with respect to these new states. We write

$$a_{\mathbf{k},\sigma} = u^{*}_{\mathbf{k}} \gamma_{\mathbf{k},0} + v_{\mathbf{k}} \gamma^{+}_{\mathbf{k},1} \tag{7.1.10a}$$

$$a^{+}_{-\mathbf{k},-\sigma} = v^{*}_{\mathbf{k}} \gamma_{\mathbf{k},0} + u_{\mathbf{k}} \gamma^{+}_{\mathbf{k},1} \quad , \tag{7.1.10b}$$

where $|u_{\mathbf{k}}|^2 + |v_{\mathbf{k}}|^2 = 1$. We insert the above into the Hamiltonian and require that terms which would not lead to a diagonalization, the terms of $\gamma_{\mathbf{k}} \gamma_{\mathbf{l}}$ or $\gamma_{\mathbf{l}} \gamma_{\mathbf{k}}$ in abbreviated notation, are zero. The end results are as follows:

$$1 - |u_{\mathbf{k}}|^2 = |v_{\mathbf{k}}|^2 = \frac{1}{2} \left(1 - \frac{\zeta_{\mathbf{k}}}{\mathcal{E}_{\mathbf{k}}} \right) \quad ; \tag{7.1.11}$$

the energies of the quasi-particle excitations are

$$\mathcal{E}_{\mathbf{k}} = \left(|\Delta_{\mathbf{k}}|^2 + \zeta^2_{\mathbf{k}} \right)^{1/2} \quad . \tag{7.1.12}$$

There is an energy minimum (for $\zeta_{\mathbf{k}} = 0$) for the excitation of the quasi-particles; this is the well known superconducting gap. We will assume that $\Delta_{\mathbf{k}}$ is real. This is not a solution yet, but we require selfconsistency by using the above coefficients to write $\Delta_{\mathbf{k}}$ in Eq. (7.1.8), and using the Fermi distribution function

$$f(\mathcal{E}_{\mathbf{k}}) = \left(\exp\left\{ \frac{\mathcal{E}_{\mathbf{k}} - \mathcal{E}_{\mathrm{F}}}{k_{\mathrm{B}}T} \right\} + 1 \right)^{-1}$$

to describe their population probability at a temperature T. This leads to the so-called gap equation:

$$\Delta_{\mathbf{k}} = -\sum_{\mathbf{l}} V_{\mathbf{k},\mathbf{l}} \Delta_{\mathbf{l}} \frac{1 - 2f(\mathcal{E}_{\mathbf{l}})}{2\mathcal{E}_{\mathbf{l}}} = -\sum_{\mathbf{l}} V_{\mathbf{k},\mathbf{l}} \Delta_{\mathbf{l}} \tanh\left\{ \frac{\mathcal{E}_{\mathbf{l}}}{2k_{\mathrm{B}}T} \right\} \quad . \tag{7.1.13}$$

We assume that $V_{\mathbf{k},\mathbf{l}}$ is a constant up to a cutoff energy $\mathcal{E}_{\mathrm{c}}$ and is zero above this energy; this yields, by converting the summation over $\mathbf{l}$ to an integral,

$$1 = D(0)V \int_0^{\mathcal{E}_{\mathrm{c}}} \frac{\tanh\{\mathcal{E}\}}{\mathcal{E}} \, \mathrm{d}\mathcal{E} \quad , \tag{7.1.14}$$

where $D(0)$ is the density of states in the normal state, i.e. the metallic state above the transition.

The end results are as follows: there is a finite transition temperature T_{c} which is related to the gap Δ (which does not depend on the momentum) by

$$2\Delta(T = 0) = 3.528 k_{\mathrm{B}} T_{\mathrm{c}} \quad ; \tag{7.1.15}$$

both parameters are non-analytical functions of V, the interaction potential which leads to the superconducting state. The gap $\Delta(T)$ is temperature dependent and approaches zero in a fashion familiar for order parameters of a second order (mean field) phase transitions; this temperature dependence is displayed in Fig. 7.1. The density of the quasi-particle state $D_{\mathrm{s}}(\mathcal{E})$ is

$$\frac{D_{\mathrm{s}}(\mathcal{E})}{D(0)} = \begin{cases} 0 & \text{if } \mathcal{E} < \Delta \\ \dfrac{\mathcal{E}}{(\mathcal{E}^2 - |\Delta|^2)^{1/2}} & \text{if } \mathcal{E} > \Delta \end{cases} \tag{7.1.16}$$

and diverges at the gap. The source of this divergence is similar to that encountered in one-dimensional semiconductors, a situation similar to perfect nesting, as every $\mathbf{k}, \sigma$ state will have its $-\mathbf{k}, -\sigma$ counterpart. The above density of states is sampled by tunneling experiments.

The procedure outlined above can also be used to discuss the other broken symmetry ground states. The gap equation, the form of the density of states, remains the same; what is different are the energy scales involved and the character of the ground states. The BCS superconducting state arises as the consequence of

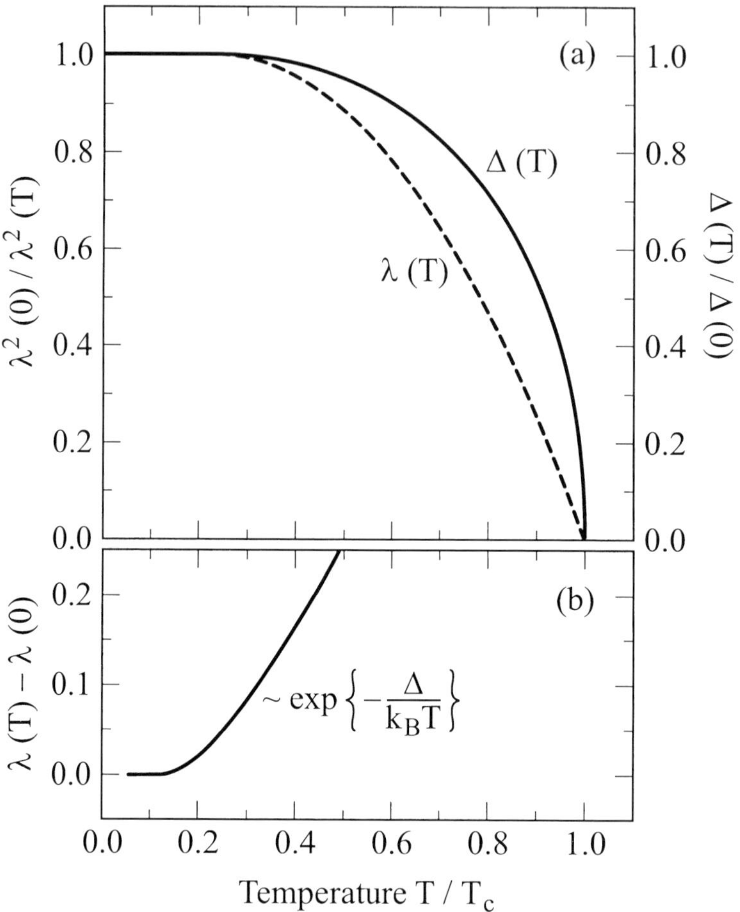

Fig. 7.1. (a) Temperature dependence of the superconducting penetration depth $\lambda(T)$ in comparison with the order parameter $\Delta(T)$. According to Eq. (7.2.10), $(\lambda(T)/\lambda(0))^{-2}$ is proportional to the density of superconducting charge carriers N_{s}. (b) Temperature dependence of the penetration depth $\lambda(T)$ calculated by Eq. (7.4.25).

(retarded) electron–phonon interaction and the cutoff is associated with the energy of the relevant phonons $\hbar\omega_{\mathbf{p}}$. In case of density wave states such retardation does not play a role, and the cutoff energy is the bandwidth of the metallic state. As this is usually significantly larger than the phonon energies, the transition temperatures for density waves are, in general, also larger than the superconducting transition temperatures. Another difference lies in the character of the state: if we calculate the electronic density for superconductors, we find that it is constant and independent of position, this is due to the observation that the total momentum of the Cooper pairs is zero – the superconducting gap opens at zero wavevector. In contrast, electron–hole condensates of the density states have a total momentum of $2\mathbf{k}_F$ (or $-2\mathbf{k}_F$); this leads to a spatial variation of the charge density given by Eq. (7.1.4). There is, however, one important difference: in the latter case, the period of the charge density is the same as the period of the underlying lattice. For density waves, this period π/k_F may be different from the lattice period; in fact this period is often incommensurate with the underlying lattice. If this is the case, the energy associated with these condensates is independent of their positions or of the phase ϕ – with interesting implications as far as the dynamics of density waves is concerned.

7.2 The response of the condensates

The ground state is a coherent superposition of Cooper pairs, all in the same quantum state with the same wavefunction, and as such the condensate can be written as a macroscopic wavefunction with an amplitude and phase

$$\Psi(\mathbf{r}, t) = [N_s(\mathbf{r}, t)]^{1/2} \exp\{i\phi(\mathbf{r}, t)\} \quad , \tag{7.2.1}$$

where $N_s(\mathbf{r}, t)$ is the condensate density. The phase $\phi(\mathbf{r}, t)$ is that of our entire macroscopic system and is therefore a physical observable – beautifully demonstrated by the Josephson effects. In order to see the significance of the phase, we calculate the current density due to the condensate[1]

$$\mathbf{J} = -\frac{e}{2m} \left[\Psi^* \mathbf{p} \Psi + \Psi \mathbf{p}^* \Psi^* \right] \quad ; \tag{7.2.2}$$

in one dimension we find

$$J_s = \frac{-N_s e\hbar}{m} \frac{\mathrm{d}\phi}{\mathrm{d}r} \quad , \tag{7.2.3}$$

i.e. the current is determined by the spatial derivative of the phase. For density waves, a somewhat different handwaving argument applies. The current density is

$$\mathbf{J}_{DW} = -N_{DW} e\mathbf{v} = -N_{DW} e \frac{\mathrm{d}\mathbf{r}}{\mathrm{d}t} \quad . \tag{7.2.4}$$

We write the phase as $\phi = 2\mathbf{k}_F \cdot \mathbf{r}$, and there are two electrons per density wave period π/k_F; this leads to

$$J_{DW} = -\frac{e}{\pi} \frac{\mathrm{d}\phi}{\mathrm{d}t} \tag{7.2.5}$$

and N_{DW} is the (two-dimensional) density in the plane perpendicular to the density wave modulation. The continuity equation gives the electron density

$$\frac{\mathrm{d}J_{DW}}{\mathrm{d}r} + e\frac{\mathrm{d}N_{DW}}{\mathrm{d}t} = 0 \quad . \tag{7.2.6}$$

---

[1] Note that the pairs have a charge $-2e$ and mass $2m$.

These differences between the superconducting and density wave ground state lead to responses to the electromagnetic fields which are different for the various ground states.

7.2.1 London equations

We write for the acceleration of the superconducting current density $\mathbf{J}_s$ in the presence of an external field $\mathbf{E}$

$$\frac{\mathrm{d}}{\mathrm{d}t}\frac{m}{N_s e^2}\mathbf{J}_s = \mathbf{E} \quad , \tag{7.2.7}$$

where N_s is the superfluid particle density, which can be taken as equal to the particle density in the normal state N. In addition, using Maxwell's equation (2.2.7a), this relation can be written as

$$\frac{\mathrm{d}}{\mathrm{d}t}\left(\nabla \times \frac{m}{N_s e^2}\mathbf{J}_s + \frac{1}{c}\mathbf{B}\right) = 0 \quad .$$

Flux expulsion, the so-called Meissner effect, is accounted for by assuming that not only the time derivative in the previous equation, but the function in the bracket itself, is zero, $\nabla \times (mc/N_s e^2)\mathbf{J}_s + \mathbf{B} = 0$. With Eq. (2.1.2) the expression reduces to

$$\mathbf{J}_s = -\frac{N_s e^2}{mc}\mathbf{A} \quad . \tag{7.2.8}$$

Thus, the superconducting current density is proportional to the vector potential $\mathbf{A}$ instead of being proportional to $\mathbf{E}$ as in the case of normal metals. We can also utilize Maxwell's equation $\nabla \times \mathbf{B} = \frac{4\pi}{c}\mathbf{J}_s$ (neglecting the displacement term and the normal current) to obtain for the magnetic induction $\mathbf{B}$

$$\nabla^2 \mathbf{B} = \frac{4\pi N_s e^2}{mc^2}\mathbf{B} = \lambda_{\mathrm{L}}^{-2}\mathbf{B} \quad , \tag{7.2.9}$$

with the so-called London penetration depth

$$\lambda_{\mathrm{L}} = \left(\frac{mc^2}{4\pi N_s e^2}\right)^{1/2} = \frac{c}{\omega_{\mathrm{p}}} \quad , \tag{7.2.10}$$

where ω_{p} is the well known plasma frequency. As this equation describes the spatial variation of $\mathbf{B}$, λ_{L} characterizes the exponential decay of the electromagnetic field. As such, λ_{L} is the equivalent to the skin depth δ_0 we encountered in metals (Eq. (2.3.16)). There are, however, important differences. δ_0 is inversely proportional to $\omega^{-1/2}$, and thus it diverges at zero frequency. In contrast, the penetration depth is frequency independent; there is a decay for $\mathbf{B}$, even for dc applied fields. The temperature dependence of the penetration depth follows the temperature dependence of the condensate density. $N_s(T)$ can be also calculated using the BCS formalism, and it is displayed in Fig. 7.1. For time-varying fields, Eq. (7.2.7) also gives the $1/\omega$ dependence of the imaginary part of the conductivity as

$$\sigma_2(\omega) = \frac{N_s e^2}{m\omega} = \frac{c^2}{4\pi \lambda_L^2 \omega} \quad . \tag{7.2.11a}$$

Through the Kramers–Kronig relation (3.2.11a) the real part of the conductivity is given by

$$\sigma_1(\omega) = \frac{\pi}{2} \frac{N_s e^2}{m} \delta\{\omega = 0\} = \frac{c^2}{8\lambda_L^2} \delta\{\omega = 0\} \quad . \tag{7.2.11b}$$

7.2.2 Equation of motion for incommensurate density waves

The energy related to the spatial and temporal fluctuations of the phase of the density waves is described by the Lagrangian density, which also includes the potential energy due to the applied electric field. The equation of motion of the phase condensate is

$$\frac{\mathrm{d}^2\phi}{\mathrm{d}t^2} - v_F^2 \frac{m}{m^*} \frac{\mathrm{d}^2\phi}{\mathrm{d}r^2} = \frac{e}{m^*} \mathbf{k}_F \cdot \mathbf{E}(\mathbf{q}, \omega) \quad , \tag{7.2.12}$$

where m^* is the mass ascribed to the dynamic response of the density wave condensates. If the interaction between electrons is solely responsible for the formation of density wave states, m^* is the free-electron mass – or the bandmass in the metallic state, out of which these density wave states develop. This is simply because the kinetic energy of the moving condensate is $\frac{1}{2}Nmv^2$, with no additional terms included. This is the case for spin density waves. For charge density waves, however, the translational motion of the condensate leads also to oscillations of the underlying lattice; this happens because of electron–phonon coupling. The motion of the underlying ions also contributes to the overall kinetic energy, and this can be expressed in terms of the increased effective mass m^*. The appropriate expression is [Gru94]

$$\frac{m^*}{m_b} = 1 + \frac{4\Delta^2}{\lambda_P \hbar^2 \omega_P^2} \quad , \tag{7.2.13}$$

where λ_P is the dimensionless electron–phonon coupling constant, and ω_P is the relevant phonon frequency, i.e. the frequency of the phonon mode which couples the electrons together with an electron–hole condensate. As the single-particle gap $2\Delta \ll \omega_P$ as a rule, the effective mass is significantly larger than the bandmass m_b.

The frequency and wavevector dependent conductivity of the collective mode $\hat{\sigma}^{\text{coll}}(\mathbf{q}, \omega)$ is obtained by a straightforward calculation, and one finds that

$$\hat{\sigma}^{\text{coll}}(\mathbf{q}, \omega) = \frac{\mathbf{J}_s(\mathbf{q}, \omega)}{\mathbf{E}(\mathbf{q}, \omega)} = \frac{1}{8\pi} \frac{m}{m^*} \frac{i\omega\omega_p^2}{\omega^2 - v_F^2 \frac{m}{m^*} q^2} \quad . \tag{7.2.14}$$

In general, non-local effects are not important and the $\mathbf{q}$ dependence can be neglected. The real part of the optical conductivity is

$$\sigma_1^{\text{coll}}(\omega) = \frac{\pi N_s e^2}{2m^*} \delta\{\omega = 0\} \quad ; \tag{7.2.15a}$$

and the imaginary part is evaluated from Eq. (7.2.14) as

$$\sigma_2^{\text{coll}}(\omega) = -\frac{2\omega}{\pi} \int_{-\infty}^{\infty} \frac{\sigma_1^{\text{coll}}(\omega)}{\omega'^2 - \omega^2} \, d\omega' = \frac{N_s e^2}{m^*\omega} \quad . \tag{7.2.15b}$$

These are the same expressions as those we derived above for the superconducting case; this becomes evident by inserting the expression for the penetration depth λ_L into Eqs (7.2.11). Note, however, that because of the (potentially) large effective mass, the total spectral weight associated with the density wave condensate can be small.

7.3 Coherence factors and transition probabilities

7.3.1 Coherence factors

One of the important early results which followed from the BCS theory was the unusual features of the transition probabilities which result as the consequence of an external perturbation. Let us assume that the external probe which induces transitions between the various quasi-particle states has the form

$$\mathcal{H}_{\text{int}} = \sum_{\mathbf{k},\mathbf{k}',\sigma,\sigma'} \langle \mathbf{k}', \sigma' | \mathcal{H}_{\text{int}} | \mathbf{k}, \sigma \rangle a_{\mathbf{k}',\sigma'}^{+} a_{\mathbf{k},\sigma} \quad . \tag{7.3.1}$$

In a metal or semiconductor, we will find that the various transitions between the states $\mathbf{k}$ and $\mathbf{k}'$ proceed independently; as a consequence in order to obtain the total transition probability we have to add the squared matrix elements. In superconductors – and also for the other broken symmetry ground states – the situation is different. Using the transformation (7.1.10) we have utilized before, one can show that two transitions, $(a_{\mathbf{k}',\sigma}^{+} a_{\mathbf{k},\sigma})$ and $(a_{-\mathbf{k},-\sigma}^{+} a_{-\mathbf{k}',-\sigma'})$, connect the same quasi-particle states. In the case where $\sigma' = \sigma$, for example,

$$\begin{aligned}
a_{\mathbf{k}',\sigma}^{+} a_{\mathbf{k},\sigma} = \;&+ v_{\mathbf{k}'}^* v_{\mathbf{k}} \gamma_{\mathbf{k}',1}^{+} \gamma_{\mathbf{k},1}^{+} + u_{\mathbf{k}'} u_{\mathbf{k}}^* \gamma_{\mathbf{k}',0}^{+} \gamma_{\mathbf{k},0} \\
&+ v_{\mathbf{k}'}^* u_{\mathbf{k}}^* \gamma_{\mathbf{k}',1}^{+} \gamma_{\mathbf{k},0} + u_{\mathbf{k}'} v_{\mathbf{k}} \gamma_{\mathbf{k}',0}^{+} \gamma_{\mathbf{k},1}^{+}
\end{aligned} \tag{7.3.2a}$$

and

$$
a^{+}_{-\mathbf{k},-\sigma} a_{-\mathbf{k}',-\sigma} = +u_{\mathbf{k}} u^{*}_{\mathbf{k}'} \gamma^{+}_{\mathbf{k}',1} \gamma_{\mathbf{k},1} + v^{*}_{\mathbf{k}} v_{\mathbf{k}'} \gamma_{\mathbf{k}',0} \gamma^{+}_{\mathbf{k},0}
$$
$$
+ v^{*}_{\mathbf{k}} u^{*}_{\mathbf{k}'} \gamma_{\mathbf{k}',0} \gamma_{\mathbf{k},1} + u_{\mathbf{k}} v_{\mathbf{k}'} \gamma^{+}_{\mathbf{k}',1} \gamma^{+}_{\mathbf{k},0} \quad . \tag{7.3.2b}
$$

The last two terms in both equations describe the creation of (or destruction of) pairs of quasi-particles, and the process requires an energy which exceeds the gap; here the combinations $\left(v^{*}_{\mathbf{k}} u^{*}_{\mathbf{k}'} - \eta u_{\mathbf{k}} v_{\mathbf{k}'}\right)$ appear in the transition probabilities. The first two terms describe the scattering of the quasi-particles, processes which are important for small energies $\hbar\omega < \Delta$. For these processes the coefficients which determine the transition probabilities are $\left(u^{*}_{\mathbf{k}'} u_{\mathbf{k}} + \eta v^{*}_{\mathbf{k}'} v_{\mathbf{k}}\right)$. These coefficients are the so-called coherence factors, with $\eta = +1$ constructive and with $\eta = -1$ destructive interference between the two processes. Whether η is positive or negative depends upon whether the interaction changes sign by going from $\mathbf{k}$ to $\mathbf{k}'$.

For electromagnetic waves, the absorption Hamiltonian is

$$
\mathcal{H}_{\text{int}} = \frac{-e\hbar}{2mc} \sum_{\mathbf{k},\mathbf{k}',\sigma} \mathbf{A}(\mathbf{k} - \mathbf{k}') \cdot (\mathbf{k} + \mathbf{k}') \, a^{+}_{\mathbf{k}',\sigma} a_{\mathbf{k},\sigma} \quad , \tag{7.3.3}
$$

and the interaction depends on the direction of the momentum. In this case, the transition probabilities add, and $\eta = +1$. The coherence factors can be expressed in terms of quasi-particle energies and gap value: after some algebra for scattering processes one obtains the form [Tin96]

$$
\left(u^{*}_{\mathbf{k}} u_{\mathbf{k}'} + \eta v_{\mathbf{k}} v^{*}_{\mathbf{k}'}\right)^{2} = \frac{1}{2}\left(1 + \frac{\zeta_{\mathbf{k}}\zeta_{\mathbf{k}'}}{\mathcal{E}_{\mathbf{k}}\mathcal{E}_{\mathbf{k}'}} + \eta\frac{\Delta^{2}}{\mathcal{E}_{\mathbf{k}}\mathcal{E}_{\mathbf{k}'}}\right) \quad . \tag{7.3.4}
$$

When summing over the $\mathbf{k}$ values, the second term on the right hand side becomes zero, and the coherence factor can be simply defined as

$$
F(\mathcal{E}, \mathcal{E}') = \frac{1}{2}\left(1 + \eta\frac{\Delta^{2}}{\mathcal{E}\mathcal{E}'}\right) \approx
\begin{cases}
0 & \text{if } \eta = -1 \quad \text{called case 1} \\
1 & \text{if } \eta = +1 \quad \text{called case 2}
\end{cases}
\tag{7.3.5}
$$

if $\hbar\omega \ll 2\Delta$. In the opposite case of $\hbar\omega \geq 2\Delta$, quasi-particles are created or annihilated and with $\left(v^{*}_{\mathbf{k}} u_{\mathbf{k}'} - \eta u_{\mathbf{k}} v^{*}_{\mathbf{k}'}\right)^{2}$ the situation is reversed, and thus we find

$$
F(\mathcal{E}, \mathcal{E}') \approx
\begin{cases}
1 & \text{if } \eta = -1 \quad \text{case 1} \\
0 & \text{if } \eta = +1 \quad \text{case 2}
\end{cases}
\tag{7.3.6}
$$

In the case of density waves a transformation similar to that performed for the superconducting case defines the quasi-particles of the system separated into two categories: the right- and left-going carriers. After some algebra, along the lines performed for the superconducting case, one finds that the case 1 and case 2 coherence factors for density waves are the opposite to those which apply for the superconducting ground state.

Similar arguments apply for other external probes. Ultrasonic attenuation, for example, is described by the interaction Hamiltonian

$$\mathcal{H} = \sum_{\mathbf{P}} (\rho_{\mathbf{P}} \omega_{\mathbf{P}})^{1/2} |\mathbf{P}| \sum_{\mathbf{k},\sigma} \left(b_{\mathbf{P}} - b^*_{-\mathbf{P}} \right) a^*_{\mathbf{k}+\mathbf{P}} a_{\mathbf{k},\sigma} \quad , \tag{7.3.7}$$

where $b^*_{\mathbf{P}}$ and $b_{\mathbf{P}}$ refer to the longitudinal phonons of frequency $\omega_{\mathbf{P}}$ and density $\rho_{\mathbf{P}}$. Here the change of the sign of $\mathbf{P}$ does not lead to the change of the sign of interactions. Consequently the transition probabilities add differently, and case 1 coherence factors apply. For processes which reverse the spin σ, the arguments are similar; such process, like nuclear magnetic resonance (NMR) relaxation, are characterized by case 2 coherences.

7.3.2 Transition probabilities

In order to derive an expression for the absorption rate W (per unit time and per unit volume) we start from Fermi's golden rule. The transition rate $W_{\mathcal{E}\mathcal{E}'}$ of absorbed energy in a transition from an occupied to an unoccupied state is proportional to the number of occupied quasi-particle states $f(\mathcal{E}_\mathrm{s})D_\mathrm{s}(\mathcal{E}_\mathrm{s})$ and to the number of unoccupied quasi-particle states $[1 - f(\mathcal{E}_\mathrm{s} + \hbar\omega)]D_\mathrm{s}(\mathcal{E}_\mathrm{s} + \hbar\omega)$. Thus we obtain for the rate of absorbed energy:

$$W_\mathrm{s} \propto \int |p_{\mathcal{E}\mathcal{E}'}|^2 F(\mathcal{E}, \mathcal{E} + \hbar\omega) D_\mathrm{s}(\mathcal{E}) D_\mathrm{s}(\mathcal{E} + \hbar\omega) \left[f(\mathcal{E}) - f(\mathcal{E} + \hbar\omega) \right] \, \mathrm{d}\mathcal{E} \quad , \tag{7.3.8}$$

where $|p_{\mathcal{E}\mathcal{E}'}|^2$ is the matrix element of the transition from the state with energy $\mathcal{E}$ to the state with energy $(\mathcal{E}' - \mathcal{E} + \hbar\omega)$, $\hbar\omega$ is the energy of the external probe, $D_\mathrm{s}(\mathcal{E})$ is the density of states given by Eq. (7.1.16), $f(\mathcal{E}) = [1 + \exp\{\mathcal{E}/k_\mathrm{B}T\}]^{-1}$ is the Fermi distribution function, and we have also included the coherence factor $F(\mathcal{E}, \mathcal{E}')$. We normalize the transition probability to the value in the normal state – just above where the broken symmetry state develops – this is $W_\mathrm{n} = |p_{\mathcal{E}\mathcal{E}'}|^2 D_\mathrm{s}^2(0)\hbar\omega$. Then

$$\frac{W_\mathrm{s}}{W_\mathrm{n}} = \frac{1}{\hbar\omega} \int_{-\infty}^{\infty} \frac{|\mathcal{E}(\mathcal{E} + \hbar\omega) + \eta\Delta^2|[f(\mathcal{E}) - f(\mathcal{E} + \hbar\omega)]}{(\mathcal{E}^2 - \Delta^2)^{1/2} [(\mathcal{E} + \hbar\omega)^2 - \Delta^2]^{1/2}} \, \mathrm{d}\mathcal{E} \quad . \tag{7.3.9}$$

As there are no states in the gap, the regions $|\mathcal{E}|$ or $|\mathcal{E} + \hbar\omega| < \Delta$ are excluded from the integration. First we consider $T = 0$, then there are no thermally excited quasi-particles, and therefore the absorption rate is zero for $\omega < \omega_\mathrm{g} = 2\Delta/\hbar$. The process allowing absorption of energy is pair breaking by creation of two quasi-particles with $\hbar\omega \geq 2\Delta$; the normalized rate takes the form:

$$\frac{W_\mathrm{s}}{W_\mathrm{n}} = \frac{1}{\hbar\omega} \int_{\Delta - \hbar\omega}^{-\Delta} \frac{[1 - 2f(\mathcal{E} + \hbar\omega)] \, \mathcal{E}(\mathcal{E} + \hbar\omega) + \eta\Delta^2}{(\mathcal{E}^2 - \Delta^2)^{1/2} [(\mathcal{E} + \hbar\omega)^2 - \Delta^2]^{1/2}} \, \mathrm{d}\mathcal{E} \tag{7.3.10}$$

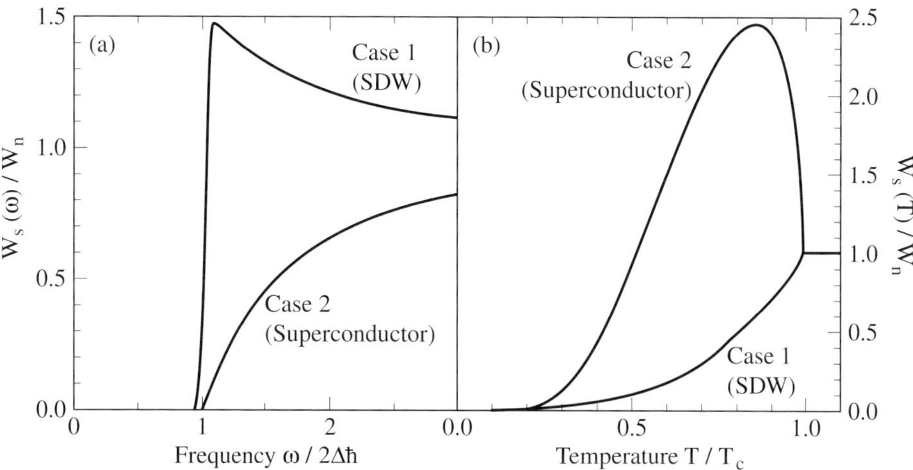

Fig. 7.2. (a) Frequency dependence of the $T \approx 0$ absorption rate $W_s(\omega)/W_n$ for case 1 and 2 coherence factors. (b) Temperature dependence of the low frequency ($\hbar\omega \approx 0.1\Delta$) absorption rate $W_s(T)/W_n$ for case 1 and 2 coherence factors evaluated from Eq. (7.3.10). Case 1 applies for ultrasonic attenuation in superconductors and case 2 describes electromagnetic absorption or nuclear relaxation in the superconducting ground state. Case 1 also applies to the electromagnetic absorption in the spin density wave ground state.

for $\hbar\omega \geq 2\Delta$. The frequency dependence of the absorption is different for case 1 and case 2 coherence factors. In the latter case, the transition probability (close to zero at the gap) effectively cancels the singularity of the density of states, and the absorption smoothly increases from zero at $\hbar\omega = 2\Delta$ to the value equal to the normal state absorption. In contrast, $F(\mathcal{E}, \mathcal{E}') = 1$ for frequencies $\hbar\omega = 2\Delta$ and is weakly frequency dependent; this leads to a peak in the absorption for case 1 coherence.

As mentioned above, one does not observe an absorption at zero temperature for frequencies $\hbar\omega < 2\Delta$ because pair breaking is not possible. Scattering of thermally excited single-particle states, however, becomes possible even at low frequencies for $T > 0$; for these the coherence effects have a different influence. At these low frequencies, the Fermi function difference $f(\mathcal{E}) - f(\mathcal{E}+\hbar\omega) \approx -\partial f/\partial\mathcal{E}$, the coherence factor, $F(\mathcal{E}, \mathcal{E}') \approx 1$, and the appropriate expression for $\omega = 0$ is

$$\frac{W_s}{W_n} = 2 \int_0^\Delta \frac{\mathcal{E}^2 + \Delta^2}{\mathcal{E}^2 - \Delta^2} \left(-\frac{\partial f}{\partial \mathcal{E}}\right) d\mathcal{E} \quad , \tag{7.3.11}$$

which diverges at T_c. At finite frequencies this divergence is removed, but one still recovers a maximum somewhat below T_c. In contrast, for case 1 coherence factors, it is easy to show that the transition rate is suppressed when compared with the normal state transition. Fig. 7.2 displays the absorption as a function of energy and as a function of temperature for both case 1 and 2 coherence factors.

7.4 The electrodynamics of the superconducting state

The electrodynamics of superconductors involves different issues and also varies from material to material. The reason for this is that different length scales play important roles, and their relative magnitude determines the nature of the superconducting state, and also the response to electromagnetic fields. The first length scale is the London penetration depth

$$\lambda_{\mathrm{L}} = \frac{c}{\omega_{\mathrm{p}}} \tag{7.4.1}$$

determined through the plasma frequency ω_{p} by the parameters of the metallic state and derived in Section 7.2.1. The second length scale is the correlation length

$$\xi_0 = \frac{\hbar v_{\mathrm{F}}}{\pi \Delta} \tag{7.4.2}$$

describing, crudely speaking, the spatial extension of the Cooper pairs. This length scale can be estimated if we consider the pair wavefunctions to be the superpositions of one-electron states within the energy region around the Fermi level; then the corresponding spread of momenta δ_p is approximately $|\Delta| = \delta(p^2/2m) \approx v_{\mathrm{F}}\delta_p/2$, where v_{F} is the Fermi velocity. This corresponds to a spatial range of $\xi_0 \approx \hbar/\delta_p$. The correct expression, the BCS coherence length, is given by Eq. (7.4.2) at zero temperature. This is the length scale over which the wavefunction can be regarded as rigid and unable to respond to a spatially varying electromagnetic field. This parameter is determined through the gap Δ, by the strength of the coupling parameter. The third length scale is the mean free path of the uncondensed electrons

$$\ell = v_{\mathrm{F}}\tau \tag{7.4.3}$$

set by the impurities and lattice imperfections at low temperatures; v_{F} is the Fermi velocity, and τ is the time between two scattering events. For typical metals, the three length scales are of the same order of magnitude, and, depending on the relative magnitude of their ratios, various types of superconductors are observed; these are referred to as the various limits. The local limit is where $\ell \ll \xi, \lambda$. More commonly it is referred to as the dirty limit defined by $\ell/\xi \to 0$. In the opposite, so-called clean, limit $\ell/\xi \to \infty$, it is necessary to distinguish the following two cases: the Pippard or anomalous limit, defined by the inequality $\lambda \ll \xi, \ell$ (type I superconductor); and the London limit for which $\xi \ll \lambda, \ell$ (type II superconductor).[2]

We first address the issues related to these length scales and describe the modifications brought about by the short mean free path; these are discussed using

---

[2] See Fig. E.6 and Appendix E.4 for further details.

spectral weight arguments. The relationship between λ_L and ξ can be cast into the form which is similar to the relationship between the skin depth δ_0 and the mean free path ℓ in metals; here also we encounter non-local electrodynamics in certain limits. Finally we recall certain expressions for the conductivity and surface impedance as calculated by Mattis and Bardeen [Mat58]; the formalism is valid in various limits, as will be pointed out.

7.4.1 Clean and dirty limit superconductors, and the spectral weight

The relative magnitude of the mean free path with respect to the coherence length has important consequences for the penetration depth, and this can be described using spectral weight arguments. The mean free path is given by $\ell \approx v_F \tau$, whereas the coherence length $\xi \approx v_F/\Delta$, and consequently in the clean limit $1/\tau < \Delta$ and $1/\tau > \Delta$ is the so-called dirty limit. In the former case the width of the Drude response in the metallic state, just above T_c, is smaller than the frequency corresponding to the gap $2\Delta/\hbar$, and the opposite is true in the dirty limit.

The spectral weight associated with the excitations is conserved by going from the normal to the broken symmetry states. While we have to integrate the real part of the normal state conductivity spectrum $\sigma_1^n(\omega)$, in the superconducting phase there are two contributions: one from the collective mode of the Cooper pairs $\sigma_1^{coll}(\omega)$, and one from the single-particle excitations $\sigma_1^{sp}(\omega)$; thus

$$\int_{-\infty}^{\infty} \left[\sigma_1^{coll} + \sigma_1^{sp}\right] d\omega = \int_{-\infty}^{\infty} \sigma_1^n d\omega = \frac{\pi}{2}\frac{Ne^2}{m} \tag{7.4.4}$$

assuming that all the normal carriers condense. Because of the difference in the coherence factors for the superconducting and density wave ground states, the conservation of the spectral weight has different consequences. The arguments for superconductors were advanced by Tinkham, Glover, and Ferrell [Glo56, Fer58], who noted that the area A which has been removed from the integral upon going through the superconducting transition,

$$\int_{0+}^{\infty} [\sigma_1^n - \sigma_1^s] d\omega = A \tag{7.4.5}$$

is redistributed to give the spectral weight of the collective mode with $\sigma_1(\omega = 0) = A\delta\{\omega\}$. A comparison between this expression and Eq. (7.2.11b) leads to

$$\lambda = \frac{c}{\sqrt{8A}} \quad , \tag{7.4.6}$$

connecting the penetration depth λ to the missing spectral weight A. This relationship is expected to hold also at finite temperatures and for various values of the mean free path and coherence length – as long as the arguments leading to the transition rate apply. In the limit of long relaxation time τ, i.e. for $\hbar/\tau \ll \Delta$, the entire Drude spectral weight of the normal carriers collapses in the collective mode, giving

$$\sigma_1(\omega = 0) = \frac{\pi N_s e^2}{2m} \delta\{\omega = 0\} \quad ;$$

leading through Eq. (7.2.11b) to the London penetration depth λ_L. With increasing $1/\tau$, moving towards the dirty limit, the spectral weight is progressively reduced, resulting in an increase of the penetration depth. In the limit $1/\tau \gg 2\Delta/\hbar$, the missing spectral weight is approximately given by $A \approx \sigma_1(2\Delta/\hbar) = 2Ne^2\tau\Delta/\hbar m$, which can be written as

$$A = \frac{c^2}{2\pi} \frac{\tau\Delta}{\hbar} \quad . \tag{7.4.7}$$

Then the relationship between the area A and the penetration depth $\lambda(\ell)$, which now is mean free path dependent as $\ell = v_F\tau$, is given by $\lambda^2(\ell) = \lambda_L^2 \hbar\pi/(4\tau\Delta)$, which leads to the approximate expression

$$\lambda(\ell) \approx \lambda_L \left(\frac{\xi_0}{\ell}\right)^{1/2} \quad . \tag{7.4.8}$$

It is easy to verify that the sum rule (3.2.28) is conserved: $\int_0^\infty \sigma_1(\omega)\,d\omega = Ne^2\pi/2m$; and it is the same in the normal and in the broken symmetry states.

7.4.2 *The electrodynamics for* $\mathbf{q} \neq 0$

We have discussed the electrodynamics and the transition processes in the $\mathbf{q} = 0$ limit, the situation which corresponds to the local electrodynamics. However, this is not always appropriate, and under certain circumstances the non-local electrodynamics of the superconducting state becomes important. For the normal state, the relative magnitude of the mean free path ℓ and the skin depth δ_0 determines the importance of non-local electrodynamics. For superconductors, the London penetration depth λ_L assumes the role of the skin depth, and the effectiveness concept introduced in Section 5.2.5 refers to the Cooper pairs; thus the relevant length scale is the correlation length ξ_0, i.e. pairs with ξ_0 much larger than λ_0 cannot be fully influenced by the electrodynamic field (here we assume that ℓ is shorter than ξ_0 and λ_L). The consequences of this argument have been explored by A. B. Pippard, and the relevant expression is similar to Chambers' formula for normal metals (5.2.27).

The electrodynamics for finite wavevector $\mathbf{q}$ can, in principle, be discussed using the formalism developed in Chapter 4, with the electronic wavefunction representing that of the superconducting case. This formalism will lead to the full $\mathbf{q}$ and ω dependence, including the response near $2k_F$.

Instead of this procedure we recall the simple phenomenological expression for the non-local response of metals, the Chambers formula (5.2.27), which reads

$$\mathbf{J}(\mathbf{r} = 0) = \frac{3\sigma_{dc}}{4\pi\ell} \int \frac{\mathbf{r}\left[\mathbf{r}\cdot\mathbf{E}(\mathbf{r}, t)\exp\{-r/\ell\}\right]}{r^4}\, d\mathbf{r} \quad .$$

Note that, in view of $\sigma_{dc} = Ne^2\tau/m = Ne^2 v_l\ell/m$, the mean free path drops out from the factor in front of the integral. For superconductors, the vector potential $\mathbf{A}$ is proportional to $\mathbf{J}$, and with $\mathbf{E} = i(\omega/c)\mathbf{A}$ we can write accordingly

$$\mathbf{J}(0) \propto \int \frac{\mathbf{r}[\mathbf{r}\cdot\mathbf{A}(\mathbf{r},\omega)]}{r^4} F(r)\, d\mathbf{r} \tag{7.4.9}$$

where the function $F(r)$ describes the spatial decay which has yet to be determined. For a vector potential of the form $\mathbf{A}(\mathbf{r}) \propto \exp\{i\mathbf{q}\cdot\mathbf{r}\}$, the $\mathbf{q}$ dependent conductivity is written as

$$\mathbf{J}(\mathbf{q}) = -\frac{c}{4\pi} K(\mathbf{q})\mathbf{A}(\mathbf{q}) \quad . \tag{7.4.10}$$

A more general definition of the kernel K is given in Appendix E.3.

Let us first assume that we deal with clean superconductors, and effects due to a finite mean free path are not important. In the London limit,

$$\mathbf{J}(\mathbf{r}) = -\frac{N_s e^2}{mc}\mathbf{A}(\mathbf{r}) = -\frac{c}{4\pi\lambda_L^2}\mathbf{A}(\mathbf{r}) \tag{7.4.11}$$

from Eq. (7.2.8). Consequently in this limit

$$\mathbf{K}(\mathbf{q}) = \mathbf{K}(0) = \frac{1}{\lambda_L^2} \quad , \tag{7.4.12}$$

independent of the wavevector $\mathbf{q}$. In general, $\mathbf{K}(\mathbf{q})$ is not constant, but decreases with increasing $\mathbf{q}$. The argument, due to Pippard, that we have advanced before suggests that

$$F(\mathbf{r}) = \exp\left\{-\frac{r}{\xi_0}\right\} \quad . \tag{7.4.13}$$

We assume that at large distances $F(r) \propto \exp\{-r/R_0\}$ with R_0 a (yet unspecified) characteristic distance. This then leads (for isotropic superconductors) to $K(0) = \lambda_L^{-2}$ at $\mathbf{q} = 0$ as seen before; for large $\mathbf{q}$ values

$$\mathbf{K}(\mathbf{q}) = \mathbf{K}(0)\frac{3\pi}{4q\xi_0} \quad . \tag{7.4.14}$$

One can evaluate the integral in Eq. (7.4.9) based on this form of $\mathbf{K}(\mathbf{q})$, but we can also resort to the same argument which led to the anomalous skin effect for normal metals. There we presented the ineffectiveness concept, which holds here for superconductors, with ξ_0 replacing ℓ, and λ replacing δ_0. The expression of the penetration depth – in analogy to the equation we have advanced for the anomalous skin effect regime – then becomes

$$\lambda = \left(\frac{mc^2}{4\pi e^2 N_s(\lambda/\xi_0)} \right)^{1/2} \tag{7.4.15}$$

leading to

$$\lambda = \left(\frac{mc^2}{4\pi e^2 N_s}\xi_0 \right)^{1/3} \approx \lambda_L^{2/3}\xi_0^{1/3} \quad ; \tag{7.4.16}$$

thus, for large coherence lengths, the penetration depth increases, well exceeding the London penetration depth λ_L.

For impure superconductors the finite mean free path can be included by a further extension of the non-local relation between $\mathbf{J}_s(\mathbf{r})$ and $\mathbf{A}(\mathbf{r})$. In analogy to the case for metals we write

$$\mathbf{J}(\mathbf{r}) \propto \int d\mathbf{r}\, \mathbf{r}\, [\mathbf{r} \cdot \mathbf{A}(\mathbf{r})]\, r^{-4} F(\mathbf{r}) \exp\{-r/\ell\} \quad .$$

This includes the effect of the finite mean path, and additional contributions to the non-local relation, dependent on the coherence length, are contained in the factor $F(\mathbf{r})$. If we take $F(\mathbf{r})$ from Eq. (7.4.13), we find that

$$\lambda_{\text{eff}} = \lambda_L \left(1 + \frac{\xi_0}{\ell} \right)^{1/2} \quad , \tag{7.4.17}$$

which reduces to Eq. (7.4.8) in the $\xi_0 > \ell$ limit. This is only an approximation; the correct expression is [Tin96]

$$\lambda_{\text{eff}}(\ell) = \lambda_L \left(1 + 0.75\frac{\xi_0}{\ell} \right)^{1/2} \quad . \tag{7.4.18}$$

7.4.3 Optical properties of the superconducting state: the Mattis–Bardeen formalism

With non-local effects potentially important, calculation of the electromagnetic absorption becomes a complicated problem. An appropriate theory has to include both the coherence factors and the non-local relationship between the vector potential and induced currents. This has been done by Mattis and Bardeen [Mat58] and by Abrikosov [Abr59]. For finite frequencies and temperatures, the relationship between $\mathbf{J}$ and $\mathbf{A}$ takes, in the presence of an ac field, the following form:

$$\mathbf{J}(0, t) \propto \exp\{-i\omega t\} \int d\mathbf{r}\, \mathbf{r}\, [\mathbf{r} \cdot \mathbf{A}(\mathbf{r})]\, r^{-4} I(\omega, \mathbf{r}, T) \exp\{-r/\ell\} \quad . \tag{7.4.19}$$

Non-local effects are included in the kernel $I(\omega, \mathbf{r}, T)$ and in the exponential function $\exp\{-r/\ell\}$. The relationship between $\mathbf{J}$ and $\mathbf{A}$ is relatively simple in two limits. For $\ell \ll \xi_0$, the local dirty limit, the integral is confined to $r \leq \ell$, and one can assume that $I(\omega, \mathbf{r}, T)$ is constant in this range of $\mathbf{r}$ values. For $\lambda \ll \xi_0$, the Pippard or extreme anomalous limit, $I(\omega, \mathbf{r}, T)$ varies slowly in space with respect to other parts in the integral, and again it can be taken as constant. The spatial variation implied by the above expression is the same as that given by the Chambers formula, the expression which has to be used when the current is evaluated in the normal state. Consequently, when the complex conductivity $\hat{\sigma}_s(\omega, T)$ in the superconducting state is normalized to the conductivity in the normal state $\hat{\sigma}_n(\omega, T)$, numerical factors drop out of the relevant equations. The expressions derived by Mattis and Bardeen, and by Abrikosov, are then valid in both limits, and they read:

$$\frac{\sigma_1(\omega, T)}{\sigma_n} = \frac{2}{\hbar\omega} \int_\Delta^\infty \frac{[f(\mathcal{E}) - f(\mathcal{E} + \hbar\omega)](\mathcal{E}^2 + \Delta^2 + \hbar\omega\mathcal{E})}{(\mathcal{E}^2 - \Delta^2)^{1/2}[(\mathcal{E} + \hbar\omega)^2 - \Delta^2]^{1/2}} \mathrm{d}\mathcal{E}$$

$$+ \frac{1}{\hbar\omega} \int_{\Delta - \hbar\omega}^{-\Delta} \frac{[1 - 2f(\mathcal{E} + \hbar\omega)](\mathcal{E}^2 + \Delta^2 + \hbar\omega\mathcal{E})}{(\mathcal{E}^2 - \Delta^2)^{1/2}[(\mathcal{E} + \hbar\omega)^2 - \Delta^2]^{1/2}} \mathrm{d}\mathcal{E} \qquad (7.4.20\mathrm{a})$$

$$\frac{\sigma_2(\omega, T)}{\sigma_n} = \frac{1}{\hbar\omega} \int_{\Delta - \hbar\omega, -\Delta}^{\Delta} \frac{[1 - 2f(\mathcal{E} + \hbar\omega)](\mathcal{E}^2 + \Delta^2 + \hbar\omega\mathcal{E})}{(\Delta^2 - \mathcal{E}^2)^{1/2}[(\mathcal{E} + \hbar\omega)^2 - \Delta^2]^{1/2}} \mathrm{d}\mathcal{E} \quad , \qquad (7.4.20\mathrm{b})$$

where for $\hbar\omega > 2\Delta$ the lower limit of the integral in Eq. (7.4.20b) becomes $-\Delta$. At $T = 0$, σ_1/σ_n describes the response of the normal carriers. The first term of Eq. (7.4.20a) represents the effects of thermally excited quasi-particles. The second term accounts for the contribution of photon excited quasi-particles; since it requires the breaking of a Cooper pair, it is zero for $\hbar\omega < 2\Delta(T)$. The expression for $\sigma_1(\omega, T)$ is the same as Eq. (7.3.9), which was derived on the basis of arguments relating to the electromagnetic absorption. This is not surprising, as

$$\frac{W_s(\omega, T)}{W_n} = \frac{\sigma_1(\omega, T)}{\sigma_n}$$

in the limits for which this equation applies. Figs 7.3 and 7.4 summarize the frequency and temperature dependence of σ_1 and σ_2 as derived from Eqs (7.4.20) assuming finite scattering effects $\ell/\pi\xi = 0.1$ discussed in [Lep83]. It reproduces the behavior predicted for case 2 coherence factors (Fig. 7.2) inferred from symmetry arguments only. Although the density of states diverges at $\pm\Delta$, the conductivity $\sigma_1(\omega)$ does not show a divergency but a smooth increase which follows approximately the dependence $\sigma_1(\omega)/\sigma_n \propto 1 - (\hbar\omega/k_B T_c)^{-1.65}$. The temperature dependent conductivity $\sigma_1(T)$ shows a peak just below T_c at low frequency. The height of the peak has the following frequency dependence: $(\sigma_1/\sigma_n)_{\max} \sim \log\{2\Delta(0)/\hbar\omega\}$.

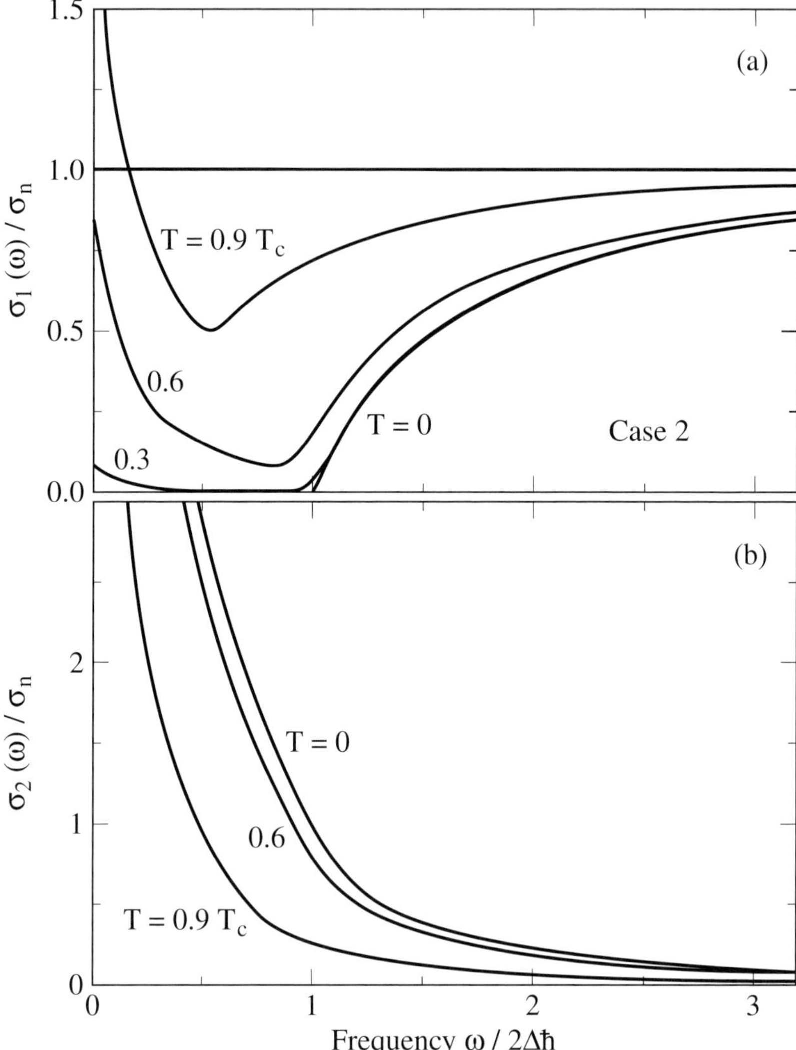

Fig. 7.3. Frequency dependence of the real and imaginary parts of the conductivity $\hat{\sigma}(\omega)$ of a superconductor (case 2) at different temperatures as evaluated from the Mattis–Bardeen expressions (7.4.20) using $\ell/\pi\xi_0 = 0.1$. (a) For $T = 0$ the conductivity $\sigma_1(\omega)$ is zero below the superconducting gap 2Δ except for the δ-peak at $\omega = 0$. For each temperature the local minimum in $\sigma_1(\omega)$ corresponds to $2\Delta(T)$. (b) The imaginary part $\sigma_2(\omega)$ diverges as $1/\omega$ in the $T = 0$ limit. The kink in $\sigma_2(\omega)$ is not a numerical artifact but indicates the gap frequency $2/\Delta(T)/\hbar$.

The peak has completely disappeared for $\hbar\omega \geq \Delta/2$ (well before 2Δ). At $T = 0$ and $\omega < 2\Delta/\hbar$ the complex part of the conductivity σ_2/σ_n describes the response of the Cooper pairs and is related to the gap parameter through the expression

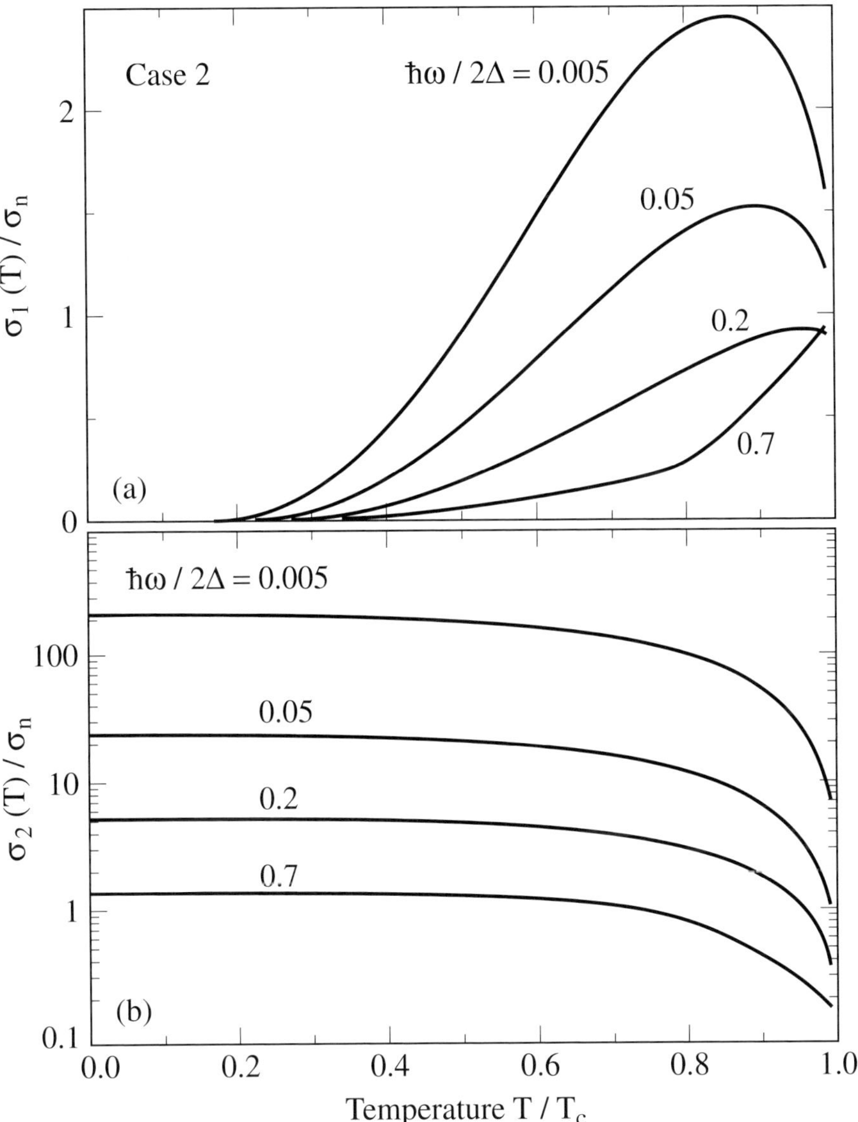

Fig. 7.4. Temperature dependence of the real and imaginary parts of the conductivity $\hat{\sigma}(T)$ of a superconductor (case 2) evaluated by Eqs (7.4.20) using $\ell/\pi\xi_0 = 0.1$. Note that the well defined coherence peak in the real part $\sigma_1(T)$ exists only at low frequencies $h\omega/2\Delta(0) < 0.1$, and $\sigma_2(T)/\sigma_n$ saturates at the value of $\pi\Delta(0)/\hbar\omega$ at low temperatures.

$$\frac{\sigma_2(T)}{\sigma_n} \approx \frac{\pi\,\Delta(T)}{\hbar\omega}\,\tanh\left\{\frac{\Delta(T)}{2k_BT}\right\} \approx \lim_{T\to 0}\frac{\pi\,\Delta(0)}{\hbar\omega}\ . \tag{7.4.21}$$

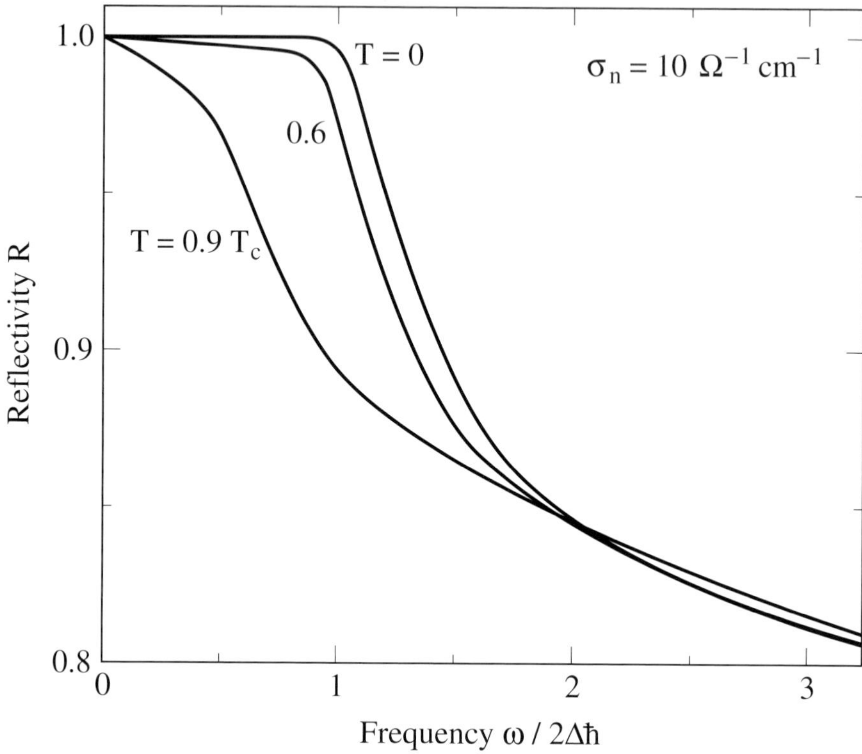

Fig. 7.5. Bulk reflectivity $R(\omega)$ of a superconducting metal as a function of frequency at different temperatures as indicated from the Mattis–Bardeen expressions (7.4.20). The figure was computed by using $\sigma_{dc} = 10^5\ \Omega^{-1}\,\mathrm{cm}^{-1}$, $\nu_p = 10^4\ \mathrm{cm}^{-1}$, and $T_c = 1$ K.

Using Eq. (2.4.17), the reflectivity $R(\omega)$ of a bulk superconductor is calculated from Eqs (7.4.20) and is shown in Fig. 7.5. We see that for frequencies below the gap energy the reflectivity goes to unity; superconductors are perfect mirrors for $\omega < 2\Delta/\hbar$. This, however, is not due to the optical conductivity since $\sigma_1(\omega, T = 0) = 0$ in the range $0 < \omega < 2\Delta$, but due to the large imaginary part of the conductivity $\sigma_2(\omega)$ which diverges as $1/\omega$. The result is a rapid phase shift near the transition.

Since in the superconducting state the imaginary part of the conductivity σ_2 cannot be neglected, the assumptions of the Hagen–Rubens limit do not apply and the general Eqs (2.3.32) and (2.3.33) have to be used to relate both complex quantities, the surface impedance $\hat{Z}_S$ and the conductivity $\hat{\sigma}$. After normalizing all quantities to their normal state properties, we obtain

$$\frac{R_S}{R_n} = \left\{ \frac{[(\sigma_1/\sigma_n)^2 + (\sigma_2/\sigma_n)^2]^{1/2} - \sigma_2/\sigma_n}{(\sigma_1/\sigma_n)^2 + (\sigma_2/\sigma_n)^2} \right\}^{1/2} \tag{7.4.22a}$$

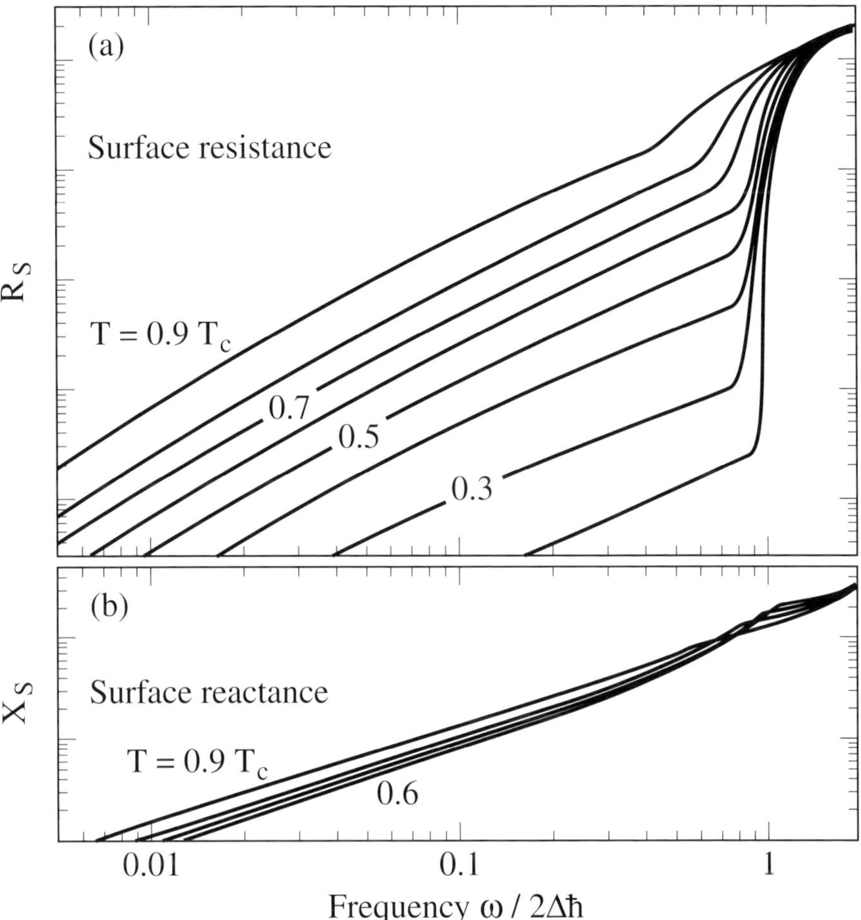

Fig. 7.6. (a) Frequency dependent surface resistance $R_{\mathrm{S}}(\omega)$ of a superconductor at various temperatures. (b) The surface reactance $X_{\mathrm{S}}(\omega)$ as a function of frequency. Both parameters were calculated using Eqs (7.4.22).

$$\frac{-X_{\mathrm{S}}}{R_{\mathrm{n}}} = \left\{\frac{[(\sigma_1/\sigma_{\mathrm{n}})^2 + (\sigma_2/\sigma_{\mathrm{n}})^2]^{1/2} + \sigma_2/\sigma_{\mathrm{n}}}{(\sigma_1/\sigma_{\mathrm{n}})^2 + (\sigma_2/\sigma_{\mathrm{n}})^2}\right\}^{1/2} . \qquad (7.4.22\mathrm{b})$$

We have to assume local electrodynamics; for a detailed discussion see Appendix E.4 and [Ric65]. In Fig. 7.6 the normalized surface resistance and surface impedance are plotted as a function of frequency for different temperatures. $R_{\mathrm{S}}/R_{\mathrm{n}}$ monotonically drops to zero for $\omega \rightarrow 0$ and for $T \rightarrow 0$ when the energy of the electromagnetic field is below the gap $2\Delta(T)$. From the BCS theory the temperature and frequency dependence of R_{S} at low temperatures ($T \ll \Delta(T)$)

and small frequencies ($\hbar\omega < \Delta(T)$) was evaluated as [Hal74]

$$R_{\mathrm{S}} \propto \frac{(\hbar\omega)^2}{k_{\mathrm{B}}T} \ln\left\{\frac{4k_{\mathrm{B}}T}{\hbar\omega}\right\} \exp\left\{-\frac{\Delta(T)}{k_{\mathrm{B}}T}\right\} \quad . \tag{7.4.23}$$

Above T_{c} the surface resistance shows the $\omega^{1/2}$ behavior of a Drude metal in the Hagen–Rubens regime (Eq. (5.1.18)). At low temperature, R_{S} is vanishingly small and approaches an approximate ω^2 dependence for $\omega \to 0$ (Fig. 7.6a). In contrast to metals, $R_{\mathrm{S}}(\omega) \neq -X_{\mathrm{S}}(\omega)$ in the case of superconductors. By reducing the temperature to close to but below T_{c}, the surface reactance $X_{\mathrm{S}}(T)$ shows an enhancement before it drops. This is typically observed as a peak of $X_{\mathrm{S}}(T)/R_{\mathrm{n}}$ right below the transition temperature. The frequency dependence of X_{S} can be evaluated from classical electrodynamics using Eq. (2.3.34b) and $\lambda = c/\omega k$ with $n \to 0$:

$$X_{\mathrm{S}} = -\frac{4\pi\omega}{c^2}\lambda \quad , \tag{7.4.24}$$

with a frequency dependence as shown in Fig. 7.6b. Equation (7.4.24) only holds for $k \gg n$, which is equivalent to $R_{\mathrm{S}} \ll |X_{\mathrm{S}}|$. It is immediately seen that the surface reactance is directly proportional to the penetration depth, which by itself is frequency independent as long as $\hbar\omega < \Delta$. This is equivalent to Eq. (2.4.27) in the normal state, where λ is replaced by $\delta/2$. The reason for the factor of 2 lies in the difference between the phase angle ϕ of the surface impedance, which is $90°$ for a superconductor (at $T \to 0$) and $45°$ for a metal in the Hagen–Rubens regime. As λ is independent of the frequency for $\hbar\omega \ll \Delta$, X_{S} is proportional to ω in this regime, as displayed in Fig. 7.6b. Even in the superconducting state the temperature dependence of the penetration depth can be calculated from the complex conductivity $\hat{\sigma}(\omega)$ using Eq. (2.3.15b). In Fig. 7.1 $(\lambda(T)/\lambda(0))^{-2}$ is plotted versus T/T_{c} for $\ell/\pi\xi_0 = 0.01$. In the London limit this expression is proportional to the density of superconducting carriers N_{s} as seen from Eq. (7.2.10). The energy gap $2\Delta(T)$ increases faster than the charge carriers condense. At low temperatures $T < 0.5T_{\mathrm{c}}$ the temperature dependence of the penetration depth $\lambda(T) - \lambda(0)$ can be well described by an exponential behavior:

$$\frac{\lambda(T) - \lambda(0)}{\lambda(0)} = \left[\frac{\pi\Delta}{2k_{\mathrm{B}}T}\right]^{1/2} \exp\left\{-\frac{\Delta}{k_{\mathrm{B}}T}\right\} \quad . \tag{7.4.25}$$

This behavior is plotted in Fig. 7.1b.

7.5 The electrodynamics of density waves

Just as for superconductors, the full electrodynamics of the density wave states includes the response of the collective mode. The collective mode, i.e. the translational motion of the entire density wave, and absorption due to the quasi-particles, ideally occurs – in the absence of lattice imperfections – at zero frequency. At zero temperature, the onset for the quasi-particle absorption is set by the BCS gap 2Δ. The differences with respect to the superconducting case are due to the different coherence factors and the possibility of a large effective mass in the case of charge density wave condensates.

Additional complications may also arise: density wave condensates with a periodic modulation of the charge and/or spin density are incommensurate with the underlying lattice only in one direction; in other directions the lattice periodicity plays an important role in pinning the condensate to the underlying lattice. If this is the case, the collective mode contribution to the conductivity is absent due to the large restoring force exercised by the lattice.

7.5.1 The optical properties of charge density waves: the Lee–Rice–Anderson formalism

Both single-particle excitations across the density wave gap and the collective mode contribute to the frequency dependent response. For charge density waves this has been examined in detail by [Lee74]. When the effective mass is large, the collective mode contribution – centered, in the absence of impurities and lattice imperfections at zero frequency, see Eq. (7.2.15a) – is small, and consequently only minor modifications from those expected for a one-dimensional semiconductor are observed. For $m^* = \infty$, the collective mode spectral weight is zero, and the expression for the conductivity is identical to that given in Section 6.3.1 for a one-dimensional semiconductor.

This can be taken into account by utilizing the formalism outlined earlier, which leads to the BCS gap equation and to the quasi-particle excitations of the superconducting and density wave states. The formalism leads to the following expressions of the single-particle contribution to the conductivity for $m^* = \infty$:

$$\hat{\sigma}^{\text{sp}}(\omega) = \frac{Ne^2}{\mathrm{i}\omega m}[F(\omega) - F(0)] \tag{7.5.1}$$

with the frequency dependent function

$$F(\omega) = -\int \frac{2\Delta^2/(\zeta_{\mathbf{k}}^2 + \Delta^2)}{(\hbar\omega)^2 - 4(\zeta_{\mathbf{k}}^2 + \Delta^2)} \, \mathrm{d}\zeta_{\mathbf{k}} \quad , \tag{7.5.2}$$

where $\zeta_{\mathbf{k}} = \mathcal{E}_{\mathbf{k}} - \mathcal{E}_{\text{F}}$. Simply integrating the expression leads to Eq. (6.3.16).

In the presence of a finite mass associated with the collective mode, the spectral function $F(\omega)$ is modified, and Lee, Rice, and Anderson [Lee74] find that

$$F'(\omega) = \frac{F(\omega)}{1 + (\frac{m}{m^*} - 1)F(\omega)} = F(\omega)\left[1 + \frac{\lambda_\mathbf{P}\hbar^2\omega_\mathbf{P}^2}{4\Delta^2}F(\omega)\right]^{-1} . \qquad (7.5.3)$$

The calculation of σ^{sp} is now straightforward, and in Fig. 7.9 the real part of the complex conductivity $\sigma_1(\omega)$ is displayed for two different values of the effective mass.

There is also a collective mode contribution to the conductivity at zero frequency, as discussed before, with spectral weight

$$\int \sigma_1^{\mathrm{coll}}(\omega)\,\mathrm{d}\omega = \int \frac{\pi Ne^2}{2m^*}\delta\{\omega = 0\}\,\mathrm{d}\omega = \frac{\pi Ne^2}{2m^*} , \qquad (7.5.4)$$

and this is taken out of the spectral weight corresponding to the single-particle excitations at frequencies $\omega > 2\Delta/\hbar$. The total spectral weight

$$\int \left[\sigma_1^{\mathrm{coll}}(\omega) + \sigma_1^{\mathrm{sp}}(\omega)\right]\,\mathrm{d}\omega = \frac{\pi}{2}\frac{Ne^2}{m} \qquad (7.5.5)$$

is of course observed, and is the same as the spectral weight associated with the Drude response in the metallic state above the density wave transition.

For an incommensurate density wave in the absence of lattice imperfections, the collective mode contribution occurs at $\omega = 0$ due to the translational invariance of the ground state. In the presence of impurities this translational invariance is broken and the collective modes are tied to the underlying lattice due to interactions with impurities [Gru88, Gru94]; this aspect of the problem is discussed in Chapter 14.

7.5.2 Spin density waves

The electrodynamics of the spin density wave ground state is different from the electrodynamics of the superconducting and of the charge density wave states. The difference is due to several factors. First, for the case of density waves, case 1 coherence factors apply, and thus the transition probabilities are different from those of a superconductor. Second, the spin density wave state arises as the consequence of electron–electron interactions. Phonons are not included here, and also the spin density wave ground state – the periodic modulation of the spin density – does not couple to the underlying lattice. Consequently, the mass which is related to the dynamics of the collective mode is simply the electron mass.

Therefore, the electrodynamics of the spin density wave ground state can be discussed along the lines developed for the superconducting state with one important difference: for the electrodynamic response, case 1 coherence factors apply.

Instead of Eqs (7.4.20) the conductivity in this case then reads

$$\frac{\sigma_1^s(\omega, T)}{\sigma_n} = \frac{2}{\hbar\omega} \int_{-\infty}^{\infty} \frac{[f(\mathcal{E}) - f(\mathcal{E} + \hbar\omega)](\mathcal{E}^2 - \Delta^2 + \hbar\omega\mathcal{E})}{(\mathcal{E}^2 - \Delta^2)^{1/2}[(\mathcal{E} + \hbar\omega)^2 - \Delta^2]^{1/2}} \, d\mathcal{E} \tag{7.5.6a}$$

$$\frac{\sigma_2^s(\omega, T)}{\sigma_n} = \frac{1}{\hbar\omega} \int_{\Delta - \hbar\omega, -\Delta}^{\Delta} \frac{[1 - 2f(\mathcal{E} + \hbar\omega)](\mathcal{E}^2 - \Delta^2 + \hbar\omega\mathcal{E})}{(\Delta^2 - \mathcal{E}^2)^{1/2}[(\mathcal{E} + \hbar\omega)^2 - \Delta^2]^{1/2}} \, d\mathcal{E} \quad . \tag{7.5.6b}$$

In Fig. 7.7 the frequency dependence of $\sigma_1(\omega)$ and $\sigma_2(\omega)$ derived from Eqs (7.5.6) is plotted assuming finite scattering effects $\ell/\pi\xi = 0.1$. This condition corresponds to the dirty limit, as introduced before. In contrast to the results obtained for the case of a superconductor, an enhancement of the conductivity $\sigma_1(\omega)$ is observed above the single-particle gap, which is similar to the results obtained from a simple semiconductor model in one dimension (Eq. (6.3.16)). As discussed above, the spectral weight in the gap region is, by and large, compensated for by the area above the single-particle gap 2Δ, leading to a small collective mode. This fact results in a low reflectivity for $\omega < 2\Delta$, as displayed in Fig. 7.8. Finite temperature effects arise here in a fairly natural fashion; thermally excited single-particle states lead to a Drude response in the gap region, with the spectral weight determined by the temperature dependence of the number of excited carriers.

7.5.3 *Clean and dirty density waves and the spectral weight*

The spectral weight arguments advanced for superconductors also apply for the density wave states with some modifications. In the density wave states the collective mode contribution to the spectral weight is

$$A^{\text{coll}} = \int_0^{\infty} \sigma_1^{\text{coll}}(\omega) \, d\omega = \frac{\pi N e^2}{2m^*} \quad , \tag{7.5.7}$$

while the single-particle excitations give a contribution

$$A^{\text{sp}} = \int_0^{\infty} \sigma_1^{\text{sp}}(\omega) \, d\omega = \int_0^{\omega_g} \sigma_1^{n}(\omega) - \sigma_1^{\text{coll}}(\omega) \, d\omega$$

$$= \frac{\pi N e^2}{2m_b} - \frac{\pi N e^2}{2m^*} = \frac{\pi N e^2}{2m_b}\left[1 - \frac{m_b}{m^*}\right] \tag{7.5.8}$$

with $\hbar\omega_g = \mathcal{E}_g$. The collective mode has a spectral weight m_b/m^*; this is removed from the excitation above the gap. For a large effective mass $m^*/m_b \gg 1$, most of the total spectral weight comes from the single-particle excitations; while for $m^*/m_b = 1$, all the spectral weight is associated with the collective mode, with no contribution to the optical conductivity from single-particle excitations. The former is appropriate for charge density waves, while the latter is valid for spin density wave transport [Gru94].

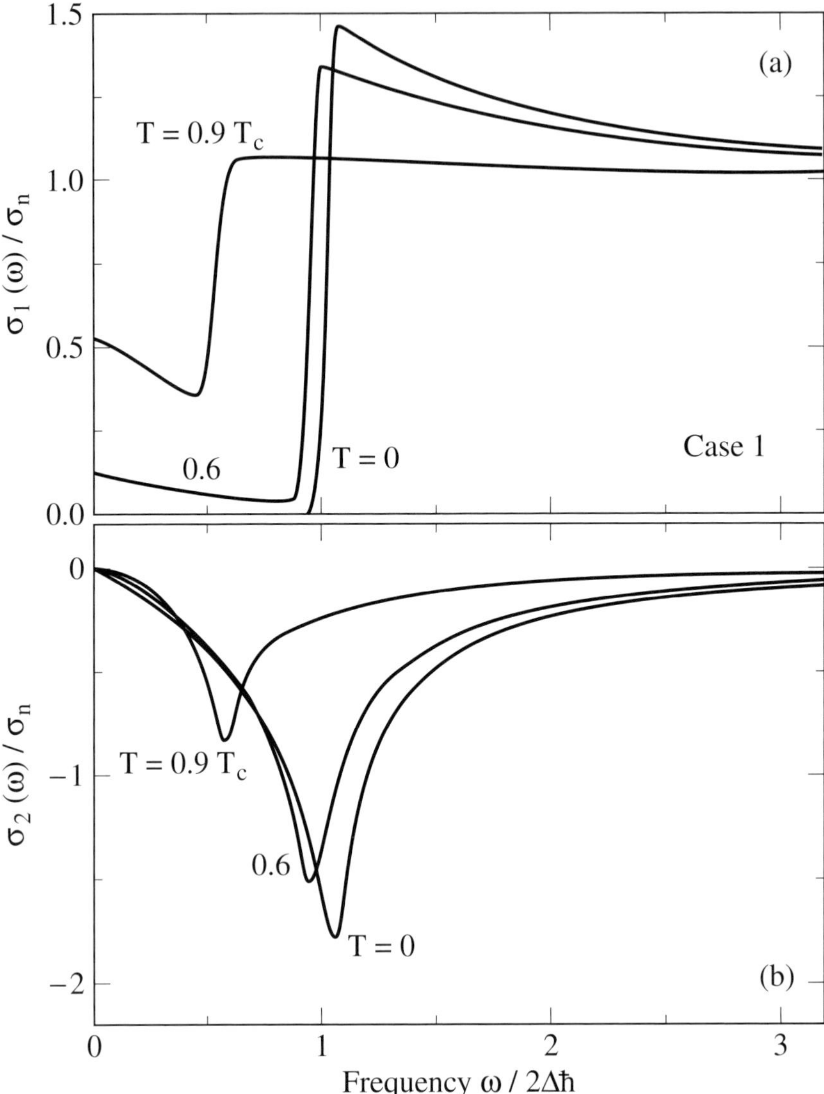

Fig. 7.7. Frequency dependence of the real and imaginary parts of the conductivity $\hat{\sigma}(\omega)$ of a spin density wave (case 1) at different temperatures as evaluated from Eqs (7.5.6) using $\ell/\pi\xi_0 = 0.1$. (a) For $T = 0$ the conductivity $\sigma_1(\omega)$ is zero below the single-particle gap 2Δ. The enhancement above 2Δ corresponds (by virtue of the Tinkham–Ferrell sum rule) to the area removed below the gap energy. (b) The imaginary part $\sigma_2(\omega)$ shows an extremum at the single-particle gap; for $\omega < 2\Delta/\hbar$ the conductivity $\sigma_2(\omega)/\sigma_n$ drops to zero.

The above arguments are appropriate in the clean limit, $1/\tau \ll \Delta/\hbar$, which is equivalent to the condition $\xi_0 \ll \ell$, where ℓ is the mean free path and ξ_0 is

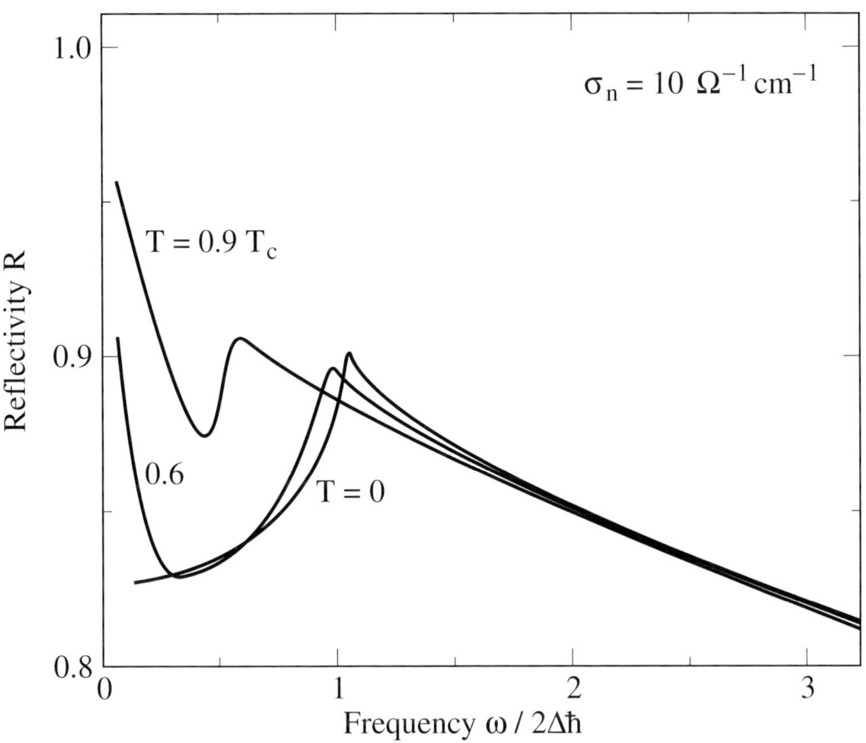

Fig. 7.8. Bulk reflectivity $R(\omega)$ of a metal in the spin density wave ground state as a function of frequency for different temperatures. The calculations are based on Eqs (7.5.6). The figure was computed by using $\sigma_{dc} = 10^5 \ \Omega^{-1} \, cm^{-1}$, $\nu_p = 10^4 \, cm^{-1}$, and $T_c = 1$ K.

the coherence length. The response for the opposite case, the so-called dirty limit $\xi_0 \gg \ell$, has not been calculated for density wave ground states. It is expected, however, that arguments advanced for superconductors, discussed in Section 7.4 and Appendix E.5, also apply for density wave ground states. In analogy to the above discussion for superconductors, the collective mode contribution to the spectral weight is given by the difference $A = \int [\sigma_1^n(\omega) - \sigma_1^{sp}(\omega)] \, d\omega$ with $\sigma_1^n(\omega)$ given by Eq. (7.4.4). This difference is approximately the area

$$A \approx \int_0^{\omega_g} \sigma_1^n(\omega) \, d\omega \approx \frac{\omega_p^2}{4\pi} \omega_g \tau \approx \frac{\omega_p^2}{2\pi^2} \left(\frac{\ell}{\xi_0} \right)^{1/2} \tag{7.5.9}$$

in the dirty limit. Consequently, the spectral weight due to the collective mode contribution is reduced, and an empirical form similar to Pippard's expression of the penetration depth [Tin96] can be anticipated:

$$\frac{A_0^{coll}}{A^{coll}} = \left(1 + \frac{\xi_0}{\alpha\ell} \right)^{1/2} \quad , \tag{7.5.10}$$

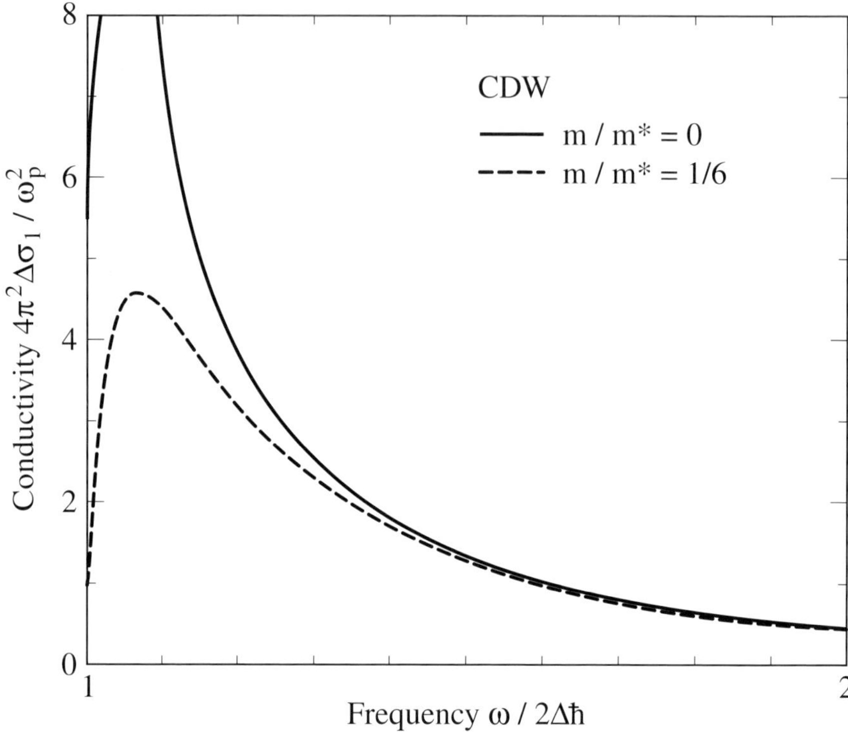

Fig. 7.9. Frequency dependence of the conductivity $\sigma_1(\omega)$ in the charge density wave state for $m^*/m = \infty$ and $m^*/m = 6$ (after [Lee74]).

where α is a numerical factor of the order of one, and A_0^{coll} is the spectral weight of the collective mode in the clean limit.

References

[Abr59]   A.A. Abrikosov, L.P. Gor'kov, and I.M. Khalatnikov, *Sov. Phys. JETP* **8**, 182 (1959)

[Bar57]   J. Bardeen, L.N. Cooper, and J.R. Schrieffer, *Phys. Rev.* **106**, 162 (1957); *ibid.* **108**, 1175 (1957)

[Fer58]   R.A. Ferrell and R.E. Glover, *Phys. Rev.* **109**, 1398 (1958)

[Glo56]   R.E. Glover and M. Tinkham, *Phys. Rev.* **104**, 844 (1956); *ibid.* **108**, 243 (1957)

[Gru88]   G. Grüner, *Rev. Mod. Phys.* **60**, 1129 (1988)

[Gru94]   G. Grüner, *Density Waves in Solids* (Addison-Wesley, Reading, MA, 1994)

[Hal74]   J. Halbritter, *Z. Phys.* **266**, 209 (1974)

[Lep83]   L. Leplae, *Phys. Rev. B* **27**, 1911 (1983)

[Lee74]   P.A. Lee, T.M. Rice, and P.W. Anderson, *Solid State Commun.* **14**, 703 (1974)

[Mat58]   D.C. Mattis and J. Bardeen, *Phys. Rev.* **111**, 561 (1958)

[Ric65]   G. Rickayzen, *Theory of Superconductivity* (John Wiley & Sons, New York, 1965)

[Tin96]   M. Tinkham, *Introduction to Superconductivity*, 2nd edition (McGraw-Hill, New York, 1996)

Further reading

[Gor89] L.P. Gor'kov and G. Grüner, eds, *Charge Density Waves in Solids*, Modern Problems in Condensed Matter Sciences **25** (North-Holland, Amsterdam, 1989)

[Mon85] P. Monceau, ed., *Electronic Properties of Inorganic Quasi-One Dimensional Compounds* (Riedel, Dordrecht, 1985)

[Par69] R.D. Parks, ed., *Superconductivity* (Marcel Dekker, New York 1969)

[Pip65] A.B. Pippard, *The Dynamics of Conduction Electrons* (Gordon and Breach, New York, 1965)

[Por92] A.M. Portis, *Electrodynamics of High-Temperature Superconductors* (World Scientific, Singapore, 1992)

[Sin77] K.P. Sinha, *Electromagnetic Properties of Metals and Superconductors*, in: *Interaction of Radiation with Condensed Matter*, **2** IAEA-SMR-20/20 (International Atomic Energy Agency, Vienna, 1977), p. 3

[Wal64] J.R. Waldram, *Adv. Phys.* **13**, 1 (1964)

Part two

Methods

Introductory remarks

The wide array of optical techniques and methods which are used for studying the electrodynamic properties of solids in the different spectral ranges of interest for condensed matter physics is covered by a large number of books and articles which focus on different aspects of this vast field of condensed matter physics. Here we take a broader view, but at the same time limit ourselves to the various principles of optical measurements and compromise on the details. Not only conventional optical methods are summarized here but also techniques which are employed below the traditional optical range of infrared, visible, and ultraviolet light. These techniques have become increasingly popular as attention has shifted from single-particle to collective properties of the electron states of solids where the relevant energies are usually significantly smaller than the single-particle energies of metals and semiconductors.

We start with the definition of propagation and scattering of electromagnetic waves, the principles of propagation in the various spectral ranges, and summarize the main ideas behind the resonant and non-resonant structures which are utilized. This is followed by the summary of spectroscopic principles – frequency and time domain as well as Fourier transform spectroscopy. We conclude with the description of measurement configurations, single path, interferometric, and resonant methods where we also address the relative advantages and disadvantages of the various measurement configurations.

General books and monographs

M. Born and E. Wolf, *Principles of Optics*, 6th edition (Cambridge University Press, Cambridge, 1999)

P.R. Griffiths and J.A. de Haseth, *Fourier Transform Infrared Spectrometry* (John Wiley & Sons, New York, 1986)

G. Grüner, ed., *Millimeter and Submillimeter Wave Spectroscopy of Solids* (Springer-Verlag, Berlin, 1996)

D.S. Kliger, J.W. Lewis, and C.E. Radall, *Polarized Light in Optics and Spectroscopy* (Academic Press, Boston, MA, 1990)

C.G. Montgomery, *Technique of Microwave Measurements*, MIT Rad. Lab. Ser. **11** (McGraw Hill, New York, 1947)

E.D. Palik, ed., *Handbook of Optical Constants of Solids* (Academic Press, Orlando, FL, 1985–1998)

8

Techniques: general considerations

The purpose of spectroscopy as applied to solid state physics is the investigation of the (complex) response as a function of wavevector and energy; here, in the spirit of optical spectroscopy, however, we limit ourselves to the response sampled at the zero wavevector, $\mathbf{q} = 0$ limit. Any spectroscopic system contains four major components: a radiation source, the sample or device under test, a detector, and some mechanism to select, to change, and to measure the frequency of the applied electromagnetic radiation. First we deal with the various energy scales of interest. Then we comment on the complex response and the requirements placed on the measured optical parameters. In the following sections we discuss how electromagnetic radiation can be generated, detected, and characterized; finally we give an overview of the experimental principles.

8.1 Energy scales

Charge excitations which are examined by optical methods span an enormous spectral range in solids. The single-particle energy scales of common metals such as aluminum – the bandwidth W, the Fermi energy $\mathcal{E}_\mathrm{F}$, together with the plasma frequency $\hbar\omega_\mathrm{p}$ – all fall into the 1–10 eV energy range, corresponding to the visible and ultraviolet parts of the spectrum of electromagnetic radiation. In band semiconductors like germanium, the bandwidth and the plasma frequency are similar to values which are found in simple metals; the single-particle bandgap $\mathcal{E}_\mathrm{g}$ ranges from 10^{-1} eV to 5 eV as we go from small bandgap semiconductors, such as InSb, to insulators, such as diamond.

Single-particle gaps which arise as the consequence of many-body interactions are typically smaller than the gaps we find in band semiconductors, such gaps – for example the superconducting gap or gaps associated with other broken symmetry states of metals – depend on the strength of the interactions which lead to the particular state; the magnitude of these gaps spans a wide range, but is usually

207

10^{-1} eV or smaller. The gap in the superconducting state of aluminum for example is 0.3 meV; in contrast this gap in high temperature superconductors or in materials with charge density wave ground states can reach energies of 100 meV. Similarly, relatively small energy scales of the order of 10 meV characterize impurity states in semiconductors and also lattice vibrations.

Electron–phonon and electron–electron interactions also lead to reduced bandwidth (while keeping the character of states unchanged), and this reduction can be substantial if these interactions are large. This is the case in particular for strong electron–electron interactions; in the so-called heavy fermion materials, for example, the typical bandwidth can be of the order of 1 meV or smaller. The response of collective modes, such as a pinned density wave, lies below this spectral range; charge excitations near phase transitions and in a glassy state extend (practically) to zero energy.

In addition to these energy scales, the frequencies which are related to the relaxation process can vary widely, and the inverse scattering time $1/\tau$ ranges from 10^{15} s^{-1} for a metal with a strong scattering process (leading to mean free paths of the order of one lattice constant) to 10^{12} s^{-1} in clean metals at low temperatures.

8.2 Response to be explored

In the case of a linear response to an electromagnetic field with a sinusoidal time variation, the response of only one individual frequency is detected, and all the other spectral components are suppressed, either on the side of the radiation source, or on the detection side. Since we consider only elastic light scattering, the frequency will not be changed upon interaction; also the response does not contain higher harmonics. The electric field $\mathbf{E}(\omega)$ is described by

$$\mathbf{E}(\omega) = \mathbf{E}_0 \sin\{-\omega t\} \quad . \tag{8.2.1}$$

The current response measured at that single frequency ω is split into an in-phase and an out-of-phase component

$$\begin{aligned}\mathbf{J}(\omega) &= \mathbf{J}_0 \sin\{-\omega t + \delta(\omega)\} \\ &= \mathbf{J}_0 \left[\sin\{-\omega t\}\cos\{\delta(\omega)\} + \cos\{-\omega t\}\sin\{\delta(\omega)\}\right] \\ &= \sigma_1 \mathbf{E}_0 \sin\{-\omega t\} + \sigma_2 \mathbf{E}_0 \cos\{-\omega t\} \quad ,\end{aligned}$$

which is then combined by the complex conductivity $\hat{\sigma}(\omega)$ at this frequency:

$$\mathbf{J}(\omega) = \mathbf{J}_0 \exp\{-i\omega t\} = [\sigma_1(\omega) + i\sigma_2(\omega)]\mathbf{E}_0 \exp\{-i\omega t\} = \hat{\sigma}(\omega)\mathbf{E}(\omega) \quad , \tag{8.2.2}$$

using the complex notation of frequency dependent material parameters. In other words, the complex notation is used to indicate the phase shift which might occur between stimulus (electric field) and response (current density). A full evaluation of the response which leads to both components of the optical conductivity obviously requires the measurement of two optical parameters; parameters which are examined by experiment and which are related – through Maxwell's equations in a medium – to the complex conductivity. At low frequencies, as a rule, both the phase and amplitude of the optical response are measured – such as the resistance and capacitance, the reflected amplitude and phase, or the surface resistance and surface reactance, respectively. Alternatively, if only one parameter is observed, such as the absorbed power, experiments in a broad frequency range have to be conducted in order to perform the Kramers–Kronig analysis (see Section 3.2) for the determination of the complex conductivity. Since a single method does not cover the entire range, it is often necessary to combine the results obtained by a variety of different techniques. The main problem here is the extrapolation to low frequencies ($\omega \to 0$) and to above the measurement range ($\omega \to \infty$).

There are other issues which have to be resolved by a particular experiment. Assume that we apply a slowly varying time dependent electric field to a specimen with contacts applied at the ends. The purpose of the contacts is to allow the electric charge to flow in and out of the specimen and thus to prevent charges building up at the boundary. The wavelength $\lambda = c/f$ for small frequencies f, say below the microwave range, is significantly larger than the typical specimen dimensions. The wavevector dependence of the problem can under such circumstances be neglected. There is a time lag between the electrical field and the current, expressed by the complex conductivity $\hat{\sigma}(\omega) = \sigma_1(\omega) + i\sigma_2(\omega)$. With increasing frequencies ω, two things happen. The alternating electric field may be screened by currents induced in the material, and the current flows in a surface layer. This layer is typically determined by the parameter called the skin depth $\delta_0 = c(2\pi\omega\sigma_1)^{-1/2}$. For typical metals at room temperature the skin depth at a frequency of 10^{10} Hz is of the order of 10^{-5} m. Whenever this length scale δ_0 is smaller than the dimension of the specimen, the skin effect has to be taken into account when the complex conductivity is evaluated. On increasing the frequency further, another point also becomes important: the wavelength of the electromagnetic field becomes comparable to the dimensions of the specimens, this typically occurs in the millimeter wave spectral range. Well above these frequencies, entirely different measurement concepts are applied. It is then assumed that in the two dimensions perpendicular to the direction of wave propagation the specimen spreads over an area which is large compared to the wavelength, and the discussion is based on the solution of Maxwell's equations for plane waves and for an infinite plane boundary.

The objective is the evaluation of the two components of the optical conductivity or alternatively the dielectric constant – the parameters which characterize the medium – and to connect the experimental observations to the behavior proposed by theory. The quantities which describe the modification of the electromagnetic wave in the presence of the specimen under study are, for example, the power reflected off or transmitted through a sample of finite thickness, and also the change of the phase upon transmission and reflection. The task at hand is either to measure two optical parameters – such as the refractive index and absorption coefficient, or the surface resistance and surface reactance, or the amplitude and phase of the reflected electromagnetic radiation. Although this is often feasible at low frequencies where, because of the large wavelength, spatial sampling of the radiation is possible, the objective, however, is difficult to meet at shorter wavelengths (and consequently at higher frequencies) and therefore another method is commonly used: one evaluates a single parameter, such as the reflected or transmitted power, and relies on the Kramers–Kronig relations to evaluate two components of the complex conductivity. Note, however, that the Kramers–Kronig relations are non-local in frequency, and therefore the parameter has to be measured over a broad frequency range, or reasonable (but often not fully justified) approximations have to be made for extrapolations to high or to low frequencies.

8.3 Sources

The spectroscopies which are discussed here depend fundamentally on the characteristics of the electromagnetic radiation utilized; consequently a discussion of some of the properties of the sources are in order. As we have seen, the electromagnetic spectrum of interest to condensed matter physics extends over many orders of magnitude in frequency. This can only be covered by utilizing a large variety of different sources and detectors; their principles, specifications, and applicable range are the subject of a number of handbooks, monographs, and articles (see the Further reading section at the end of this chapter).

We distinguish between four different principles of generating electromagnetic waves (Fig. 8.1). At low frequencies mainly solid state electronic circuits are used; they are monochromatic and often tunable over an appreciably wide range. Above the gigahertz frequency range, electron beams are modulated to utilize the interaction of charge and electric field to create electromagnetic waves from accelerated electrons. Thermal radiation (black-body radiation) creates a broad spectrum, and according to Planck's law typical sources have their peak intensity in the infrared. Transition between atomic levels is used in lasers and discharge lamps. The radiation sources might operate in a continuous manner or deliver pulses as short as a few femtoseconds.

In recent years significant efforts have been made to extend the spectral range of synchrotron radiation down to the far infrared and thus to have a powerful broadband source available for solid state applications. Also free-electron lasers

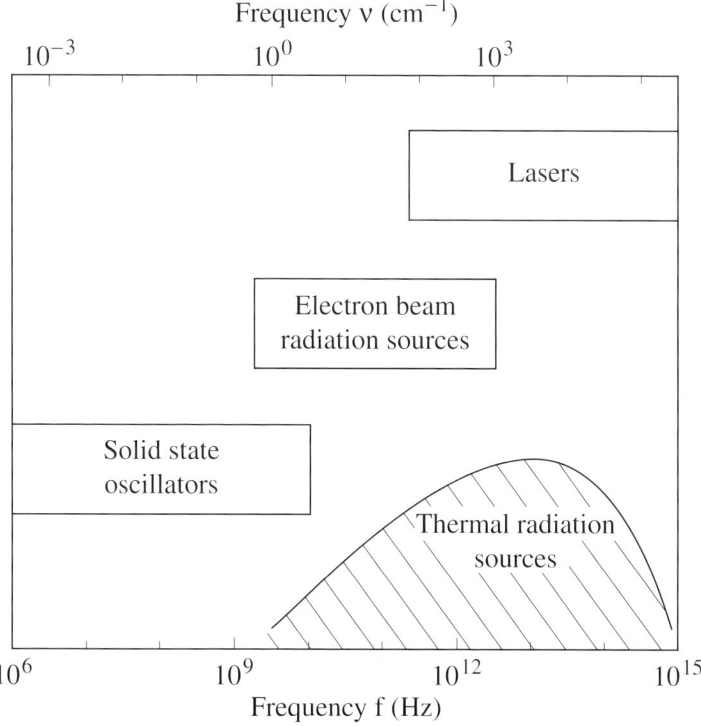

Fig. 8.1. Ranges of the electromagnetic spectrum in which the different radiation sources are applicable. At low frequencies solid state devices such as Gunn oscillators or IMPATT diodes are used. Up to about 2 THz, coherent monochromatic sources are available which generate the radiation by modulation of an electron beam (e.g. clystron, magnetron, backward wave oscillator). White light sources deliver a broad but incoherent spectrum from the far-infrared up to the ultraviolet; however, at both ends of the range the intensity falls off dramatically according to Planck's law. In the infrared, visible, and ultraviolet spectral ranges, lasers are utilized; in some cases, the lasers are tunable.

are about to become common in solid state spectroscopy since they deliver coherent but tunable radiation, and also short pulses.

For sources of electromagnetic radiation the bandwidth is an important parameter. Radiation sources produce either a broad spectrum (usually with a frequency dependent intensity), or they are limited (ideally) to a single frequency (a very narrow bandwidth) which in some cases can be tuned. A quantitative way of distinction is the coherence of the radiation, which is defined as a constant phase relation between two beams [Ber64, Mar82]. Time coherence, which has to be discriminated from the spatial coherence, is linked to the monochromaticity of the radiation, since only light which is limited to a single frequency and radiates over an infinite period of time is fully time coherent. Thus in reality any radiation exhibits only partial coherence as it originates from an atomic transition between levels with finite lifetime leading to a broadening. Monochromatism is determined by the bandwidth of the power spectrum, while the ability to form an interference pattern measures the time coherence. If light which originates from one source is split into two arms, with electric fields E_1 and E_2, of which one can be delayed by the time period τ – for instance by moving a Michelson interferometer out of balance by the distance $\delta = \tau/2c$ as displayed in Fig. 10.6 – the intensity of superposition is found to vary from point to point between maxima, which exceed the sum of the intensities in the beam, and minima, which may be zero; this fact is called interference. For the quantitative description a complex auto-coherence function is defined

$$\Gamma(\tau) = \lim_{T \to \infty} \frac{1}{T} \int_{-T/2}^{T/2} E_1^*(t) E_2(t + \tau)\, dt \quad , \tag{8.3.1}$$

where E^* denotes the complex conjugate of the electric field; the period of the sinusoidal wave with frequency $f = \omega/2\pi$ is defined by $T = 1/f$. The degree of coherence is then given by $\gamma(\tau) = \Gamma(\tau)/\Gamma(0)$, and we call $|\gamma(\tau)| = 1$ totally coherent, $|\gamma(\tau)| \leq 1$ partially coherent, and $\gamma(\tau) = 0$ totally incoherent. By definition, the coherence time τ_c is reached when $|\gamma(\tau)| = 1/e$; the coherence length l_c is then defined as $l_c = c\tau_c$, where c is the speed of light. l_c can be depicted as the average distance over which the phase of a wave is constant. For a coherent source it should be at least one hundred times the period of the oscillation; the wavepackages are then in phase over a distance more than one hundred times the wavelength. If the radiation is not strictly monochromatic, the coherence decreases rapidly as τ_c increases. A finite bandwidth Δf reduces the coherence length to $l_c = c/2\Delta f$. In general, the finite bandwidth is the limitation of coherence for most radiation sources.

The spatial coherence, on the other hand, refers to a spatially extended light source. The coherence of light which originates from two points of the source decreases as the distance δ between these positions increases. In analogy to the time coherence, we can define the auto-correlation function $\Gamma(\delta)$ and the degree of coherence $\gamma(\delta)$. Spatial coherence is measured by the interference fringes of light coming through two diaphragms placed in front of the radiation source. The absence of coherence becomes especially important for arc lamps, but also for some lasers with large beam diameters.

8.4 Detectors

Electromagnetic radiation is in general detected by its interaction with matter. The most common principles on which devices which measure radiation are based are

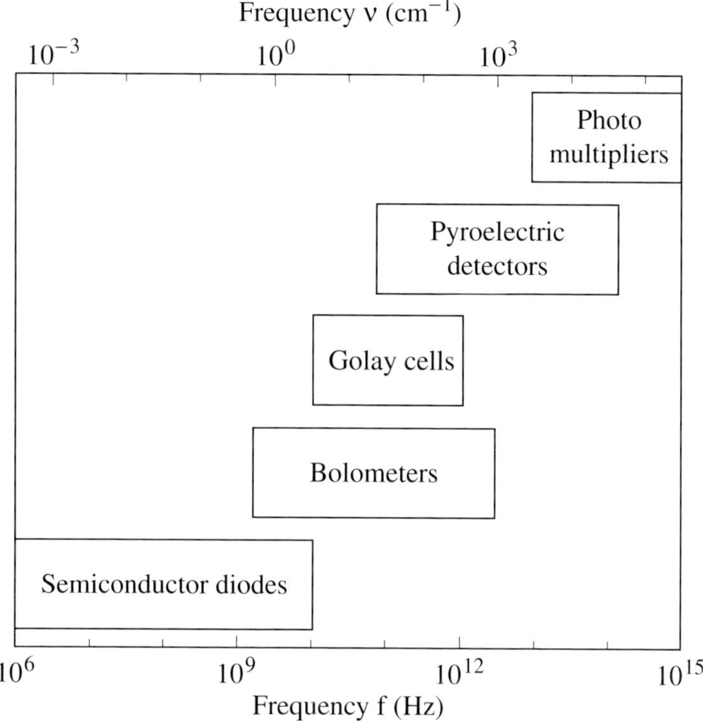

Fig. 8.2. Operating range of detectors. Semiconductor devices, such as Schottky diodes, can be used well into the gigahertz range of frequency. Thermal detectors, such as Golay cells and bolometers, operate up to a few terahertz; the infrared range is covered by pyroelectric detectors. Photomultipliers are extremely sensitive detectors in the visible and ultraviolet spectral ranges.

the photoelectric effect and the thermal effect (heating); less important are luminescence and photochemical reactions. In the first category, photons of frequency ω excite carriers across a gap $\mathcal{E}_g$ if $\hbar\omega > \mathcal{E}_g$ and thus the conductivity increases. The second class is characterized by a change in certain properties of the material due to an increase in temperature arising from the absorption of radiation; for example, the resistivity decreases (semiconductor bolometer) or the material expands (Golay cell). Most detectors are time averaging and thus probe the beam intensity, but non-integrating detectors are also used for measuring the power of the radiation; the time constant for the response can be as small as nanoseconds. Details of the different detector principles and their advantages are discussed in a large number of books [Den86, Der84, Kin78, Key80, Rog00, Wil70]. The various detectors commonly used for optical studies are displayed in Fig. 8.2.

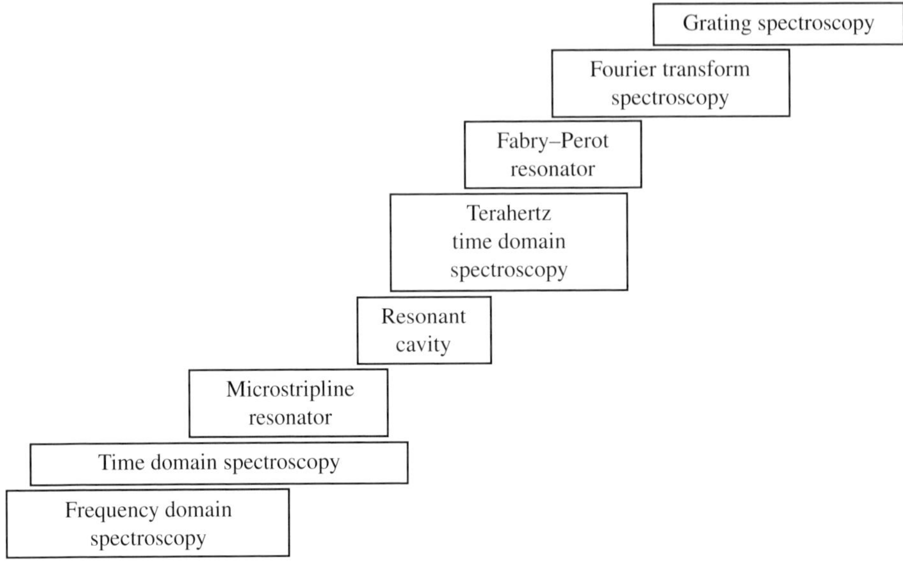

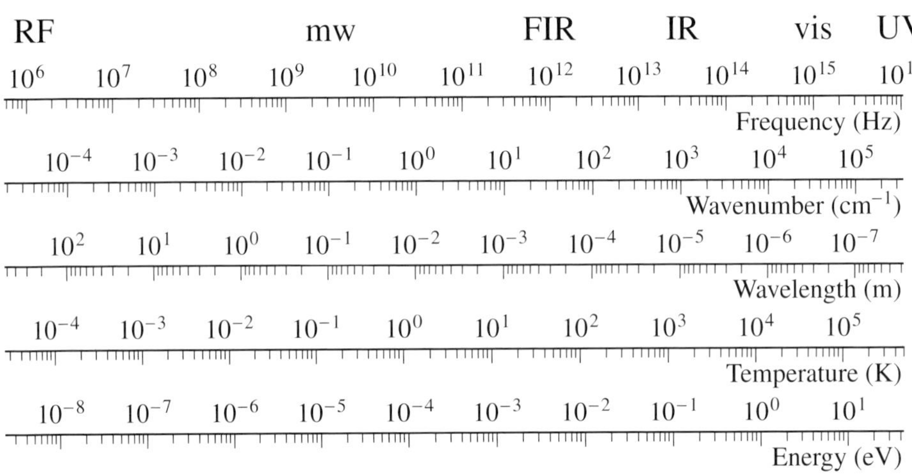

Fig. 8.3. Commonly employed methods of exploring the electrodynamics of metals in a wide spectral range. Also shown are the various energy units used. RF refers to radio frequencies; mw to microwaves; FIR and IR mean far-infrared and infrared, respectively; vis stands for visible, and UV stands for ultraviolet.

8.5 Overview of relevant techniques

The broad range of energy scales we encounter in solids implies that various techniques which are effective at vastly different parts of the electromagnetic spectrum are all of importance, and in fact have to be combined if a full account of the physics under question is attempted. This explains the variety of methods which are used, the variety of the hardware, the means of propagation of the electromagnetic radiation, and the relevant optical parameters which are measured.

The methods which have been utilized are of course too numerous to review. Some of the most commonly used experimental techniques are displayed in Fig. 8.3 together with the electromagnetic spectrum, as measured by different units. All these units have significance and Table G.4, p. 464, may be helpful. The s^{-1} or hertz scale is the natural unit for the radiation with a sinusoidal time variation; attention has to be paid not to confuse frequency f and angular frequency $\omega = 2\pi f$. The energy associated with the angular frequency is $\hbar\omega$; the unit commonly used is electron-volts. The temperature scale is in units for which $k_B T$ is important, as it establishes a correspondence between temperature driven and electromagnetic radiation induced charge response. Often this response is fundamentally different in the so-called quantum limit $k_B T < \hbar\omega$ as opposed to the $k_B T > \hbar\omega$ classical limit. Finally, the wavelength λ of the electromagnetic radiation is important, not least because it indicates its relevance with respect to typical sample dimensions.

The various methods applied in different ranges of the electromagnetic spectrum have much in common, in particular as far as the principles of light propagation and the overall measurement configurations are concerned. Interferometric techniques work equally well in the infrared and in the microwave spectral ranges – although the hardware is vastly different. Resonant techniques have been also employed in different spectral ranges – the advantages and disadvantages of these techniques do not depend on the frequency of the electromagnetic radiation.

Finally, one should keep in mind that the optical spectroscopy covered in this book is but one of the techniques which examine the charge excitations of solids; complementary techniques, such as electron energy loss spectroscopy, photo-emission, or Raman and Brillouin scattering, are also widely used. The response functions which are examined are different for different methods, and often a comparison of information offered by the different experimental results is required for a full characterization of the charge excitations of solids.

References

[Ber64] M.J. Beran and G.B. Parrent, *Theory of Partial Coherence* (Prentice-Hall, Englewood Cliffs, NJ, 1964)

[Den86] P.N.J. Dennis, *Photodetectors* (Plenum Press, New York, 1986)

[Der84] E.L. Dereniak and D.G. Crowe, *Optical Radiation Detectors* (J. Wiley, New York, 1984)

[Kin78] *Detection of Optical and Infrared Radiation*, edited by R.H. Kingston, Springer Series in Optical Sciences **10** (Springer-Verlag, Berlin, 1978)

[Key80] *Optical and Infrared Detectors*, edited by R.J. Keyes, 2nd edition, Topics in Applied Physics **19** (Springer-Verlag, Berlin, 1980)

[Mar82] A.S. Marathay, *Elements of Optical Coherence Theory* (Wiley, New York, 1982)

[Rog00] A. Rogalski, *Infrared Detectors* (Gordon and Breach, Amsterdam, 2000)

[Wil70] *Semiconductors and Semimetals*, **5**, edited by R.K. Willardson and A.C. Beer (Academic Press, New York, 1970)

Further reading

[Gra92] R.F. Graf, *The Modern Oscillator Circuit Encyclopedia* (Tab Books, Blue Ridge Summit, PA, 1992)

[Hec92] J. Hecht, *The Laser Guidebook*, 2nd edition (McGraw-Hill, New York, 1992)

[Hol92] E. Holzman, *Solid-State Microwave Power Oscillator Design* (Artech House, Boston, MA, 1992)

[Hen89] B. Henderson, *Optical Spectroscopy of Inorganic Solids* (Clarendon Press, Oxford, 1989)

[Kin95] R.H. Kingston, *Optical Sources, Detectors, and Systems* (Academic Press, San Diego, CA, 1995)

[Kuz98] H. Kuzmany, *Solid-State Spectroscopy* (Springer-Verlag, Berlin, 1998)

[Pei99] K.-E. Peiponen, E.M. Vartiainen, and T. Asakura, *Dispersion, Complex Analysis and Optical Spectroscopy*, Springer Tracts in Modern Physics **147** (Springer-Verlag, Berlin, 1999)

[Sch90] *Dye Lasers*, edited by P. Schäfer, 3rd edition, Topics in Applied Physics **1** (Springer-Verlag, Berlin, 1990)

[Sch00] W. Schmidt, *Optische Spektroskopie*, 2nd edition (Wiley-VCH, Weinheim, 2000)

[Yng91] S.Y. Yngversson, *Microwave Semiconductor Devices* (Kluwer Academic Publisher, New York, 1991)

9

Propagation and scattering of electromagnetic waves

A large variety of structures are used to propagate and guide electromagnetic radiation, with their applicability depending on the spectral range. From zero frequency up to the radio frequency range, the current can be supplied through electrical leads. Coaxial cables of different sizes are the most common arrangement in the kilohertz and megahertz range of frequency, and recent progress has made them available up to 100 GHz. In the microwave and millimeter wave range, in general, the propagation of electromagnetic radiation takes place via striplines or waveguides. From the infrared through the visible up to the ultraviolet spectral range, the light is transmitted via free space or, as in the case of optical fibers, guided through a dielectric material. Although the terminology used in these areas might differ vastly, the general principles of wave propagation are always the same. All structures which guide electromagnetic waves and even free space can be discussed within the concept of transmission lines as far as the wave propagation is concerned. Obviously the propagation of electromagnetic waves in a transmission line is fully given by Maxwell's equations; however, it is not necessary actually to solve the wave equations with the appropriate boundary conditions each time. Instead, we can use the impedance $\hat{Z}$ as the characteristic parameter, and then the wave propagation can be discussed in a manner which is independent of the particular kind of guiding structure.

After presenting the principles of wave propagation, we discuss the scattering from boundaries which terminate a transmission line or which serve as part of the line; this can for instance be a sample in a waveguide, termination of a coaxial cable, or a mirror in an optical setup. All these cases can be treated in the same way: as a change of the impedance at the interface. Here we evaluate the reflection and transmission at a single impedance mismatch, and subsequently we examine what happens for two interfaces, for example the front and back of a specimen.

Finally we look at resonant structures, which can be considered as isolated parts of a transmission line which are terminated by two impedance mismatches, and

217

thus which are weakly coupled to the exterior. This includes different arrangements such as RLC circuits, enclosed cavities, or Fabry–Perot resonators.

9.1 Propagation of electromagnetic radiation

In the absence of free charge and current, the propagation of electromagnetic radiation in free space or in a homogeneous medium is described by the wave equations (2.1.19) and (2.2.20), respectively. In the presence of conducting material, such as a wire or a waveguide, the appropriate boundary conditions (2.4.4) have to be taken into account; i.e. the tangential component of the electric field $\mathbf{E}$ and the normal component of the magnetic induction $\mathbf{B}$ are zero at the surface of a good conductor: $\mathbf{n_s} \times \mathbf{E} = 0$ and $\mathbf{n_s} \cdot \mathbf{B} = 0$, where $\mathbf{n_s}$ denotes the unit vector perpendicular to the boundary. The solutions for many simple, mainly symmetrical, configurations (a pair of parallel wires, microstriplines, coaxial cables, and rectangular waveguides, for example) are well documented, and we do not elaborate on them here.

The important overall concept which emerges is that the structure which is used to propagate the electromagnetic radiation can be characterized by a quantity called the impedance $\hat{Z}_c$, a complex quantity which depends on the particular structure used. The parameter also appears in the equation which describes the reflection from or transmission through a test structure (the sample to be measured) which is placed in the path of the electromagnetic radiation; the structure is again described by the characteristic impedance.

9.1.1 Circuit representation

Instead of solving Maxwell's equations for the description of wave propagation, for cables and wires it is more appropriate to use the circuit representation, eventually leading to the telegraphist's equation. This approach is common to electrical engineering, and is a convenient way of describing and calculating the wave propagation in transmission lines. Although the description can be applied to all waveguiding structures, it is best explained if we look at a pair of wires or a coaxial cable. Let us assume that the leads have a certain resistance and inductance, and that between the two wires there is some capacitance and maybe even losses (expressed as conductance) due to a not perfectly insulating dielectric. The circuit which includes these four components can be used to write down the relationship between currents and voltages at both ends of a transmission line segment displayed in Fig. 9.1. We define the wave to be traveling in the z direction and therefore expand $V(z + \Delta z, t)$ and $I(z + \Delta z, t)$ in a Taylor series ignoring the

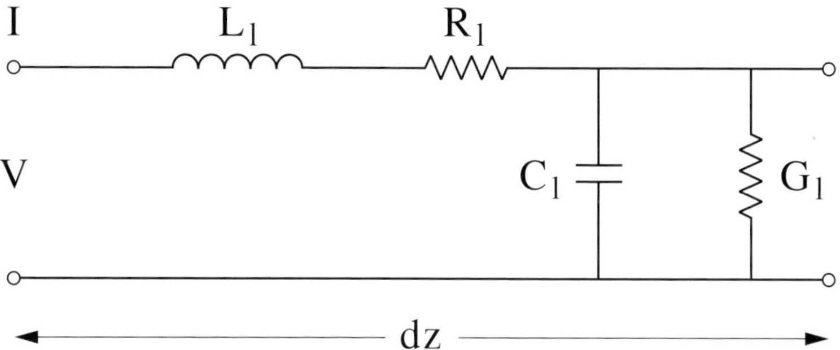

Fig. 9.1. Circuit representing a transmission line: the losses in the wires are given by the resistance $R_1\,dz$, the inductance of the wires is $L_1\,dz$, the capacitance between them is given by $C_1 dz$, and the losses of the dielectric between the wires are described by the conductance $G_1 dz$. If a voltage V is applied between the two contacts on the left hand side of the circuit, $V + \frac{\partial V}{\partial z}\,dz$ is obtained at the right hand side. A current I on one side becomes $I + \frac{\partial I}{\partial z} dz$.

terms which contain second and higher powers. Kirchhoff's laws yield

$$-\frac{\partial V(z,t)}{\partial z} = R_1 I(z,t) + L_1 \frac{\partial I(z,t)}{\partial t} \tag{9.1.1a}$$

$$-\frac{\partial I(z,t)}{\partial z} = G_1 V(z,t) + C_1 \frac{\partial V(z,t)}{\partial t} , \tag{9.1.1b}$$

where R_1, L_1, G_1, and C_1 are the resistance, inductance, conductance, and capacitance per unit length, respectively. Combining the two equations, the propagation of electromagnetic waves in a transmission line can now be described by the so-called telegraphist's equations of voltage $V(z,t)$ and current $I(z,t)$

$$\frac{\partial^2 V(z,t)}{\partial z^2} = R_1 G_1 V(z,t) + (R_1 C_1 + L_1 G_1)\frac{\partial V(z,t)}{\partial t} + L_1 C_1 \frac{\partial^2 V(z,t)}{\partial t^2} \tag{9.1.2a}$$

$$\frac{\partial^2 I(z,t)}{\partial z^2} = R_1 G_1 I(z,t) + (R_1 C_1 + L_1 G_1)\frac{\partial I(z,t)}{\partial t} + L_1 C_1 \frac{\partial^2 I(z,t)}{\partial t^2} . \tag{9.1.2b}$$

One set of solutions of these equations is given by traveling attenuated waves[1]

$$V(z,t) = V_0 \exp\{-i\omega t\} \exp\{\pm\hat{\gamma} z\} \tag{9.1.3}$$

(a similar equation exists to describe the current), where we obtain for the propagation constant

$$\hat{\gamma} = \frac{\alpha}{2} + i\beta = [(R_1 - i\omega L_1)(G_1 - i\omega C_1)]^{1/2} \tag{9.1.4}$$

---

[1] In engineering textbooks generally $\exp\{j\omega t\}$ is used, leading to equations apparently different from the one given here. In cases where the time dependence of waves is concerned, the replacement of i by $-j$ leads to identical results; there is never any difference in the physical results.

where the parameter α is the attenuation constant and β is the phase constant. The complex propagation constant $\hat{\gamma}$ is the common replacement for the wavevector $\mathbf{q} = \frac{\omega}{c}\hat{N}\mathbf{n_q}$ as defined in Eq. (2.3.2), where $\hat{q} = i\hat{\gamma}$; it fully characterizes the wave propagation. Hence the solution of the telegraphist's equation is reduced to a static problem which has to be calculated for each specific cross-section. Knowing the parameters of our circuit, we can therefore calculate how the electromagnetic waves propagate in the transmission line. From Eq. (9.1.1a) we find the voltage decay

$$-\frac{\partial V}{\partial z} = (R_1 - i\omega L_1)I = \pm\hat{\gamma}V \quad ,$$

and the decay of the current follows from Eq. (9.1.1),

$$I = \left[\frac{G_1 - i\omega C_1}{R_1 - i\omega L_1}\right]^{1/2} V = \hat{Y}_\mathrm{c}V \quad ,$$

where the electromagnetic fields propagating to the left (positive $\hat{\gamma}$ in Eq. (9.1.3)) are neglected, and $\hat{Y}_\mathrm{c}$ is called the admittance. The ratio of voltage to current

$$\hat{Z}_\mathrm{c} = \left[\frac{R_1 - i\omega L_1}{G_1 - i\omega C_1}\right]^{1/2} = \frac{1}{\hat{Y}_\mathrm{c}} \tag{9.1.5}$$

is called the characteristic impedance of the transmission line; it has the units of a resistance and is in accordance with the definition of the characteristic wave impedance as the ratio of electric field and magnetic field given by Eq. (2.3.27). This impedance fully characterizes the wave propagation in a transmission line and contains all information necessary for applications. It is the basic parameter used in the next section to calculate the reflectivity off or transmission through a material placed in the path of the electromagnetic wave.

Losses along the line and between the leads are assumed to be relatively small, i.e. $R_1 < \omega L_1$ and $G_1 < \omega C_1$. This condition allows (roughly speaking) the propagation of waves. The phase constant β becomes

$$\beta \approx \omega\,(L_1C_1)^{1/2} \left(1 - \frac{R_1G_1}{4\omega^2 L_1C_1} + \frac{G_1^2}{8\omega^2 C_1^2} + \frac{R_1^2}{8\omega^2 L_1^2}\right) \quad , \tag{9.1.6}$$

while the attenuation constant α

$$\alpha \approx R_1 \left(\frac{C_1}{L_1}\right)^{1/2} + G_1 \left(\frac{L_1}{C_1}\right)^{1/2} = \frac{R_1}{\hat{Z}_\mathrm{c}} + G_1\hat{Z}_\mathrm{c} \tag{9.1.7}$$

describes the exponential decay of the electric field maxima; i.e. the power dissipation along the line $P(z) = P_0 \exp\{-\alpha z\}$ according to Lambert–Beer's law (2.3.17). Without dissipation $G_1 = R_1 = 0$, $\alpha = 0$, and $\beta = \omega(L_1C_1)^{1/2}$. The characteristic impedance $\hat{Z}_\mathrm{c}$ is then always a real quantity with $\hat{Z}_\mathrm{c} = R_\mathrm{c} = (L_1/C_1)^{1/2}$. The losses of a transmission line are determined by the conductance G_1 and the resistance R_1 per unit length; in most common cases, $G_1 \approx 0$ and R_1 is small. The phase velocity v_{ph} of an electromagnetic wave is given by $v_{ph} = \omega/\beta \approx (L_1 C_1)^{-1/2}$ for a lossless transmission line. The group velocity $v_{gr} = \left.\frac{d\omega}{d\beta}\right|_0$ describes the velocity of the energy transport and cannot be expressed in a simple form. On the other hand, for $\beta = 0$ no wave propagates ($v_{ph} = \infty$), and the field decays exponentially even without damping by the line (evanescent waves).

9.1.2 Electromagnetic waves

Whereas in free space only transverse electric and magnetic (TEM) waves propagate, transmission lines may also support the propagation of transverse electric (TE) and transverse magnetic (TM) modes. As a point of fact this is not always disturbing: the modes excited in the transmission line influence the attenuation and may also be important when these structures are used for investigations of condensed matter.

TEM waves

As we have seen in Section 2.1.2, the electric and magnetic field components of electromagnetic waves in free space are perpendicular to the direction of propagation. Due to the boundary conditions this does not hold in the case of conducting media. However, with ohmic losses along the line negligible, guiding structures which contain two or more conductors are in general capable of supporting electromagnetic waves that are entirely transverse to the direction of propagation; the electric and the magnetic field have no longitudinal components ($E^L = E_z = 0$ and $H^L = H_z = 0$). They are called transverse electromagnetic (TEM) waves. The transverse electric field is $\mathbf{E}^T = -\nabla^T \Phi \exp\{i(\hat{q}z - \omega t)\}$; we want to suppress the time dependence in the following. The propagation can be expressed as

$$[(\nabla^T)^2 + \hat{\gamma}^2 + \hat{q}^2]\mathbf{E}^T \Phi \exp\{-\hat{\gamma}z\} = 0 \quad .$$

The magnetic field is given by Eq. (2.2.21), and with $\mathbf{n_q}$ the unit vector along the propagation direction $\mathbf{q}$:

$$\mathbf{H}^T = \frac{\hat{q}c}{\omega\mu_1'}\mathbf{n_q} \times \mathbf{E}^T = \frac{4\pi}{c}\frac{1}{\hat{Z}_{TEM}}\mathbf{n_q} \times \mathbf{E}^T \quad ; \tag{9.1.8a}$$

we find that the characteristic impedance of the TEM waves is given by

$$\hat{Z}_{TEM} = \frac{4\pi}{c}\left(\frac{\mu_1'}{\epsilon_1'}\right)^{1/2} = Z_0 \left(\frac{\mu_1'}{\epsilon_1'}\right)^{1/2} \tag{9.1.8b}$$

with the free space impedance Z_0. There is only a single TEM mode possible. Note that $\hat{Z}_{\text{TEM}}$ decreases as the dielectric constant of the transmission line ϵ_1' becomes larger. As mentioned above, in the absence of guiding structures (i.e. in a vacuum) TEM waves also propagate. There is no cutoff frequency (i.e. the electromagnetic radiation can propagate at any frequency) for TEM waves. This is particularly important for coaxial cables which are used down to zero frequency. The electric field distribution of TEM modes in coaxial cables is radial; the magnetic field is circular. The upper frequency limits for the use of coaxial cables is determined by the increasing losses of the conducting wires but also by the dielectric material between. At higher frequencies, however, the propagation of TE and TM modes also becomes possible.

TM waves

Lossless (or weakly dissipating) guiding structures with one or more conductors and a homogeneous dielectric can support electromagnetic waves in which the magnetic field is entirely transverse to the direction of propagation ($H^{\text{L}} = H_z = 0$) but in which the electric field has in addition a longitudinal component ($E^{\text{L}} = E_z \neq 0$). These waves are called TM or E waves. The electric and magnetic fields have the form $\mathbf{E} = (\mathbf{E}^{\text{T}} + \mathbf{E}^{\text{L}}) \exp\{-\hat{\gamma}z\}$ and $\mathbf{H} = \mathbf{H}^{\text{T}} \exp\{-\hat{\gamma}z\}$, and the spatial part of the wave equation is given by

$$[(\nabla^{\text{T}})^2 + \hat{\gamma}^2 + q^2](\mathbf{E}^{\text{T}} + \mathbf{E}^{\text{L}}) \exp\{-\hat{\gamma}z\} = 0 \quad ,$$

where the longitudinal component is $[(\nabla^{\text{T}})^2 + \hat{q}_{\text{c}}^2]E^{\text{L}} = 0$ and the cutoff wavenumber $\hat{q}_{\text{c}}$ is defined as

$$\hat{q}_{\text{c}}^2 = \hat{\gamma}^2 + \hat{q}^2 = \left(\frac{2\pi}{\lambda_{\text{c}}}\right)^2 \quad . \tag{9.1.9}$$

There is no wave propagation possible for wavelength λ exceeding the cutoff wavelength λ_{c}; crudely speaking, the wave has to be shorter than twice the distance between the two conductors which constitute the transmission line or the opening of the waveguide. For longer waves only evanescent waves are possible where the fields die off exponentially in space. Even without damping no energy transport takes place; a fact which is utilized if leakage has to be avoided through openings. The relations for TM waves are:

$$\mathbf{H}^{\text{T}} = -\frac{i\omega\epsilon_1'}{c\hat{\gamma}}\mathbf{n_q} \times \mathbf{E}^{\text{T}} = \frac{i\omega\epsilon_1'}{c\hat{q}_{\text{c}}^2}\mathbf{n_q} \times \nabla^{\text{T}}E^{\text{L}} \tag{9.1.10a}$$

$$\mathbf{E}^{\text{T}} = \frac{c}{4\pi}\hat{Z}_{\text{TM}}\mathbf{n_q} \times \mathbf{H}^{\text{T}} \tag{9.1.10b}$$

$$\hat{Z}_{\text{TM}} = -\frac{4\pi\hat{\gamma}}{i\omega\epsilon_1'} = -\frac{c\hat{\gamma}}{i\omega\epsilon_1'}Z_0 \quad , \tag{9.1.10c}$$

where $\hat{\gamma}$ is determined by the particular propagating mode. This implies that the characteristic impedance of a transmission line is different for different modes. Although not commonly done, in principle TM waves can be used if the longitudinal response of a material is studied. If we place a material of interest between two plates, we can excite a TM wave which has also a longitudinal component of the electric field. Thus σ_1^L and ϵ_1^L can be measured and evaluated at finite frequencies, which is not possible for free space wave propagation.

TE waves

In analogy to TM waves, lossless guiding structures that contain one or more conductors and a homogeneous dielectric are also capable of supporting electromagnetic waves in which the electric field is entirely transverse to the direction of propagation ($E^L = E_z = 0$) but the magnetic field may also have a longitudinal component ($H^L = H_z \neq 0$). These waves are called TE or H waves. The propagation of electric fields $\mathbf{E} = \mathbf{E}^T \exp\{-\hat{\gamma}z\}$ and magnetic fields $\mathbf{H} = (\mathbf{H}^T + \mathbf{H}^L)\exp\{-\hat{\gamma}z\}$. The spatial propagation can be written as

$$[(\nabla^T)^2 + \hat{\gamma}^2 + \hat{q}^2](\mathbf{H}^T + \mathbf{H}^L)\exp\{-\hat{\gamma}z\} = 0 \quad ,$$

where the longitudinal component is $[(\nabla^T)^2 + \hat{q}_c^2]H^L = 0$. The equations for TE waves are:

$$\mathbf{E}^T = \frac{i\omega\mu_1'}{c\hat{\gamma}}\mathbf{n_q} \times \mathbf{H}^T = \frac{-i\omega\mu_1'}{c\hat{q}_c^2}\mathbf{n} \times \nabla^T H^L \tag{9.1.11a}$$

$$\mathbf{H}^T = \frac{4\pi}{c}\frac{1}{\hat{Z}_{TE}}\mathbf{n_q} \times \mathbf{E}^T \tag{9.1.11b}$$

$$\hat{Z}_{TE} = -\frac{4\pi}{c}\frac{i\omega\mu_1'}{c\hat{\gamma}} = \frac{i\omega\mu_1'}{c\hat{\gamma}}Z_0 \quad . \tag{9.1.11c}$$

As in the case of TM waves the characteristic impedance of the guiding structure is defined in analogy to the wave impedance of free space (Eqs (2.3.27) to (2.3.29)). The characteristic impedance describes the resistance of the transmission line; in contrast to TM waves, the impedance of TE waves does not depend on the dielectric material the structure is filled with. The cutoff wavenumber is defined in the same way as for TM waves: $\hat{q}_c^2 = \hat{\gamma}^2 + \hat{q}^2 = (2\pi/\lambda_c)^2$; for smaller wavevectors no wave propagation is possible.

9.1.3 Transmission line structures

Next we evaluate $\hat{Z}_c$, the parameter which is needed to describe the scattering problem: the scattering of electromagnetic radiation on a sample which, as we will see, is described by an impedance $\hat{Z}_S$.

In the case that metallic boundaries are present, the characteristic impedance $\hat{Z}_c$ depends not only on the medium, but also on the geometry and the mode which is excited. The task at hand now is to evaluate the geometrical factor for the transmission lines of interest and then see how the propagation parameters are modified compared to free space propagation.

The best known examples of transmission lines (Fig. 9.2) are two parallel wires (Lecher line), parallel plates (microstripline and stripline), coaxial cables, and hollow (rectangular) waveguides; the respective circuit representations are given in [Ell93, Gar84, Poz90, Ram93] and in numerous handbooks [Dix91, Mag92, Mar48, Smi93]. There is a simple concept which applies in general: the calculation of the four circuit parameters (capacitance, inductance, conductance, and resistance) is reduced to two geometrical parameters (called A and A') and the knowledge of the material parameters which form the transmission line.[2] In this chapter we denote the material parameters of the surrounding dielectric by a prime, and unprimed symbols refer to the properties of the actual transmission line. The transmission line is assumed to be filled by a material with the dielectric constant ϵ_1' and permeability μ_1'; the ohmic losses, due to the conductance of the material, are described by σ_1'. The capacitance per unit length of a system is given by

$$C_1 \;=\; \frac{1}{4\pi}\frac{\epsilon_1'}{V_0^2}\int_S \mathbf{E}\cdot\mathbf{E}^*\,\mathrm{d}s \;=\; \frac{1}{4\pi}\epsilon_1'\frac{1}{A} \quad, \tag{9.1.12}$$

where the integral is taken over the cross-section S of the transmission line. A is a constant which solely depends on the geometry of the particular setup; we will give some examples below. In a similar way the inductance, impedance, and conductance can be calculated:

$$L_1 \;=\; \frac{4\pi}{c^2}\frac{\mu_1'}{I_0^2}\int_S \mathbf{H}\cdot\mathbf{H}^*\,\mathrm{d}s \;=\; \frac{4\pi}{c^2}\mu_1' A \quad, \tag{9.1.13}$$

$$Z_c \;=\; \left(\frac{L_1}{C_1}\right)^{1/2} \;=\; \frac{4\pi}{c}\left(\frac{\mu_1'}{\epsilon_1'}\right)^{1/2} A = Z_0\left(\frac{\mu_1'}{\epsilon_1'}\right)^{1/2} A \tag{9.1.14}$$

and

$$G_1 \;=\; \frac{1}{4\pi}\frac{\sigma_1'}{V_0^2}\int_S \mathbf{E}\cdot\mathbf{E}^*\,\mathrm{d}s \;=\; \frac{1}{4\pi}\sigma_1'\frac{1}{A} \quad. \tag{9.1.15}$$

The ohmic losses per unit length R are proportional to the surface resistance R_S of the guiding material and can be calculated by integrating over the conductor

---

[2] Optical fibers commonly used in the visible and near-infrared range of frequency are not covered by the description due to different boundary conditions since there is no conducting material used, but total reflection at oblique incidence is used [Ada81, Mar91]. Of course total reflection is also covered by the impedance mismatch approach.

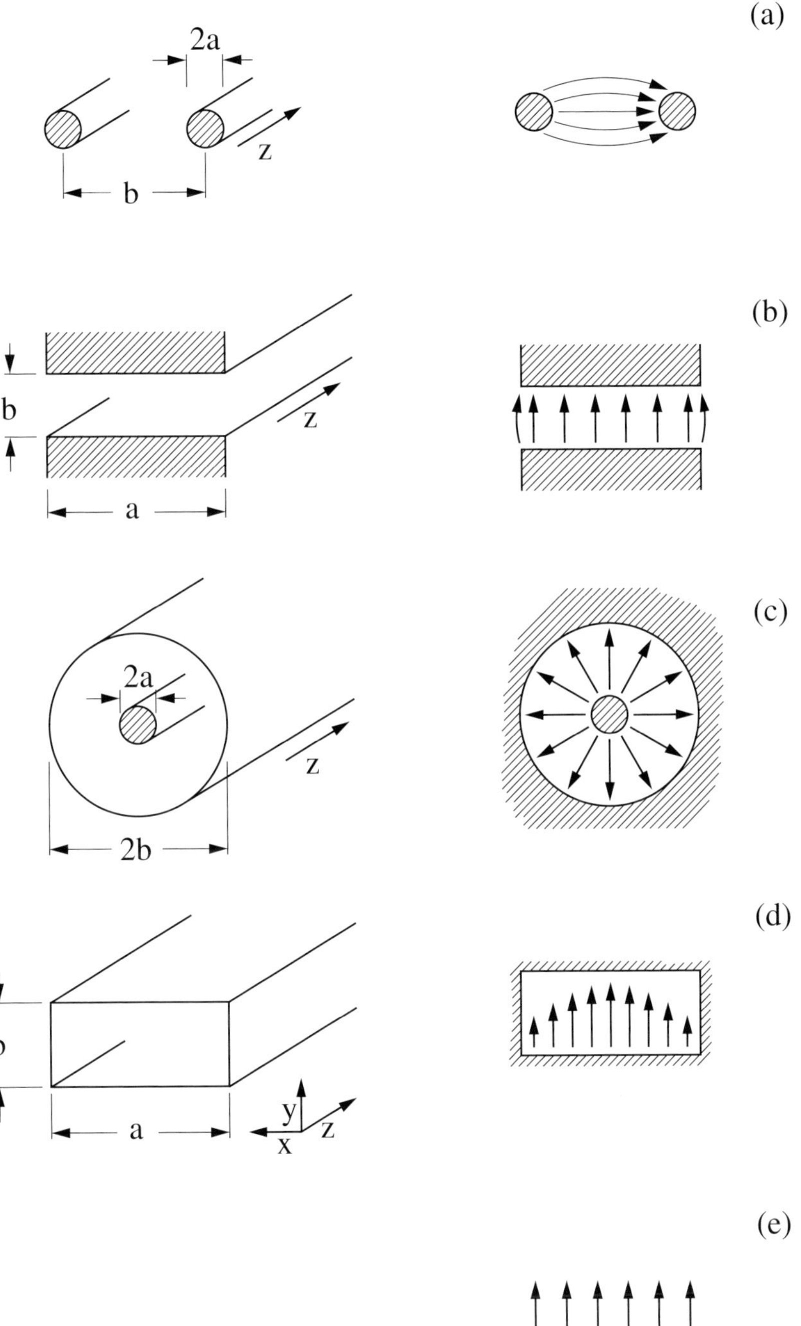

Fig. 9.2. Different transmission line configurations: (a) two wires, (b) parallel plate, (c) coaxial cable, and (d) rectangular waveguide. On the right hand side are the corresponding electric field configurations, where the free space (e) is added.

boundaries:

$$R_1 = \frac{R_S}{I_0^2} \int_C \mathbf{H} \cdot \mathbf{H}^* \, dl = A' R_S = \frac{A'}{\sigma_1 \delta_0} = A' \left(\frac{2\pi\omega}{c^2 \sigma_1} \right)^{1/2} = A' \frac{2\pi}{c} \left(\frac{f}{\sigma_1} \right)^{1/2} \quad ,$$

$$(9.1.16)$$

where the proportionality constant A' also depends on the geometry of the transmission line. σ_1 describes the conductivity of the metal which constitutes the line, $f = \omega/2\pi$ is the frequency of the transmitted waves, and δ_0 is the skin depth. From this equation we see that the losses of a transmission line increase with frequency f. In the infrared range this becomes a problem, in particular since $\sigma_1(\omega)$ of the metal decreases and $\sigma_1'(\omega)$ of the dielectric in general increases. This then makes the use of free space propagation or the utilization of optical fibers advantageous. The power losses per unit length are in general due to both the finite conductivity of the metallic conductors σ_1 and the lossy dielectric σ_1' with which the line is filled. The attenuation α along the line is defined by the power dissipation in direction z; it can easily be calculated by Eq. (9.1.7), if it is only due to the metallic surface

$$\alpha = \frac{A'}{A} \frac{R_S}{Z_0} \quad . \tag{9.1.17}$$

Integrating the Poynting vector over the cross-section yields the power propagation along the line:

$$P = \frac{1}{2} \frac{c}{4\pi} \int_S E_s H^T \, ds = \frac{1}{2} \left(\frac{c}{4\pi} \right)^2 \frac{A}{A'} Z_0 (H^T)^2 \quad . \tag{9.1.18}$$

For low-loss transmission lines A should be large and A' should be small. From Eq. (9.1.6) we see that for a transmission line with losses $R_1 = G_1 = 0$ the phase velocity $v_{ph} = (L_1 C_1)^{-1/2} = c/(\mu_1 \epsilon_1)^{1/2}$ independent of the particular geometry.

The problem of wave propagation in a transmission line is now reduced to finding the geometrical factors A and A' for special cases of interest; this means evaluating the field arrangement and solving the integrals in the above equations. Note that the particular field distribution does not only depend on the geometry, but also on the mode which is excited, as discussed in Section 9.1.2.

Free space and medium

Before we discuss special configurations of metallic transmission lines, let us recall the propagation of a plane wave along the z direction in an isotropic medium (with ϵ_1, μ_1 but no losses) or free space ($\epsilon_1 = \mu_1 = 1$), in order to demonstrate that our approach of circuit representation also covers this case. Since $\mathbf{H} = (\epsilon_1/\mu_1)^{1/2} \mathbf{n}_z \times \mathbf{E}$ and $\mathbf{n}_z$ is the unit vector along z, we find that only transverse electromagnetic waves propagate. The wavevector $q = 2\pi/\lambda$ is given by

$$q = \frac{\omega}{c} (\mu_1 \epsilon_1)^{1/2} = i\hat{\gamma} = -\beta \quad ,$$

and a plane wave is not attenuated in free space. Obviously there are no losses associated with free space propagation. The impedance of free space can be evaluated to be

$$Z_0 = \left(\frac{L_1}{C_1}\right)^{1/2} = \frac{4\pi}{c} = 377 \ \Omega \tag{9.1.19}$$

in SI units (which equals $Z_0 = 4\pi/c = 4.19 \times 10^{-10}$ s cm^1 in cgs units), as we have derived in Section 2.3.2. If the wave propagates in a dielectric medium, characterized by $\hat{\epsilon}$ and μ_1, without boundaries we obtain

$$\hat{Z}_c = \frac{4\pi}{c} \left(\frac{\mu_1}{\hat{\epsilon}}\right)^{1/2} = Z_0 \left(\frac{\mu_1}{\hat{\epsilon}}\right)^{1/2} \tag{9.1.20}$$

for the impedance according to Eq. (2.3.30). Depending on the imaginary part of the dielectric constant $\hat{\epsilon}$ of the medium, losses become important.

Two wire transmission line

Perhaps the simplest transmission line, the so-called Lecher system shown in Fig. 9.2a, consists of two parallel wires separated by an insulating material (ϵ_1', μ_1') – which might also be air. Let us assume two perfectly conducting wires of diameter $2a$ spaced a distance b apart. The geometrical constant A of this system can be calculated, and we find $A = (1/\pi)\text{arccosh}\,\{2b/a\} \approx (1/\pi)\ln\{b/a\}$ where the approximation is valid for $a \ll b$; typical values are $b/a \approx 10$. Assuming $R_1 \to 0$ and $G_1 \to 0$ the characteristic impedance Z_c of a two wire transmission line is given by:

$$Z_c = Z_0 \left(\frac{\mu_1'}{\epsilon_1'}\right)^{1/2} \frac{1}{\pi}\text{arccosh}\left\{\frac{2b}{a}\right\} \approx Z_0 \left(\frac{\mu_1'}{\epsilon_1'}\right)^{1/2} \frac{1}{\pi}\ln\left\{\frac{b}{a}\right\} \ . \tag{9.1.21}$$

The resistance R_1 per unit length is given by

$$R_1 = \frac{4\pi}{c} \left(\frac{f}{\sigma_1}\right)^{1/2} \frac{b}{4\pi a^2} \left[\left(\frac{b}{2a}\right)^2 - 1\right]^{-1/2}, \tag{9.1.22}$$

where σ_1 is the conductivity of the (usually copper) wires.

Parallel plate transmission line

Similar considerations hold for a parallel plate transmission line of spacing b and width a (Fig. 9.2b) which in general may be filled with a dielectric, to first approximation lossless material (ϵ_1' and μ_1', $\sigma_1' = 0$). For wide transmission lines $a \gg b$, when the effect at the edges is negligible, the geometrical factor is approximately

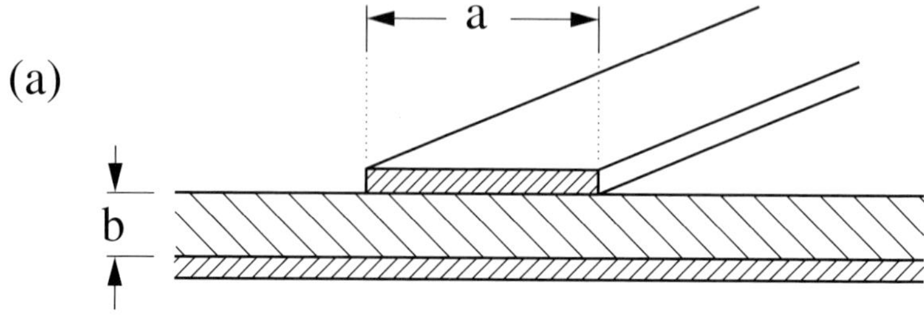

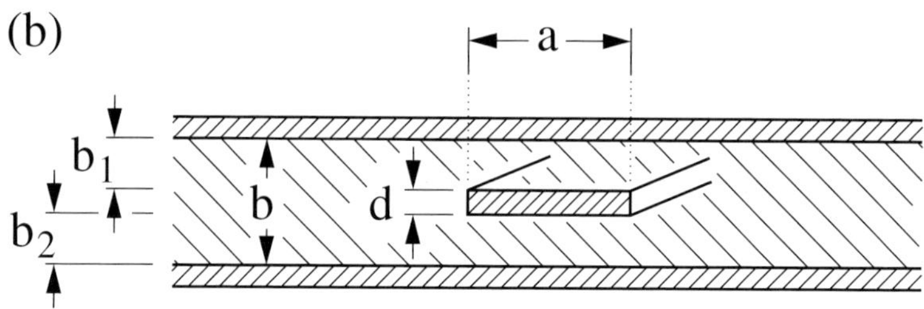

Fig. 9.3. (a) Cross-section through a microstripline; the line of width a is usually made from copper and is separated from the metallic ground plate by an dielectric spacer of thickness b. (b) Design of a stripline where the metallic line is embedded in the dielectric material with the ground plates at the bottom and on the top; in most cases $b_1 = b_2 \gg d$.

$A \approx b/a$. The characteristic impedance is then evaluated as

$$Z_c = Z_0 \left(\frac{\mu_1'}{\epsilon_1'}\right)^{1/2} \frac{b}{a} \left[1 + \frac{2b}{\pi a} + \frac{2b}{\pi a} \ln\left\{\frac{\pi a}{2b} + 1\right\}\right]^{-1} \approx Z_0 \left(\frac{\mu_1'}{\epsilon_1'}\right)^{1/2} \frac{b}{a} \quad. \tag{9.1.23}$$

The resistance per unit length is independent of b and is given by

$$R_l = \frac{4\pi}{c} \left(\frac{f}{\sigma_1}\right)^{1/2} \frac{1}{a} \quad. \tag{9.1.24}$$

The parallel plate transmission line is the model for microstriplines and striplines (Fig. 9.3) [Bha91, Gar94].

Coaxial cable

Coaxial cables are the preferred transmission lines in the microwave and in the lower millimeter wave spectral range. The geometrical constant of a lossless coaxial cable, with a and b the radii of the inner and the outer conductor as displayed in Fig. 9.2c, is $A = (1/2\pi) \ln\{b/a\}$. The characteristic impedance of a lossless coaxial cable can be written as:

$$Z_c = Z_0 \left(\frac{\mu_1'}{\epsilon_1'}\right)^{1/2} \frac{1}{2\pi} \ln\left\{\frac{b}{a}\right\} \quad . \tag{9.1.25}$$

Note, that – as this solution implies – coaxial cables can have different dimensions with the same impedance. Besides the losses of dielectric material ($\sigma_1' \neq 0$), the attenuation is determined by the metal; in the latter case the resistance per unit length is given by

$$R_1 = \frac{4\pi}{c} \left(\frac{f}{\sigma_1}\right)^{1/2} \frac{1}{4\pi} \left(\frac{1}{a} + \frac{1}{b}\right) \quad , \tag{9.1.26}$$

which sets the higher frequency limitation for the use of coaxial cables. A second restriction is the occurrence of higher modes as seen in Section 9.1.2. A commonly used coaxial cable has an impedance of 50 Ω.

Rectangular waveguide at TE_{10} mode

The most important application of transverse electrical waves is in standard rectangular metal waveguides commonly operated at the basic TE_{10} mode. In the case of a waveguide with (almost) no losses, $\hat{q} = \frac{\omega}{c}\left(\epsilon_1'\mu_1'\right)^{1/2}$ and $\hat{\gamma} = i\beta$ with $\beta^2 = (\omega^2\epsilon_1'\mu_1'/c^2) - (\pi/a)^2$ and $Z_{TE} = Z_0(\hat{q}/\beta)$ for frequencies larger than the cutoff frequency $f_c = c/\lambda_c = c/(2a)$. As usual a is the larger side of the rectangular cross-section of the waveguide as displayed in Fig. 9.2d; it is about half the wavelength. The length of the guided wave λ is defined as the distance between two equal phase planes along the waveguide,

$$\lambda = \frac{2\pi}{\beta} = 2\pi \left[\frac{\omega^2}{c^2}\epsilon_1'\mu_1' - \left(\frac{\pi}{a}\right)^2\right]^{-1/2} > \lambda_0 \quad ,$$

where λ_0 is the wavelength in free space. In the most common case of an empty waveguide ($\epsilon_1' = \mu_1' = 1$), the relations for Z_{TE} and λ reduce to:

$$Z_{TE} = Z_0 \left[1 - \left(\frac{\lambda_0}{2a}\right)^2\right]^{-1/2} \quad \text{and} \quad \lambda = \lambda_0 \left[1 - \left(\frac{\lambda_0}{2a}\right)^2\right]^{-1/2} \quad . \tag{9.1.27}$$

The frequency dependences of the phase velocity and group velocity behave similarly. Below the cutoff frequency $\omega_c = \pi c/a$, no wave propagation is possible: the attenuation α, the impedance Z_{TE}, and the phase velocity v_{ph} diverge, while the group velocity v_{gr} and $\beta = \text{Re}\{\hat{q}\}$ (the real part of the wavevector) drop to zero.

The attenuation due to the wall losses is given by Eq. (9.1.7);

$$
\begin{aligned}
\alpha_{\mathrm{TE}_{01}} &= \frac{2R_{\mathrm{S}}}{Z_0 b}\frac{1 + \frac{2b}{a}\,(f_{\mathrm{c}}/f)^2}{\left[1 - (f_{\mathrm{c}}/f)^2\right]^{1/2}} \\
&= \frac{1}{2b}\left[\frac{f}{\sigma_1}\right]^{1/2}\left[1 - \left(\frac{f_{\mathrm{c}}}{f}\right)^2\right]^{-1/2}\left[1 + \frac{2b}{a}\left(\frac{f_{\mathrm{c}}}{f}\right)^2\right] \quad , \quad (9.1.28)
\end{aligned}
$$

and increases with frequency.

9.2 Scattering at boundaries

As we have discussed in the previous section, the propagation of electromagnetic radiation through transmission lines (and even free space) is fully described by the characteristic impedance of this structure. Hence we can disregard the particular arrangement of a transmission line and only consider its impedance. As we have seen, the concept of the characteristic impedance is quite general, and it allows us to present a unified description of wave propagation and scattering for low frequency problems as well as for typical optical arrangements. Any change in the characteristic impedance leads not only to a variation in the propagation parameter $\hat{\gamma}$ as we have discussed in the previous section, but also to a partial reflection of the electromagnetic wave at the boundary or interfaces of media with different $\hat{Z}_{\mathrm{S}}$. The scattering of the wave at the interface of two impedances is the subject of this section.

A large number of treatises deal with changes of the geometry of transmission lines or obstacles in the transmission line which modify the characteristic impedance [Mar48, Sch68]. Here we are less interested in a variation of the geometry than in the effects associated with specimens placed within the structures. We consider the change in the characteristic impedance if a specimen (of dielectric properties under consideration) is placed at an appropriate position in the transmission structure. The simplest case is light reflected off a mirror: the propagation in free space is described by the impedance Z_0, and the mirror by the surface impedance $\hat{Z}_{\mathrm{S}}$. From these two quantities we can immediately evaluate the complex reflection coefficient as demonstrated in Section 2.4.5. Similarly, the sample could be a thin rod inside a waveguide, some material used to replace the dielectric of a coaxial cable, or a device under test connected to the ports of a network analyzer. In these examples the simple change of the material properties described by the surface impedance $\hat{Z}_{\mathrm{S}}$ has to be supplemented by geometrical considerations leading to the concept of the load impedance $\hat{Z}_{\mathrm{L}}$.

In many practical cases, a specimen of finite size is placed inside a transmission line, implying that there is a second boundary at the back of the sample after

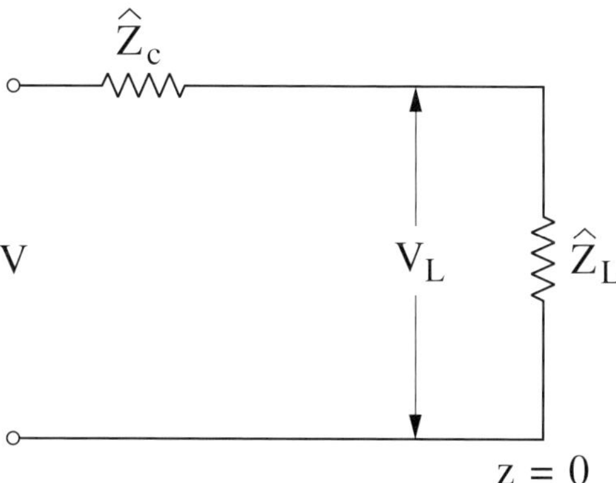

Fig. 9.4. Circuit representation of a transmission line with characteristic impedance $\hat{Z}_c$ which is terminated at the position $z = 0$ by the load impedance $\hat{Z}_L$. V_L denotes the voltage across the load impedance.

which the transmission line continues. A thin, semitransparent film in the path of light propagating in free space is the simplest example. If the wave is not fully absorbed by the sample, part of the incident power will be transmitted through both boundaries and continues to propagate in the rear transmission line. Transparent samples of finite thickness are discussed in Appendix B at length, here we will present only the main ideas pertaining to the scattering problem.

9.2.1 Single bounce

Let us start with the circuit representation introduced in Section 9.1.1 to illustrate the approach taken. The equivalent circuit of this general scattering problem is shown in Fig. 9.4, where z is the direction along the transmission line. In the case of no attenuation of the undisturbed transmission line ($\alpha = 0$), the solutions of the second order differential equations (9.1.2) for the voltage and current are the following (considering the spatial variation only):

$$V(z) = V_+ \exp\{-i\beta z\} + V_- \exp\{i\beta z\} \tag{9.2.1a}$$

$$I(z) = I_+ \exp\{-i\beta z\} - I_- \exp\{i\beta z\} \tag{9.2.1b}$$

$$= \frac{1}{\hat{Z}_c}[V_+ \exp\{-i\beta z\} + V_- \exp\{i\beta z\}] \quad ;$$

as defined in the previous section, β is the phase constant and $\hat{Z}_c$ is the characteristic impedance of the transmission line which also depends on the mode which is excited. Assuming that we terminate one end of the line (at the position $z = 0$) with a load impedance $\hat{Z}_L = R_L + iX_L$, part of the electromagnetic wave is reflected back, and the ratio of the amplitude of the reflected V_- and incident waves V_+ is called the reflection coefficient $r = V_-/V_+$. The voltage at the load resistance is $V_L = V_+ + V_-$, and the current going through is $I_L = I_+ - I_-$. Together with $V_L = \hat{Z}_L I_L$ this yields the complex reflection coefficient

$$\hat{r} = |\hat{r}|\exp\{i\phi_r\} = \frac{\hat{Z}_L - \hat{Z}_c}{\hat{Z}_L + \hat{Z}_c} \tag{9.2.2}$$

which also includes the phase difference between the incident and reflected waves. The transmission coefficient

$$\hat{t} = |\hat{t}|\exp\{i\phi_t\} = \frac{2\hat{Z}_L}{\hat{Z}_L + \hat{Z}_c} \tag{9.2.3}$$

describes the ratio of the voltage (or electric field) which passes the boundary to the incident one; it also takes the phase change into account. These two are the main equations which fully characterize the behavior of a wave as it hits a boundary. These equations also imply that the electromagnetic wave is sensitive only to the impedance in the guiding structure, meaning that changes in geometry or material cannot be observed if the complex impedances on both sides of the interface match. Note that we assumed implicitly that the same wave type can propagate on both sides of the boundary. This, for example, does not hold if a metallic waveguide where a TE mode propagates is terminated by an open end, since only TEM waves are possible in free space; in this case the reflection coefficient $\hat{r} = 1$.

In practice, the approach is the opposite: from the measurement of both components of the complex reflection (or transmission) coefficient and knowing the characteristic line impedance of the transmission line, it is straightforward to calculate the complex load impedance $\hat{Z}_L$ of the specimen under test:

$$\hat{Z}_L = \hat{Z}_c \frac{1+\hat{r}}{1-\hat{r}} \qquad \text{and} \qquad \hat{Z}_L = \hat{Z}_c \frac{\hat{t}}{2-\hat{t}} \ . \tag{9.2.4}$$

$\hat{Z}_L$ depends on the geometry used and on the optical parameters of the material forming the load. For more complicated and multiple scattering events, the representation by scattering matrices S is advantageous and is widely used in engineering textbooks [Ell93, Gar84, Poz90]. The impedance is the ratio of the total voltage to the total current at one point; the scattering matrix relates the electric field of the incident wave to the field of the wave reflected from this point. If the S matrix is known, the impedance $\hat{Z}$ can be evaluated, and vice versa.

This general concept holds for scattering on any kind of impedance mismatch experienced by the traveling wave in a transmission line; it is equally appropriate when we terminate a coaxial line by some material of interest, if we fill the waveguide with the sample, or replace a wire or plate in a Lecher line or parallel plate transmission line by the unknown metal. In our simple example of a large (compared to the wavelength and beam diameter) and thick ($d \gg \delta_0$) mirror placed in the optical path of a light beam, it is the surface impedance of the mirror $\hat{Z}_S$ which solely determines the reflection coefficient according to the equation

$$\hat{r} = \frac{\hat{Z}_S - Z_0}{\hat{Z}_S + Z_0} \quad .$$

In this case, and also when the material acts as a short for a coaxial line or waveguide (where the free space impedance Z_0 is replaced by the characteristic impedance $\hat{Z}_c$), the load impedance $\hat{Z}_L$ is equal to the surface impedance $\hat{Z}_S$. In general a geometrical factor enters the evaluation; this is, as a rule, difficult to calculate. The main problem is the following: if we know the load impedance $\hat{Z}_L$ which causes the scattering of the wave as it propagates along the transmission line, it is no trivial task to evaluate the surface impedance $\hat{Z}_S$ of the material from which the obstacle is made because of depolarization effects. Having obtained the surface impedance, we then evaluate the complex conductivity or the complex dielectric constant of the specimen as discussed in Chapter 2.

A remark on the difference between measuring conductivity (or admittance) and resistivity (or impedance) is in order here. In the first case, the applied voltage is kept constant and the induced current is measured; or the applied electric field is constant and the current is evaluated by observing the dissipated power due to losses (Joule heat) within the material. In the second case when the impedance is probed, the current flowing through the device is kept constant and the voltage drop studied: a typical resistance measurement. This implies that absorption measurements look for dissipation and thus the admittance $\hat{Y} = 1/\hat{Z}$; the same holds for reflection experiments off a thick ($d \gg \delta_0$) material which probe the power not absorbed $R = 1 - A$. Optical transmission experiments through thin films, on the other hand, actually measure the impedance $\hat{Z}$.

9.2.2 Two interfaces

In general it is not possible to evaluate the transmission of the electromagnetic wave through a single interface, but we can measure the transmission through a sample of finite thickness; thus we have to consider two interfaces. This can be done, for instance, if we replace a certain part (length d) of a copper waveguide (characterized by an impedance $\hat{Z}_c$) by a metal (of the same geometry) with a load impedance $\hat{Z}_L$ and the transmission line continues with its characteristic impedance $\hat{Z}_c$. Alternatively we can fill a microstripline with some unknown dielectric over a length d, or we can shine light through a slab of material of thickness d. The latter case can be simplified for thick samples: when the skin depth δ_0 is much smaller than the thickness d, the problem is reduced to a single bounce. Hence the load impedance $\hat{Z}_L$ is equivalent to the surface impedance $\hat{Z}_S$ of the material as defined in Eq. (2.4.23); no power is transmitted through the material: it is either reflected or absorbed. However, if this is not the case, the second impedance mismatch (at the back of the sample) also causes reflection of the wave and thus reduces the power transmitted into the rear part of the transmission line; the reflection coefficients for both sides of the sample are the same (though with opposite sign) if the line continues with $\hat{Z}_c$. For arbitrary systems (but assuming local and linear electrodynamics) the load impedance $\hat{Z}_L$ can be calculated by

$$\hat{Z}_L = \hat{Z}_S \frac{\hat{Z}_c \cosh\{-i\hat{q}d\} + \hat{Z}_S \sinh\{-i\hat{q}d\}}{\hat{Z}_S \cosh\{-i\hat{q}d\} + \hat{Z}_c \sinh\{-i\hat{q}d\}} \quad , \tag{9.2.5}$$

if we assume that the surrounding structure has the same characteristic impedance $\hat{Z}_c$ before and after the load, and the wavevector in the material (with impedance $\hat{Z}_S$) is given by $\hat{q} = \frac{\omega}{c}(\mu_1\hat{\epsilon})^{1/2}$. Again, the simplest example is a thin (free standing) dielectric film probed by optical techniques. In the $d \to 0$ limit, $\hat{Z}_L = \hat{Z}_c$ and no reflection or absorption takes place; for $d \to \infty$ we observe no transmission and are left with a single bounce. In all other cases part of the wave is reflected off the front surface, and part is reflected at the rear surface of the material. This wave, however, is again partially reflected at the front, and the process repeats itself *ad infinitum,* as depicted in Fig. 9.5. Thus we have to take all these contributions to the totally reflected and totally transmitted wave into account and sum them up using the proper phase. The optical properties of media with finite thickness (and also of systems with many layers) are discussed in Appendix B in more detail. The main features of the reflected and transmitted radiation are as follows. Whenever the thickness d of the slab is a multiple of half the wavelength in the material λ, we observe a maximum in the transmitted power. Hence, the absorption is enhanced at certain frequencies (and suppressed at others) due to multireflection; an effect which will be utilized by resonant techniques discussed below.

If the material does not fill the entire waveguide or coaxial line, and hence part of the radiation is transmitted beyond the obstacle, the analysis is more complicated because geometrical factors enter the appropriate relations. Many simple cases, however, such as a rod of a certain diameter in a waveguide or a coaxial cable (see Section 11.1) can be analyzed analytically [Joo94, Kim88, Mar91, Sri85].

9.3 Resonant structures

In the previous sections we have discussed the propagation of electromagnetic waves in various structures (transmission lines) together with the reflection off

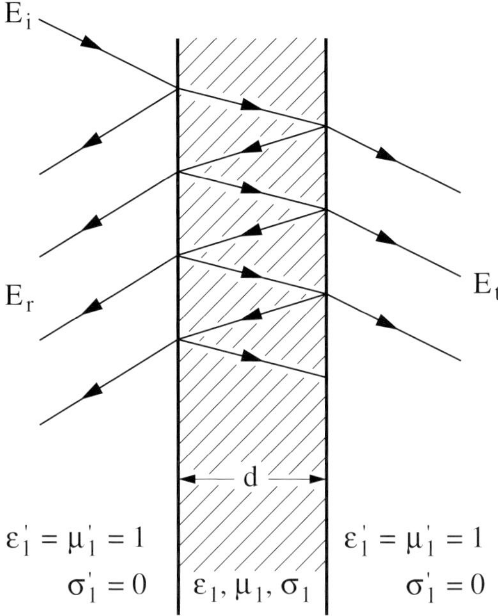

Fig. 9.5. Reflection off and transmission through a thin semitransparent material with thickness d and optical parameters ϵ_1, σ_1, and μ_1. The multiple reflections cause interference. $\mathbf{E}_i$, $\mathbf{E}_t$, and $\mathbf{E}_r$ indicate the incident, transmitted, and reflected electric fields, respectively. The optical properties of the vacuum are given by $\epsilon'_1 = \mu'_1 = 1$ and $\sigma'_1 = 0$.

boundaries. If a transmission line contains two impedance mismatches at a certain distance (a distance of the order of the wavelength, as will be discussed below), as sketched in Fig. 9.6, the electromagnetic wave can be partially trapped between these discontinuities by successive reflections. If the distance is roughly a multiple of half the wavelength, the fields for each cycle add with the proper phase relations and thus are enhanced: we call the structure to be at resonance. In any real system, losses cannot be avoided and a (generally small) fraction of the energy per cycle dissipates.

The resonant system becomes useful for measurements of materials when part of a well characterized structure can be replaced by a specimen of interest. If the resonance characteristics are modified only weakly by the specimen, this modification can be considered to be a perturbation of the resonant structure and analyzed accordingly. If the geometry is known, the change in the width and shift in resonant frequency allow the evaluation of the impedance and finally the complex conductivity of the material under consideration. Crudely speaking, the losses of the material (for example given by the surface resistance) determine the quality factor Q or the width of the resonance, whereas the refractive index

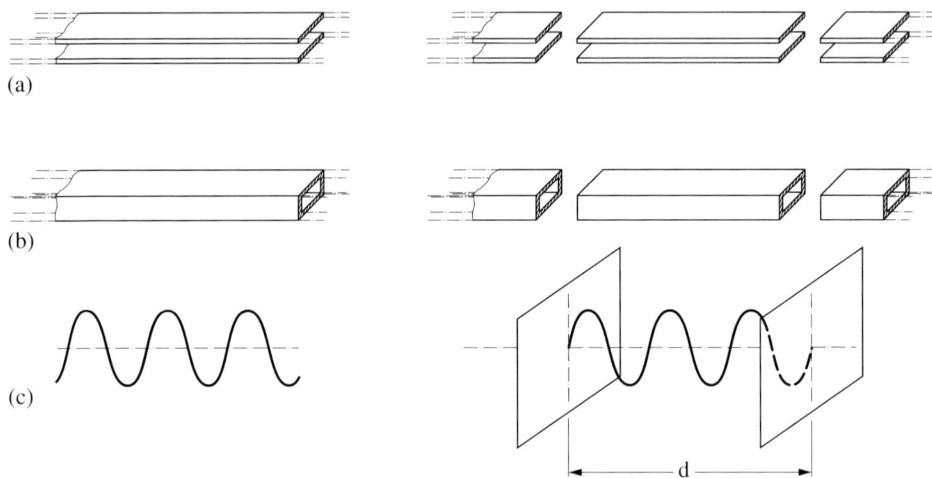

Fig. 9.6. Transformation of a transmission line to a resonant structure by introducing two impedance mismatches (discontinuities) at a certain distance d. (a) Parallel plate transmission line, (b) rectangular waveguide, and (c) free space becoming an open resonator.

(or surface reactance) determines the resonance frequency. The advantage when compared with a simple reflection or transmission experiment is the fact that the electromagnetic wave bounces off the material which forms the resonance structure many times (roughly of the order of Q); this then enhances the interaction, and thus the sensitivity, significantly.

9.3.1 Circuit representation

In Section 9.1 we discussed the characteristics of transmission lines in terms of an electrical circuit analogy and have established the relevant electrical parameters such as the characteristic impedance. A similar approach is also useful for discussing the various resonant structures which are employed to study the electrodynamics of solids. We consider a series RLC circuit as shown in Fig. 9.7a; similar considerations hold for parallel RLC circuits shown in Fig. 9.7b. The impedance $\hat{Z}$ of a resonant structure is in general given by

$$\hat{Z} = R - i\omega L + \frac{i}{\omega C} \ . \tag{9.3.1}$$

If the system is lossless ($R = 0$), the impedance is purely imaginary and the phase angle $\phi = \arctan\{(1/\omega C - \omega L)/R\} = \pi/2$. The structure resonates at the angular frequency

$$\omega_0 = 2\pi f_0 = (LC)^{-1/2} \ . \tag{9.3.2}$$

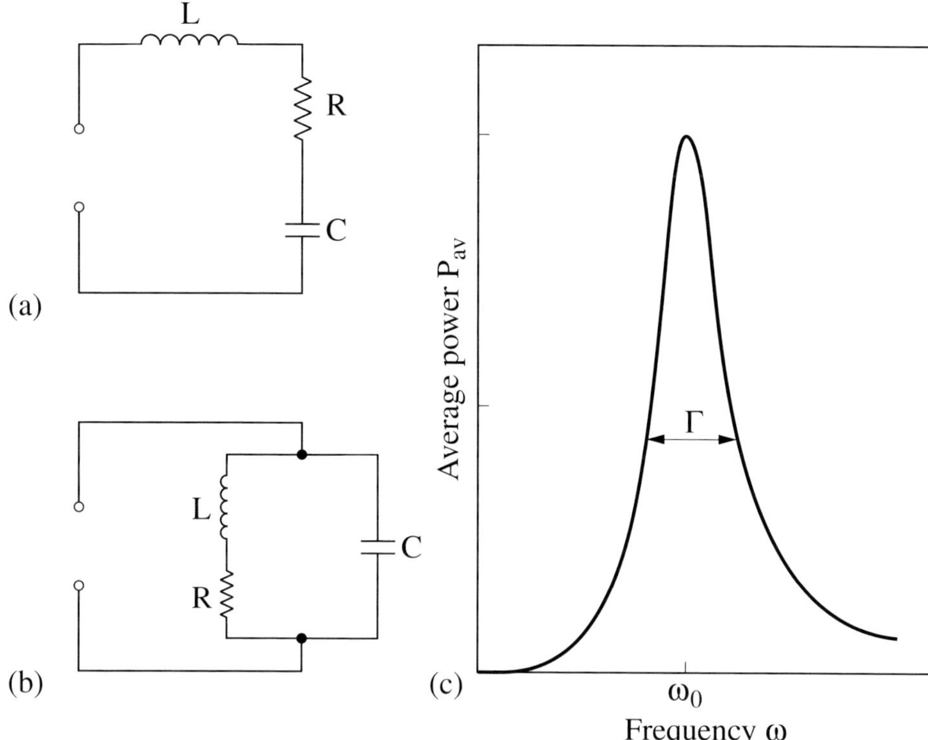

Fig. 9.7. (a) Series *RLC* circuit, (b) parallel *RLC* circuit, and (c) the average power absorbed as a function of frequency.

The absorbed power is due to the dissipation by the resistor R, and its time average is

$$P_{\text{av}} = \frac{V_{\text{rms}}^2 R \omega^2}{L^2(\omega_0^2 - \omega^2)^2 + \omega^2 R^2}$$

where the root mean square value $V_{\text{rms}} = V_{\text{max}}/\sqrt{2}$, and it shows a Lorentzian frequency dependence as displayed in Fig. 9.7c. The power is half of its maximum value when $L(\omega_0^2 - \omega^2) = \omega R$, and the full width of the resonance curve is $\Gamma = 2|\omega_0 - \omega| = R/L$ at that point. This defines the quality factor Q as the ratio of the resonant frequency to the full width of the resonance at half its maximum (FWHM, often called halfwidth):

$$Q = \frac{\omega_0}{\Gamma} = \frac{\omega_0 L}{R} = \frac{1}{\omega_0 RC} \quad . \tag{9.3.3}$$

Thus the quality factor can be evaluated by measuring the relative bandwidth Γ/ω_0 of the resonance. Quality factors of 10^3 can easily be achieved in circuits built of metallic components. Another way of expressing the quality factor is in terms of stored and absorbed power:

$$Q = \frac{\omega_0 W_{\text{stored}}}{P_{\text{lost}}} \quad ,$$

where W_{stored} refers to the time average of the stored energy $W_{\text{stored}} = \frac{1}{2} I_{\text{rms}}^2 L$, and the dissipated power, i.e. the average Joule heat lost per second, is $P_{\text{lost}} = \frac{1}{2} I^2 R$, leading to Eq. (9.3.3). This means that the time dependence of the stored energy – if no power is added from the outside – has the form $W_{\text{stored}} = W_0 \exp\{-\omega_0 t / Q\}$, an exponential decay with Q/ω_0 being the characteristic time scale of the damping.

9.3.2 Resonant structure characteristics

Resonant structures come in different shapes and forms, these being determined by the spectral range in which they are employed and also by the objectives of the experiment. In the upper end of the radio frequency spectrum, where guiding structures are employed to propagate the electromagnetic fields, transmission line resonators are commonly used. The resonant structures can be formed by terminating transmission lines such as a coaxial line or parallel plate lines at two points as shown in Fig. 9.6. In general two impedance mismatches are needed to form a resonator; depending on the impedance associated with the mismatch, the appropriate electrical analog is a parallel or a series RLC circuit [Poz90].

As a rule waveguides are utilized in the microwave spectral range; here, enclosing part of the waveguide can form the resonant structures. An antenna or a coupling hole usually provides the coupling to this structure (called a hollow resonator); both provide access of the electromagnetic field to the resonant structure. Various types of enclosed cavities have been designed and used, the simplest form being a rectangular cavity such as depicted in Fig. 9.8; often however circular cavities are employed [Gru98].

Instead of enclosing a volume by conducting walls, a dielectric body of appropriate shape can also be utilized as a resonator. As the dielectric constant of the structure $\epsilon_1 > 1$, the dielectric constant of a vacuum, there is an impedance mismatch at the surface; this in turn leads to a resonance, the frequency of which depends on the size, the dielectric constant ϵ_1, and the geometry of the structure. Of course, significant impedance mismatch is required to lead to well confined electromagnetic fields. Again, coupling to the structure is through an appropriate antenna arrangement or simply by placing the dielectric in the proximity of a transmission line [Kaj98]. Note also that dielectric resonators have a reduced size (roughly by $\sqrt{\epsilon_1}$), compared with hollow metal resonators – the reason for this is the reduction of the wavelength of light in a dielectric according to Eq. (2.3.9).

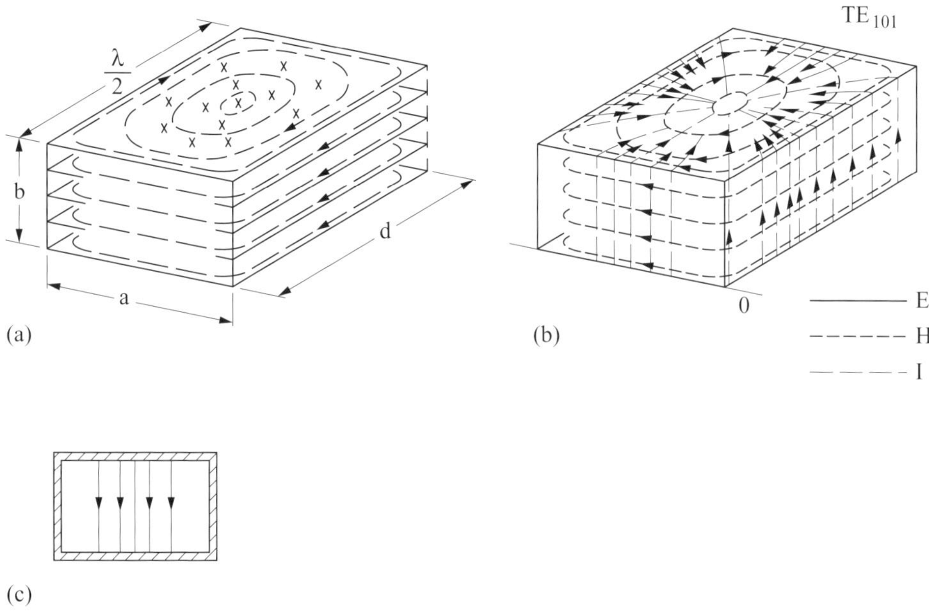

Fig. 9.8. Rectangular TE$_{101}$ cavity. (a) The electric field lines **E** span the space between the top and bottom wall while the magnetic field **H** circles around. (b) The current I runs up on the side walls and to the center of the top plate; (c) the electric field has its highest density in the center of the cavity.

With increasing frequency, the resonators discussed above become progressively smaller and, consequently, impractical. Thus for the millimeter wave spectral range and above, typically for frequencies above 100 GHz, open, so-called Fabry–Perot resonators are employed where no guiding structure for the electromagnetic wave is utilized [Afs85, Cul83]. The simplest form is that of two parallel plates separated by a distance d, as shown in Fig. 9.9.

The resonators support various modes, as the appropriate boundary conditions – such as the vanishing of the electric field component, perpendicular to the surface at the two end walls in the case of hollow resonators – are satisfied for different wavelengths of the electromagnetic field. In most cases, however, only the fundamental mode (corresponding to the largest wavelength for which the boundary condition is satisfied) is utilized, and then the resonant frequency to first approximation is written as

$$\omega_0 = \frac{2\pi c}{\lambda} g \quad , \tag{9.3.4}$$

where λ is the wavelength of the electromagnetic wave in the structure, and g is the geometrical factor of the order of one, which can be evaluated for the particular

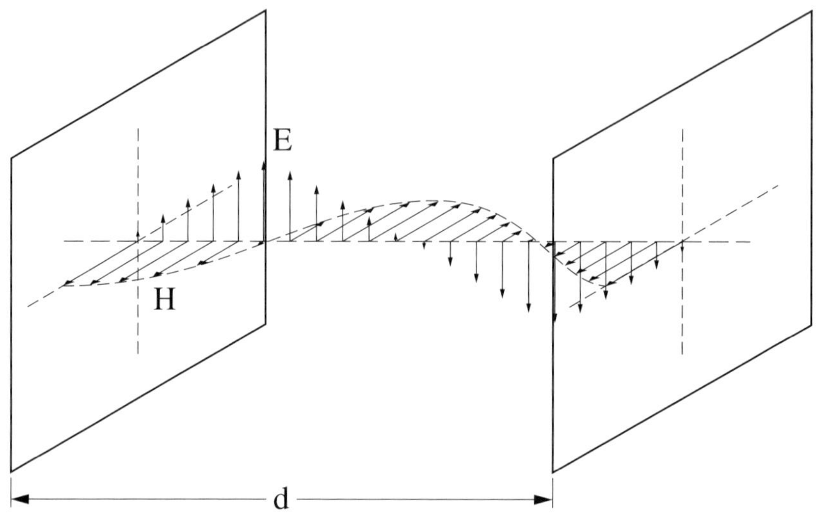

Fig. 9.9. Simple open resonator consisting of two parallel plates at a distance d between which the electromagnetic waves bounce back and forth.

structure. For the simplest case of a waveguide section of width a and cavity of length d, the resonant frequency is given by $\pi c \left(a^{-2} + d^{-2}\right)^{1/2}$.

The loss, which determines the quality factor Q of the structure, has various components. Radiation losses, either intrinsic to the structure – as for a transmission line or for a dielectric resonator – or due to coupling to the guiding structures which connect the resonator to the source and detector, contribute to a decrease in the quality factor. This can be evaluated by examining the electromagnetic fields associated with the resonant structure. As a rule, increased coupling to the guiding structures leads to increased loss, and thus to smaller Q. Another source of the loss is the so-called ohmic loss, arising from currents induced in the structure itself, either in the (metallic) cavity walls, or in the dielectric which forms the resonant structure. In all cases this latter contribution to the loss can be written as

$$Q = \frac{Z_0}{R} g' \quad .$$

(9.3.5)

Here Z_0 is the impedance in free space and R is the resistance responsible for the ohmic losses. If the resonant structure is formed of metallic components, R is the surface resistance R_S. The geometrical factor g' is determined by the integration of the currents over the surface of the resonator, and by the integration of the electric field inside the resonator. g' is approximately given by

$$g' = \frac{d}{\lambda} \quad ,$$

where d is the characteristic length of the structure.[3] As $d \approx \lambda$ for a typical structure, g' is of the order of unity, and the quality factor

$$Q \approx \frac{Z_0}{R_S}$$

for the cavities, where currents induced in the metallic part of the structure are the source of the loss. In the microwave spectral range, where $R_S \ll Z_0$, Q values of the order of 10^3 are easily obtainable. As the surface impedance increases with increasing frequency (note that $R_S \propto \omega^{1/2}$, see Eq. (9.1.16)), the quality factor decreases with decreasing frequency. Extremely high quality factors can be achieved in resonators composed of superconducting parts; this is due to the small surface impedance associated with the superconducting state.

It is important to note that the resonant frequency ω_0 is also influenced by the conducting characteristics of the structure. Equation (9.3.4) refers to the situation where it is assumed that the electric or magnetic fields do not penetrate the structure itself. This is, however, not the case. For metals the magnitude of the penetration of the electromagnetic field into the walls of the resonator depends on the skin depth δ_0, the length scale which increases with increasing resistance of the material which forms the (conducting) walls. This, however, is a small effect: a typical cavity dimension d is around 1 cm, while the skin depth is of the order of 1 μm for good metals of microwave frequencies; thus the change in the resonant frequency, given approximately by δ_0/d, is small.

9.3.3 Perturbation of resonant structures

The resonant structures discussed above can be used to evaluate the optical parameters of solids. The commonly used method is to replace part of the structure by the material to be measured or, alternatively, to insert the specimen into the resonator (in the case of metallic, enclosed resonators) or to place the sample in close proximity to the resonator (in the case of dielectric resonators).

Two parameters, center frequency and halfwidth, fully characterize the resonant structure; and the impedance, Eq. (9.3.1), can be expressed as

$$
\begin{aligned}
\hat{Z} &= R - i\omega L \left(1 - \frac{\omega_0^2}{\omega^2}\right) \approx \frac{\omega_0 L}{Q} - 2iL(\omega - \omega_0) \\
&= -2iL\left[\omega - \omega_0\left(1 - \frac{i}{2Q}\right)\right] = -2iL(\omega - \hat{\omega}_0) \quad , \quad (9.3.6)
\end{aligned}
$$

where the approximation used is valid for $\omega_0 \gg |\omega - \omega_0|$. We now define a complex

---

[3] This characteristic length, of course, depends on the form of the structure used. For a cubic cavity, for example, $g' = 0.76$.

frequency

$$\hat{\omega}_0 = \omega_0 - \frac{\mathrm{i}\omega_0}{2Q} = \omega_0 - \mathrm{i}\frac{\Gamma}{2} \quad , \tag{9.3.7}$$

which contains the resonant frequency in its real part and the dissipation described by the halfwidth in the imaginary part (note the factor of 2). The effects of a specimen which forms part of the structure can now be considered as a perturbation of the complex frequency.[4] The change in the impedance due to the material leads to a complex frequency shift $\Delta\hat{\omega}$ from the case of the sample being absent: $\Delta\hat{\omega} = \hat{\omega}_S - \hat{\omega}_0$.

Let us look at a lossless system with a characteristic impedance $Z_c = (L/C)^{1/2}$ according to Eq. (9.1.5). The introduction of a specimen or replacement of parts of the resonant structure by the sample causes a shift in the resonance frequency from ω_0 to ω_S and a decrease of the quality factor from Q_0 to Q_S, summarized in $\Delta\hat{\omega}$. The impedance of the sample $\hat{Z}_S = R_S + \mathrm{i}X_S$ is related to the complex frequency change by Eq. (9.3.6)

$$\hat{Z}_S = \mathrm{i}g_0 L \Delta\hat{\omega} \tag{9.3.8}$$

with a geometrical factor g_0 depending on the geometries of the specimen and of the structure. Thus we obtain

$$\hat{Z}_S = \mathrm{i}g_0 L \omega_0 \left(\frac{\Delta\omega}{\omega_0} - \frac{\mathrm{i}}{2}\frac{\Delta\Gamma}{\omega_0} \right) = \left(\frac{\Delta\Gamma}{2\omega_0} + \mathrm{i}\frac{\Delta\omega}{\omega_0} \right) g_0 \hat{Z}_c \quad , \tag{9.3.9}$$

where the real and imaginary parts of $\hat{Z}_S$ are related to the change in Q and shift in frequency $(\omega_S - \omega_0)$ separately:

$$\frac{R_S}{Z_c} = g_0\frac{\Delta\Gamma}{2\omega_0} = \frac{g_0}{2}\left(\frac{1}{Q_S} - \frac{1}{Q_0} \right) \quad , \tag{9.3.10a}$$

$$\frac{X_S}{Z_c} = g_0\frac{\Delta\omega}{\omega_0} = g_0\frac{\omega_S - \omega_0}{\omega_0} \quad . \tag{9.3.10b}$$

Thus the resistance (i.e. the loss) determines the change in the width of the resonance or the Q factor, while from the shift in the resonance frequency we recover the reactance X_S. Both parameters are similarly dependent on the geometry, determined by g_0. This geometrical factor can either be evaluated experimentally or can be obtained analytically in some simple cases. Thus, evaluating the change in Q and the resonant frequency allows the evaluation of $\hat{Z}_S$, and this parameter then can be used to extract the components of the complex conductivity.

---

[4] The assumption of this approach is that the perturbation is so small that the field configuration inside the resonant structure is not changed significantly.

References

[Ada81] M.J. Adams, *An Introduction to Optical Waveguides* (Wiley, Chichester, 1981)

[Afs85] M.N. Afsar and K.J. Button, *Proc. IEEE* **73**, 131 (1985)

[Bha91] B. Bhat and S.K. Koul, *Stripline-Like Transmission Lines for Microwave Integrated Circuits* (John Wiley & Sons, New York, 1991)

[Cul83] A.L. Cullen, *Millimeter-Wave Open Resonator Techniques*, in: *Infrared and Millimeter Waves* **10**, edited by K.J. Button (Academic Press, Orlando, FL, 1983), p. 233

[Dix91] M.W. Dixon, ed., *Microwave Handbook* (Radio Society of Great Britain, Potters Bar, 1991–92)

[Ell93] R.S. Elliott, *Introduction to Guided Waves and Microwave Circuits* (Prentice-Hall, Englewood Cliffs, NJ, 1993)

[Gar84] F.E. Gardiol, *Introduction to Microwaves* (Artech House, Dedham, 1984)

[Gar94] F.E. Gardiol, *Microstrip Circuits* (John Wiley & Sons, New York, 1994)

[Gru98] G. Grüner, ed., *Waveguide Configuration Optical Spectroscopy*, in: *Millimeter and Submillimeter Wave Spectroscopy of Solids* (Springer-Verlag, Berlin, 1998), p. 111

[Joo94] J. Joo and A.J. Epstein, *Rev. Sci. Instrum.* **65**, 2653 (1994)

[Kaj98] D. Kajfez and P. Guillon, eds, *Dielectric Resonators* (Noble Publishing, Atlanta, 1998)

[Kim88] T.W. Kim, W. Beyermann, D. Reagor, and G. Grüner, *Rev. Sci. Instrum.* **59**, 1219 (1988)

[Mag92] P.C. Magnusson, G.C. Alexander, and V.K. Tripathi, *Transmission Lines and Wave Propagation*, 3rd edition (CRC Press, Boca Raton, FL, 1992)

[Mar48] M. Marcuvitz, *Waveguide Handbook*, MIT Rad. Lab. Ser. **10** (McGraw-Hill, New York, 1948)

[Mar91] D. Marcuse, *Theory of Dielectric Optical Waveguides*, 2nd edition (Academic Press, Boston, MA, 1991)

[Poz90] D.M. Pozar, *Microwave Engineering* (Addison Wesley, Reading, MA, 1990)

[Ram93] S. Ramo, J.R. Whinnery, and T.v. Duzer, *Fields and Waves in Communication Electronics*, 3rd edition (John Wiley & Sons, New York, 1993)

[Sch68] J. Schwinger and D.S. Saxon, *Discontinuities in Waveguides* (Gordon and Breach, New York, 1968)

[Smi93] B.L. Smith, ed., *The Microwave Engineering Handbook* (Chapman & Hall, London, 1993)

[Sri85] S. Sridhar, D. Reagor, and G. Grüner, *Rev. Sci. Instrum.*, **56**, 1946 (1985)

Further reading

[Coh68] S.B. Cohn, *IEEE Trans. Microwave Theory Tech.* **MTT-16**, 218 (1968)

[Col92] R.E. Collin, *Foundations for Microwave Engineering*, 2nd edition (McGraw-Hill, New York, 1992)

[Her86] G. Hernandez, *Fabry–Perot Interferometers* (Cambridge University Press, Cambridge, 1986)

[Hip54] A.R.v. Hippel, *Dielectrics and Waves* (John Wiley & Sons, New York, 1954)

[Hop95] D.J. Hoppe and Y. Rakmat-Samii, *Impedance Boundary Conditions in Electromagnetics* (Taylor & Francis, London, 1995)

[Ish95] T. K. Ishii, ed., *Handbook of Microwave Technology* (Academic Press, San Diego, CA, 1995)

[Ito77] T. Itoh and R.S. Rudokas, *IEEE Trans. Microwave Theory Tech.* **MTT-25**, 52 (1977)

[Tab90] R.C. Taber, *Rev. Sci. Instrum.* **61,** 2200 (1990)

[Whe77] H.A. Wheeler, *IEEE Trans. Microwave Theory Tech.* **MTT-25**, 631 (1977); **MTT-26**, 866 (1978)

10

Spectroscopic principles

Different approaches are commonly utilized to measure the frequency dependent optical properties of solids, with three principles being distinguished. Investigations of the electrodynamic response as a function of frequency can be performed in a straightforward manner by applying monochromatic radiation and measuring the amplitude and phase of the response at one frequency ω: this might be the resistance R and capacitance C, the complex reflection coefficient $\hat{r}$, or the amplitude drop and phase shift upon transmission T and ϕ_t. In the next step the complex refractive index $\hat{N}$ or the complex surface impedance $\hat{Z}_S$ can be calculated from these equations. Finally the complex conductivity $\hat{\sigma}$ or dielectric constant $\hat{\epsilon}$ is calculated. In order to evaluate $\hat{\sigma}(\omega)$, the measurement has to be repeated for each frequency of interest.

The following two approaches do not involve monochromatic radiation but rather an excitation with a well defined time or spectral dependence, thus they lead to the determination of the response in a wide frequency range. Time domain spectroscopy applies voltage steps or pulses with a short rise time, and from the time dependent response of the material under investigation, i.e. from the broadening and delay of the pulse, for example, the frequency dependence of the complex conductivity or complex dielectric constant is calculated using an appropriate mathematical approach. The high frequency limit of the accessible spectral range is determined by the inverse rise time of the pulse. Typical rise times of a few nanoseconds are commercially available to evaluate the response up to 1 GHz. Optical femtosecond pulses allow this method to be extended up to a few terahertz.

Fourier transform spectroscopy utilizes two beams of one broadband source and records the response of the material as a function of the difference in path lengths between the two arms; a Fourier transformation then leads to the frequency dependent response. The spectral limitations on the high frequency side are the timely stability of the setup and the intensity of the source. On the lower frequency side, diffraction of the beam due to the large wavelength limits the performance, and also the spectral intensity of the light source diminishes rapidly due to Planck's law. Hence Fourier transform spectroscopy is mainly used in the infrared range.

We introduce the basic ideas rather than focus on the technical details of particular arrangements; these depend mainly on the individual task at hand, and further information can be found in the vast literature we selectively refer to at the end of the chapter.

10.1 Frequency domain spectroscopy

Frequency domain measurements are straightforward to analyze as each single frequency point can be evaluated separately. Due to the large variety of methods employed, which often probe different responses (such as the resistance, the loss tangent, the surface impedance, the reflectivity, etc.), it is necessary to combine the results using common material parameters, such as the real and imaginary parts of the conductivity $\hat{\sigma}$ or the dielectric constant $\hat{\epsilon}$. If a certain experimental method delivers only one parameter, for example the surface resistance R_S or the reflectivity R, a Kramers–Kronig analysis is required in order to obtain the complex response. In this case, data from the complementary methods have to be transformed to one selected parameter, for example $R(\omega)$; this common parameter is then used as the input to the Kramers–Kronig transformation.

10.1.1 Analysis

Typically, both components of the complex conductivity can be determined in the radio frequency range and below, since vector network analyzers are available, whereas sometimes only the surface resistance can be measured in the microwave range; in most cases only the reflectivity of metals is accessible experimentally in the optical range of frequencies. The combination of different techniques then allows one to calculate the reflectivity $R(\omega)$ over a broad spectral range. Beyond the spectral range in which measurements have been performed, suitable extrapolations are necessary, the choice being guided by the information available on the material. For metals a Hagen–Rubens behavior of the reflectivity $[1 - R(\omega)] \propto \omega^{-1/2}$ is a good approximation at low frequency. Semiconductors and insulators have a constant reflectivity for $\omega \to 0$. At very high frequencies the reflectivity eventually has to fall to zero; usually a ω^{-2} or ω^{-4} behavior is assumed in the spectral range above the ultraviolet. Then with $R(\omega)$ known over the entire range of frequencies, the Kramers–Kronig analysis can be performed.

10.1.2 Methods

A large number of different techniques are available and also necessary to cover the entire frequency range of interest. We divide these into three major groups. This division is based on the way in which the electromagnetic signal is delivered, which also determines the measurement configurations: the low frequency range where leads and cables are used, the optical range where the radiation is delivered via free space, and the intermediate range where transmission lines and waveguides are employed.

Audio and radio frequency arrangements

For low noise, low frequency signals up to approximately 1 GHz, the driving signal is provided by a commercial synthesizer which can be tuned over several orders of magnitude in frequency; the signal is usually guided to the sample and to the detector by coaxial cables with negligible losses. Phase sensitive detection (lock-in amplifier) can be used up to 100 MHz; precautions have to be taken for the proper phase adjustment above 1 MHz.

In recent years, fully automated and computer controlled test and measurement instruments have been developed and by now are well established, making the kilohertz, megahertz, and lower gigahertz range easily accessible. An RLC meter measures the resistance R, and the inductance L or the capacitance C of a device at one frequency which then can be continuously varied in the range from 5 Hz up to 1 GHz. All of these techniques require that contacts are attached to the sample. Due to the high precision of sources and detectors, the low frequency techniques are mainly used in a single path method; however, for highest sensitivity, interferometric methods and resonant arrangements have been utilized, as discussed in Chapter 11. These methods are limited at higher frequencies by effects associated with the finite cable length and stray capacitance.

Microwave and millimeter wave measurement techniques

Based on solid state high frequency generators and the stripline technique, the upper frequency limit of commercial test equipment (currently at 100 GHz) is continuously rising as technology improves. Impedance analyzers measure the complex impedance of materials or circuits in a certain frequency range; up to about 2 GHz the current and voltage are probed. Network analyzers, which are commercially available for an extremely wide range of frequency from well below 1 kHz up to 100 GHz, often utilize a bridge configuration as described in Section 11.2.1. They are combined driver/response test systems which measure the magnitude and phase characteristics of linear networks by comparing the incident signal (which is always a sine wave) with the signal transmitted or reflected by the device under test, and provide a complete description of linear network behavior in the frequency domain. Whereas scalar network analyzers evaluate only the magnitude of the signal, vector network analyzers deliver magnitude and phase, i.e. the complex impedance, or alternatively the so-called scattering matrix or transfer function.

The sample is usually placed between the two conductors of a coaxial line or the walls of a waveguide. Calibration measurements are required to take effects into account, for example the resistance, stray capacitance, and inductance of the leads. For special requirements – concerning the accuracy, output power, or stability – monochromatic solid state sources (such as Gunn diodes or IMPATT sources) are utilized as generators in custom made setups instead of network analyzers.

Optical measurements

If the wavelength λ is much smaller than the sample size a, the considerations of geometrical optics apply and diffraction and boundary effects become less important; by rule of thumb, diffraction and edge effects can be neglected for $a > 5\lambda$. In contrast to lower frequencies, phase information cannot be obtained. Standard power reflection measurements are performed in an extremely wide frequency range, from millimeter waves up to the ultraviolet (between 1 cm^{-1} and 10^6 cm^{-1}); they are probably the single most important technique of studying the electrodynamic properties of solids.

Although tunable, monochromatic radiation sources (such as backward wave oscillators, infrared gas lasers, solid state and dye lasers) are available in the infrared, visible, and ultraviolet frequency range, it is more common that broadband sources are used and dispersive spectrometers (grating and prism spectrometers) are employed to select the required frequency. For the latter case, either a certain frequency of the radiation is selected (either by dispersing prisms or by utilizing diffraction gratings) which is then guided to the specimen, or the sample is irradiated by a broad spectrum but only the response of a certain frequency is analyzed. Combined with suitable detectors the relative amount of radiation at each frequency is then recorded.

Because of their great light efficiency, single order spectrum, ruggedness, and ease of manufacture, prisms have long been favored as a dispersing medium in spectrographs and monochromators in the visible spectral range. The disadvantages of prisms are their non-linear dispersion and the limited frequency range for which they are transparent. If high resolution is required, grating spectrometers are commonly used as research instruments. Two or more monochromators may be employed in series to achieve higher dispersion or greater spectral purity; stray light is also greatly reduced [Dav70].

Optical experiments are in general performed in a straightforward manner by measuring, at different frequencies, either the bulk reflectivity or both the trans-

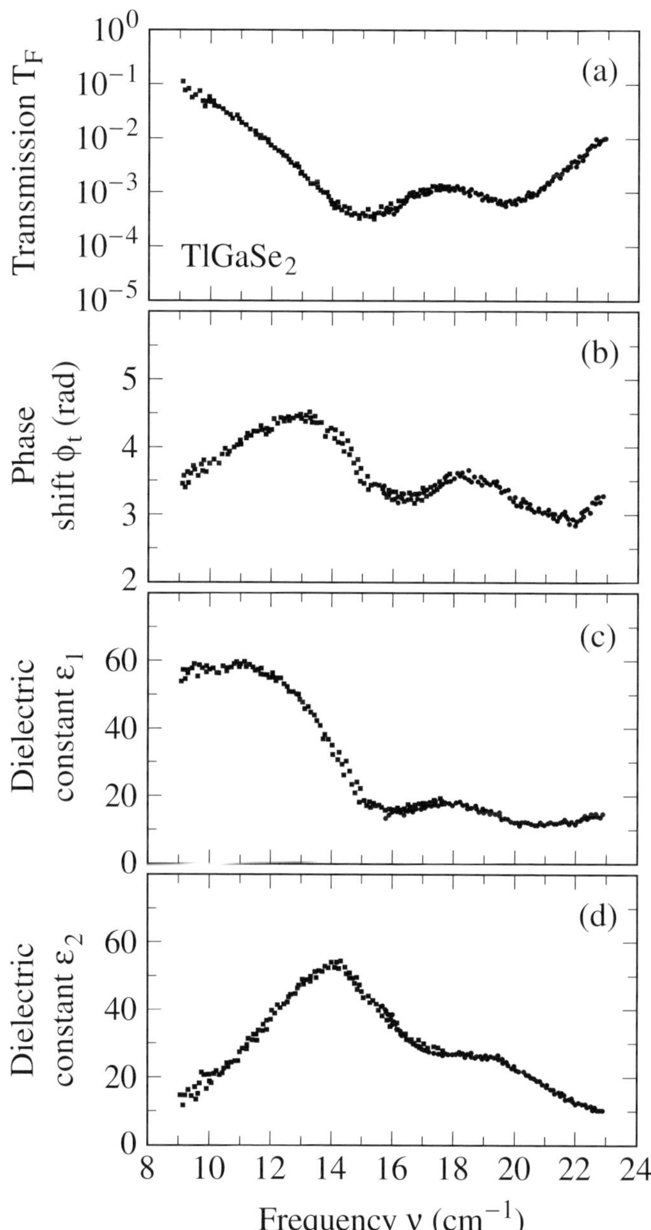

Fig. 10.1. Frequency dependence of the optical properties of TlGaSe$_2$ measured at $T = 300$ K by using a submillimeter wave Mach–Zehnder interferometer. (a) The transmission $T_F(\omega)$ and (b) the phase $\phi_t(\omega)$ spectra of a 0.079 mm thick plate are recorded. (c) The real part $\epsilon'(\omega)$ and (d) the imaginary part $\epsilon''(\omega)$ of the dielectric constant are calculated at each frequency according to Eqs (B.9), (B.10b), and (B.12) on the basis of the $T_F(\nu)$ and $\phi_t(\nu)$ spectra given in (a) and (b).

mission and reflectivity. If the investigations are conducted on samples of different thickness, the real and imaginary parts of the response can be evaluated. Measurements of the complex optical properties are also possible using Mach–Zehnder interferometers or Fabry–Perot resonators as discussed in Section 11.3.4.

As an example of an optical experiment performed in the frequency domain, data from a room-temperature transmission experiment on $TlGaSe_2$ are displayed in Fig. 10.1. Fig. 10.1a shows the frequency dependence of the power transmission T_F through a 1 mm slab of $TlGaSe_2$; the corresponding phase shift ϕ_t – as obtained by a Mach–Zehnder interferometer using a coherent radiation source (Section 11.2.3) – is plotted in Fig. 10.1b. Following the analysis laid out in Section 2.4.2, for each frequency point these two parameters T_F and ϕ_t allow evaluation of both components of the complex dielectric constant $\hat{\epsilon} = \epsilon_1 + i\epsilon_2$ as shown in Fig. 10.1c,d. There a strong absorption peak centered at $14\ \mathrm{cm}^{-1}$ and a shoulder around $19\ \mathrm{cm}^{-1}$; the maximum of ϵ_2 corresponds to a minimum of transmission. The phase shift is mainly determined by the real part of the dielectric constant. A peak in $\epsilon_2(\omega)$ leads to a step in $\epsilon_1(\omega)$ as we have already seen in the simple model of a Lorentzian oscillator (Fig. 6.3).

10.2 Time domain spectroscopy

Instead of measuring the response of a solid by applying monochromatic radiation varied over a wide frequency range, the optical constants can also be obtained by performing the experiment in the time domain using a voltage pulse with a very short rising time. Such a pulse – via the Laplace transformation presented in Appendix A.2 – contains all the frequencies of interest. In general, a linear time invariant system can be described unequivocally by its response to an applied pulse. Two basic setups are employed depending on the frequency range of importance: either the sample is placed in a capacitor and the charging or discharging current is observed; or a voltage pulse is applied at a transmission line and the attenuation and broadening of the pulse upon passing through the sample placed in this line is observed. The latter method can also be used in free space by utilizing an optical setup. In both cases, from this information ϵ_1 and σ_1 of the material can be determined by utilizing the Laplace transformation. Crudely speaking, the losses of the material cause the attenuation of the pulse while the real part of the dielectric constant is responsible for the broadening.

The experimental problem lies in the generation and detection of suitable pulses or voltage steps. Short pulses of electromagnetic radiation are generated by fast switches or by electro-optical configurations. Although with a voltage step of rise time t the entire spectrum up to approximately $1/t$ can be covered, time domain experiments are preferably performed in those parts of the spectrum where no standard techniques are available in the frequency domain.

10.2.1 Analysis

The objective of time domain spectroscopy is to evaluate the complex response function $\hat{\chi}(\omega)$ by analyzing the time dependent response $P(t)$. To this end the mathematical methods of the Laplace transformation are used as described in Appendix A.2. In a linear and causal system, the general response $P(t)$ to a perturbation $E(t)$ can be expressed by

$$P(t) = \int_{-\infty}^{t} \chi(t - t') \, E(t') \, dt' \quad , \tag{10.2.1}$$

where $\chi(t)$ is the response function in question. Note that χ is not local in time, meaning that $P(t)$ depends on the responses to the perturbation $E(t)$ for all times t' prior to t. The equation describes a convolution $P(t) = (\chi * E)(t)$ as defined in Eq. (A.2b), leading to the Fourier transform:

$$\begin{aligned} P(\omega) &= \int_{-\infty}^{\infty} P(t) \, \exp\{-i\omega t\} \, dt = \int_{-\infty}^{\infty} (\chi * E)(t) \, \exp\{-i\omega t\} \, dt \\ &= \chi(\omega) \, E(\omega) \quad . \end{aligned} \tag{10.2.2}$$

Thus for a known time dependence of the perturbation $E(t)$ we can measure the response as a function of time $P(t)$ and immediately obtain the frequency dependent response function $\chi(\omega)$:

$$\chi(\omega) = \frac{\int_{-\infty}^{\infty} P(t) \, \exp\{-i\omega t\} \, dt}{\int_{-\infty}^{\infty} E(t) \, \exp\{-i\omega t\} \, dt} = \frac{1}{E(\omega)} \int_{-\infty}^{\infty} P(t) \, \exp\{-i\omega t\} \, dt \quad . \tag{10.2.3}$$

If the system is excited by a Dirac delta function $E(t) = \delta(t)$, the response directly yields $\chi(t)$. Then the convolution simplifies to $P(t) = (\chi * \delta)(t) = \chi(t)$, and the frequency dependence is the Fourier transform of the pulse response given by

$$\chi(\omega) = \int_{-\infty}^{\infty} P(t) \, \exp\{-i\omega t\} \, dt \quad , \tag{10.2.4}$$

which can also be obtained from Eq. (10.2.3) with the help of Eq. (A.1): $E(\omega) = 1$.

Another commonly used perturbation is a step function

$$E(t) = \begin{cases} 0 & t < 0 \\ E_0 & t > 0 \end{cases} \quad ,$$

where we have to replace the Fourier transform by the Laplace transform in order to avoid mathematical complications. From Eq. (10.2.1) we see that

$$P(t) \;=\; \int_0^t \chi(t')E_0 \, dt' \tag{10.2.5a}$$

$$\chi(t) \;=\; \frac{1}{E_0}\frac{d}{dt}P(t) \;; \tag{10.2.5b}$$

and the differentiation theorem:

$$\chi(\omega) = \frac{i\omega}{E_0}\int_{-\infty}^{\infty} P(t)\,\exp\{-i\omega t\}\, dt \tag{10.2.6}$$

as discussed in Appendix A.1. The same result is obtained from Eq. (10.2.3) by using

$$E(\omega) = 2\pi\,\delta(\omega)E_0 - \frac{i}{\omega}E_0$$

and by neglecting the dc response.

Due to the finite rise time of the electronic instrumentation, the step function is better written as

$$E(t) = \begin{cases} 0 & t < 0 \\ E_0(1 - \exp\{-\eta t\}) & t > 0 \end{cases}$$

leading to

$$E(\omega) = 2\pi\,\eta(\omega)E_0 - E_0\left[\frac{\eta}{\eta^2 + \omega^2} + i\frac{\eta^2}{\omega(\eta^2 + \omega^2)}\right] \quad,$$

i.e. Lorentzian broadening of the dc response with a halfwidth of η, and narrowed high frequency response:

$$\chi(\omega) = \frac{1}{E_0}\frac{\eta^2 + \omega^2}{\eta}\int_{-\infty}^{\infty} P(t)\,\exp\{-i\omega t\}\, dt \quad.$$

Since the time domain response function $\chi(t)$ should be real, we can split the Fourier transform into its real and imaginary parts by using the sine and cosine transforms according to our discussion in Section 8.2:

$$\begin{aligned}
\hat{\chi}(\omega) &= \mathrm{Re}\{\hat{\chi}(\omega)\} + i\,\mathrm{Im}\{\hat{\chi}(\omega)\} \\
&= \int_0^{\infty} \chi(t)\cos\{-\omega t\}\, dt + i\int_0^{\infty} \chi(t)\sin\{-\omega t\}\, dt \quad. \tag{10.2.7}
\end{aligned}$$

The real and imaginary parts of $\hat{\chi}(\omega)$ are also related by the Kramers–Kronig relation (Section 3.2). This implies that all the assumptions made in the derivation of the Kramers–Kronig relations also have to be observed in the analysis of time domain spectroscopic data.

10.2.2 Methods

Time domain spectroscopy is mainly used in frequency ranges where other techniques discussed in this chapter – frequency domain spectroscopy and Fourier transform spectroscopy – have significant disadvantages. At very low frequencies (microhertz and millihertz range), for instance, it is not feasible to measure a full cycle of a sinusoidal current necessary for the frequency domain experiment. Hence time domain experiments are typically conducted to cover the range from 10^{-6} Hz up to 10^6 Hz. Of course, also in the time domain, data have to be acquired for a long period of time to obtain information on the low frequency side since both are connected by the Fourier transformation, but the extrapolation for $t \rightarrow \infty$ with an exponential decay, for example, is easier and often well sustained.

As we have seen in Section 8.3, from 10 GHz up to the terahertz range, tunable and powerful radiation sources which allow experiments in the frequency domain are not readily available. Even for the use of Fourier transform techniques, broadband sources with sufficient power below infrared frequencies are scarce. To utilize time domain spectroscopy in this range requires extremely short pulses with an inverse rise time comparable to the frequency of interest. Due to recent advances in electronics, pulses of less than one nanosecond are possible; using short pulse lasers, pulses with a rise time of femtoseconds are generated. In these spectral ranges time domain methods are well established.

We consider three cases, employed in three different ranges of the electromagnetic spectrum. First, the sample is placed in a capacitor and the characteristic discharge is determined; this is a suitable setup for experiments on dielectrics at the lower frequency end of the spectrum. In a second arrangement, a short electrical pulse travels along a coaxial cable or waveguide which contains the sample at one point; the change in intensity and shape of the pulse due to reflection off or transmission through the sample is measured. The third setup utilizes a similar principle, but is an optical arrangement with light pulses of only a few picoseconds duration traveling in free space; here the transmission and time delay are probed.

Audio frequency time domain spectroscopy

Time domain spectroscopy is useful for studying slow processes in the frequency range down to microhertz, where the frequency domain method faces inherent limits such as stability or long measurement times [Hil69, Hyd70, Sug72]. To determine the complex dielectric constant, a voltage or current step is applied and the time decay of the charging or discharging current or voltage, respectively, is monitored. Fig. 10.2 shows the basic measurement circuit which can be used to investigate low-loss samples at very low frequencies. Voltage steps of equal magnitude but opposite polarity are applied simultaneously to the two terminal capacitors C_x and C_r. C_r is an air filled reference capacitor, and C_x contains the specimen

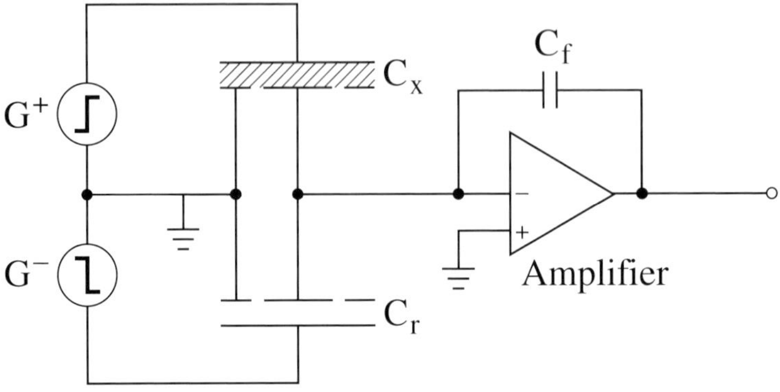

Fig. 10.2. Measurement circuit of a dielectric spectrometer for time domain investigations in the low frequency range. The generators G$^+$ and G$^-$ apply positive and negative voltage steps across the sample capacitor C_x and the reference capacitor C_r, respectively. The charge detector, formed by the operational amplifier and feedback capacitor C_f, provides an output proportional to the net charge introduced by the step voltage.

of the material to be studied. The symmetric design compensates the switching spike and hence considerably reduces the dynamic-range requirements. For better conducting samples the influence of the contacts becomes more pronounced and the discharge is faster.

In the special case of a step voltage, $dP_\mathrm{step}(t)/dt$ is related to the charging current density by Eq. (10.2.5b), and finally we find for the real and imaginary parts of the dielectric constant of the sample

$$\epsilon_1(\omega) \;=\; 1 + 4\pi \int_0^\infty \frac{J(t)}{E_0} \cos\{-\omega t\}\, dt \qquad (10.2.8a)$$

$$\epsilon_2(\omega) \;=\; 4\pi \int_0^\infty \frac{J(t)}{E_0} \sin\{-\omega t\}\, dt \quad . \qquad (10.2.8b)$$

One assumption of this analysis is the linearity and causality of the response. Due to the fact that the Kramers–Kronig relations are inherently utilized, both components of the dielectric response are obtained by the measurement of only one parameter. Equally spaced sampling presents an experimental compromise between the resolution needed near the beginning of the detected signal and the total sampling time necessary to capture the entire response. As mentioned above, the short spacing at the start is needed for the high frequency resolution, and the long-time tail dominates the low frequency response. Non-uniform sampling, i.e. stepping the intervals logarithmically, however, precludes the applicability of well developed algorithms, such as the fast Fourier transformation.

As an example, the time dependent polarization $P(t)$ of propylene carbonate after switching off the applied electric field is displayed in Fig. 10.3a. Applying

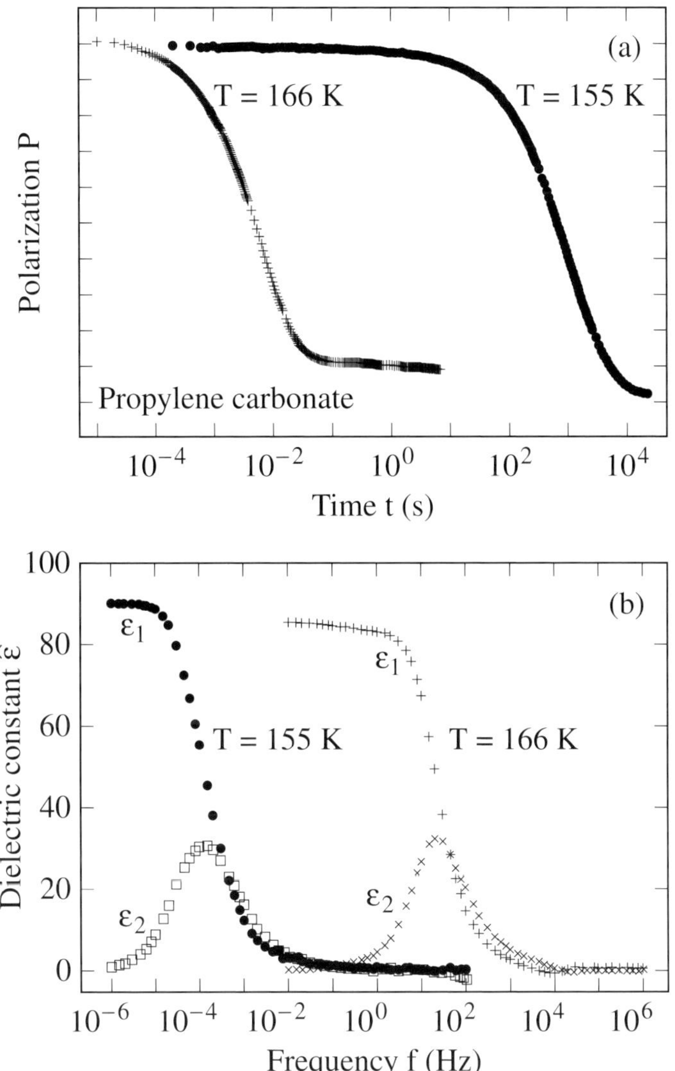

Fig. 10.3. (a) Time dependent polarization $P(t)$ of propylene carbonate at two temperatures $T = 155$ K and 166 K measured after switching the electric field off. (b) Real and imaginary parts of the frequency dependent dielectric constant $\hat{\epsilon}(f)$ obtained by Laplace transformation of the time response shown in (a) (after [Boh95]).

Eqs (10.2.8), the frequency dependent dielectric constant is calculated and plotted in Fig. 10.3b. The data were taken from [Boh95]. The dielectric relaxation strongly depends on the temperature, and the time dependence deviates from a simple exponential behavior. The slow response at low temperatures (155 K) corresponds to a low frequency relaxation process.

Radio frequency time domain spectroscopy

The time dependence of a voltage step or pulse which travels in a waveguide or a coaxial cable line, and is reflected at the interface between air and a dielectric medium, can be used to determine the high frequency dielectric properties of the material. From the ratio of the two Fourier transforms, the incident and the reflected signals, the scattering coefficient is obtained; and eventually the complex response spectrum of the sample can be evaluated using the appropriate expressions derived in Section 9.2. Similar considerations hold for a pulse being transmitted through a sample of finite thickness. Fast response techniques permit measurements ranging from a time resolution of less than 100 ps, corresponding to a frequency range up to 10 GHz.

A time domain reflectometer consists of a pulse generator, which produces a voltage step with a rise time as fast as 10 ps, and a sampling detector, which transforms the high frequency signal into a dc output. The pulse from the step generator travels along the coaxial line until it reaches a point where the initial voltage step is detected for reference purposes; the main signal travels on. At the interface of the transmission line and the sample, part of the step pulse will be reflected and then recorded. Comparing both pulses allows calculation of the response function of the sample. The time dependent response to a step function perturbation is widely used for measurements of the dielectric properties, mainly in liquids, biological materials, and solutions where long-time relaxation processes are studied. A detailed discussion of the different methods and their analysis together with their advantages and disadvantages can be found in [Fel79, Gem73].

Terahertz time domain spectroscopy

In order to expand the time domain spectroscopy to higher frequencies in the upper gigahertz and lower terahertz range, two major changes have to be made. Due to the high frequencies, guided transmission by wires or waveguides is not feasible and a quasi-optical setup is used. Furthermore, the short switching time required is beyond the capability of an electronic device, thus short optical pulses of a few femtoseconds furnished by lasers are utilized. Here the sudden discharge between two electrodes forms a transient electric dipole and generates a short electromagnetic wave which contains a broad range of frequencies; its center frequency corresponds to the inverse of the transient time. Terahertz spectroscopy measures two electromagnetic pulse shapes: the input (or reference) pulse and the propagated pulse which has changed shape owing to its passage through the sample under study. We follow the discussion of Section 10.2.1, but as the response function $\hat{\chi}(\omega)$ we now use the complex refractive index $\hat{N}(\omega)$, which describes the wave propagation, or the conductivity $\hat{\sigma}(\omega)$, which relates the current density to

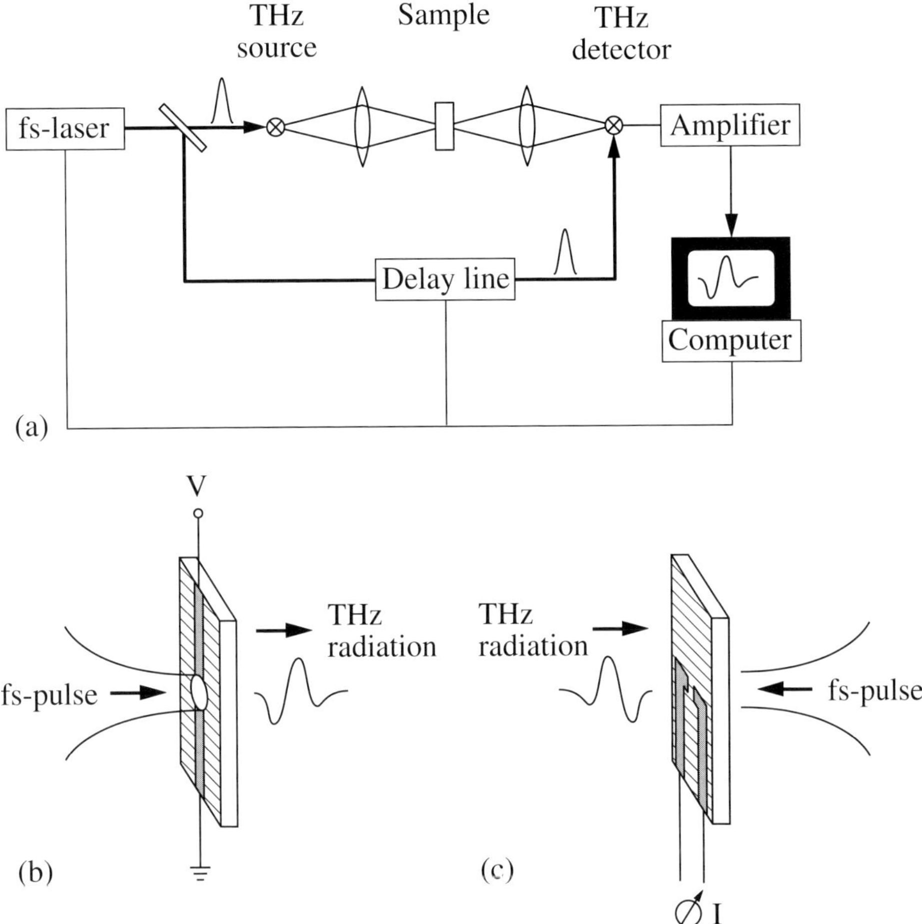

Fig. 10.4. (a) Schematic of a terahertz (THz) time domain spectrometer. The femtosecond laser pulse drives the optically gated THz transmitter and – via beamsplitter and delay line – the detector. The THz radiation is collimated at the sample by an optical arrangement. (b) The radiation source is an ultrafast dipole antenna which consists of a charged transmission line. The beam spot of the femtosecond pulses is focused onto the gap in the transmission line and injects carriers into the semiconductor leading to a transient current flowing across the gap, which serves as a transient electric dipole. (c) In order to detect the THz radiation, a similar arrangement is used. The current flows though the switch only when both the THz radiation and photocarriers created by the femtosecond laser beam are present.

the electric field applied. Equation (10.2.1) then reads:

$$\mathbf{J}(t) = \int \hat{\sigma}(t - t')\mathbf{E}(t')\,\mathrm{d}t' \quad .$$

A terahertz time domain spectrometer is displayed diagrammatically in

Fig. 10.4a; between the terahertz source and detector a standard optical arrangement of lenses or parabolic mirrors guides the radiation and focuses the beam onto the sample. The most suitable design of a terahertz radiation transmitter is a charged transmission line shorted by a laser pulse of a few femtoseconds duration. Fig. 10.4b shows a terahertz radiation source. The subpicosecond electric dipoles of micrometer size are created by photoconductive shorting of the charged transmission line with the femtosecond pulses from a dye or Ti–sapphire laser. The detection segment is designed similarly to the radiation source (Fig. 10.4c) via the transmission line; one side of the antenna is grounded and a current amplifier is connected across the antenna. During operation, the antenna is driven by the incoming terahertz radiation pulse which causes a time dependent voltage across the antenna gap. The induced voltage is measured by shorting the antenna gap with the femtosecond optical pulse in the detection beam and monitoring the collected charge (current) versus time delay of the detection laser pulses with respect to the excitation pulses [Ext89, Ext90]. If the photocarrier lifetime is much shorter than the terahertz pulse, the photoconductive switch acts as a sampling gate which samples the terahertz field. In principle, reflection measurements are also possible, although they are rarely conducted because the setup is more sensitive to alignment. Details of the experimental arrangement and the analysis can be found in [Nus98].

As an example of an experiment performed in the time domain, Fig. 10.5a shows the amplitude of the terahertz radiation transmitted through a thin niobium film deposited onto a quartz substrate at two different temperatures, above and below the superconducting transition temperature of niobium [Nus98]. Both the real and imaginary parts of the complex conductivity $\hat{\sigma}(\omega)$ of the niobium film can be obtained directly from these terahertz waveforms without the use of the Kramers–Kronig relations. The frequency dependence of $\sigma_1(\omega)$ and $\sigma_2(\omega)$ normalized to the normal state value σ_n are shown in Fig. 10.5b for $T = 4.7$ K. These results may be compared with the analogous experiments performed in the frequency domain by the use of a Mach–Zehnder interferometer (see Fig. 14.5).

10.3 Fourier transform spectroscopy

The concept of a Fourier transform spectrometer is based on Michelson's design of an interferometer in which a beam of monochromatic light is split into two approximately equal parts which follow different paths before being brought together again. The intensity of the recombined light is a function of the relative difference in path length between the two arms; i.e. the light shows an interference pattern. Recording the combined intensity as a function of the delay δ of one of these beams allows the spectral distribution of the light by Fourier transformation (Appendix A.1) to be recovered.

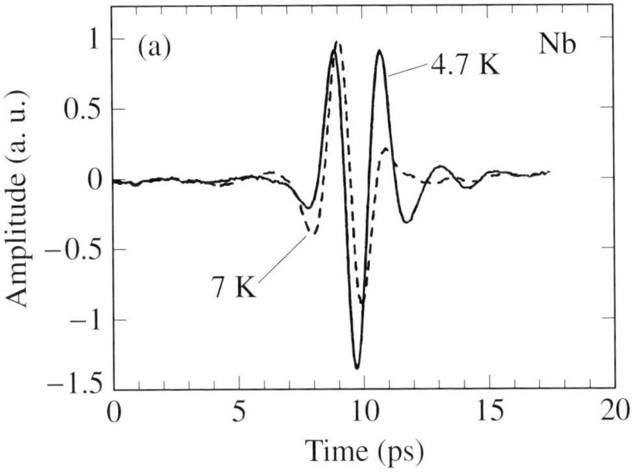

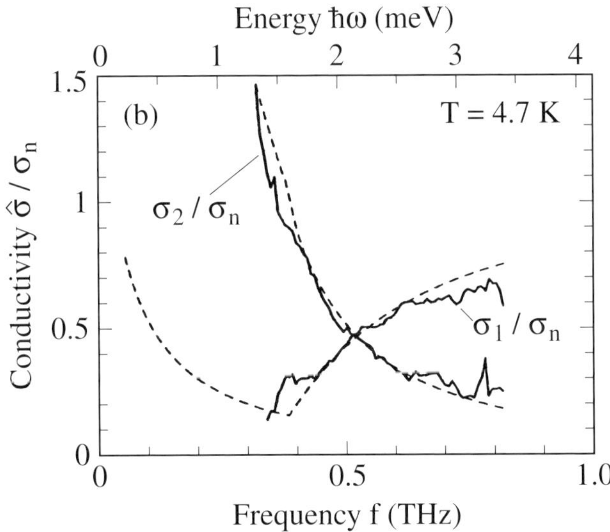

Fig. 10.5. (a) Terahertz transients transmitted through a thin niobium film ($T_c \approx 7$ K) on quartz in the normal state (dashed line) and the superconducting state (solid line). (b) Real and imaginary parts of the normalized complex conductivity, $\sigma_1(\omega)$ and $\sigma_2(\omega)$, of niobium in the superconducting state evaluated from the above data. The dashed line indicates the predictions by the BCS theory using the Mattis–Bardeen equations (7.4.20) (after [Nus98]).

Modern Fourier transform spectrometers mainly operate in the infrared spectral range (10–10 000 cm^{-1}); however, Fourier transform spectrometers have been built in the microwave frequency range as well as in the visible spectral range. A large number of detailed and excellent monographs on Fourier transform spectroscopy are available [Bel72, Gri86, Gen98]; only a short description is given here.

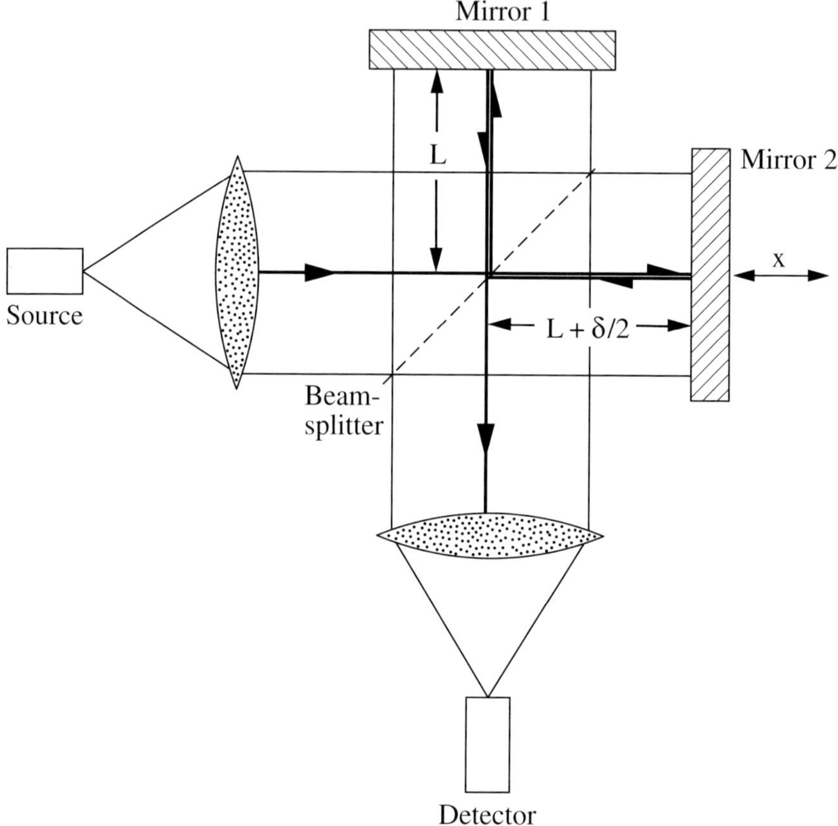

Fig. 10.6. Basic outline of a Michelson interferometer. The beam of the source is divided by the beamsplitter to the fixed mirror 1 and the moving mirror 2, at distances L and $L + \delta/2$, respectively. The recombined beam is focused onto the detector, which measures the intensity I as a function of displacement $\delta/2$.

10.3.1 Analysis

In order to discuss the principles of Fourier transform spectroscopy, the examination of a simple diagram of a Michelson interferometer shown in Fig. 10.6 is useful. Mirror 1 is fixed at a distance L from the beamsplitter, and the second mirror can be moved in the x direction. If mirror 2 is at a distance $L \pm \delta/2$ from the beamsplitter, then the difference in path length between the two arms is δ. If δ is an integer multiple of the wavelengths ($\delta = n\lambda$, $n = 1, 2, 3, \ldots$), we observe constructive interference and the signal at the detector is at a maximum; conversely, if $\delta = (2n + 1)\lambda/2$, the beams interfere destructively and no light is detected. Hence the instrument measures $I(\delta)$, the intensity of the recombined beam as a function of optical path difference. In other words, the interferometer

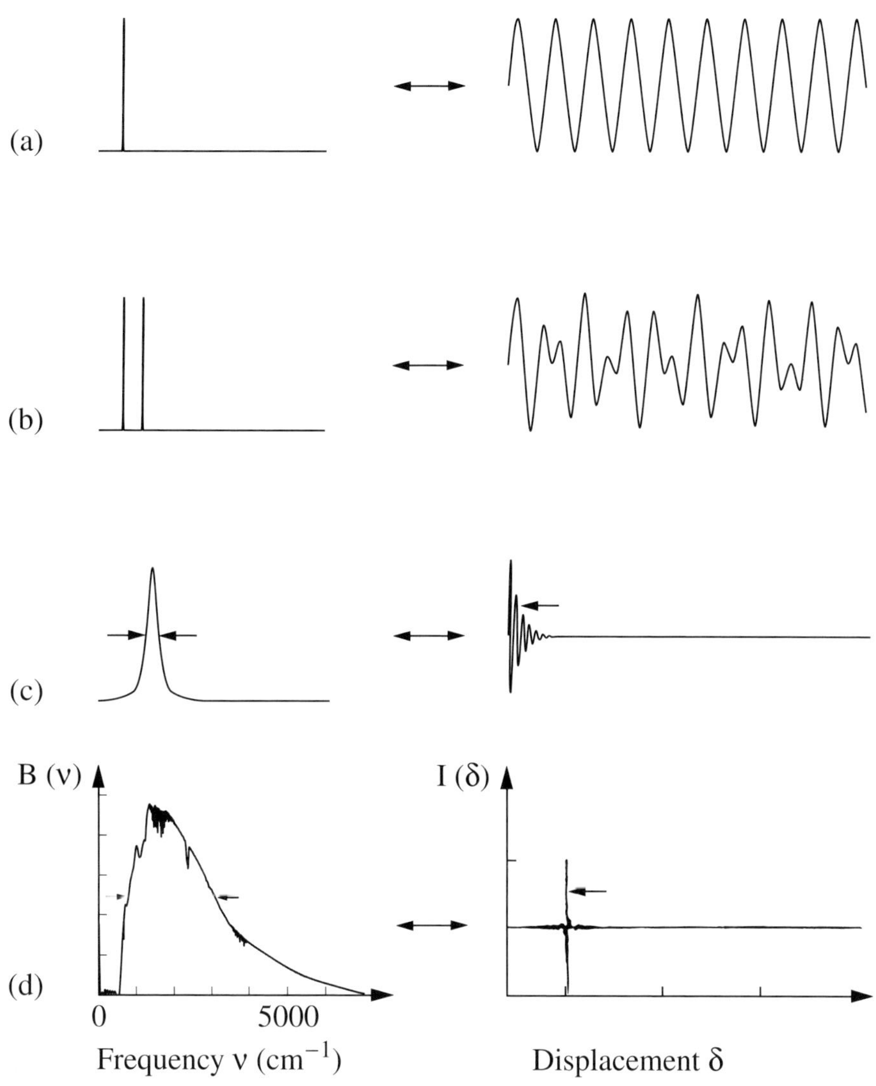

Fig. 10.7. Different spectra recorded by the interferometer and their Fourier transforms:
(a) monochromatic light; (b) two frequencies; (c) Lorentzian peak; (d) typical spectrum in
the mid-infrared range.

converts the frequency dependence of the spectrum $B(\omega)$ into a spatial dependence
of the detected intensity $I(\delta)$. From this information, it is possible to reconstruct
mathematically the source spectrum $B(\omega)$ no matter what form it has. In Fig. 10.7
we display such interference patterns: a single frequency obviously leads to a
signal with a $\cos^2$ dependence of the path length difference; interferograms from
non-monochromatic sources are more complicated.

The mathematical background of a Fourier transform spectrometer is laid out in Appendix A.1; here we apply it to the Michelson interferometer. We can write the electric field at the beamsplitter as

$$E(x, v)\, dv = E_0(v) \exp\{i(2\pi x v - \omega t)\}\, dv \quad , \tag{10.3.1}$$

where $v = 1/\lambda = \omega/2\pi c$ is the wavenumber of the radiation. Each of the two beams which reach the detector have undergone one reflection and one transmission at the beamsplitter, and therefore we can consider their amplitudes to be equal. If one light beam travels a distance $2L$ to the fixed mirror and the other a distance $2L + \delta$, then we can write the reconstructed field as

$$E_R(\delta, v)\, dv = |\hat{r}||\hat{t}| E_0(v)\big[\exp\{i(4\pi v L - \omega t)\} + \exp\{i[2\pi v(2L + \delta) - \omega t]\}\big]\, dv,$$

where we have assumed that both beams have the same polarization; $\hat{r}$ and $\hat{t}$ are the complex reflection and transmission coefficients of the beamsplitter. For a given spectral range, the intensity is proportional to the complex square of the electric field ($E_R E_R^*$). The preceding equation then gives

$$I(\delta, v)\, dv \propto E_0^2(v)[1 + \cos\{2\pi v \delta\}]\, dv \quad ,$$

and the total intensity from all wavenumbers at a particular path difference δ is

$$I(\delta) \propto \int_0^\infty E_0^2(v)\big[1 + \cos\{2\pi v \delta\}\big]\, dv \quad .$$

This is usually written in a slightly different form:

$$\left[I(\delta) - \frac{1}{2}I(0)\right] \propto \int_0^\infty E_0^2(v) \cos\{2\pi v \delta\}\, dv \quad , \tag{10.3.2}$$

often referred to as the interferogram. For a broadband source the intensity at infinite path difference $I(\infty)$ corresponds to the average intensity of the incoherent radiation which is exactly half the intensity obtained at equal paths: $I(\infty) = I(0)/2$; the interferogram is actually the deviation from this value at infinite path difference. Finally, from Eq. (10.3.2) and using the fact that $B(v) \approx E_0^2(v)$, we can use the inverse Fourier transformation to write

$$B(v) \propto \int_0^\infty \left[I(\delta) - \frac{1}{2}I(0)\right] \cos\{2\pi v \delta\}\, d\delta \quad . \tag{10.3.3}$$

Thus, $I(\delta)$ can be measured by the interferometer, and it is theoretically a simple task to perform the Fourier transform to arrive at $B(\omega)$, the power spectrum of the signal.

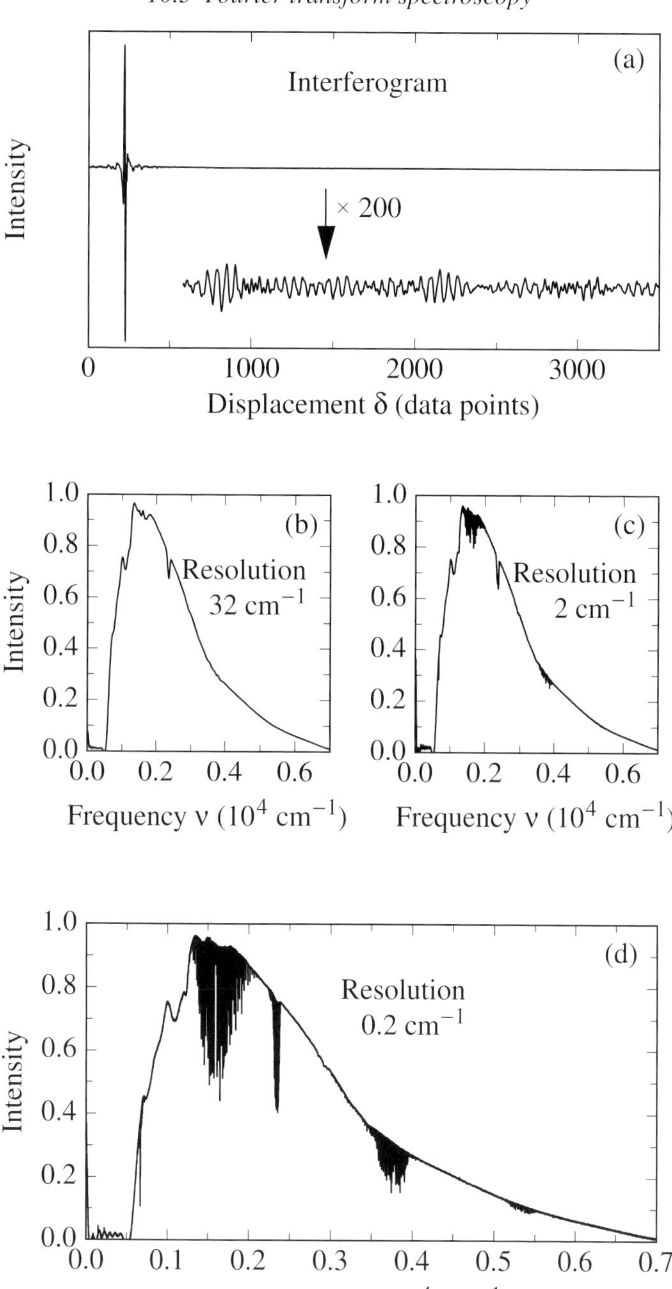

Fig. 10.8. (a) The first 3500 points of an interferogram. The signal in the wings is amplified 200 times; (b) Fourier transform of the first 512 points of the interferogram in (a), corresponding to a resolution of 32 cm^{-1}; (c) Fourier transform of 8196 points of the interferogram, corresponding to a resolution of 2 cm^{-1}; (d) taking all 81 920 points leads to a resolution of 0.2 cm^{-1}.

The integral in Eq. (10.3.3) is infinite, but for obvious reasons the interferogram is measured only over a finite mirror displacement $(\delta/2) \leq (\delta_{max}/2)$. The problem of truncating the interferogram is solved by choosing an appropriate apodization, instead of cutting it abruptly off.[1] As demonstrated in Fig. 10.8, this restriction limits the resolution to $\Delta\nu = 1/\delta_{max}$ – as can be easily seen from the properties of the Fourier transformation explained in Appendix A.1 – but it can also lead to errors in the calculated spectrum depending on the extrapolation used. In addition, an interferogram is in general not measured continuously but at discrete points, which might lead to problems like picket-fence effects and aliasing, i.e. the replication of the original spectrum and its mirror image on the frequency axis. However, this allows one to use fast computational algorithms such as the fast Fourier transformation [Bel72, Gri86].

10.3.2 Methods

One limitation of the spectral range of Fourier transform spectrometers is the availability of broadband sources. For a typical black-body radiation source, the peak of the intensity is typically somewhat below the visible. The spectral power is small above and well below this frequency and falls to zero as $\omega \to 0$. In general, three different sources are used to cover the range from the far-infrared up to the visible (Fig. 10.9). A set of filters, beamsplitters, windows, and detectors are necessary to cover a wide spectral range. At the low frequency end, in the extreme far-infrared, standing waves between the various optical components are of importance and diffraction effects call for dimensions of the optical components to be larger than a few centimeters. The limitation at higher frequencies is given by the mechanical and thermal stability of the setup and by the accuracy of the mirror motion; this is – as a rule – limited to a fraction of a micrometer. In practice, interferometers are used in two different ways depending on the scanning mode of the moving mirror. For a slow scanning interferometer or a stepped scan interferometer, on the one hand, the light is modulated with a mechanical chopper and lock-in detection is employed. The advantage of a stepped scan interferometer is that it can accumulate a low signal at one position of the mirror over a long period of time; it also can be used for time dependent experiments. A rapid scan interferometer, on the other hand, does not use chopped light because the fast mirror movement itself modulates the source radiation at audio frequency; here the velocity of the mirror is limited by the response time of the detector used. During the experiment, the recombined light from the interferometer is reflected off or transmitted through a sample before being focused onto the detector. Any

---

[1] The interferogram is usually multiplied by a function, the apodizing function, which removes false sidelobes introduced into transformed spectra because of the finite optical path displacement.

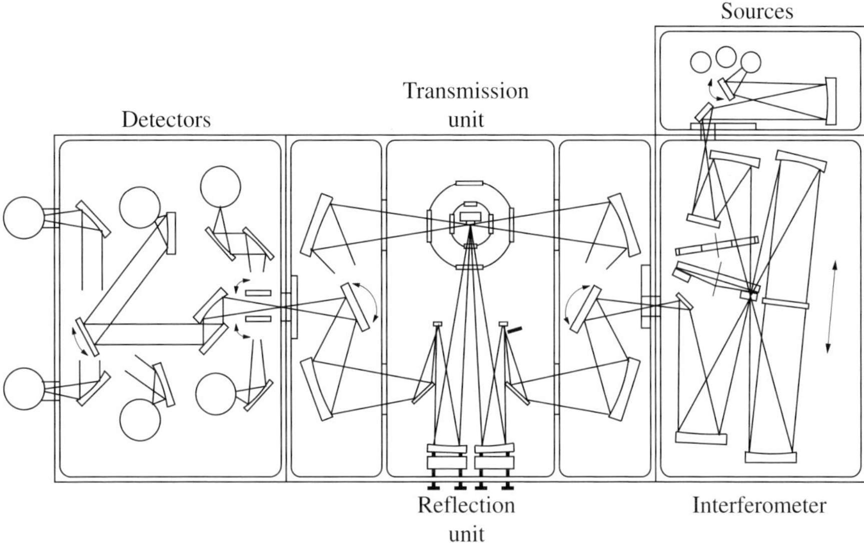

Fig. 10.9. Optical layout of a modified Bruker IFS 113v Fourier transform interferometer. The radiation is selected from three sources, guided through an aperture and a filter to the Michelson interferometer with six beamsplitters to select. A switching chamber allows transmission or reflection measurements of the sample inside the cryostat. The light is detected by one of six detectors.

arrangement used in optical measurements can also be utilized in combination with Fourier transform spectroscopy; in the following chapter we discuss the measurement configurations in detail.

A major disadvantage of the standard Fourier transform measurement – common to other optical techniques such as grating spectrometers – is that in general only one parameter is measured.[2] As discussed above, this means that the Kramers–Kronig relations must be employed in order to obtain the complex optical parameters such as $\hat{N}(\omega) = n(\omega) + \mathrm{i}k(\omega)$ or $\hat{\epsilon}(\omega)$ or $\hat{\sigma}(\omega)$.

In Fig. 10.10 experimental data obtained by a Fourier transform spectrometer are shown as an example. The polarized optical reflectivity ($\mathbf{E} \parallel a$) of $Sr_{14}Cu_{24}O_{41}$ was measured at room temperature and at $T = 5$ K over a wide spectral range. The highly anisotropic material becomes progressively insulating when the temperature decreases, as can be seen by the drop in low frequency reflectivity. Above 50 cm^{-1} a large number of well pronounced phonon modes dominate the spectra; they become sharper as the temperature decreases. In order to obtain the optical conductivity $\sigma_1(\omega)$ via Eq. (11.1.1b), the data are extrapolated by a Hagen–Rubens

---

[2] In order to measure both components, the sample must be placed in one of the active arms of the interferometer, e.g. replacing the fixed mirror in the case of a highly reflective sample or in front of a mirror in the case of dielectric samples. This arrangement is also called dispersive Fourier transform spectroscopy.

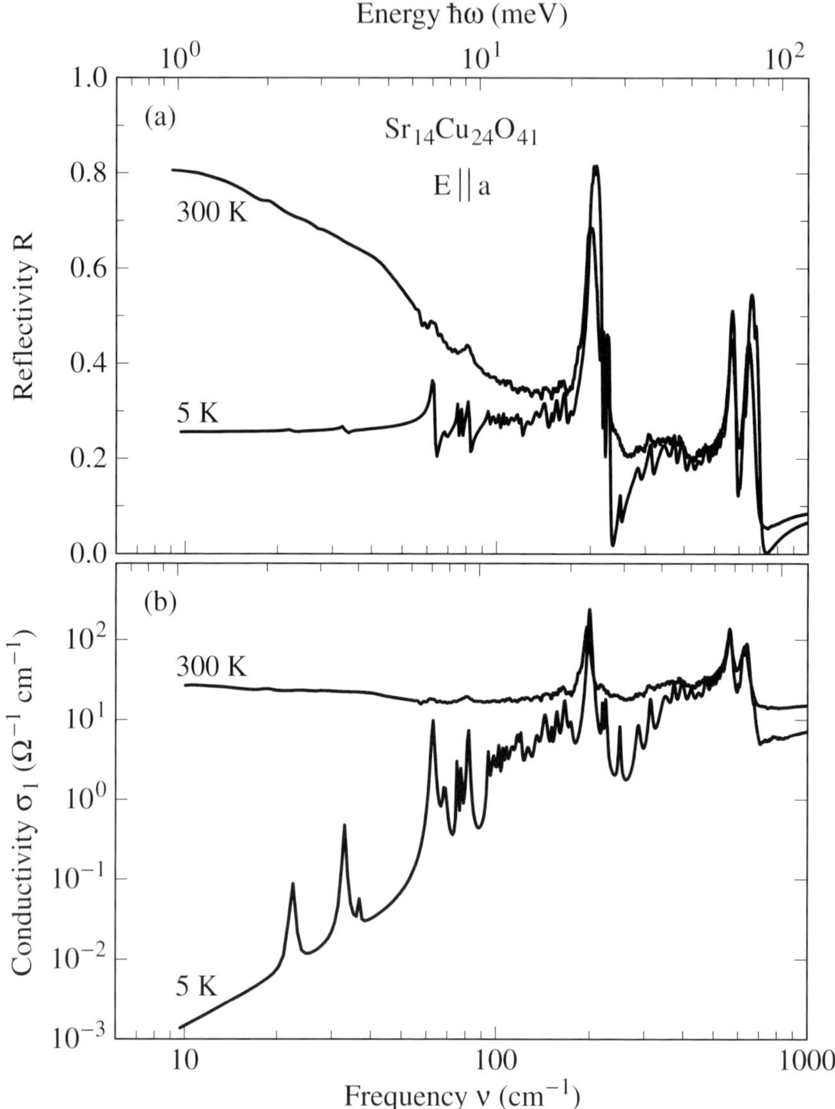

Fig. 10.10. (a) Frequency dependent reflectivity $R(\omega)$ of $Sr_{14}Cu_{24}O_{41}$ measured at two different temperatures for the electric field **E** oriented parallel to the a axis. (b) Optical conductivity $\sigma_1(\omega)$ of $Sr_{14}Cu_{24}O_{41}$ obtained by the Kramers–Kronig analysis of the reflection data [Gor00].

behavior (5.1.17) – for a metal as in the case of 300 K – or constant reflectivity – for an insulator as in the case of 5 K – in the limit $\omega \to 0$ and by assuming a smooth decrease of the reflectivity with the functional dependence of $R \propto \omega^{-2}$ for $\omega \to \infty$.

References

[Bel72] R.J. Bell, *Introductory Fourier Transform Spectroscopy* (Academic Press, New York, 1972)

[Boh95] R. Böhmer, B. Schiener, J. Hemberger, and R.V. Chamberlin, *Z. Phys. B* **99**, 91 (1995)

[Dav70] S.P. Davis, *Diffraction Grating Spectrographs* (Holt Rinehart & Winston, New York, 1970)

[Ext89] M. van Exter, Ch. Fattinger, and D. Grischkowsky, *Appl. Phys. Lett.* **55**, 337 (1989)

[Ext90] M. van Exter and D. Grischkowsky, *Phys. Rev. B* **41**, 12 140 (1990)

[Fel79] Yu.D. Fel'dman, Yu.F. Zuev, and V.M. Valitov, *Instrum. Exp. Techn.*, **22**, 611 (1979)

[Gem73] M.J.C. van Gemert, *Philips Res. Rep.* **28**, 530 (1973)

[Gen98] L. Genzel, *Far-Infrared Fourier Transform Spectroscopy*, in: *Millimeter and Submillimeter Wave Spectroscopy of Solids*, edited by G. Grüner (Springer-Verlag, Berlin, 1998), p. 169

[Gri86] P.R. Griffiths and J.A. de Haseth, *Fourier Transform Infrared Spectrometry* (John Wiley & Sons, New York, 1986)

[Gor00] B. Gorshunov, P. Haas, and M. Dressel, *unpublished*

[Hec92] J. Hecht, *The Laser Guidebook*, 2nd edition (McGraw-Hill, New York, 1992)

[Hil69] N.E. Hill, W.E. Vaughan, A.H. Price, and M. Davies, *Dielectric Properties and Molecular Behaviour* (Van Nostrand Reinhold, London, 1969)

[Hyd70] P.J. Hyde, *Proc. IEE* **117**, 1891 (1970)

[Key80] Keyes, R.J., ed., *Optical and Infrared Detectors*, 2nd edition, Topics in Applied Physics **19** (Springer-Verlag, Berlin, 1980)

[Nus98] M.C. Nuss and J. Orenstein, *Terahertz Time-Domain Spectroscopy*, in: *Millimeter and Submillimeter Wave Spectroscopy of Solids*, edited by G. Grüner (Springer-Verlag, Berlin, 1998)

[Sug72] A. Suggett, in: *Dielectric and Related Molecular Processes*, Vol. I (Chemical Society, London, 1970), p. 100

Further reading

[Cha71] G.W. Chantry, *Submillimetre Spectroscopy* (Academic Press, London, 1971)

[Fly87] M. O'Flynn and E. Moriarty, *Linear Systems: Time Domain and Transform Analysis* (Harper & Row, New York, 1987)

[Han01] P.Y. Han, M. Tani, M. Usami, S. Kono, R. Kersting, and X.C. Zhang, *J. Appl. Phys.* **89**, 2357 (2001)

[Kaa80] U. Kaatze and K. Giese, *J. Phys. E: Sci. Instrum.* **13**, 133 (1980)

[Mac87] J.R. Macdonald, *Impedance Spectroscopy* (John Wiley & Sons, New York, 1987)

[Mil86] E.K. Miller, ed., *Time-Domain Measurements in Electromagnetic* (Van Nostrand Reinhold, New York, 1986)

[Mit96] D.M. Mittleman, R.H. Jacobsen, and M.C. Nuss, *IEEE J. Sec. Topics Quantum Electron.* **2**, 679 (1996)

[Rie94] G.H. Rieke, *Detection of Light: From the Ultraviolet to the Submillimeter* (Cambridge University Press, 1994)

[Rze75] M.A. Rzepecka and S.S. Stuchly, *IEEE Trans. Instrum. Measur.* **IM-24**, 27 (1975)

[Sch95] B. Schrader, ed., *Infrared and Raman Spectroscopy* (VCH, Weinheim, 1995)

[Smi96] B.C. Smith, *Fundamentals of Fourier Transform Infrared Spectroscopy* (CRC Press, Boca Raton, FL, 1996)

[Sob89] M.I Sobhy, *Time-Domain and Frequency-Domain Measurements of Microwave Circuits* in: *Microwave Measurements*, edited by A.E. Bailey, 2nd edition, IEEE Electrical Measurement Series **3** (Peter Peregrinus Ltd, London, 1989)

[Ste84] S. Stenholm, *Foundations of Laser Spectroscopy* (John Wiley & Sons, New York, 1984)

11

Measurement configurations

The configurations which guide the electromagnetic radiation and allow the interaction of light with matter vary considerably, with a wide variation of techniques developed over the years and employed today. We can distinguish between single-path and interferometric arrangements, and – as far as the interaction of light with the material under study is concerned – single-bounce from multiple-bounce, so-called resonant techniques. In this chapter we present a short summary.

Single-path arrangements sample the change of the electromagnetic wave if only one interaction with matter takes place; for instance if the light is reflected off the sample surface or is transmitted through the specimen in a single path. In general, part of the radiation is absorbed, and from this the optical properties of the material can be evaluated. However, only at low frequencies (i.e. at long wavelengths) does this simple configuration allow the determination of the phase change of the radiation due to the interaction; for higher frequencies only the attenuation in power is observed.

Interferometric techniques compare one part of the radiation, which undergoes the interaction with the material (i.e. reflection from or transmission through the material), with a second part of the signal, which serves as a reference. In this comparative approach – the so-called bridge configuration – the mutual coherence of the two beams is crucial. The interference of the two beams is sensitive to both the change in amplitude and in phase upon interaction, and thus allows calculation of the complex response of the specimen.

Resonant techniques enhance the sensitivity of the measurement because the radiation interacts with the material multiple times and the electromagnetic fields are added with the proper phase relation (interference). By observing the two parameters which characterize the resonance, i.e. the resonance frequency and its quality factor, both components of the electrodynamic response of the sample are evaluated.

All of the above methods have advantages and disadvantages. Single-path arrangements are simple and are the preferred method if significant changes in the measured parameters are encountered. Interferometric techniques are, as a rule, more elaborate, but offer increased sensitivity and precision. Both of these are broadband techniques, and – at the same time – use a single bounce of the light at the interface of the specimen. Resonant techniques offer high sensitivity due to the multiple interaction of the radiation with the sample (broadly speaking, for a resonant structure of quality Q, the electromagnetic radiation bounces off approximately Q times from the surface of the specimen), but at the expense of a narrow bandwidth. Of course, the different optical path arrangements and the different – resonant or non-resonant – techniques can sometimes be combined.

In order to obtain the frequency dependent optical properties, the three measurement configurations have to be used in combination with one of the three spectroscopic principles discussed in the previous chapter. This then leads to a variety of measurement arrangements. The quality factor of a resonator, for instance, can be obtained in the time domain by cavity ring-down methods as well as in the frequency domain by measuring the width of the absorption curve. As another example, using the Fourier transform technique, we can study the transmission through a slab or simple reflection off a bulk sample – typical single-path configurations. Fourier transform spectroscopy also allows us to observe multireflection within a thin slab or within the substrate of a film – the sample then acts as a resonator with interference effects becoming important.

Of course, only a summary of the main principles can be given here, together with a short description of the commonly used arrangements in the different ranges of frequency. Sophisticated setups developed over the years in order to enhance sensitivity and accuracy and to address particular problems are beyond the scope of this book, but are covered by a vast literature that we extensively refer to.

11.1 Single-path methods

In the simplest measurement configuration, the radiation interacts with the material only once; for example, dc current flows though the specimen or light is reflected from the surface of the sample. These configurations can be regarded as transmission lines in which the specimen is inserted; in a single-bounce experiment either the reflection or the transmission (and in some cases both) is measured. In general, four data points have to be acquired to obtain the complex response: two parameters determine the amplitude and the phase of the signal before and two parameters determine these quantities after the interaction with the sample. With both amplitude and phase information available, the complex impedance $\hat{Z}$ of the sample – and then its conductivity $\hat{\sigma}$ – are evaluated. If only the ratio of

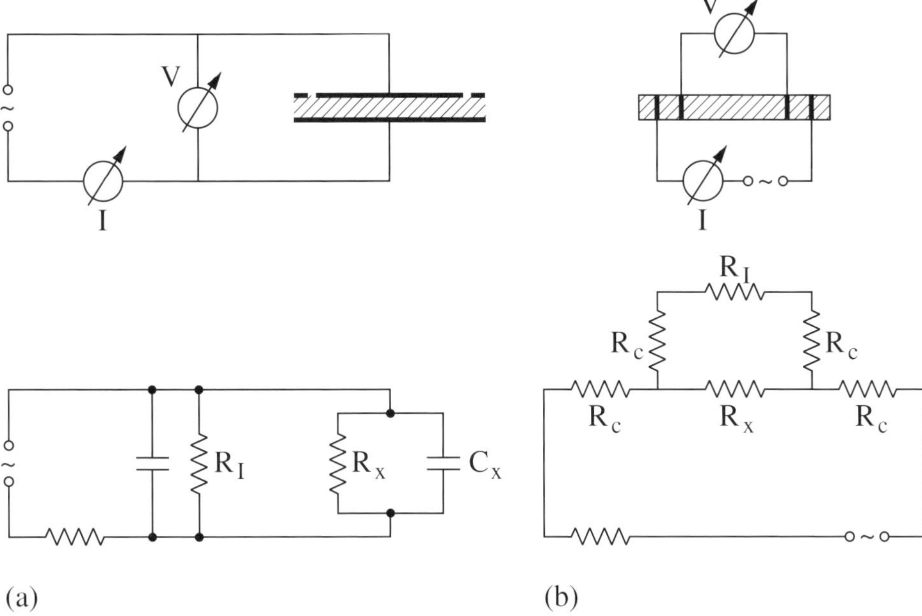

Fig. 11.1. Experimental arrangement for measuring the low frequency properties of materials and the corresponding equivalent circuit. (a) Guard-ring capacitor filled with material: R_x corresponds to the losses of the sample, and C_x describes the capacitance which changes due to the dielectric constant of the sample. (b) Four-lead technique used to measure the resistance R_x of the sample. The contact resistance is R_c, and R_I is the internal resistance of the voltmeter.

the power before and after the interaction is determined, certain assumptions are required to analyze further the reflectivity or transmission. Single-bounce methods, while less sensitive than the interference and resonant configuration techniques, nevertheless offer advantages, for example the simple arrangement, straightforward analysis, and, in general, a broad bandwidth in which they are applicable. For these reasons the single-path configuration is widely used and is perhaps the single most important technique.

Ellipsometry is a single-bounce method of a special kind which utilizes the dependence of the reflected power upon the polarization for oblique incidence in order to evaluate both the real and imaginary parts of the electrodynamic response function. This method is widely used and important in the visible spectral range.

11.1.1 Radio frequency methods

In the spectral range up to radio frequencies, the dielectric and transport properties of materials are measured either by placing the specimen in a parallel plate ca-

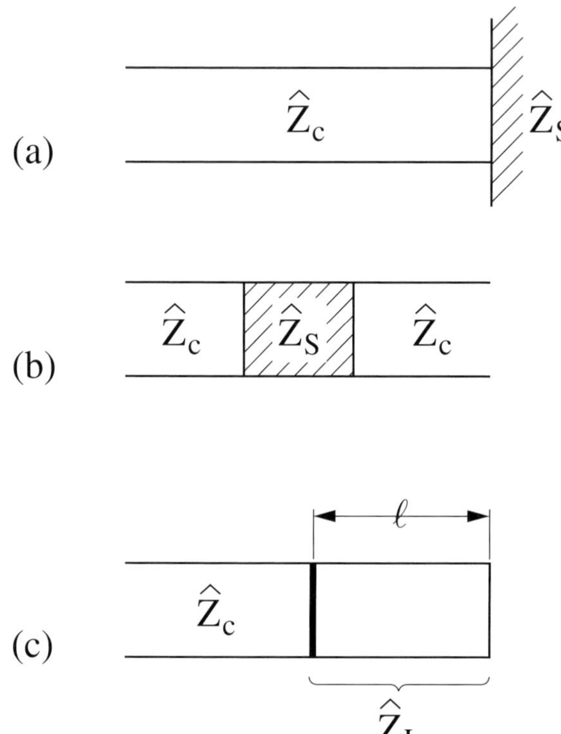

Fig. 11.2. Sample with impedance $\hat{Z}_S$ in a waveguide characterized by the impedance $\hat{Z}_c$. (a) A conducting sample terminates the transmission line. (b) The waveguide is filled with an insulating material. (c) A sample is placed at a distance $\ell = \frac{2n+1}{4}\lambda$ from the shorted end of a waveguide; at this position the electric field is at its maximum. For a well conducting material, the load impedance $\hat{Z}_L$ depends on the surface impedance of the sample $\hat{Z}_S$ and the geometrical configuration.

pacitor or by attaching four leads to the sample – the actual arrangement depends on the conductivity of the particular sample. If the material under investigation is insulating, a capacitor is filled as depicted in Fig. 11.1a; the dielectric constant and the conductivity are then calculated from the capacitance and the conductance provided the geometry of the capacitor is known. For materials with an appreciably large conductivity, the standard four-point technique is preferable where the outer two contacts supply the current and the inner two contacts probe the voltage drop (Fig. 11.1b).

Performing time domain measurements of dielectric materials, a voltage step is applied (Fig. 10.2) and the charging and discharging of the capacitor is studied, as discussed in Section 10.2.2. For measurements in the frequency domain, the current and voltage amplitude and the phase difference between them is determined; usually by employing a sine wave generator and oscilloscope or lock-in amplifier for detection. Stray capacitance and the inductance of the electrical wires determine the upper frequency limit (in the megahertz range) for the use of electrical leads to measure materials.

11.1.2 Methods using transmission lines and waveguides

In the frequency range of a few megahertz up to about 50 GHz, transmission lines are employed to study the electrodynamic properties of conducting as well as insulating materials [Boh89, Jia93]. In the simplest arrangement the sample terminates the transmission line for reflection measurements, as depicted in Fig. 11.2. Since (for a straightforward analysis) the sample dimensions have to exceed the skin depth, this is the preferential arrangement for conducting materials. On the other hand, if the material under consideration is insulating, the sample is placed inside the transmission line – between the inner and the outer conductors of a coaxial line, for instance. Similar arrangements are commonly used in the microwave and millimeter wave spectral range where rectangular waveguides are employed; the sample then replaces the end wall of the waveguide (reflection setup for conductors) or is positioned inside the waveguide (transmission setup for insulating materials). With microstrip and stripline techniques, the electrodynamic properties of the dielectric material placed between the conducting strips can be obtained; the method can also be used to fabricate the microstrip of a metal which is to be investigated.

At frequencies above the radio frequency range, only the reflected or transmitted power is probed. Using frequency domain spectroscopy, the phase information can still be obtained if a bridge configuration is utilized, as discussed in the following section; in the simplest case of a reflecting sample, the standing wave pattern in front of the specimen is measured and (by position and ratio of minimum and maximum) yields both components of the response. In the time domain, both parameters of the response function are evaluated from the delay and the dephasing of the reflected or transmitted pulse by utilizing the Laplace transformation (Section 10.2).

The electrodynamic properties of the conductors and the dielectric medium which constitute the transmission line – for example the inner and outer leads and the dielectric spacer in the case of a coaxial cable – alter the attenuation α and the phase velocity v_{ph} of the traveling wave in the transmission lines. In Eq. (9.1.4) we arrived at an expression for the phase velocity v_{ph} which, for a lossless transmission line, simplifies to $v_{\mathrm{ph}} = (L_1 C_1)^{-1/2}$; as usual, L_1 and C_1 are the inductance and capacitance per unit length, respectively. The attenuation constant α was derived in Section 9.1 and is given by $\alpha \approx R_1/Z_{\mathrm{c}} + G_1 Z_{\mathrm{c}}$. Thus this technique yields

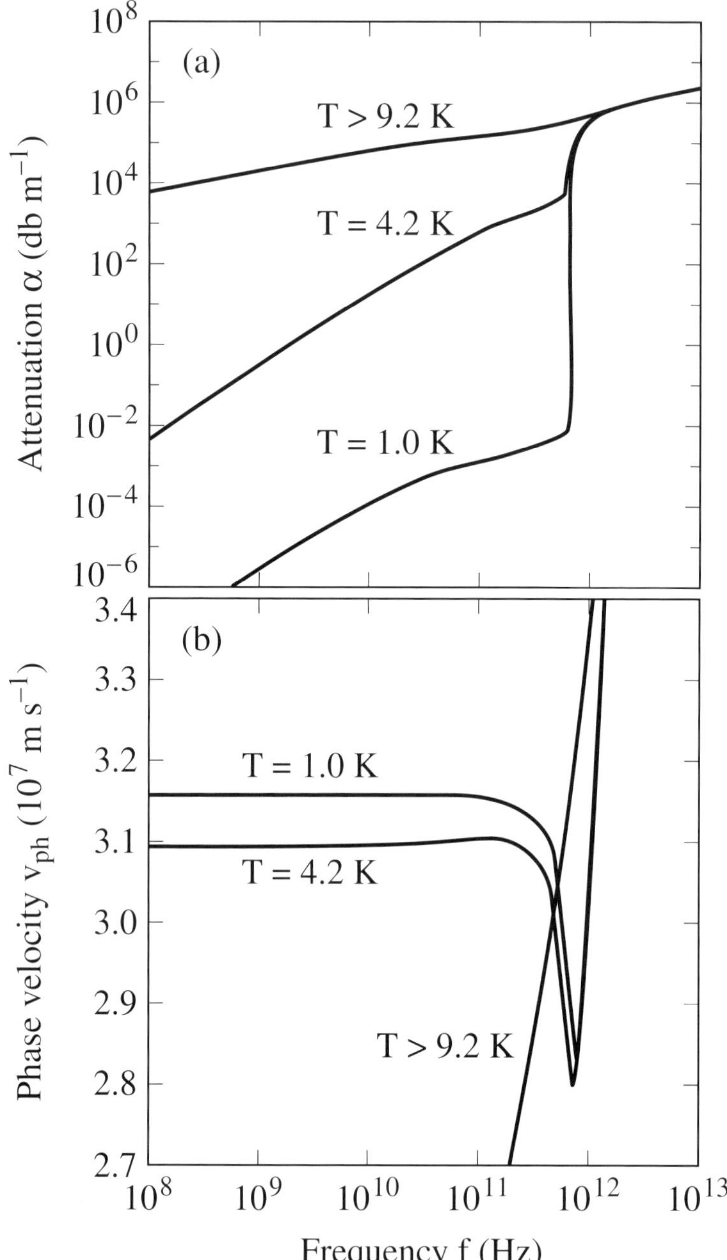

Fig. 11.3. (a) Attenuation and (b) phase velocity as a function of frequency for a niobium stripline (after [Kau78]). The calculations are based on the Mattis–Bardeen theory, Eqs (7.4.20). For $T > 9.2$ K niobium is metallic; at lower temperatures it is superconducting. While the attenuation α is proportional to the surface resistance, the phase velocity v_{ph} is related to the surface reactance and thus to the penetration depth.

information on ohmic losses in the conductor (given by R_l) and on the skin depth of a conductor or the penetration depth (both influence the phase velocity) and losses in a superconductor. This technique, however, can also be used to measure the dielectric constant and the losses of the material (determined by C_l and G_l) surrounding the transmission line.

As an example, let us consider the propagation of electromagnetic waves in a superconducting microstrip structure. If short pulses are sent along a parallel plate transmission line of superconducting niobium on a dielectric substrate, the attenuation and the phase velocity change as the material properties vary; the frequency dependences of the propagation parameters as calculated by Kautz [Kau78] are displayed in Fig. 11.3. In the superconducting phase, $T \ll T_\mathrm{c}$, the attenuation α is strongly reduced for frequencies $\omega < 2\Delta/\hbar$, with 2Δ the superconducting energy gap, and follows a ω^2 behavior. The phase velocity v_ph changes because of the influence of the skin depth and penetration depth on the effective geometry. These experiments – performed by using frequency domain as well as time domain techniques – are widely employed to study the electrodynamic properties of high temperature superconductors [Gal87, Lan91].

11.1.3 Free space: optical methods

Wave propagation in free space is preferred as soon as the wavelength becomes smaller than roughly a millimeter, i.e. the frequency exceeds a few hundred gigahertz; in this spectral range the assumptions of geometrical optics are valid, and the analysis is straightforward because the wavelength is much smaller than the sample size (and any other relevant dimension). Thus we can safely neglect diffraction effects and – since the surface is assumed to be infinite – the build up of charges at the edges of the sample.

Standard power reflection measurements are the most important technique from the submillimeter waves up to the ultraviolet spectral region. In Fig. 11.4a a typical reflection setup is shown; in order to separate the incident and reflected beams, the light has an angle of incidence ψ_i of $45°$. According to the Fresnel formulas (2.4.7), the reflectivity for oblique incidence depends on the polarization of the light – this angular dependence is used by the ellipsometric method discussed in Section 11.1.4. Since this is often a disturbing effect, strictly normal incidence is sought by using a beamsplitter in front of the sample (Fig. 11.4b). Reflectivity measurements are commonly performed in such a way that the sample is replaced by a reference mirror (evaporated aluminum or gold) to determine the incident power. Much effort is needed to prepare an exactly flat sample, which is particularly important in the spectral range above the visible. Another crucial point is the reproducible interchange between sample and reference; this becomes increasingly important

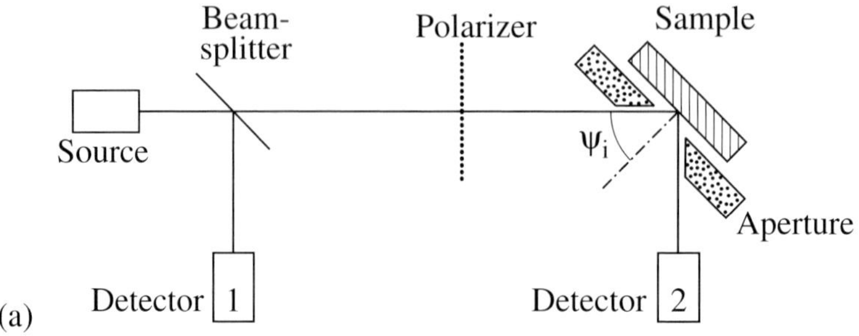

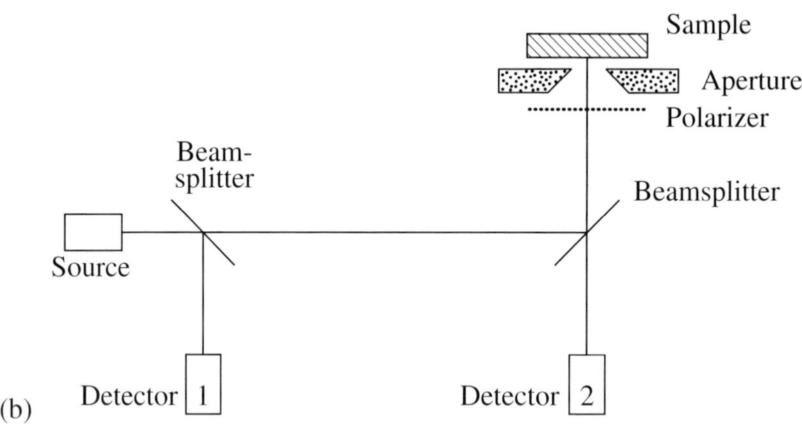

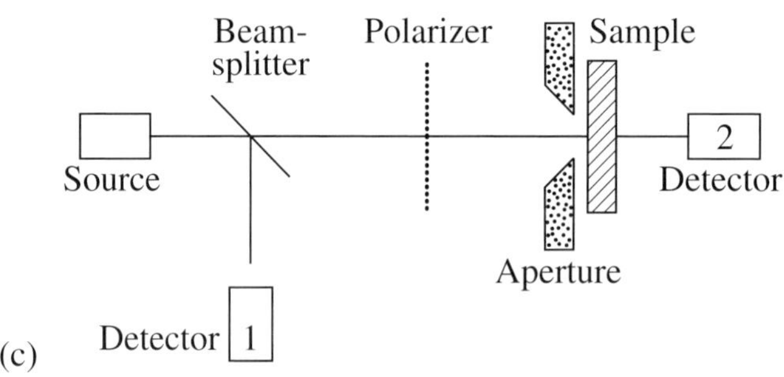

Fig. 11.4. (a) Experimental setup for measuring the reflectivity. A beamsplitter directs part of the radiation from a monochromatic source to a detector which monitors the output power of the generator; after passing through a polarizer the beam hits the sample at an angle $\psi_i = 45°$; a second detector measures the reflected intensity. (b) Using a beamsplitter allows reflection experiments at normal incidence. (c) Experimental setup for transmission measurements.

for highly conducting samples at low frequencies where the reflectivity follows the Hagen–Rubens law (5.1.17) and is close to unity [Ben60, Eld89, Hom93].

The optical arrangements do not vary significantly between the experiments performed in the frequency domain and in the time domain. While the latter principle allows the evaluation of both components of the optical response, in the frequency domain, for both transmission and reflection measurements, the power, but not the phase change upon interaction, is determined by a straightforward single-path method. Nevertheless, with the help of the Kramers–Kronig analysis we can obtain the complex electrodynamic response if the reflected power $R(\omega)$ is measured over a large range of frequency. The relations give the phase shift $\phi_r(\omega)$ of the reflected signal as

$$\phi_r(\omega) = \frac{\omega}{\pi} \int_0^\infty \frac{\ln\{R(\omega')\} - \ln\{R(\omega)\}}{\omega^2 - \omega'^2} \, d\omega' \quad , \qquad (11.1.1a)$$

which then allows the calculation of the complex conductivity

$$\sigma_1(\omega) = \frac{\omega}{4\pi}\epsilon_2(\omega) = \frac{\omega}{4\pi} \frac{4\sqrt{R(\omega)}[1 - R(\omega)]\sin\phi_r}{[1 + R(\omega) - 2\sqrt{R(\omega)}\cos\phi_r]^2} \qquad (11.1.1b)$$

$$\sigma_2(\omega) = \frac{\omega}{4\pi}[1 - \epsilon_1(\omega)] = \frac{\omega}{4\pi}\left(1 - \frac{[1 - R(\omega)]^2 - 4R(\omega)\sin^2\phi_r}{[1 + R(\omega) - 2\sqrt{R(\omega)}\cos\phi_r]^2}\right) \qquad (11.1.1c)$$

using the well known relations between the conductivity and the reflectivity. As mentioned in Chapter 10, the extrapolations at the low and high frequency ends of the experimental data are of great significance. For the simple analysis just presented, the skin depth (or the penetration depth) has to be smaller than the thickness of the sample in the direction of wave propagation to avoid transmission through the specimen. These types of reflectance studies are often referred to as bulk reflectivity measurements in contrast to optical experiments on thin films; in this case, as discussed in Appendix B in full detail, multireflection of the front and back of the material has to be included in the overall reflectivity.

In the cases of insulating materials or thin films, transmission experiments are more suitable and accurate than reflection measurements, as no reference mirror is needed (Fig. 11.4c); simply removing the sample from the optical path serves as a reference. Since part of the radiation is reflected off the surface of the material, additional information is needed in order to obtain the transmission coefficient; either the bulk reflectivity has to be known or two samples with different thicknesses are probed. The accurate expression, which also takes multireflection inside the specimen into account, is given by Eq. (B.12a); simpler relations for limiting cases are derived by [Pot85]. Other common techniques used to determine the

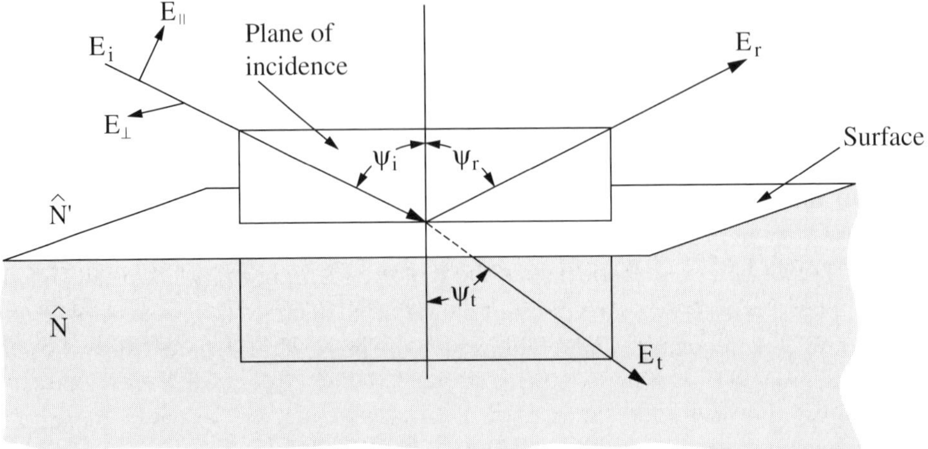

Fig. 11.5. Reflection and transmission of a light beam at a surface. The plane of incidence contains the incoming beam E_i, the outgoing beam E_r, and the transmitted beam E_t. The ratio of the angles of incidence ψ_i and transmission ψ_t at a single interface between two media is characterized by $\hat{N}'$ and $\hat{N}$. The amplitudes of the reflected and transmitted fields depend on the polarization with respect to the plane of incidence, $E_\parallel$ and $E_\perp$, and are given by Fresnel's equations (2.4.3).

optical properties of solids, such as frustrated and attenuated total reflectance, are not presented here, but are discussed in handbooks, for example [Pal85].

11.1.4 Ellipsometry

Ellipsometry or polarimetry measures the polarization of an electromagnetic wave in order to obtain information about an optical system which modifies the polarization; in general it is used for reflection measurements. While most of the early work was concentrated in the visible spectral range, recent advances have made ellipsometric investigations possible from the far-infrared to the ultraviolet. In contrast to standard reflectivity studies which only record the power reflectance, two independent parameters are measured, thus allowing a direct evaluation of the complex optical constants. Furthermore, as the magnitude of the reflected light does not enter the analysis, ellipsometric studies are not sensitive to surface roughness and do not require reference measurements. Several reviews of this topic and a selection of important papers can be found in [Azz87, Ros90, Tom93].

As discussed in Section 2.4.1, an electromagnetic wave reflected at a surface can be decomposed into a wave with the polarization lying in the plane of incidence (subscript $\parallel$) and a part which is polarized perpendicular to the plane of incidence (subscript $\perp$) according to Fig. 11.5. The two components of the electric field generally experience a different attenuation and phase shift upon reflection even for isotropic media, and hence the state of polarization changes as described by Fresnel's equations (2.4.7d) and (2.4.7b). Ellipsometry is a measurement of the complex ratio of both reflection coefficients:

$$\hat{\rho} = \frac{\hat{r}_\|}{\hat{r}_\perp} = \tan\{\theta\}\exp\{i\Delta\} \quad . \tag{11.1.2}$$

The two angles which have to be determined are related to the amplitude ratio θ and the phase $\Delta = \phi_{r\|} - \phi_{r\perp}$, where $\phi_{r\|}$ and $\phi_{r\perp}$ are the phase shifts induced by the sample upon the reflection. These coefficients depend upon the angle of incidence ψ_i, upon the material properties (the complex conductivity $\hat{\sigma}$ or dielectric constant $\hat{\epsilon}$), and typically also upon the frequency ω. As expected, for normal incidence the ellipsometric effect disappears because $R_\| = |\hat{r}_\||^2 = |\hat{r}_\perp|^2 = R_\perp$.

It is important to note that ellipsometers measure θ and Δ, and not the optical constants: to evaluate these, models are used with certain assumptions. With the simplest model (two semi-infinite dielectric materials with an abrupt discontinuity in $\hat{N}$ at the interface as shown in Fig. 11.5), both the real and imaginary components of the dielectric constant are determined using the following:

$$\epsilon_1 = \sin^2\{\psi_i\}\left[1 + \frac{\tan^2\{\psi_i\}(\cos^2\{2\theta\} - \sin^2\{\Delta\}\sin^2\{2\theta\})}{(1 + \sin 2\{\theta\}\cos\{\Delta\})^2}\right] \tag{11.1.3}$$

$$\epsilon_2 = \sin^2\{\psi_i\}\frac{\tan^2\{\psi_i\}\sin\{4\theta\}\sin\{\Delta\}}{(1 + \sin\{2\theta\}\cos\{\Delta\})^2} \quad . \tag{11.1.4}$$

Two configurations are often used to obtain the ellipsometric angles. The nulling technique determines the angle of the analyzer for which the reflected signal vanishes; while in the case of the photometric technique the transmitted signal is measured as a function of the polarization angle. The scheme of a null ellipsometer is displayed in Fig. 11.6: the linearly polarized light is reflected from the surface of the sample in an angle larger than $60°$ and becomes elliptically polarized; it is then converted back to a linearly polarized beam by a suitable rotation of the quarter wave ($\lambda/4$) plate. The intensity of the light at the detector is finally minimized by rotating the analyzer until it is perpendicular to the axis of polarization. The rotation angles of both the $\lambda/4$ plate and the analyzer allow the determination of θ and Δ. In other words, in general the transmitted light is elliptically polarized so that the major and minor axes are aligned with the directions of fast and slow propagation. The angle γ of the major axis is adjusted by rotating the $\lambda/4$ plate; by rotating the polarizer in front of the $\lambda/4$ plate, we account for the eccentricity $e = b/a$ – the ratio of the minor axis b to the major axis a.

In general no attempt is made to reconvert the elliptically polarized light to

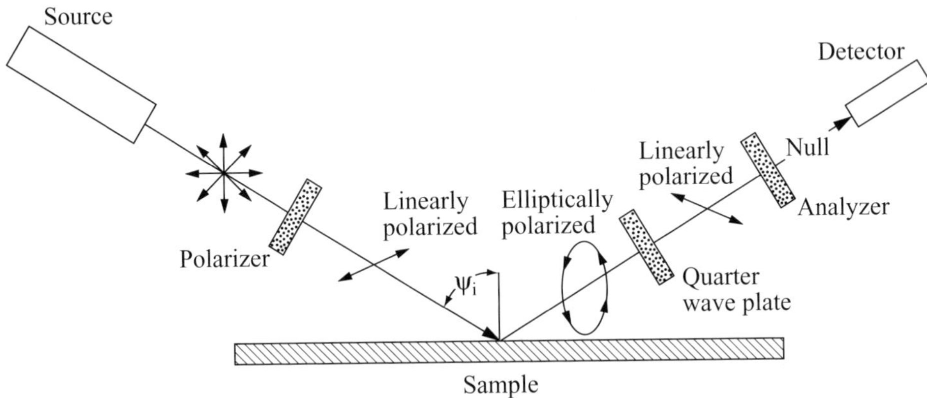

Fig. 11.6. Schematic arrangement of a null ellipsometer. A monochromatic light (polarized 45° out of the plane of incidence) hits the sample at an angle ψ_i. The reflected light is elliptically polarized; by passing through a $\lambda/4$ (quarter wave) plate it is transformed back to linear polarization. A second polarizer determines the polarization. The angular settings of the quarter wave plate and the analyzer are used to determine the phase shift Δ and attenuation ratio θ (after [Tom93]).

linearly polarized light; instead the first polarizer is set at a fixed value – usually 45° with respect to the plane of incidence – and the reflected intensity of the light is measured as a function of the analyzer angle (so-called rotating analyzer ellipsometer). The intensity as a function of analyzer angle is given by the expression

$$I(\alpha) = I_0 + I_1 \cos\{2\alpha\} + I_2 \sin\{2\alpha\} \quad ,$$

where I_1 and I_2 depend on both γ and e; here α is the position of the analyzer. The coefficients I_1 and I_2 can be found from the measured data since the azimuth is

$$2\gamma = \frac{\pi}{2} - \arctan\left\{\frac{I_1}{I_2}\right\} \tag{11.1.5}$$

and the eccentricity is

$$e = \left(\frac{I_0 + \left[I_1^2 + I_2^2\right]^{1/2}}{I_0 - \left[I_1^2 + I_2^2\right]^{1/2}}\right)^{1/2} \quad . \tag{11.1.6}$$

Thus, as with the nulling technique, we determine θ and Δ from the measured parameters; the material properties are finally evaluated from these values.

Ellipsometric investigations which cover a wide spectral range are cumbersome because the experiments are performed in the frequency domain; in the infrared range Fourier transform ellipsometers have become available [Ros90]. At each analyzer position an interferogram is recorded and a Fourier transform of the interferogram generates the frequency spectra of the reflected light at this position of the analyzer. The measurements are then repeated for the other analyzer position, and thus for a certain frequency the intensity is obtained as a function of angle. Measurements with these techniques have recently been made in the far-infrared frequency range using synchrotrons as highly polarized light sources [Kir97].

11.2 Interferometric techniques

In order to obtain the complex electrodynamic response of the material of interest, single-path configurations – as just described – have to record the absolute values of four different quantities; for example, magnitude and phase of the signal before the interaction and magnitude and phase after the interaction of the light with the sample. If the variations caused by the material are small, it is advantageous not to probe the signal itself, but to compare it with a (well characterized) reference and measure only the difference. The basic idea of interferometric measurements is therefore to compare the parameters of interest with known parameters, and to analyze the difference; this technique is often called the bridge method.

An interferometer splits the monochromatic radiation coming from one source into two different paths, with the sample being introduced in one arm and a reference into the other (no reference is needed for transmission measurements); the radiation is eventually combined and guided to a detector. If the coherence of the source (introduced in Section 8.3) is larger than the path difference, the recombined beam shows interference. From the two parameters, the phase difference and the attenuation caused by the sample, the material properties (e.g. complex conductivity $\hat{\sigma}$) are evaluated.

Since monochromatic radiation is required, interferometric techniques can only be utilized in the frequency domain. In the following we discuss three different setups which cover the spectral range where circuits, transmission lines, and optical arrangements are utilized.

11.2.1 Radio frequency bridge methods

The arrangement of a low frequency bridge is best explained by the Wheatstone bridge (Fig. 11.7). An element, the electrical properties of which we intend to measure and which is described by a complex impedance $\hat{Z}_1$, is inserted into a network of known impedances $\hat{Z}_2$ and $\hat{Z}_3$. Two points in the network are connected to an alternating current source, while a detecting instrument bridges the other two points. The fourth impedance $\hat{Z}_4$ is then adjusted until the two bridged points are at the same potential and phase – leading to a null reading at the detector

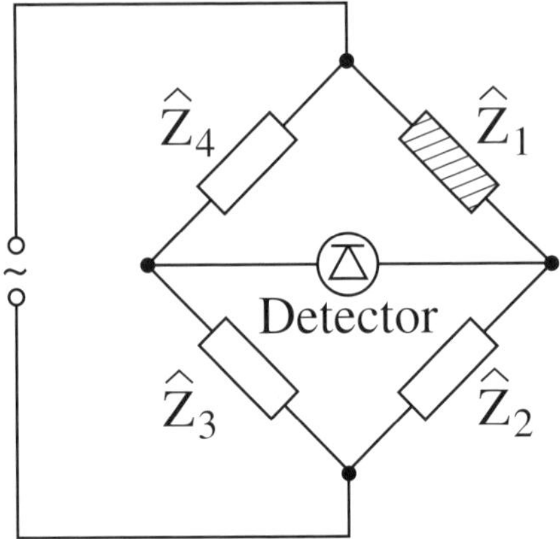

Fig. 11.7. Wheatstone bridge circuit with four impedances. The impedance $\hat{Z}_1$ is determined by adjusting $\hat{Z}_4$ in such a way that no signal is recorded at the detector. In the simplest case the two elements $\hat{Z}_2$ and $\hat{Z}_3$ are equal.

[Hag71]. The experimental determination of the electromagnetic properties is then reduced to the measurement of the values of an impedance $\hat{Z}_4$. In the simplest case the Wheatstone bridge contains only resistors, but the general layout also works for complex impedances, consisting of ohmic and capacitive contributions, for example. Then two parameters have to be adjusted to bring the detector signal to zero, corresponding to the phase and to the amplitude of the signal. Knowing the geometry of the specimen, and from the values of the resistance and the capacitance, the complex conductivity of the material can be calculated. Bridge configurations as depicted in Fig. 11.7 are available in a wide frequency range up to a few hundred megahertz. The sample is either placed in a capacitor or has wires attached (Fig. 11.1).

11.2.2 Transmission line bridge methods

A large variety of waveguide arrangements have been developed which follow the principles of bridge techniques [Gru98] to measure the complex conductivity of a material in the microwave and millimeter wave range (10–200 GHz). In general the beam is split into two arms with the sample placed in one and the second serving as a reference; both beams are finally recombined and the interference is observed. Up to approximately 50 GHz coaxial components are also employed for similar bridge configurations.

The technique is used for transmission and for reflection measurements. Since the former arrangement is similar to the Mach–Zehnder interferometer discussed in Section 11.2.3, we only treat the reflection bridge here. This configuration

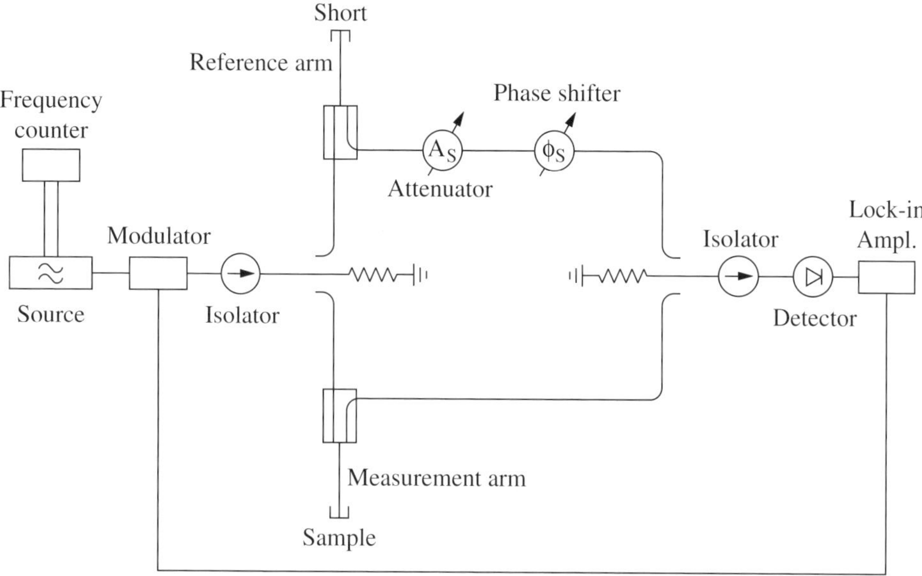

Fig. 11.8. Diagram of the millimeter wave impedance bridge for reflection measurements (after [Sri85]). First the bridge is nulled by terminating the measurement arm with a metal of known $\hat{Z}_S$. Then, with the sample replacing the metal, the interferometer is readjusted by changing the phase shift ϕ_S and attenuation A_S. From these two readings, ϕ_S and A_S, the load impedance terminating the transmission line and eventually the complex conductivity of sample are determined.

measures the complex reflection coefficient (or the scattering parameter $\hat{S}$) of the sample, which is written in the form

$$\hat{r} = -10^{A_S/20} \exp\{i\phi_S\} \quad , \tag{11.2.1}$$

where A_S is the change of the attenuation given in decibels and ϕ_S is the change in the phase. Once this quantity is known, using Eq. (9.2.4) the impedance of the sample is extracted, and from that the complex conductivity can be evaluated. The arrangement of a microwave reflection bridge [Joo94, Kim88, Sri85] is shown in Fig. 11.8. The sample either terminates the measurement arm (in this case the specimen has to be metallic or at least thicker than the skin depth – $\hat{Z}_L$ then simplifies to the surface impedance $\hat{Z}_S$ of the material); or the sample is placed at a distance ℓ from a short end of the transmission line, where the electric field is at a maximum. This is the case for $\ell = (2n+1)\lambda/4$ (with $n = 0, 1, 2, \ldots$) and thus the position depends on the particular frequency used (Fig. 11.2); for other frequencies appropriate transformations have to be made [Ram94].

As an example of a measurement performed by a millimeter wave impedance bridge operating at 109 GHz, we present results from TaSe$_3$ [Sri85]. The phase

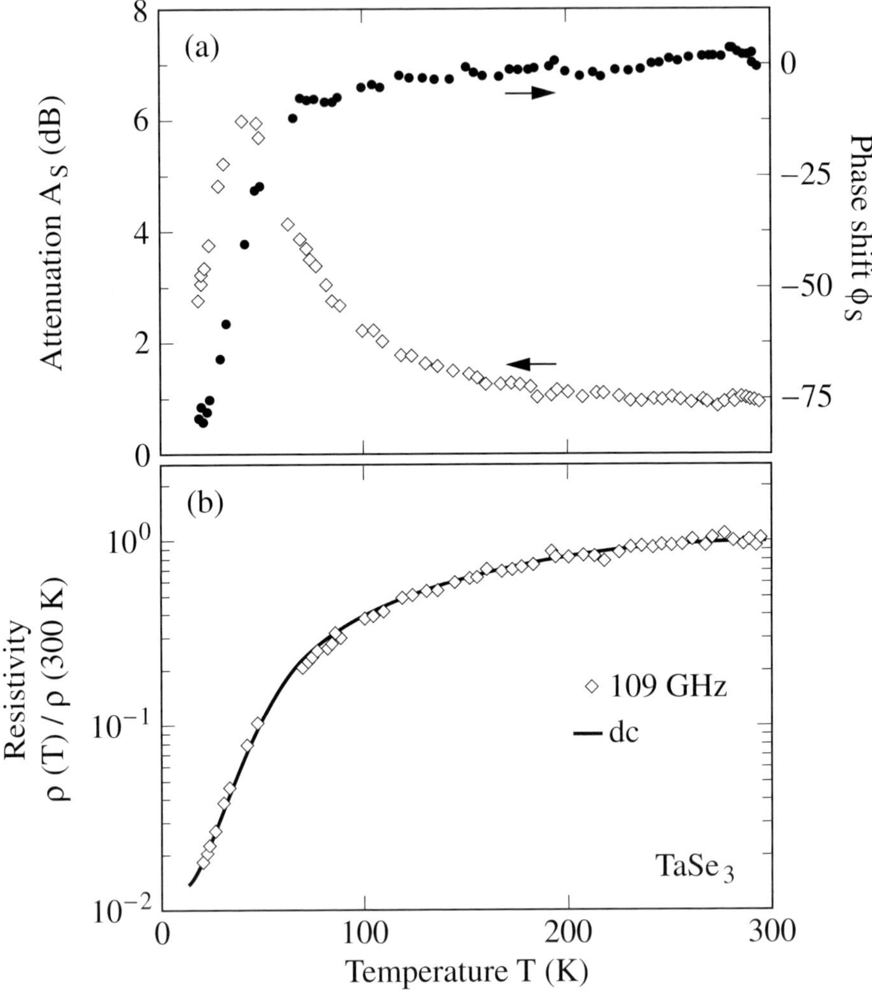

Fig. 11.9. (a) Attenuation (open diamonds) and phase shift (full circles) measured as a function of temperature on a sample of $TaSe_3$ with a millimeter wave bridge at 109 GHz. (b) From the data shown in (a) the temperature dependent resistivity of $TaSe_3$ is obtained at 109 GHz and compared with dc measurements. The sample was a thin needle with an approximate cross-section of 1 μm $\times$ 1 μm. At room temperature, the dc resistivity is 500 $\mu\Omega$ cm and the microwave results were normalized to this value (after [Sri85]).

shift and the attenuation due to the needle shaped crystal placed in the maximum of the electric field are shown in Fig. 11.9a as a function of temperature. For this configuration the load impedance and from that the complex conductivity can be evaluated; the calculated resistivity $\rho(T)$ is displayed in Fig. 11.9b. The solid line is the four-probe dc resistivity measured – the experiments provide evidence that the conductivity is independent of frequency in the measured spectral range.

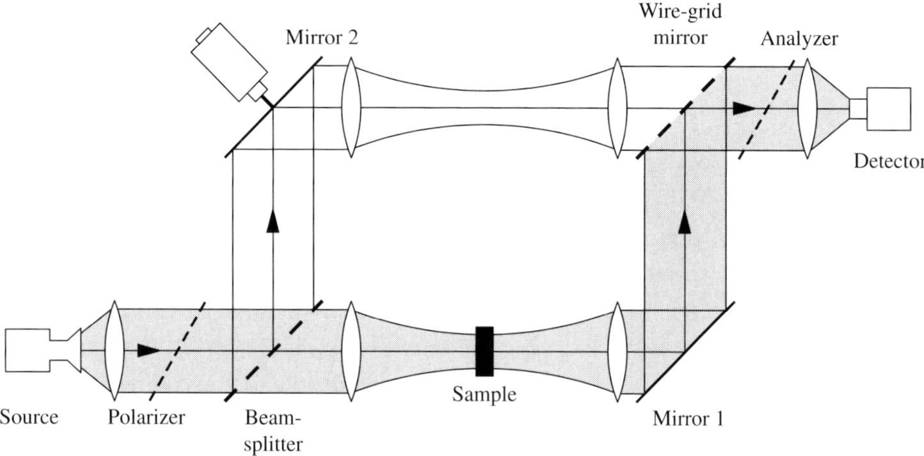

Fig. 11.10. Mach–Zehnder type interferometer used for quasi-optical transmission measurements in the submillimeter wave range. The coherent radiation is split by wire grids. The length of the reference arm can be adjusted for destructive interference by moving mirror 2. After the sample is introduced, the interferometer is readjusted in order to obtain ϕ_t.

11.2.3 Mach–Zehnder interferometer

The Mach–Zehnder interferometer is arranged following the outline of a transmission bridge; it is common in the optical range (the frequency range from millimeter waves up to the visible spectrum) where the electromagnetic waves propagate in free space. The monochromatic beam is split into two paths which are finally recombined. The sample is placed in one arm, and the changes in phase and the attenuation are measured by compensation; this procedure is repeated at each frequency. Provided the sample thickness is known, the refractive index n of the material is evaluated from the change in phase. The absorbed power is measured by a transmission measurement in a single-bounce configuration as discussed in Section 11.1.3, for instance. From the absorption coefficient α of the sample, the extinction coefficient k is determined using Eq. (2.3.18). Other material parameters, such as the complex conductivity $\hat{\sigma}$ or dielectric constant $\hat{\epsilon}$, are evaluated from n and k using the relations given in Section 2.3.

Fig. 11.10 shows the outline of a Mach–Zehnder type interferometer developed for the submillimeter wave region, i.e. from 2 cm^{-1} to 50 cm^{-1}, based on backward wave oscillators as tunable and coherent sources [Koz98]. As an example of a measurement conducted using this interferometer, the transmission $T_F(\omega)$ and phase shift $\phi_t(\omega)$ obtained on a semiconducting TlGaSe$_2$ sample (thickness 0.1 mm) is displayed in Fig. 10.1a and b. For each frequency, T_F and ϕ_t are obtained by separate measurements. From these two quantities both components of the complex dielectric constant $\hat{\epsilon}(\omega)$ are determined and are depicted in Fig. 10.1c and d. The close relation between the phase shift $\phi_t(\omega)$ and the dielectric constant $\epsilon_1(\omega)$ or the refractive index $n(\omega) \approx [\epsilon_1(\omega)]^{1/2}$ is clearly seen.

In contrast to resonant techniques, which are limited to a narrow range of frequency, interferometric methods can in general be used in a broad frequency range. The most significant advantage of interferometric arrangements compared to single-path methods is the possibility – in addition to the attenuation of the radiation – of determining the phase shift introduced by the sample. Furthermore, the method has increased sensitivity by directly comparing the electromagnetic wave with a reference wave in a phase sensitive way. The interferometric method can be combined with resonant techniques to enhance the sensitivity further.

Fig. 11.11 displays the results of a transmission experiment [Pro98] performed on a metal film on a substrate. The niobium film (thickness 150 Å) was deposited on a 0.45 mm thick sapphire substrate, which acts as a Fabry–Perot resonator due to multireflection. The transmission through this arrangement is measured by a Mach–Zehnder interferometer in order to also determine the phase shift. In Fig. 11.11a the transmission T_F through this composite sample is shown as a function of frequency, and the phase shift is displayed in Fig. 11.11b. As the temperature decreases below the superconducting transition $T_c = 8.3$ K, the transmitted power and phase shift are modified significantly. Since the properties of the dielectric substrate do not vary in this range of frequency and temperature, the changes observed are due to the electrodynamic properties of the superconductor. The change of the electrodynamic properties at the superconducting transition strongly changes the transmission and the phase of the composite resonator [Pro98]. The data can be used to calculate directly the real and imaginary parts of the complex conductivity $\hat{\sigma}(\omega)$. The results, together with the theoretical prediction by the Mattis and Bardeen formulas (7.4.20), are displayed in Fig. 14.5.

11.3 Resonant techniques

Resonant methods utilize multiple reflection to increase the interaction of the electromagnetic radiation with the material under investigation. The fundamental concept of resonant structures were discussed in Section 9.3 where the technical aspects of these measurement configurations are summarized. The quality factor Q of a resonant structure – as defined in Section 9.3.1 – indicates the number of times the wave bounces back and forth in the resonator, and, roughly speaking, the sensitivity of the measurement by a resonant technique is Q times better than the equivalent non-resonant method. The increase in sensitivity is at the expense of the bandwidth – the major drawback of resonant techniques. In general the applicability of resonant structures is limited to a single frequency, and only in

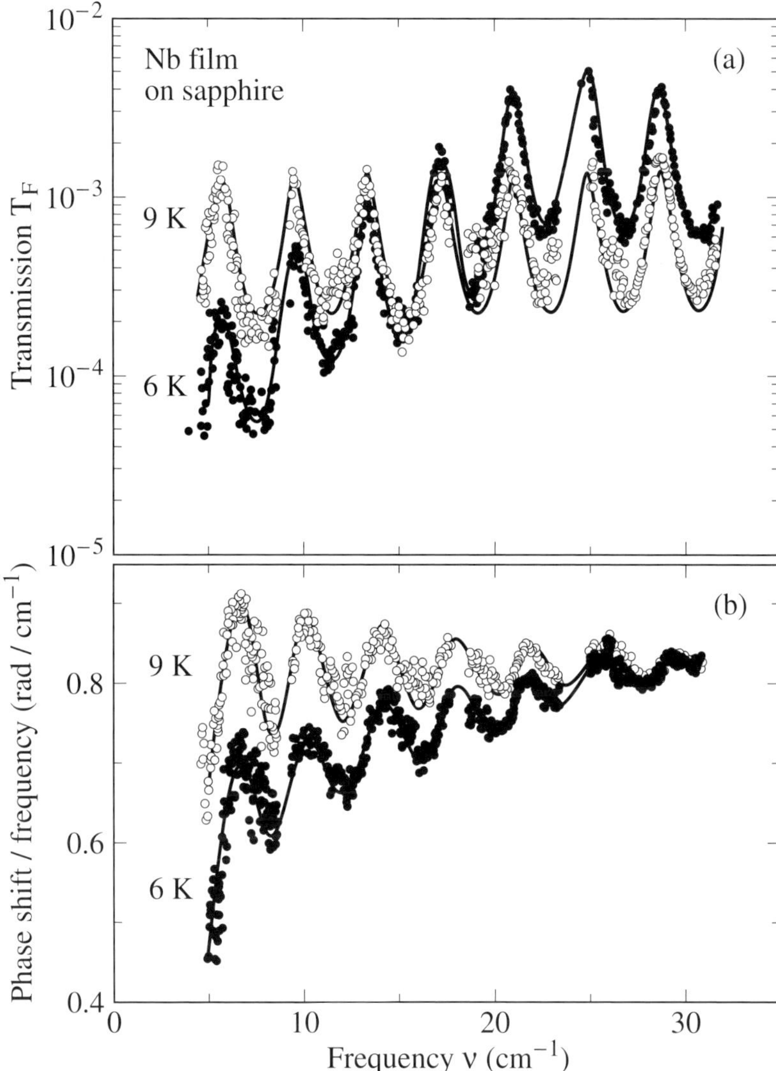

Fig. 11.11. Frequency dependent transmission $T_{\mathrm{F}}(\omega)$ and phase shift $\phi(\omega)$ spectra of a niobium film on a sapphire substrate (0.45 mm) at two temperatures above and below the superconducting transition $T_{\mathrm{c}} = 8.3$ K (after [Pro98]).

some cases are higher harmonics or different modes utilized. Resonant techniques are widely used in the gigahertz range (microwaves up to submillimeter waves) where non-resonant measurement techniques lack sensitivity. The experiments are usually performed in the frequency domain but the resonator can also be excited by a short pulse, and then the timely decay of the signal is observed. This so-called cavity ring-down method is preferable for extremely large Q factors.

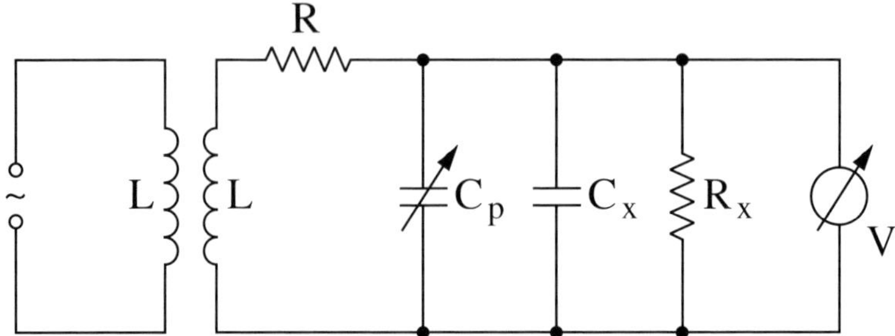

Fig. 11.12. Schematic of lumped resonance circuit. The source is weakly coupled to the resonator by the inductance L. R_x and C_x represent the sample. The tunable capacitance C_p allows coverage of a wide range of frequency.

11.3.1 Resonant circuits of discrete elements

In the audio and radio frequency range, up to approximately 100 MHz [Hil69], *RLC* resonant circuits are used to enhance the sensitivity of dielectric measurements. The complex dielectric constant is evaluated from the change in the resonance frequency and the decrease of the quality factor upon introducing the sample. The sample is usually contained between two parallel plates and is modeled by a parallel circuit of a capacitive part C_x and a resistive part R_x. Fig. 11.12 shows a simple corresponding circuit; as the capacitance C_p is varied, the signal detected by the voltmeter goes through a maximum when the resonance condition is passed (cf. Section 9.3). Roughly speaking, the resonance frequency depends on the dielectric constant ϵ_1 of the material introduced into the capacitor, while the width of the resonance curve increases as the losses of the sample (described by ϵ_2) increase. Following this arrangement, resonant circuits are designed to operate over a large range of frequencies; they are capable of high accuracy, provided the losses are low.

11.3.2 Microstrip and stripline resonators

As we have seen in Section 9.3, any transmission line between two impedance mismatches forms a resonant structure. Microstrip and stripline resonators utilize the fact that the resonance frequency and bandwidth of the transmission line resonator depend upon the electrodynamic properties of the conductors and of the dielectric media comprising the transmission line. Consequently, measurement configurations using microstrip resonators offer the following: if a conductor has to be studied, the stripline itself is made out of the material of interest; insulating material, on the other hand, is placed as dielectric spacers between the metallic

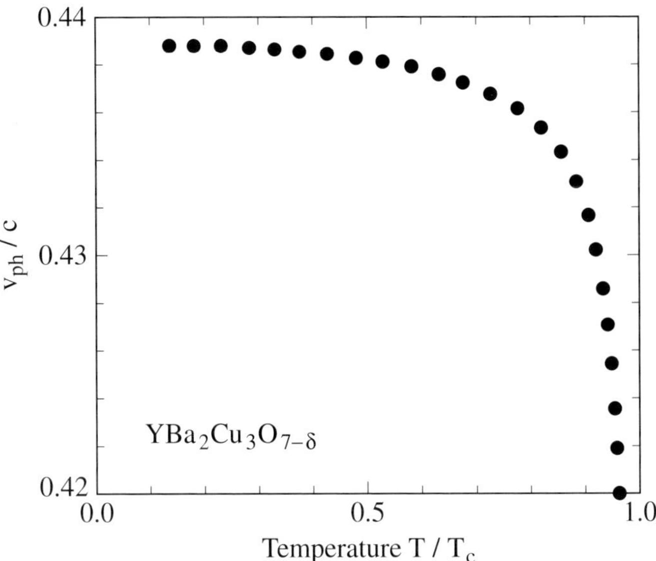

Fig. 11.13. Phase velocity v_{ph} data (normalized to the speed of light c) obtained for a YBa$_2$Cu$_3$O$_{7-\delta}$ superconducting microstrip resonator, plotted versus the reduced temperature (after [Lan91]).

transmission line. Microstrip and stripline resonators are used in the frequency range from 100 MHz to 10 GHz.

The change in the two characteristic parameters – the resonance frequency ω_0 and the quality factor Q – upon introducing a sample allows the determination of the dielectric constant and conductivity of the material if the geometry and the mode remain constant. The experiments are performed by recording the transmitted power as a function of frequency; whenever the resonance condition (9.3.4) is fulfilled, a maximum in transmission is observed. Alternatively, if a reflection arrangement is used, a minimum in the reflected power is recorded.

If the dielectric properties of an insulating material are investigated, the material of interest forms the dielectric spacer separating the ground plate and the microstrip pattern which constitutes the resonator. Here the resonant frequency increases with the dielectric constant ϵ_1' of the specimen; the dielectric losses ϵ_2' determine the width of the resonance curve. In an alternative configuration microstrip resonators are utilized to investigate the properties of a conducting (or even superconducting) material which forms the ground plate and/or the microstrip pattern. The ohmic losses of this metal lead to a broadening of the resonance curve; the shift of the resonance frequency allows the determination of the skin depth or the penetration depth, respectively.

Microstrip resonators are a well established technique used to study superconducting films which are deposited onto a substrate to form a resonant microstrip pattern; and in Fig. 11.13 measurements on a $YBa_2Cu_3O_{7-\delta}$ superconducting stripline resonator operating in the vicinity of 3 GHz [Lan91] are displayed. The phase velocity v_{ph}, a parameter proportional to the resonance frequency ω_0, is plotted as a function of the temperature; this parameter also contains the penetration of the electromagnetic field into the structure – and thus is a measure of the penetration depth λ. The phase velocity $v_{ph}(T)$ increases for decreasing temperature because $\lambda(T)$ becomes larger [Zho94].

11.3.3 Enclosed cavities

In the frequency range from 1 to 300 GHz enclosed cavities are employed to measure the dielectric properties of materials; the sample is introduced into the cavity and the changes of the resonance parameters are observed [Don93, Dre93, Kle93, Sch95]. Enclosed cavities are not limited by diffraction problems which other methods face if the wavelength λ becomes comparable to the sample size – a common occurrence in this range of frequency. Often it is sufficient to consider the sample (placed in the cavity) as a small perturbation of the resonator; the material parameters are evaluated from this perturbation as outlined in Section 9.3.3. The major disadvantage of cavities – like any resonant technique – is that they are (with a few exceptions) limited to a single frequency.

If the material under study is a low-loss dielectric, the entire cross-section of a cavity can be filled by the sample, as shown in Fig. 11.14a. The evaluation of the complex dielectric constant $\hat{\epsilon}$ is straightforward [Har58, Hip54a] since the geometrical factor is particularly simple. If the sample is a good conductor (and significantly thicker than the skin depth), it can replace the wall of a cavity. In Fig. 11.14b a cylindrical TE_{011} cavity is displayed where the copper end plate is replaced by a sample. The advantage of these two configurations is that the analysis has no geometrical uncertainties. Instead of filling the entire cavity or replacing part of it, a small specimen (with dimensions significantly smaller than the dimensions of the cavity) can be placed inside the cavity. This also leads to a modification of the cavity characteristics, which can be treated in a perturbative way. For a quantitative evaluation of the electrodynamic properties of the material, however, the geometry of the sample is crucial. Details of the analysis and discussion of particular sample shapes are found in [Kle93, Osb45, Pel98].

The experimental results obtained from a specimen which undergoes a metal–insulator transition at 135 K are shown in Fig. 11.15. The needle shaped crystal of α-(BEDT-TTF)$_2$I$_3$ was placed in the electric field antinode of a 12 GHz cylindrical TE_{011} reflection cavity. In Fig. 11.15a we display the temperature dependence of

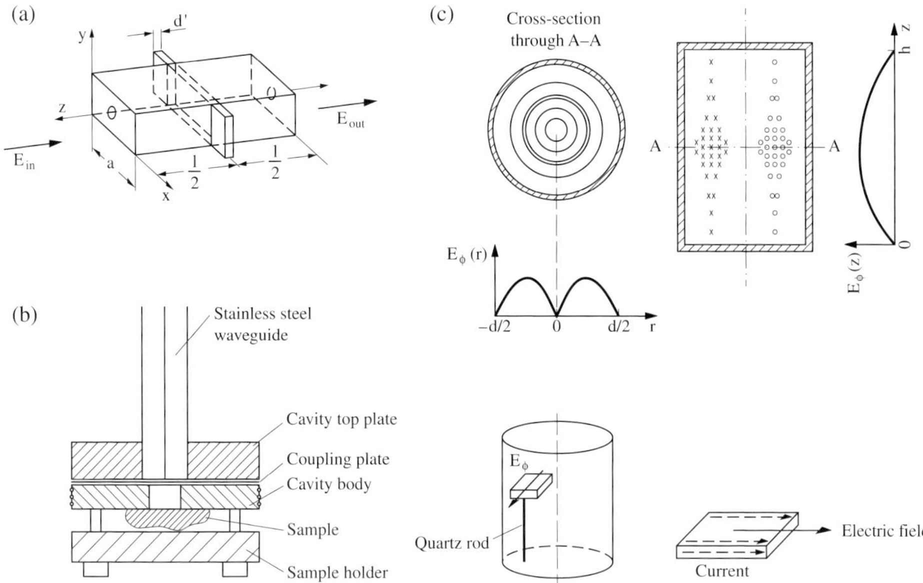

Fig. 11.14. (a) Arrangement for the measurement of slab-like samples in a rectangular TE_{101} cavity. (b) End plate method where the sample replaces parts of the cavity walls. (c) Electrical field configuration in a cylindrical TE_{011} cavity. The sample is located at the maximum of the electric field; see lower right (after [Don93, Dre96]).

the change in both the width $\Delta\Gamma/2\omega_0$ and resonance frequency $\Delta\omega/\omega_0$. The phase transition causes a large and rapid change in the frequency shift (about one order of magnitude) and a maximum in the bandwidth. These features indicate that a crossover from the metallic to the insulating regime occurs [Dre93, Dre94] as seen in Fig. 11.15b; and an appropriate analysis leads to the temperature dependence of the conductivity.

11.3.4 Open resonators

The spectral range from 1 cm^{-1} to 100 cm^{-1} is a transition region between microwave techniques (utilizing waveguides or coaxial cables) and optical methods (free space wave propagation). In this regime, open resonators are employed for the measurement of electromagnetic properties of materials [Afs85, Cha82, Cul83]. These resonators (often called Fabry–Perot resonators) consist of two mirrors separated by a distance which in general is considerably larger than the wavelength. A large number of higher order modes can be utilized for measurement purposes and thus open resonators are usually not as restricted to a single measurement frequency as enclosed cavities.

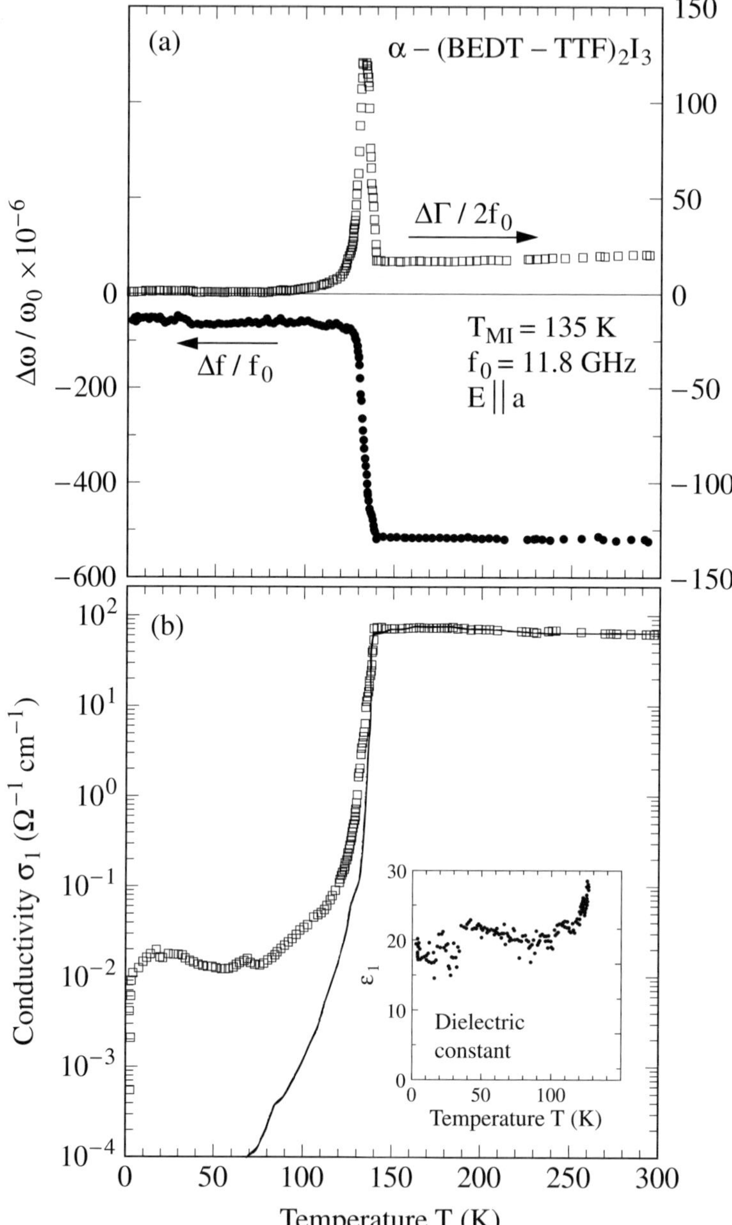

Fig. 11.15. (a) Temperature dependence of $\Delta\Gamma/2f_0$ and $\Delta f/f_0$ obtained on a sample of α-(BEDT-TTF)$_2$I$_3$ in an 11.8 GHz cavity. Near the phase transition, T_{MI}=135 K, a large change in the frequency shift $\Delta f/f_0$ occurs together with a peak in the bandwidth $\Delta\Gamma/2f_0$. (b) The a axis conductivity observed in the microwave range is displayed together with the dc conductivity. Below the metal–insulator phase transition, the high frequency conductivity develops a plateau while the dc values continue to decrease. The temperature dependence of the dielectric constant is displayed in the inset (after [Dre93]).

The reflected and transmitted intensities of a symmetrical Fabry–Perot resonator are described by the Airy function plotted in Fig. B.3. Instead of the quality factor, the finesse $\mathcal{F}$ defined as

$$\mathcal{F} = \frac{\pi}{2}\sqrt{F} = \frac{\pi\sqrt{R}}{(1-R)} = \frac{2\pi}{\Gamma} \tag{11.3.1}$$

is commonly used to describe the resonator; here Γ is the half-intensity full width of the transmission maxima of the Airy function $1/[1 + F\sin^2\{\beta\}]$ which peak at $\beta = 2\pi d/\lambda = \pm m\pi$, with $m = 0, 1, 2, \ldots$ accounting for higher harmonics. The finesse is related to the quality factor by $Q = 2\nu d\mathcal{F}$ (with the frequency $\nu = \omega/(2\pi c) = 1/\lambda$) and can be as large as 1000.

The simplest open resonator is built of two plane parallel mirrors; to reduce radiation losses the mirrors have to be much larger than the wavelength used. Spherical and hemispherical Fabry–Perot resonators were also developed [Cla82, Cul83] which overcome limitations due to these diffraction effects and due to alignment problems. In a good approximation the quality factor is given by $Q = \omega_0 d/(1 - R_{\text{eff}})c$, where d is the distance between the mirrors and R_{eff} is the effective reflectivity of the open resonator. At 150 GHz, for example, quality factors up to 3×10^5 have been achieved, making this a very sensitive arrangement.

We distinguish between three different arrangements which are used to measure material properties by open resonators. First, a slab of a low-loss material is placed inside the resonating structure. Second, the dielectric sample itself forms a Fabry–Perot resonator where the light is reflected at the front and at the back of the specimen. Third, one of the mirrors is replaced by the (highly conducting) sample.

In the first arrangement, the electrodynamic properties of a low-loss material are determined by introducing it into the resonator and measuring the change in frequency and halfwidth of the resonance; some of the possible experimental setups are shown in Fig. 11.16. For these arrangements the evaluation of the complex conductivity by the perturbation method is less accurate compared to that in enclosed cavities because in general radiation losses cannot be neglected.

In the second case, the sample itself forms the resonant structure – for example, a slab with the opposite sides being plane parallel; due to the impedance mismatch at the boundaries, multireflection within the sample occurs. The experiments, which are conducted either in transmission or reflection, map the interference pattern in a finite frequency range; fitting this pattern yields electromagnetic properties of the dielectric material using the expressions given in Appendix B. This method requires that the sample dimensions significantly exceed the wavelength – the thickness being a multiple of $\lambda/2$. The upper frequency limit is given by the ability to prepare plane surfaces which are parallel within a fraction of a wavelength. Due to these limitations, the technique of using the sample as a Fabry–Perot

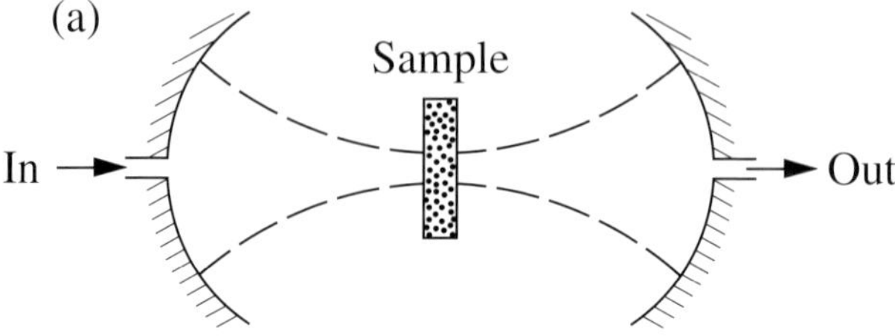

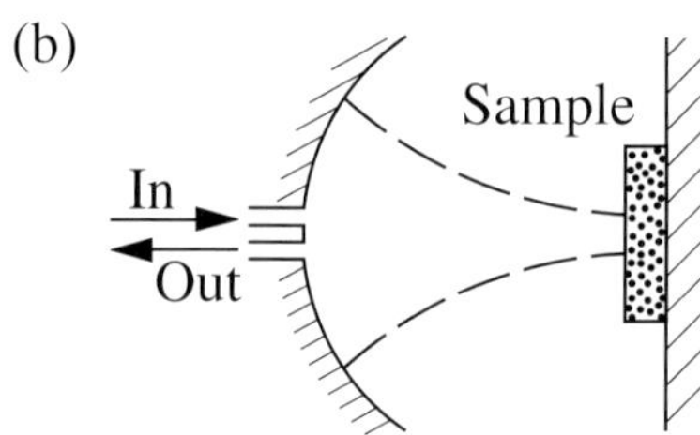

Fig. 11.16. Open resonator setups for the measurement of the optical conductivity: (a) confocal transmission resonator with low-loss specimen in the center; (b) hemispherical type for highly conducting samples.

resonator is mainly employed in the submillimeter and infrared frequency range. The measurements are performed by placing the Fabry–Perot arrangement in an optical spectrometer; the data are taken in the frequency domain or by Fourier transform technique. In combination with a Mach–Zehnder interferometer, the real and imaginary parts of the response can be measured independently, as shown in Fig. 10.1.

In order to measure highly conducting samples – the third case – the specimen is used as part of the resonant structure (e.g. as one mirror); most important is the case of thin films deposited on low-loss dielectric substrates. The (usually transparent) substrate acts as an asymmetric Fabry–Perot interferometer with one mirror made of the thin film. The interference pattern depends on the real and imaginary parts of the electrodynamic response of the film and of the substrate. If the latter parameters are known (for example by an independent measurement of the bare substrate), the complex conductivity $\hat{\sigma}$ of the conducting film is evaluated from the position and the height of the absorption minima. The detailed analysis of the optical properties of a sandwich structure is given in Appendix B. When the optical properties of a bulk sample – instead of a thin film – are investigated, a transparent material is brought in contact with the specimen; the measurements are performed in reflection configuration. Interference minima appear when the optical thickness of the dielectric is roughly equal to an integer number of a half

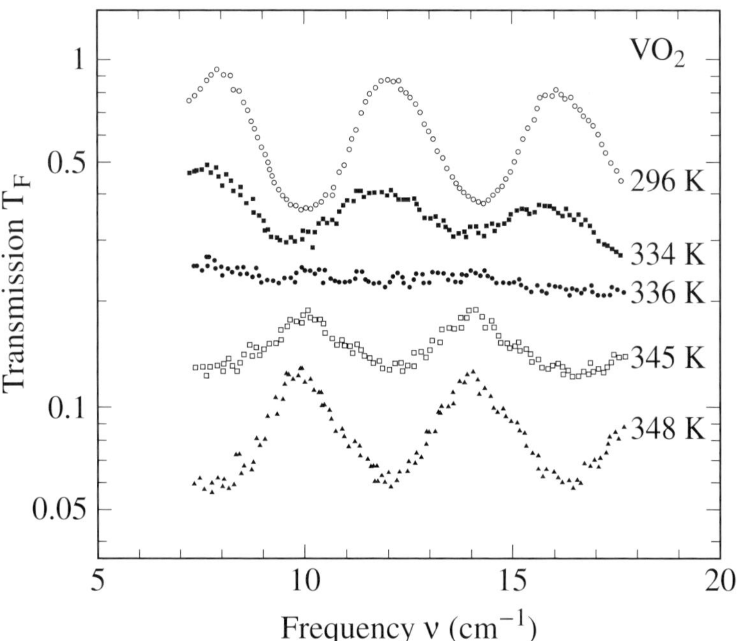

Fig. 11.17. Transmission spectra $T_F(\omega)$ of a polycrystalline VO_2 film deposited on a sapphire substrate. The thicknesses of the film and the substrate are 0.1 μm and 0.383 mm, respectively. The shape of the $T_F(\omega)$ spectra depends on the values of n and k for the film, going from an insulating to a metallic state with increasing temperature (measured by B. Gorshunov).

wavelength of the radiation; the interface pattern also sensitively depends on the impedance mismatch between dielectric and metal [Sch95].

The influence of the thin film impedance on the interference pattern which occurs due to multireflection within the substrate is seen in Fig. 11.17, where the transmission spectra $T_F(\nu)$ of a thin film of polycrystalline vanadium dioxide (VO_2) deposited on a sapphire plate are displayed. VO_2 is an insulator at low temperatures and shows a phase transition at 336 K to a metallic state. This transition clearly manifests itself in the transmission spectra by a strong decrease in $T_F(\nu)$ above the transition temperature and by a phase change in the interference oscillations by π in the course of the insulator-to-metal transition.

References

[Afs85] M.N. Afsar and K.J. Button, *Proc. IEEE* **73**, 131 (1985)

[Azz87] R.M.A. Azzam and N.M. Bashara, *Ellipsometry and Polarized Light*, 2nd edition (North-Holland, Amsterdam, 1987)

[Ben60] H.E. Bennett and W.F. Koehler, *J. Opt. Soc. Am.* **50**, 1 (1960)

[Boh89] R. Böhmer, M. Maglione, P. Lunkenheimer, and A. Loidl, *J. Appl. Phys.* **65**, 901 (1989)

[Cha82] G.W. Chantry, *J. Phys. E: Sci. Instrum.* **15**, 3 (1982)

[Cla82] R.N. Clarke and C.B. Rosenberg, *J. Phys. E: Sci. Instrum.* **15**, 9 (1982)

[Cul83] A.L. Cullen, *Millimeter-Wave Open Resonator Techniques*, in: *Infrared and Millimeter Waves* **10**, edited by K.J. Button (Academic Press, Orlando, FL, 1983), p. 233

[Don93] S. Donovan, O. Klein, M. Dressel, K. Holczer, and G. Grüner, *Int. J. Infrared and Millimeter Waves* **14**, 2459 (1993)

[Dre93] M. Dressel, O. Klein, S. Donovan, and G. Grüner, *Int. J. Infrared and Millimeter Waves* **14**, 2489 (1993)

[Dre94] M. Dressel, G. Grüner, J. P. Pouget, A. Breining, and D. Schweitzer, *J. Phys. I (France)* **4**, 579 (1994)

[Dre96] M. Dressel, O. Klein, S. Donovan, and G. Grüner, *Ferroelectrics*, **176**, 285 (1996)

[Eld89] E. Eldridge and C.C. Homes, *Infrared Phys.* **29**, 143 (1989)

[Gal87] W.J. Gallagher, CC. Chi, I.N. Duling, D. Grischkowsky, N.J. Halas, M.B. Ketchen, and A.W. Kleinsasser, *Appl. Phys. Lett.* **50**, 350 (1987)

[Gru98] G. Grüner, ed., *Waveguide Configuration Optical Spectroscopy*, in: *Millimeter and Submillimeter Wave Spectroscopy of Solids* (Springer-Verlag, Berlin, 1998), p. 111

[Hag71] B. Hague, *Alternating Current Bridge Methods*, 6th edition (Pitman, London, 1971)

[Har58] L. Hartshorn and J.A. Saxton, *The Dispersion and Absorption of Electromagnetic Waves*, in: *Handbuch der Physik* **16**, edited by S. Flügge (Springer-Verlag, Berlin, 1958), p. 640

[Hil69] N.E. Hill, W.E. Vaughan, A.H. Price, and M. Davies, *Dielectric Properties and Molecular Behaviour* (Van Nostrand Reinhold, London, 1969)

[Hip54a] A.v. Hippel, *Dielectrics and Waves* (Wiley, New York, 1954)

[Hom93] C.C. Homes, M. Reedyk, D.A. Cradles, and T. Timusk, *Appl. Opt.* **32**, 2976 (1993)

[Jia93] G.Q. Jiang, W.H. Wong, E.Y. Raskovich, W.G. Clark, W.A. Hines, and J. Sanny, *Rev. Sci. Instrum.* **64**, 1614 and 1622 (1993)

[Joo94] J. Joo and A.J. Epstein, *Rev. Sci. Instrum.* **65**, 2653 (1994)

[Kau78] R.L. Kautz, *J. Appl. Phys.* **49**, 308 (1978)

[Kim88] T.W. Kim, W. Beyermann, D. Reagor, and G. Grüner, *Rev. Sci. Instrum.* **59**, 1219 (1988)

[Kir97] J. Kircher, R. Henn, M. Cardona, P.L. Richards, and G.O. Williams, *J. Opt. Soc. Am. B* **104**, 705 (1997)

[Kle93] O. Klein, S. Donovan, M. Dressel, and G. Grüner, *Int. J. Infrared and Millimeter Waves* **14**, 2423 (1993)

[Koz98] G.V. Kozlov and A.A. Volkov, *Coherent Source Submillimeter Wave Spectroscopy*, in: *Millimeter Wave Spectroscopy of Solids*, edited by G. Grüner (Springer-Verlag, Berlin, 1998), p. 51

[Lan91] B.W. Langley, S.M. Anlage, R.F.W. Pease, and M.R. Beasley, *Rev. Sci. Instrum.* **62**, 1801 (1991)

[Osb45] J.A. Osborn, *Phys. Rev.* **67**, 351 (1945)

[Pal85] E.D. Palik, ed., *Handbook of Optical Constants of Solids* (Academic Press, Orlando, FL, 1985, 1991, and 1998)

[Pel98] D.N. Peligrad, B. Nebendahl, C. Kessler, M. Mehring, A. Dulcic, M. Pozek, and D. Paar, *Phys. Rev. B* **58**, 11 652 (1998)

[Pot85] R.F. Potter, *Basic Parameters for Measuring Optical Properties*, in: *Handbook of Optical Constants of Solids*, Vol. 1, edited by E.D. Palik (Academic Press, Orlando, FL, 1985), p. 3

[Pro98] A.V. Pronin, M. Dressel, A. Pimenov, A. Loidl, I. Roshchin, and L.H. Greene, *Phys. Rev. B* **57**, 14 416 (1998)

[Ram93] S. Ramo, J.R. Whinnery, and T.v. Duzer, *Fields and Waves in Communication Electronics*, 3rd edition (John Wiley & Sons, New York, 1993)

[Ros90] A. Röseler, *Infrared and Spectroscopic Ellipsometry* (Akademie-Verlag, Berlin, 1990)

[Sch95] A. Schwartz, M. Dressel, A. Blank, T. Csiba, G. Grüner, A.A. Volkov, B.P. Gorshunov, and G.V. Kozlov, *Rev. Sci. Instrum.* **60** (1995)

[Sri85] S. Sridhar, D. Reagor, and G. Grüner, *Rev. Sci. Instrum.*, **56**, 1946 (1985)

[Tom93] H.G. Tompkins, *A User's Guide to Ellipsometry* (Academic Press, Boston, MA, 1993)

[Zho94] S.A. Zhou, *Physica C* **233**, 379 (1994)

Further reading

[Alt63] H.M. Altschuler, *Dielectric Constant*, in: *Handbook of Microwave Measurements*, edited by M. Sucher and J. Fox, 3rd edition (Polytechnic Press, New York, 1963), p. 495

[Azz91] R.M.A. Azzam, ed., *Selected Papers on Ellipsometry*, SPIE Milestone Series **MS 27** (SPIE, Bellingham, 1991)

[Ben66] H.E. Bennett and J.M. Bennett, *Validity of the Drude Theory for Silver, Gold and Aluminum in the Infrared*, in: *Optical Properties and Electronic Structure of Metals and Alloys*, edited by F. Abelès (North-Holland, Amsterdam, 1966)

[Cha85] S.H. Chao, *IEEE Trans. Microwave Theo. Techn.* **MTT-33**, 519 (1985)

[Col92] R.E. Collin, *Foundations For Microwave Engineering* (McGraw-Hill, New York, 1992)

[Coo73] R.J. Cook, in: *High Frequency Dielectric Measurements*, edited by J. Chamberlain and G.W. Chantry (IPC Science and Technology Press, Guildford, 1973), p. 12

[Duz81] T. van Duzer and T.W. Turner, *Principles of Superconductive Devices and Circuits* (Elsevier, New York, 1981)

[Gri96] J. Grigas, *Microwave Dielectric Spectroscopy of Ferroelectrics and Related Materials* (Gordon and Breach, Amsterdam, 1996)

[Her86] G. Hernandez, *Fabry–Perot Interferometers* (Cambridge University Press, Cambridge, 1986)

[Hip54b] A.v. Hippel, *Dielectric Materials and Applications* (Technology Press M.I.T., Wiley, Cambridge, 1954)

[Jam69] J.F. James and R.S. Sternberg, *The Design of Optical Spectrometers* (Chapman and Hall Ltd, London, 1969)

[Kei84]   Keithley Instruments, *Low Level Measurements* (Keithley, Cleveland, 1984)

[Mac87]   J.R. Macdonald, *Impedance Spectroscopy* (John Wiley & Sons, New York, 1987)

[Mol00]   M. Mola, S. Hill, P. Goy, and M. Gross, *Rev. Sci. Instrum.* **71**, 186 (2000)

[Mul39]   J. Müller, *Z. Hochfrequenztech. Elektroak.* **54**, 157 (1939)

[Sch68]   J. Schwinger and D. Saxon, *Discontinuities in Waveguides* (Gordon and Breach, New York, 1968)

[Sla46]   J.C. Slater, *Rev. Mod. Phys.* **18**, 441 (1946)

[Vau89]   J.M. Vaughan, *The Fabry–Perot Interferometer* (A. Hilger, Bristol, 1989)

[Vol85]   A.A. Volkov, Yu.G. Goncharov, G.V. Kozlov, S.P. Lebedev, and A.M. Prokhorov, *Infrared Phys.* **25**, 369 (1985)

[Vol89]   A.A. Volkov, G.V. Kozlov, and A.M. Prokhorov, *Infrared Phys.* **29**, 747 (1989)

Part three

Experiments

Introductory remarks

The theoretical concepts of metals, semiconductors, and the various broken symmetry states were developed in Part 1. Our objective in this part is to subject these theories to test by looking at some examples, and thus to check the validity of the assumptions which lie behind the theories and to extract some important parameters which can be compared with results obtained by utilizing other methods.

We first focus our attention on simple metals and simple semiconductors, on which experiments have been conducted since the early days of solid state physics. Perhaps not too surprisingly the comparison between theory and experiment is satisfactory, with the differences attributed to complexities associated with the electron states of solids which, although important, will not be treated here. We also discuss topics of current interest, materials where electron–electron, electron–phonon interactions and/or disorder are important. These interactions fundamentally change the character of the electronic states – and consequently the optical properties. These topics also indicate some general trends of condensed matter physics.

The discussion of metals and semiconductors is followed by the review of optical experiments on various broken symmetry ground states. Examples involving the BCS superconducting state are followed by observations on materials where the conditions of the weak coupling BCS approach are not adequate, and we also describe the current experimental state of affairs on materials with charge or spin density wave ground states.

General books and monographs

F. Abelès, ed., *Optical Properties and Electronic Structure of Metals and Alloys* (North Holland, Amsterdam, 1966)

G. Grüner, *Density Waves in Solids* (Addison Wesley, Reading, MA, 1994)

S. Nudelman and S.S. Mitra, eds, *Optical Properties of Solids* (Plenum Press, New York, 1969)

E.D. Palik, ed., *Handbook of Optical Constants of Solids* (Academic Press, Orlando, 1985–98)

D. Pines, *Elementary Excitations in Solids* (Addison Wesley, Reading, MA, 1963)

A.V. Sokolov, *Optical Properties of Metals* (American Elsevier, New York, 1967)

J. Tauc, ed., *The Optical Properties of Solids*, Proceedings of the International School of Physics 'Enrico Fermi' **34** (Academic Press, New York, 1966)

P.Y. Yu and M. Cardona, *Fundamentals of Semiconductors* (Springer-Verlag, Berlin, 1996)

12

Metals

Optical investigations have contributed much to our current understanding of the electronic state of conductors. Early studies have focused on the behavior of simple metals, on the single-particle and collective responses of the free-electron gas, and on Fermi-surface phenomena. Here the relevant energy scales are the single-particle bandwidth W, the plasma frequency ω_p, and the single-particle scattering rate $1/\tau$, all lying in the spectral range of conventional optics. Consequently, when simple metals are investigated standard optical studies are of primary importance. Recent focus areas include the influence of electron–electron and electron–phonon interactions on the electron states, the possibility of non-Fermi-liquid states, the highly anisotropic, in particular two-dimensional, electron gas, together with disorder driven metal–insulator transition. Here, because of renormalization effects and low carrier density, and also often because of close proximity to a phase transition, the energy scales are – as a rule – significantly smaller than the single-particle energies. Consequently the exploration of low energy electrodynamics, i.e. the response in the millimeter wave spectral range or below, is of central importance.

12.1 Simple metals

In a broad range of metals – most notably alkaline metals, but also metals like aluminum – the kinetic energy of the electrons is large, significantly larger then the potential energy created by the periodic underlying lattice. Also, because of screening the strength of electron–electron and electron–phonon interactions is small; they can for all practical purposes be neglected. Consequently it is expected that for these metals the free-electron theory is an excellent approximation, and this has indeed been confirmed by a wide range of transport and thermodynamic experiments. We refer to these materials as simple metals, and proceed to explore their optical properties first.

12.1.1 Comparison with the Drude–Sommerfeld model

The fundamental prediction of the Drude–Sommerfeld model discussed in Section 5.1 is a frequency dependent complex conductivity, reproduced here

$$\hat{\sigma}(\omega) = \frac{Ne^2\tau}{m_{\mathrm{b}}} \frac{1}{1 - i\omega\tau} = \sigma_1(\omega) + i\sigma_2(\omega) \quad ; \tag{12.1.1a}$$

with

$$\sigma_1(\omega) = \frac{Ne^2\tau}{m_{\mathrm{b}}} \frac{1}{1 + \omega^2\tau^2} \qquad \text{and} \qquad \sigma_2(\omega) = \frac{Ne^2\tau}{m_{\mathrm{b}}} \frac{\omega\tau}{1 + \omega^2\tau^2} \quad . \tag{12.1.1b}$$

The ingredients of the model are the frequency independent relaxation rate $1/\tau$, and the mass m_{b} which loosely can be defined as the bandmass,[1] also assumed to be frequency independent; N is the electron density – more precisely the number of conduction electrons per unit volume. In Chapter 5 we derived the complex conductivity as given above, using the Kubo and also the Boltzmann equations. In addition to the frequency independent relaxation rate and mass, the underlying condition for these formalisms to apply is that the elementary excitations of the system are well defined, a condition certainly appropriate for good metals which can be described as Fermi liquids.

Let us first estimate the two fundamental frequencies of the model, the inverse relaxation time $1/\tau$ and the plasma frequency

$$\omega_{\mathrm{p}} = \left(\frac{4\pi Ne^2}{m_{\mathrm{b}}}\right)^{1/2} \tag{12.1.2}$$

for a typical good metal at room temperature. This can be done by using the measured dc conductivity

$$\sigma_{\mathrm{dc}} = \frac{Ne^2\tau}{m_{\mathrm{b}}} \quad ,$$

which is of the order of 10^6 Ω^{-1} cm^{-1}. For a number of free electrons per volume of approximately 10^{23} cm^{-3} and a bandmass which is assumed to be the same as the free-electron mass, we find $\gamma = 1/(2\pi c\tau) \approx 150$ cm^{-1}, while the plasma frequency so obtained is $\nu_{\mathrm{p}} = \omega_{\mathrm{p}}/(2\pi c) = 9.5 \times 10^4$ cm^{-1}, i.e. well in the ultraviolet.[2] Therefore for simple metals typically $1/\tau \ll \omega_{\mathrm{p}}$ and the

---

[1] The bandmass is usually defined by the curvature of the energy band in **k** space as given by Eq. (12.1.17), and this definition is used to account, in general, for the thermodynamic and magnetic properties by assuming that this mass enters – instead of the free-electron mass – in the various expressions for the specific heat or (Pauli-like) magnetic susceptibility.

[2] Some clarification of the units used is in order here. We will frequently use the inverse wavelength, the cm^{-1} scale, which is common in infrared spectroscopy to describe optical frequencies. The correct unit is frequency per velocity of light (f/c); however, this is usually omitted and we refer to ω_{p} or ν_{p}, for example, as frequencies, but using the cm^{-1} scale. We will follow this notation and refer to Table G.4 for conversion.

three regimes where the optical parameters have well defined characteristics – the Hagen–Rubens, the relaxation, and the transparent regimes – are well separated. These characteristics and the expected behavior of the optical constants in the three regimes are discussed in detail in Section 5.1.2.

When exploring the optical properties of metals in a broad frequency range, not only the response of the free conduction carriers described by the Drude model are important, but also excitations between different bands. The interband transitions can be described along the same lines as for semiconductors; in general the analysis is as follows: we assume a Drude form for the intraband contributions, with parameters which are fitted to the low frequency part of the electromagnetic spectrum. With this contribution established, the experimentally found optical parameters – say the dielectric constant $\epsilon_1(\omega)$ – can, according to Eq. (6.2.17), be decomposed into intra- and inter-band contributions,

$$\epsilon_1(\omega) = \epsilon_1^{\text{intra}}(\omega) + \epsilon_1^{\text{inter}}(\omega) \quad . \tag{12.1.3}$$

The interband contribution has, as expected, a finite ϵ_1 at low frequencies, with a peak at a frequency which corresponds to the onset of interband absorption, features which will be discussed in more detail in the next chapter.

The optical parameters evaluated for gold at room temperature from transmission and reflectivity measurements over a wide range of frequency are displayed in Fig. 12.1; the data are taken from [Ben66] and [Lyn85]. The overall behavior of the various optical constants is in broad qualitative agreement with those calculated, as can clearly be demonstrated by contrasting the experimental findings with the theoretical curves displayed in Figs 5.2–5.5. We observe a high reflectivity up to about 2×10^4 cm^{-1}, the frequency where the sudden drop of reflectivity R signals the onset of the transparent regime. The reflectivity does not immediately approach zero above this frequency because of interband transitions – not included in the simple Drude response which treats only one band, namely the conduction band. The onset of transparency can be associated with the plasma frequency ω_{p}; with a carrier concentration $N = 5.9 \times 10^{22}$ cm^{-3} (assuming that one electron per atom contributes to the conduction band) we obtain a bandmass significantly larger than the free-electron mass. This may be related to the narrow conduction band or to the fact that interband transitions are neglected. For the conduction band (where the carriers have the bandmass m_{b}) the sum rule of the conductivity is

$$\int \sigma_1^{\text{intra}}(\omega) \, d\omega = \frac{\pi N e^2}{2m_{\text{b}}} \quad , \tag{12.1.4}$$

and by including the contributions from all bands we obviously have

$$\int_0^{\infty} \sigma_1(\omega) \, d\omega = \int \sigma_1^{\text{intra}}(\omega) \, d\omega + \int \sigma_1^{\text{inter}}(\omega) \, d\omega = \frac{\pi N e^2}{2m} \quad , \tag{12.1.5}$$

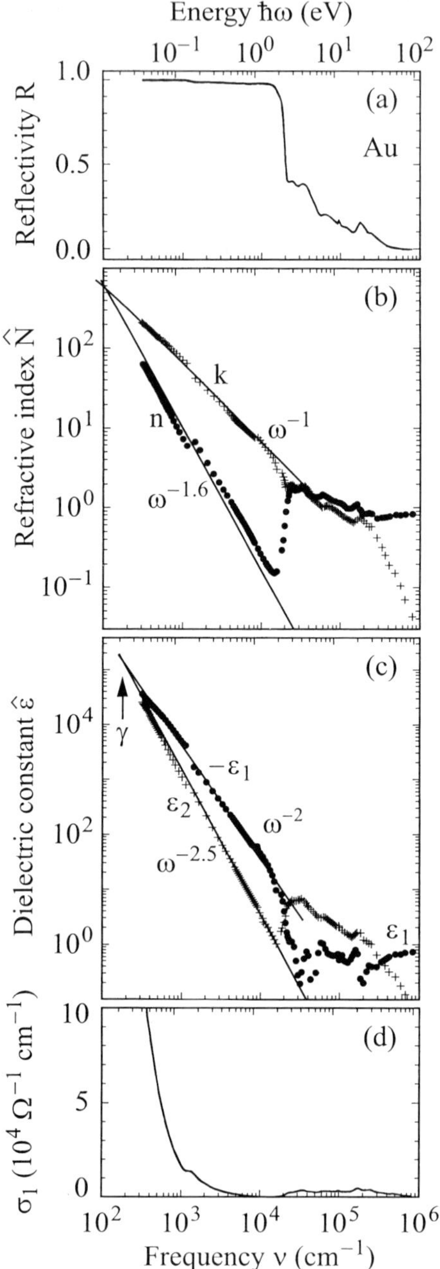

Fig. 12.1. Frequency dependence of the optical properties of gold at room temperature. (a) Reflectivity data $R(\omega)$ obtained by [Ben66, Lyn85]. In panels (b)–(d) the refractive index, $n(\omega)$ and $k(\omega)$, the dielectric constant, $\epsilon_1(\omega)$ and $\epsilon_2(\omega)$, and the conductivity $\sigma_1(\omega)$ as obtained by Kramers–Kronig analysis of the reflectivity are displayed. The discontinuity at around 1000 cm^{-1} is due to a mismatch in the data sets of two different groups.

where σ_1^{inter} refers to the interband contribution to the conductivity. The deviation of the bandmass m_b from the free-electron mass m is therefore directly related to the spectral weight of the conductivity coming from the interband transitions[3]

$$\int_0^\infty \sigma_1^{\text{inter}}(\omega)\,d\omega = \frac{\pi N e^2}{2}\left[\frac{1}{m} - \frac{1}{m_b}\right] \quad . \tag{12.1.6}$$

The interband transitions are usually evaluated by fitting the optical parameters to the Drude response at low frequencies (where intraband contributions dominate) and plotting the difference between experiment and the Drude contribution at high frequencies. This procedure can also be applied to the real part of the dielectric constant $\epsilon_1(\omega)$ as displayed in Fig. 12.2, where the dielectric constant of gold is decomposed into an interband contribution and an intraband contribution according to Eq. (12.1.3). Such decomposition and fit of the intraband contribution gives a plasma frequency ω_p of 7.0×10^4 cm^{-1} from the zero-crossing of $\epsilon_1(\omega)$; this compares favorably with the plasma frequency of 7.1×10^4 cm^{-1} calculated from Eq. (5.1.5) assuming free-electron mass and, as before, one electron per atom in the conduction band. The form of $\epsilon_1^{\text{inter}}(\omega)$ is that of a narrow band with a bandgap of 2×10^4 cm^{-1} between valence and conduction bands, in broad qualitative agreement with band structure calculations. Instead of a detailed analysis, such as in Fig. 12.2, it is frequently assumed that interband contributions lead only to a finite high frequency dielectric constant ϵ_∞, and we incorporate this into the analysis of the Drude response of the conduction band. The condition for the zero-crossing of $\epsilon_1(\omega)$ is given by

$$\epsilon_1(\omega) = \epsilon_1^{\text{intra}}(\omega) + \epsilon_\infty \quad , \tag{12.1.7}$$

and thus we recover a plasma frequency renormalized by the interband transitions

$$\omega_p^+ = \frac{4\pi N e^2}{m_b \epsilon_\infty} = \frac{\omega_p}{\sqrt{\epsilon_\infty}} \tag{12.1.8}$$

for $1/\tau \ll \omega_p$. Over an extended frequency region, below ω_p, power law behaviors of various optical parameters are found, with the exponent indicated on Fig. 12.1. Both $k(\omega)$ and $\epsilon_1(\omega)$ are well described with expressions appropriate for the relaxation regime (see Eqs (5.1.20) and (5.1.21)), although the frequency dependence for $n(\omega)$ and $\epsilon_2(\omega)$ is somewhat different from that predicted by these equations. The reason for this probably lies in the contributions coming from interband transitions.[4] The fact that $n > k$ observed over an extended spectral range also upholds

---

[3] The argument here is different from the spectral weight argument developed in Appendix D. Here, collisions (included in the damping term) lead to momentum relaxation and to a conductivity for which the sum rule (12.1.4) applies. In contrast, a collisionless electron gas is considered in Appendix D, and under this condition there is no absorption at finite frequencies.

[4] The experimental data are taken from two different sources [Ben66, Lyn85]; they seem to have some offset with respect to each other.

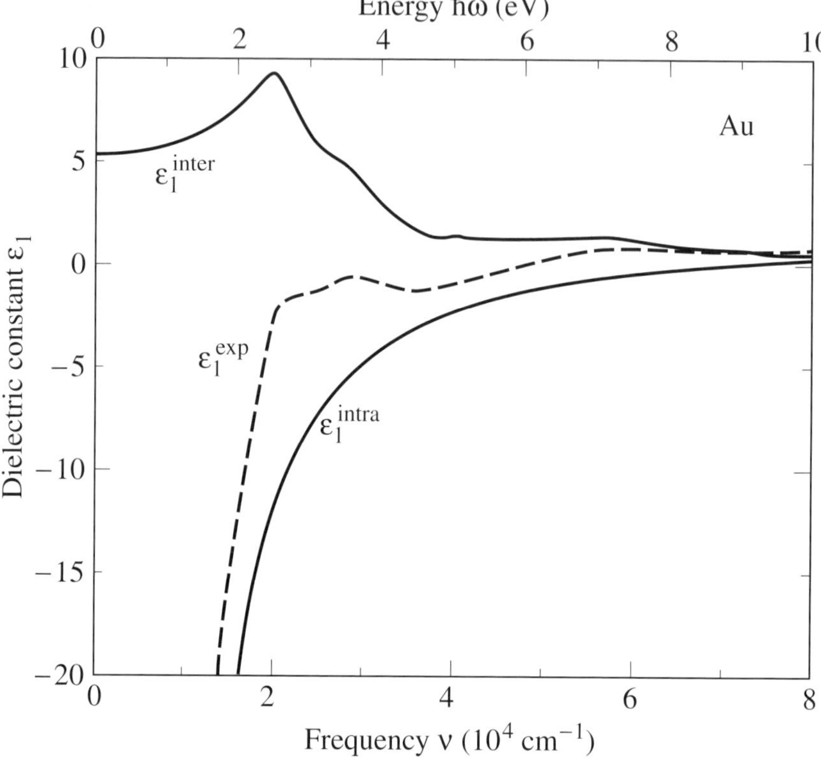

Fig. 12.2. Frequency dependence of the real part of the dielectric constant of gold at high frequencies; data from [Ehr63, Ehr66]. The experimental data $\epsilon_1^{\text{exp}}(\omega)$ are decomposed into an intraband contribution $\epsilon_1^{\text{intra}}(\omega)$ of the quasi-free electrons and a remaining interband contribution $\epsilon_1^{\text{inter}}(\omega)$, as described in the text.

$1/\tau < \omega$ at these frequencies, and there is a wide relaxation regime. Note that the relaxation rate is given by the frequency when $n = k$ or alternatively $1 - \epsilon_1 = \epsilon_2$. In our example, this leads to $1/(2\pi c\tau) \approx 200$ cm^{-1}, although the crossover at this frequency is not directly measured because of obvious technical difficulties: the reflectivity is close to unity for $\omega < 1/\tau$. Now we are in a position to evaluate the dc conductivity using Eq. (5.1.6):

$$\sigma_{\text{dc}} = \frac{\omega_{\text{p}}^2 \tau}{4\pi} \quad .$$

The values obtained from the analysis of the optical data, $\nu_{\text{p}} = 7 \times 10^4$ cm^{-1} and $\gamma = 200$ cm^{-1}, give $\sigma_1(\omega \to 0) = 3.3 \times 10^6$ Ω^{-1} cm^{-1}, a value somewhat larger than that obtained by direct resistivity measurement $\sigma_{\text{dc}} = 4.9 \times 10^5$ Ω^{-1} cm^{-1}. Such analysis, when performed for experiments conducted at different temperatures, leads, in general, to a temperature independent plasma frequency (not surprising since the number of carriers is constant and factors which determine the bandmass also do not depend on the temperature) and to a relaxation rate which compares well with the temperature dependent relaxation rate as evaluated from the dc resistivity since $\rho(T) \propto 1/\tau(T)$.

The uncertainties are typical of what is observed in simple metals: they are the consequence of band structure effects, and the influence of electron–electron and electron–phonon interactions. These effects can often be incorporated into an effective bandmass which leads to an enhanced specific heat and magnetic susceptibility (note that within the framework of the nearly free-electron model and also the tight binding model, both are inversely proportional to the band-mass), and, through Eq. (12.1.1b), to a reduced plasma frequency. However, further complications may also arise. The mass which enters into the expression of the dc conductivity may be influenced by electron–phonon interactions, although such interactions are not important above the phonon frequencies (typically located in the infrared range). Similar frequency dependent renormalizations occur for electron–electron interactions. Interband transitions may also lead to contributions to the conductivity at low frequencies, thus interfering with the Drude analysis, and anisotropy effects (discussed later) can also be important.

It is clear from the analysis given above that the Hagen–Rubens regime cannot easily be explored; this is mainly due to the fact that, for a highly conducting metallic state, the reflectivity is close to unity, and at the same time the range of validity of the Hagen–Rubens regime is limited to low frequencies where conducting the experiments is not particularly straightforward. The situation is different for so-called bad metals, materials for which σ_{dc} is small, $1/\tau$ is large, and consequently the reflectivity deviates well from unity even in this regime, where

$$R(\omega) = 1 - \left(\frac{2\omega}{\pi \sigma_{dc}} \right)^{1/2} . \tag{12.1.9}$$

Stainless steel is a good example, and the reflectivity as measured by a variety of experimental methods is displayed in Fig. 12.3. It is evident that Eq. (12.1.9) is well obeyed, and the full line is calculated using $\sigma_{dc} = 1.4 \times 10^4 \ \Omega^{-1} \ \mathrm{cm}^{-1}$ in excellent agreement with the directly measured dc conductivity. For short relaxation times additional complications occur. In the high frequency limit, the complex dielectric constant is

$$\hat{\epsilon}(\omega) = 1 - \frac{\omega_p^2}{\omega^2 - i\omega/\tau}$$

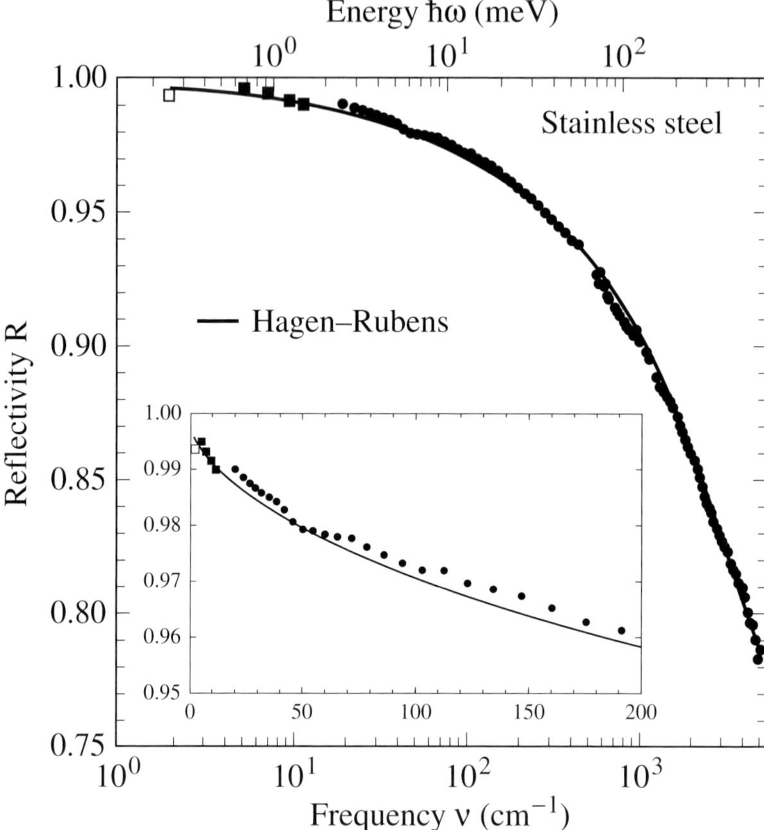

Fig. 12.3. Reflectivity of stainless steel versus frequency obtained at $T = 300$ K. The infrared data were measured with a Fourier transform spectrometer. In the submillimeter wave range the reflectivity, using single-frequency radiation sources, at 60 GHz cavity perturbation technique was used. The solid line is a fit by the Hagen–Rubens formula (12.1.9) with $\sigma_{dc} = 1.38 \times 10^4\ \Omega^{-1}\ \text{cm}^{-1}$ (after [Dre96]). The inset shows the same data plotted on a linear scale.

and the zero-crossing of the real part occurs to first approximation at

$$\omega \approx \left(\omega_p^2 - \frac{1}{\tau^2}\right)^{1/2}\quad ;$$

this is, for $1/\tau$ comparable to ω_p, different from what one would observe in an electron energy loss experiment, the standard method for measuring the plasma frequency [Rae80].

Electron energy loss experiments give a wealth of information about plasmons, including their energy, damping, and dispersion relation, and the dependence of the plasmon energy on its wavevector. As displayed in the inset of Fig. 12.5, a

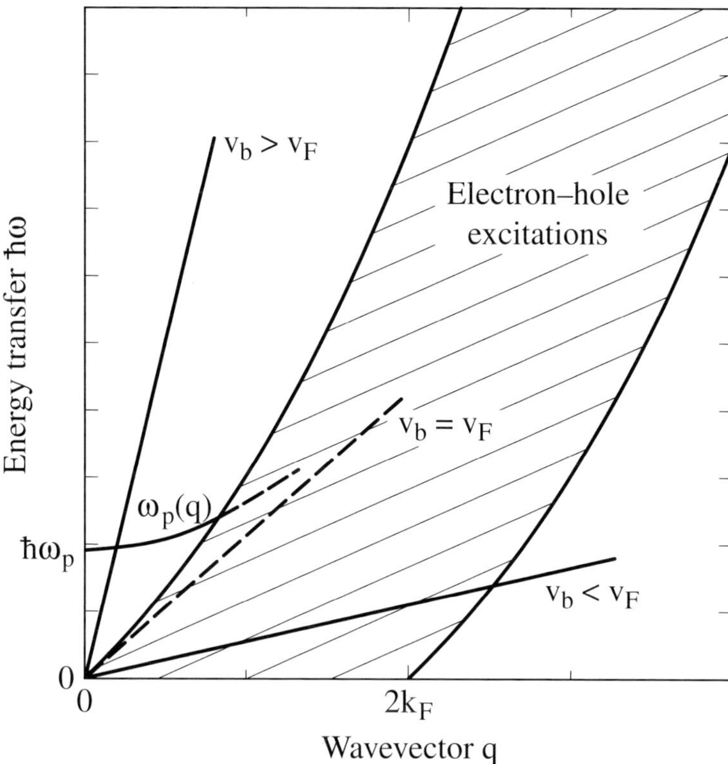

Fig. 12.4. The interaction of high velocity electrons with excitations of a three-dimensional free-electron gas. For $v_b > v_F$ plasmons are excited, while for $v_b < v_F$ single-particle excitations are responsible for the electron energy loss.

beam of high energy electrons is passed through a thin film, and the energy loss (and momentum transfer) of the electrons is measured. This loss is described by the so-called loss function

$$W(\mathbf{q}, \omega) = \frac{\omega}{8\pi} \mathrm{Im}\left\{ \frac{1}{\hat{\epsilon}(\mathbf{q}, \omega)} \right\} \ . \qquad (12.1.10)$$

Here $\hat{\epsilon}(\mathbf{q}, \omega)$ refers to the wavevector and frequency dependent longitudinal dielectric constant. As discussed in Section 3.1, this can be different from the transverse dielectric constant, sampled by an optical experiment. Both single-particle and collective plasmon excitations contribute to the loss; which of these contributions is important depends on the velocity v_b of the electron beam with respect to the Fermi velocity of the electron gas v_F. The situation is shown in Fig. 12.4. If v_b is large and exceeds the Fermi velocity v_F, single-particle excitations cannot occur, and the exchange of energy between the electron beam and the electron gas is possible only by creating plasma excitations; here we can use the expression of $\hat{\epsilon}(\mathbf{q}, \omega)$ obtained

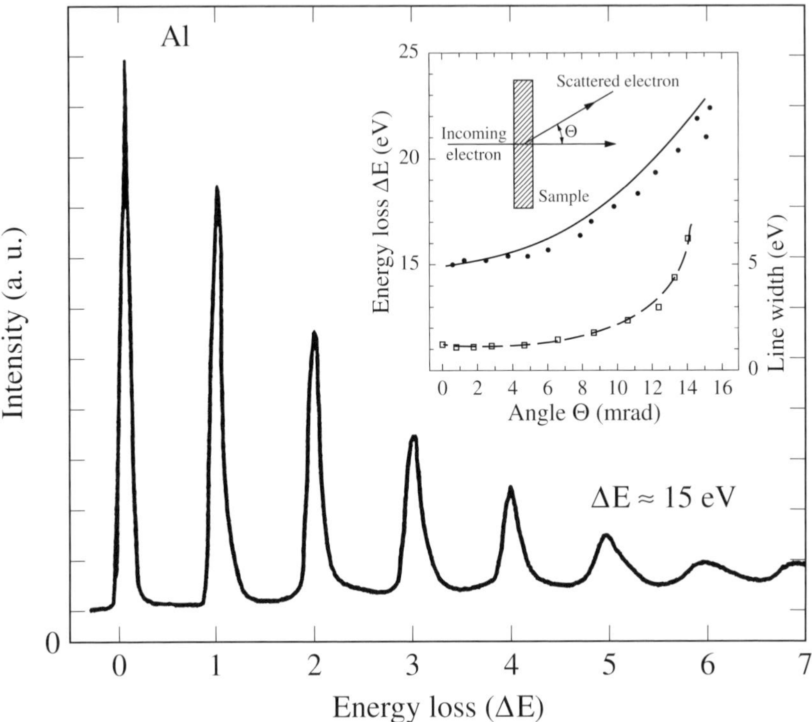

Fig. 12.5. Energy loss spectra for a beam of electrons (approximately 20 keV) passing through an aluminum foil. The units of energy loss $\Delta\mathcal{E}$ is 15.0 eV, the plasmon excitation energy of aluminum (after [Mar62]). The fundamental refers to a single excitation; the peaks at multiples of $\Delta\mathcal{E}$ are due to multiple absorption processes. The inset shows the angular variation of plasmon frequency and line width (after [Kun62]). The solid line is the theoretical prediction of a parabolic angular dependence, and the dashed line corresponds to a theoretical description of the line width.

in the $\mathbf{q} \to 0$ limit to see what happens. In this limit the longitudinal and transverse dielectric responses are identical. By utilizing the Drude expression of the $(\mathbf{q} = 0)$ dielectric constant (or conductivity) we find that

$$W(\omega) = \frac{\omega}{8\pi}\mathrm{Im}\left\{\frac{\omega^2 - i\omega/\tau}{\omega^2 - \omega_{\mathrm{p}}^2 - i\omega/\tau}\right\} \quad , \qquad (12.1.11)$$

which, in the absence of damping, reduces to

$$W(\omega) \propto \frac{\pi}{2}\left[\delta\{\omega - \omega_{\mathrm{p}}\} + \delta\{\omega_{\mathrm{p}} - \omega\}\right] \quad ; \qquad (12.1.12)$$

i.e. delta functions at the plasma frequency $\pm\omega_{\mathrm{p}}$. It is also straightforward to see that the width of the electron energy loss spectrum is determined by the relaxation rate $1/\tau$ for velocities which well exceed the Fermi velocity of the electron gas.

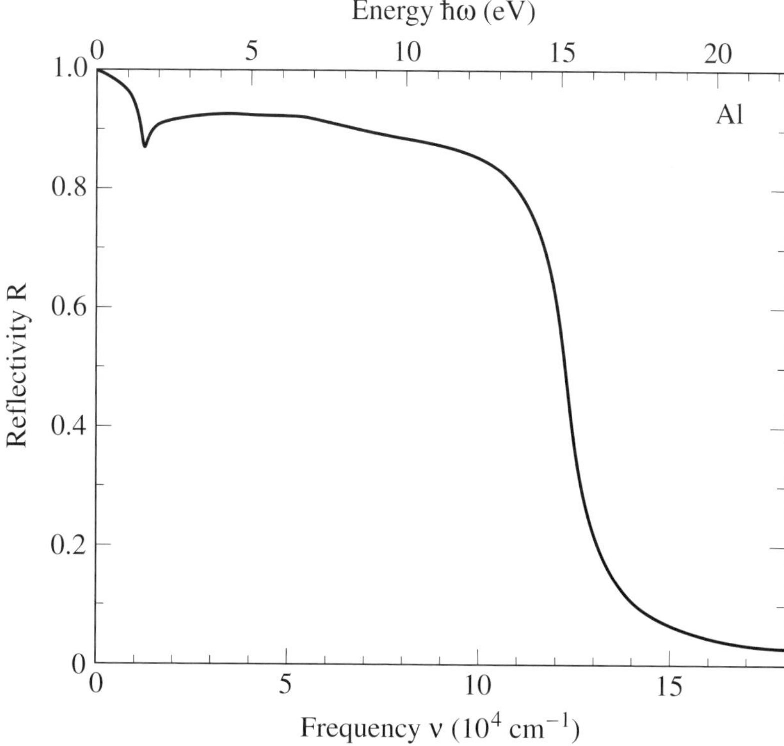

Fig. 12.6. Reflectivity of aluminum over a wide frequency range compiled from different measurements of thin films (after [Ehr63]). The dip in the reflectivity at 1.2×10^4 cm^{-1} is caused by an interband transition.

One can also examine the dispersion relation, such as that given by Eq. (5.4.28), together with the onset of damping by the single-particle excitations, the so-called Landau damping. The (conceptually) simplest way to do that would be to change the velocity of the electron beam; however, this is not practical and therefore the angular dependence of the energy loss spectrum is examined [Kun62]. Such experiments establish the onset of Landau damping for critical $\mathbf{q}$ values, the approximate expression for which was given in Section 5.4.4. All this applies for high beam velocities $v_b > v_F$, as in this case the $\mathbf{q}$ dependence of the conductivity of the electron gas can be neglected. For small beam velocities $v_b < v_F$, the situation is fundamentally different, as also shown in Fig. 12.4. In this latter case, the energy loss comes mainly from electron–hole excitations with velocities the same as the beam velocity, and plasmon excitations do not occur. The calculation of the loss function is somewhat difficult, as the full $\mathbf{q}$ dependent conductivity has to be utilized in order to evaluate $W(\omega)$.

Table 12.1. *Plasma frequencies of simple metals, as obtained from the onset of transparency, from electron energy loss (EEL), and from theory [Kit63, Rae80]. The values are given in energy $\hbar\omega_p$ or in wavenumber $\nu_p = \omega_p/2\pi c$.*

Material	Number of electrons in conduction band	Optics		EEL		Calculated	
		ν_p (cm^{-1})	$\hbar\omega_p$ (eV)	ν_p (cm^{-1})	$\hbar\omega_p$ (eV)	ν_p (cm^{-1})	$\hbar\omega_p$ (eV)
Li	1	6.4×10^4	8.0	7.7×10^4	9.5	6.6×10^4	8.2
Na	1	4.6×10^4	5.9	4.4×10^4	5.4	4.6×10^4	5.7
Ca	1	3.1×10^4	3.9	3.1×10^4	3.8	3.1×10^4	3.9
Au	1	7.0×10^4	8.7	6.3×10^4	7.8	7.3×10^4	9
Al	3	12.1×10^4	15	12.1×10^4	15.0	12.7×10^4	15.8
Si	4			13.3×10^4	16.5	13.4×10^4	16.6

A typical result, obtained for aluminum at high electron beam velocities, is shown in Fig. 12.5. The value of the plasma frequency is obtained from the spacing of the peaks at $\Delta\mathcal{E} = 15.0$ eV. Assuming three electrons per atom and free-electron mass leads to a plasma frequency $\omega_p/2\pi c = 1.3 \times 10^5$ cm^{-1}, corresponding to 16 eV, again showing excellent agreement between theory and experiment. This value also compares well with what is obtained from reflectivity, the data for aluminum being displayed in Fig. 12.6. The plasma edge is not particularly sharp, and this can be interpreted as damping; however, the notoriously bad surface characteristics of aluminum may be responsible for this feature of the reflectivity data. Nevertheless, the drop in R occurs around 1.2×10^5 cm^{-1}, in full agreement with the value derived from the electron energy loss spectroscopy. In principle, the loss function can be calculated from the dielectric constant as evaluated from the optical experiments, and may be compared with the loss measured directly with electron energy loss spectroscopy. This has been done for certain semiconductors, but not for metals.

Some plasma frequency values obtained from both reflectivity and electron energy loss spectroscopy are collected in Table 12.1, together with the values calculated assuming free electrons, with the number of electrons per atom as given in the table. The excellent agreement between theory and experiment is perhaps one of the most powerful arguments for applying the free-electron theory to metals, where the bandwidth, i.e. the kinetic energy of electrons is large and exceeds all other energy scales of the problem.

12.1.2 The anomalous skin effect

The prevailing notion that we have relied on in the previous sections is the local response to the applied electromagnetic field, the assumption that the current at a particular position is determined by the electric field at the same position only, and hence that the conductivity is independent of the position at which it is examined. Of course this does not mean that the currents and fields are independent of position, as the exponential decay of both $\mathbf{J}$ and $\mathbf{E}$ at the surface of a conducting medium – the examination of which leads to the normal skin effect described by Eq. (2.3.16) – so clearly demonstrates. The various consequences of this wavevector independent response are well known and were discussed in the previous section. The local response also leads, via Eq. (5.1.18), to a surface impedance $\hat{Z}_S = R_S + iX_S$ where – in the Hagen–Rubens regime – the components R_S and X_S are proportional to $\omega^{1/2}$ and $R_S = -X_S$. This is all confirmed by experiments on various simple metals, where the mean free path is not extremely large. This approximation progressively breaks down if the mean free path of the electrons ℓ becomes longer, and in the limit where ℓ exceeds the skin depth δ_0 the non-local response has to be taken into account. Utilizing the Chambers formula (5.2.27) to examine what happens near to the surface of a metal for which $\ell > \delta_0$ leads to the so-called anomalous skin effect, and the fundamental expression is given by Eq. (5.2.32). Both the normal and the anomalous skin effect have been derived in the Hagen–Rubens regime $\omega\tau < 1$, although it is straightforward to develop appropriate expressions in the opposite, so-called relaxation, regime (see Appendix E).

Let us estimate where the gradual transition from normal to anomalous skin effect occurs if we cool down a good metal such as copper. At room temperature the dc conductivity is typically 1×10^5 Ω^{-1} cm^{-1}, and the number of carriers (assuming that each copper atom donates approximately one electron into the conduction band) $N = 8.5 \times 10^{22}$ cm^{-3} leads to a relaxation time of $\tau = 2 \times 10^{-14}$ s. With a Fermi velocity of v_F approximately 5×10^7 cm s^{-1} the mean free path is $\ell = v_F\tau \approx 1000$ Å. The skin depth

$$\delta_0 = \left(\frac{c}{2\pi\omega\sigma_{\mathrm{dc}}}\right)^{1/2}$$

with the same parameters at 1 GHz, for instance (the upper end of the radio frequency spectral region), is 20 000 Å. As the skin depth is much larger than the mean free path, copper at this frequency is in the normal skin effect regime at room temperature. On cooling, the conductivity increases and consequently the mean free path increases while the skin depth decreases. At liquid nitrogen temperature the resistivity is about one order of magnitude larger than at room temperature, and estimations similar to those given above lead to $\ell \approx 10\,000$ Å and $\delta_0 = 7000$ Å, placing the material in the anomalous limit. There must be therefore a smooth crossover from the normal to the anomalous regime at relatively high temperatures. This has indeed been found by Pippard [Pip57, Pip60], and the experimental results are displayed in Fig. 12.7. For small conductivities, R_S^{-2}

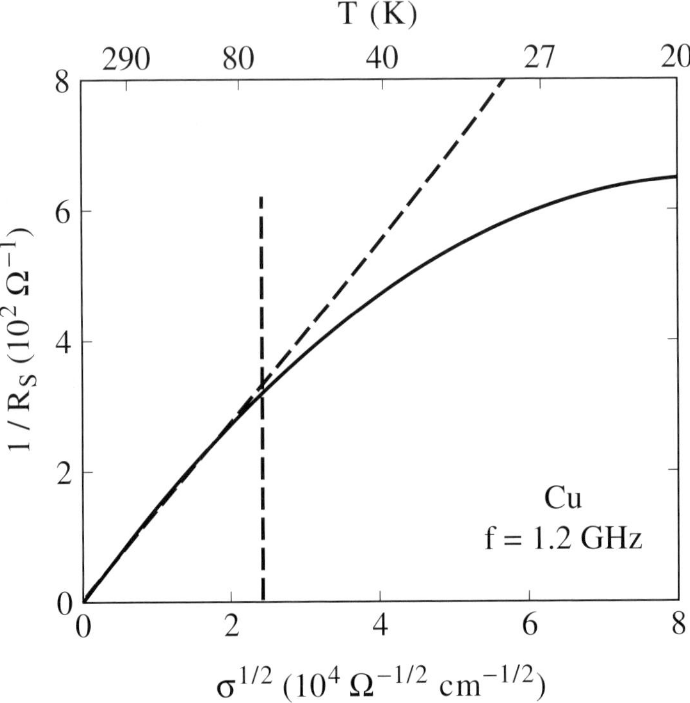

Fig. 12.7. Inverse surface resistance of copper measured at 1.2 GHz as a function of the square root of the conductivity. The upper axis shows the temperature corresponding to the conductivity value. In the Hagen–Rubens regime, and assuming local electrodynamics, (Eq. (5.1.18)), a linear dependence is expected. Below about 70 K the surface resistance is larger than expected from the theory of normal skin effect, indicating a crossover to the anomalous regime (after [Cha52]).

is found to be proportional to the dc conductivity, as expected for the surface resistance in the normal skin effect region (see Eq. (5.1.18)), but R_S saturates and becomes independent of the conductivity (or alternatively of the mean free path) at low temperatures. As predicted by the expression (5.2.32) for the anomalous skin effect, R_S in this limit reads

$$R_S = \left[\left(\frac{2\pi\omega}{c^2} \right)^2 \frac{m v_F}{N e^2} \right]^{1/3} \; . \tag{12.1.13}$$

By using the previous parameters, we can estimate $1/R_S$ at 1.2 GHz (the frequency at which the experiments were conducted), and we find that it agrees well with the value towards which the data tend to extrapolate for large conductivities.

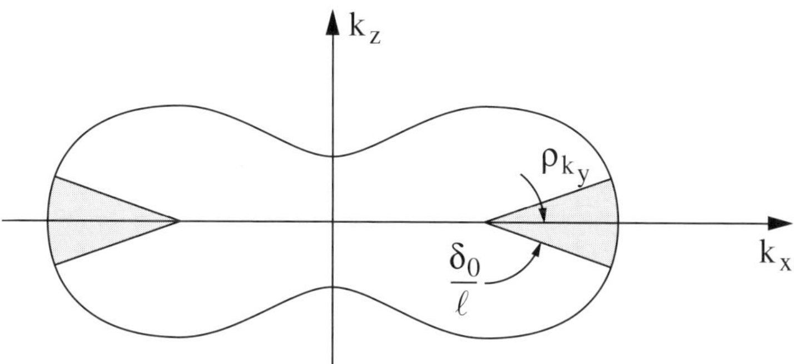

Fig. 12.8. Section of an anisotropic Fermi surface in the xz plane, with the shaded region representing electrons which are affected by the electric field in the $\delta_0 \ll \ell$ limit.

The ineffectiveness concept summarized in Section 5.2.5, and which led to the expression displayed above, can be extended to anisotropic Fermi surfaces. Such an anisotropic Fermi surface is shown in Fig. 12.8, where a cross-section in the xz plane is displayed for a particular value of the y component k_y. The radius of curvature $\rho(k_y)$ at any given point is wavevector dependent. The slices in Fig. 12.8 indicate electrons with velocities which lie within the angle δ_0/ℓ of the surface. The current is proportional to

$$J_x \propto \int e E_x \tau v_{\mathrm{F}} \, \mathrm{d}S \quad . \tag{12.1.14}$$

The surface $\mathrm{d}S$ over which the integration must be performed is given by this slice defined by a constant k_y, and consequently we find

$$J_x \propto \int e E_x \tau v_{\mathrm{F}} \rho(k_y) \frac{\delta_0}{\ell} \mathrm{d}k_y \quad . \tag{12.1.15}$$

The integral over $\mathrm{d}k_y$ has to be taken around the line where the Fermi surface is cut by the xz plane. Inserting this effective conductivity, defined through Eq. (12.1.15) by J_x/E_x, into the expression of the anomalous skin effect, we find that the surface resistance

$$R_{\mathrm{S}} \propto \left[\omega^2 \left(\oint \rho(k_y) \, \mathrm{d}k_y \right)^{-1} \right]^{1/3} \quad . \tag{12.1.16}$$

Although this is a qualitative argument, analytical results have been obtained for ellipsoidal Fermi surfaces by Reuter and Sondheimer [Reu48]. The importance of this result lies in the fact that the main contribution to the integral comes from regions of large curvature; these correspond to the flat regions of the Fermi surface. Therefore – at least in principle – the anisotropy of the Fermi surface can be mapped out by surface resistance measurements with the electromagnetic

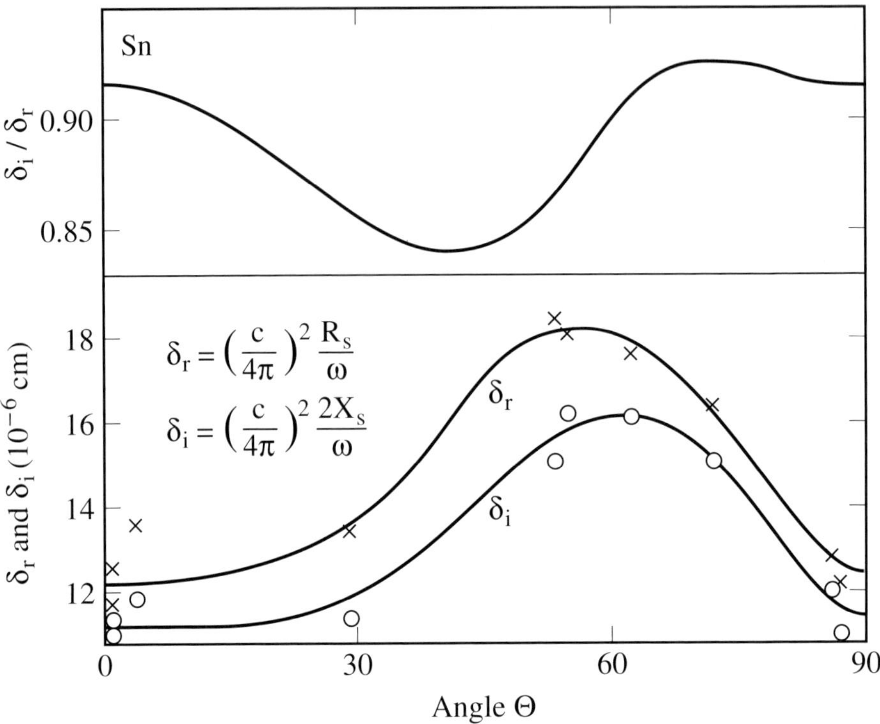

Fig. 12.9. Surface resistance and surface reactance of tin measured at different orientations of the electric field with respect to the crystallographic axis [Pip50]. The normalized values of the real and imaginary parts of the surface impedance $\delta_r = (c/4\pi)^2 R_S/\omega$ and $\delta_i = (c/4\pi)^2 2X_S/\omega$ are plotted in the lower panel. The upper panel shows the ratio of the two components.

field pointing in different directions with respect to the main crystallographic axes. A typical result obtained for tin at low temperatures by Chambers [Cha52] is displayed in Fig. 12.9 with both the surface resistance and surface reactance displaying substantial anisotropy. One has to note, however, that the evaluation of the characteristics of the Fermi surface from such data is by no means straightforward, and other methods of studying the Fermi-surface phenomena, like Subnikov–de Haas oscillations, and the de Haas–van Alphen effect, or cyclotron resonance, have proven to be more useful.

12.1.3 Band structure and anisotropy effects

The Drude model, as it was used before, applies for an isotropic, three-dimensional medium where subtleties associated with band structure effects are fully neglected. Such effects enter into the various expressions of the conductivity in different ways, and usually the Boltzmann equation in its variant forms where the electron velocities and the applied electric field appear in a vector form can be used to explore such band structure dependent phenomena.

A particularly straightforward modification occurs when the consequence of band structure effects can be absorbed into a dispersion relation which retains a parabolic form. If this is the case

$$\frac{1}{m_{\mathrm{b}}} = \frac{1}{\hbar^2}\frac{\partial^2 \mathcal{E}(\mathbf{k})}{\partial \mathbf{k}^2} \neq \frac{1}{m} \quad ; \tag{12.1.17}$$

i.e. the electron mass m is not the same as the bandmass m_{b}. This parameter depends on the scattering of electrons on the periodic potential, and may be anisotropic. When the above expression applies, all features of the interband transitions remain unchanged, for example the plasma frequency is given by $\omega_{\mathrm{p}} = (4\pi N e^2/m_{\mathrm{b}})^{1/2}$, and, through Eq. (12.1.17), is dependent on the orientation of the electric field with respect to the crystallographic axes. Such effects are particularly important when the band structure is highly anisotropic; an example is displayed in Fig. 12.10. This material, $(\mathrm{TMTSF})_2\mathrm{ClO}_4$, is an anisotropic metal, and band structure calculations suggest rather different single-particle transfer integrals in the two directions, with $t_a \approx 200$ meV and $t_{\mathrm{b}} \approx 20$ meV. The small bandwidth also suggests that a tight binding approximation is appropriate. The resistivity is metallic in both directions, and its anisotropy $\rho_{\mathrm{b}}/\rho_a \approx 10^2$ is in full agreement with the anisotropic band structure as determined by the above transfer integrals. The metallic, but highly anisotropic, optical response leads to different plasma frequencies in the two directions; and the expression for ω_{p} given above, together with the known carrier concentration, leads to bandmass values which, when interpreted in terms of a tight binding model, are in full agreement with the transfer integrals.

We encounter further complications if the approximation in terms of an effective bandmass as given above is not appropriate and we have to resort to the full Boltzmann equation as given by Eq. (5.2.16). The relevant integral which has to be examined is

$$\hat{\sigma}(\omega) \propto \int \mathbf{n_E} \cdot \mathbf{v_k} \frac{\tau}{1 - i\omega\tau} \, dS_{\mathrm{F}} \quad , \tag{12.1.18}$$

where $\mathbf{n_E}$ is the unit vector in the direction of the electric field $\mathbf{E}$; through $\mathbf{n_E}$ and $\mathbf{v_k}$ it leads to a complex dependence on the band structure. As $\sigma = N e^2 \tau/m$, one can define an effective mass

$$\frac{1}{m} \propto \int \mathbf{n_E} \cdot \mathbf{v_k} \, dS_{\mathrm{F}} \quad ,$$

which, in general, will depend also on the orientation of the applied electromag-

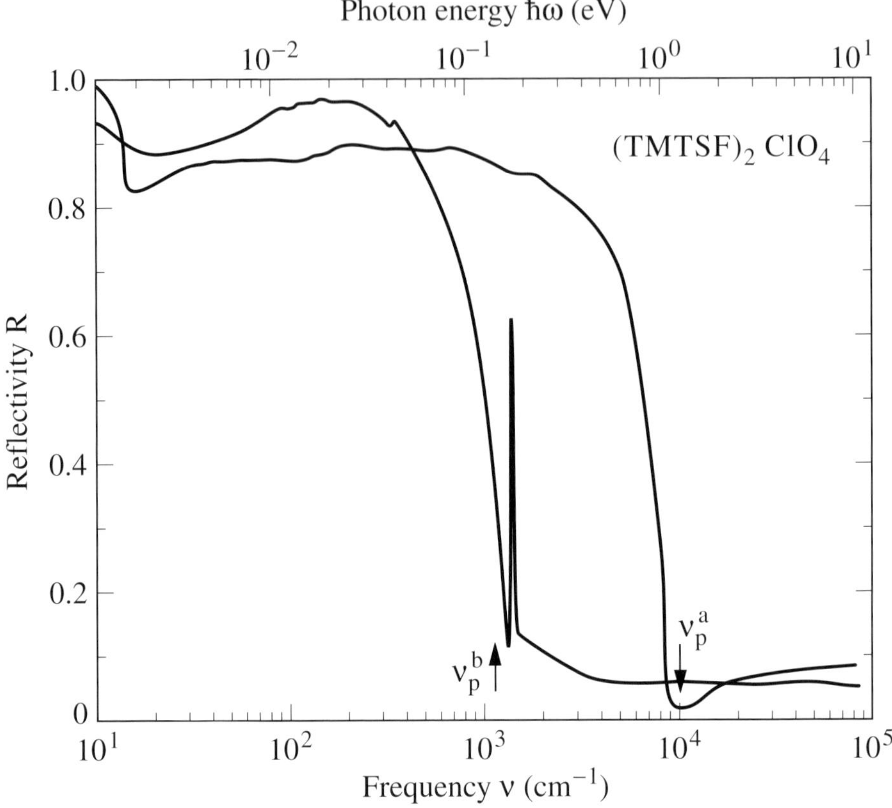

Fig. 12.10. Frequency dependence of the optical reflectivity $R(\omega)$ of the anisotropic metal (TMTSF)$_2$ClO$_4$ measured at 10 K with the electric field polarized along two different crystallographic directions a and b (after [Ves98]). The anomalies observed at low frequencies for both directions, while important, do not affect the analysis of the plasma frequencies in terms of an anisotropic bandmass.

netic field **E**. These subtleties are usually not considered in detail but absorbed into an isotropic effective optical bandmass, a mass which is the average of the mass given above, integrated over all directions – such an average occurs, for example, if the optical properties of polycrystalline materials are examined. This mass is in general compared with the mass, called the thermal mass, which is obtained from the thermodynamic quantities such as the electronic contribution (linear in T) to the specific heat, the so-called Sommerfeld coefficient γ. The parameter is – within the framework of the nearly free-electron approximation – given by $\gamma \propto N m_b$. Such a correspondence taken at face value, however, is highly misleading as the different experiments sample different averages over the Fermi surface. The specific heat, and thus the thermal mass, is proportional to the density of state and therefore to the average

$$\int \frac{\mathrm{d}S_\mathrm{F}}{\mathbf{v}_\mathrm{F}} \quad ;$$

this average being distinctively different from the average sampled by $\hat{\sigma}(\omega)$ given above. Therefore the thermal bandmass is, as a rule, different from the bandmass determined by optical experiments, the optical mass.

12.2 Effects of interactions and disorder

The optical experiments summarized in the previous section provide powerful evidence that the main assumptions which lie behind the Drude–Sommerfeld model are appropriate: for simple metals we can neglect the interaction of the electrons with lattice vibrations and with other electrons. This picture has to be modified by both interaction and disorder effects. We expect these to occur: (i) when the energy scale which describes the strength of the electron–phonon or electron–electron interactions and (ii) when – for disordered metals – the overall strength of the random potentials, are comparable to the kinetic energy of the electrons. This depends on many factors: electron–electron interactions are important for narrow bands, electron–phonon interactions are expected to be strong for soft lattices; while the strength of the random potential can be modified by alloying and/or by removing the underlying lattice periodicity, by making an amorphous solid.

12.2.1 Impurity effects

Most of the optical effects associated with impurities in metals can be absorbed into an impurity induced relaxation rate $1/\tau_{\mathrm{imp}}$ which affects the response in conjunction with the relaxation rate as determined by phonons $1/\tau_\mathbf{P}$. This, together with Matthiessen's rule, leads to the relaxation rate $1/\tau = 1/\tau_{\mathrm{imp}} + 1/\tau_\mathbf{P}$ which enters into the relevant expressions of the Drude model. In addition to an increased relaxation rate, the plasma frequency can also be influenced by impurities which are non-isoelectric with the host matrix; for small impurity concentrations this modification is expected to be small. While the above effects can be incorporated into a Drude model, more significant are the effects due to impurities with unoccupied d or f orbitals. In this case the (originally) localized orbitals at energy position $\mathcal{E}_d$ (or $\mathcal{E}_f$) away from the Fermi level are, due to interaction with the broadened conduction band; this broadening is described by a transition matrix element V_{sd} between the localized orbitals and the conduction band. These states then acquire a finite width W – estimated through a simple golden-rule argument which gives $W = 2\pi |V_{sd}|^2 D(\mathcal{E}_\mathrm{F})$ with $D(\mathcal{E}_\mathrm{F})$ the density of s states at the Fermi

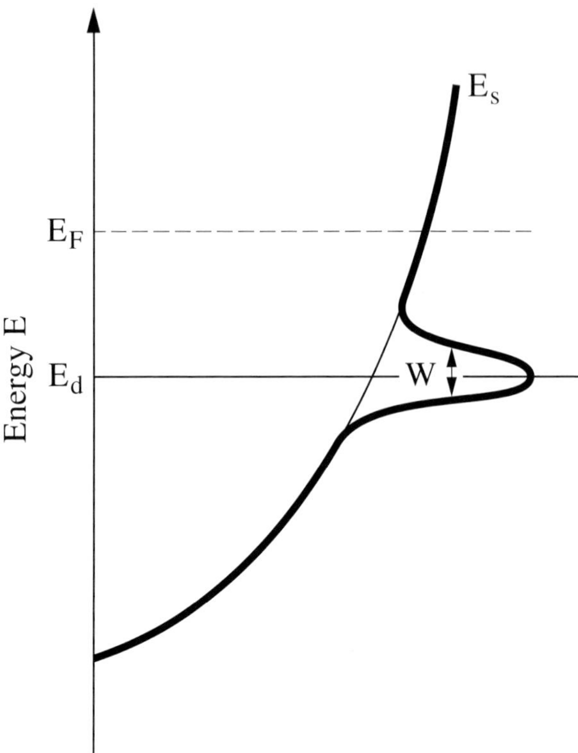

Fig. 12.11. Density of states of a free-electron metal, alloyed with an impurity which has a localized level at $\mathcal{E}_d$. Mixing between the s electrons and d localized states, described by a matrix element V_{sd}, leads to the broadening of this so-called virtually bound state, with the width of the peak $W = 2\pi|V_{sd}|^2 D(\mathcal{E}_F)$.

level; the density of states is displayed in Fig. 12.11. The notion of virtually bound states has been adopted for this situation [Fri56]. Because the symmetry of these virtually bound states is different from the electron states in the conduction band, optical transitions from these states to empty states just above the Fermi level $\mathcal{E}_F$ (or the reverse process, if $\mathcal{E}_d > \mathcal{E}_F$) may take place.

Such virtually bound states have indeed been observed in noble metals alloyed with transition metals such as palladium or manganese [Abe66, Mye68]. The interband part of the optical conductivity $\sigma_1^{\text{inter}}(\omega)$, measured on a AgPd alloy with 10% and 23% of palladium [Mye68], is displayed in Fig. 12.12 together with the results for pure silver. The data were obtained by measuring both the reflection and transmission of films. The peak between 2 and 3 eV clearly corresponds to the transitions from this nearly bound d state, about 2.6 eV below the Fermi level, to the unoccupied conduction electron states. As expected, photoemission experiments [Nor71] also provide additional evidence for such virtually bound states.

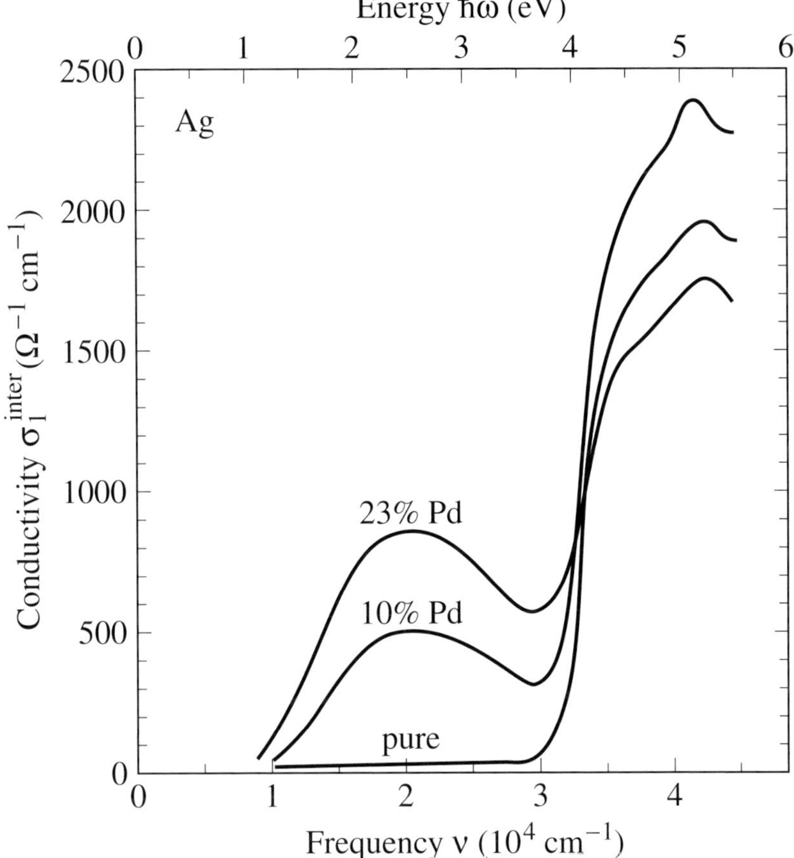

Fig. 12.12. Frequency dependence of the conductivity due to interband transitions $\sigma_1^{\text{inter}}(\omega)$ of silver films doped with different concentrations of palladium (after [Mye68]).

Together with Coulomb interaction, these states are described by Anderson's impurity model [And61], and such interactions lead to the Kondo problem, with all of its implications on many-body and single-particle resonances [Gru74a, Gru74b]. Interesting optical effects, associated in particular with the many-body resonance pinned to the Fermi level, have not been explored to date.

12.2.2 Electron–phonon and electron–electron interactions

Electron–electron and electron–phonon interactions are more a rule than an exception in metals. This is particularly true for cases where bands of d and f electrons are formed, and when scattering of s electrons on other, localized or narrow band, states occurs. Many metals of current interest belong to this group of materials, and optical studies have greatly contributed to the field by elucidating the important aspects of these interactions. In general the optical properties are treated by extending the formalism we have used in Part 1 in its simple form. This leads to a frequency dependent relaxation rate and mass (the two quantities are related by the Kramers–Kronig transformation), which in turn can be extracted from the measured optical parameters. In some cases the frequency dependencies can also be calculated both for electron–electron and electron–phonon interactions. This will not be done here; we confine ourselves to the main aspects of this fascinating field.

Without specifying the underlying mechanism, we assume that the relaxation rate $\hat{\Gamma}$ which appears in the Drude response of metals is frequency dependent:

$$\hat{\Gamma}(\omega) = \Gamma_1(\omega) + i\Gamma_2(\omega) \quad . \tag{12.2.1}$$

$\hat{\Gamma}(\omega)$ can be regarded as the memory function; when this is used as the (complex) relaxation rate in Eq. (5.1.4), the model referred to is the extended Drude model. If we define the dimensionless quantity $\lambda(\omega) = -\Gamma_2(\omega)/\omega$, then the complex conductivity can be written as

$$\hat{\sigma}(\omega) = \sigma_1(\omega) + i\sigma_2(\omega) = \frac{\omega_{\mathrm{p}}^2}{4\pi[1 + \lambda(\omega)]} \frac{1}{\frac{\Gamma_1(\omega)}{1+\lambda(\omega)} - i\omega} \quad . \tag{12.2.2}$$

It is clear that this has the same form as the simple Drude expression (5.1.4) but with a renormalized and frequency dependent scattering rate

$$\frac{1}{\tau^*(\omega)} = \frac{\Gamma_1(\omega)}{1 + \lambda(\omega)} \tag{12.2.3}$$

which approaches the constant $1/\tau$ as $\lambda \to 0$. The parameter which demonstrates this renormalization can be written in terms of a renormalized, enhanced mass $m^*/m_{\mathrm{b}} = 1 + \lambda(\omega)$, which then leads to another form of the conductivity:

$$\hat{\sigma}(\omega) = \frac{\omega_{\mathrm{p}}^2}{4\pi} \frac{1}{\Gamma_1(\omega) - i\omega[m^*(\omega)/m_{\mathrm{b}}]} \quad . \tag{12.2.4}$$

By rearranging this equation, we can write down expressions for $\Gamma_1(\omega)$ and $m^*(\omega)$ in terms of $\sigma_1(\omega)$ and $\sigma_2(\omega)$ as follows:

$$\Gamma_1(\omega) = \frac{\omega_{\mathrm{p}}^2}{4\pi} \frac{\sigma_1(\omega)}{|\hat{\sigma}(\omega)|^2} \quad , \tag{12.2.5a}$$

$$\frac{m^*(\omega)}{m_{\mathrm{b}}} = \frac{\omega_{\mathrm{p}}^2}{4\pi} \frac{\sigma_2(\omega)/\omega}{|\hat{\sigma}(\omega)|^2} \quad . \tag{12.2.5b}$$

Because $\hat{\Gamma}(\omega)$ obeys causality, $\Gamma_1(\omega)$ and $m^*(\omega)$ are related through the Kramers–Kronig relation.

There are several reasons for such a frequency dependent relaxation rate: both electron–electron and electron–phonon interactions (together with more exotic mechanisms, such as scattering on spin fluctuations) may lead to a renormalized and frequency dependent scattering process. Such renormalization is not effective at high frequencies, but is confined to low frequencies; typically below the relevant phonon frequency for electron–phonon interactions or below an effective, reduced bandwidth in the case of electron–electron interactions. Hence the frequency dependent relaxation rate – and via Kramers–Kronig relations also a frequency dependent effective mass – approaches for high frequencies the unrenormalized, frequency independent value.

Electron–electron interactions leading to an enhanced thermodynamic mass are known to occur in various intermetallic compounds, usually referred to as heavy fermions. The thermodynamic properties are that of a Fermi liquid, but with renormalized coefficients [Gre91, Ott87, Ste84]. Within the framework of a Fermi-liquid description, the Sommerfeld coefficient γ, which accounts for the electronic contribution to specific heat $C(T) = \gamma T + \beta T^3$ at low temperatures where the phonon contribution can be ignored, is given by

$$\gamma = \frac{C(T)}{T} = \frac{1}{3}\left(\frac{k_B}{\hbar}\right)^2 k_F m^* V_m = \frac{N m^* k_F}{6} \quad , \qquad (12.2.6)$$

where V_m is the molar volume, k_F is the Fermi wavevector, m^* is the effective mass, and N is the number of (conduction) electrons per unit volume. The same formalism gives the magnetic susceptibility χ_m as

$$\chi_m(T \to 0) = \frac{N m^* k_F}{18} \quad , \qquad (12.2.7)$$

which has the same form as for a non-interacting Fermi glass, but with m^* replacing the free-electron mass m. As for non-interacting electrons, the so-called Wilson ratio $\gamma(0)/\chi_m(0)$ is independent of the mass enhancement. The enhanced χ_m and γ imply a large effective mass, hence the name heavy fermions. Because of the periodic lattice, the dc resistivity ρ_{dc} of metals is expected to vanish at $T = 0$. The finite values observed at zero temperature are attributed to lattice imperfections and impurities. At low temperatures, the temperature dependence of the resistivity is best described by

$$\rho_{dc} = \rho_0 + AT^2, \qquad (12.2.8)$$

where ρ_0 is the residual resistivity due to impurities, the T^2 contribution is a consequence of electron–electron scattering, and A is proportional to both γ^2 and χ_m^2. This relation holds for a broad variety of materials spanning a wide range of mass enhancements [Dre97, Miy89]. In most cases, the dc resistivity has a broad maximum at a temperature T_{coh}, which is often referred to as the

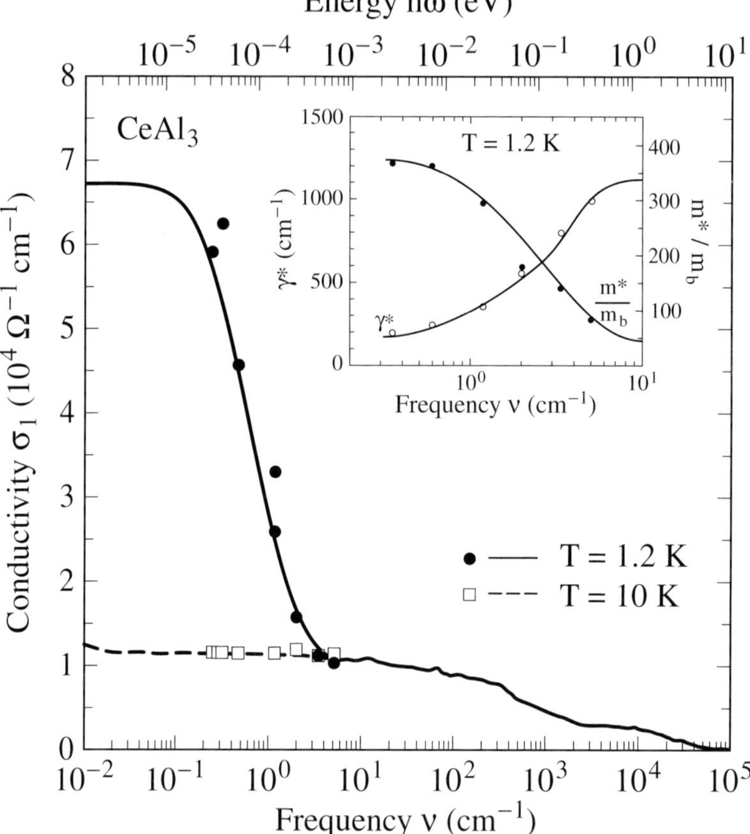

Fig. 12.13. Frequency dependent conductivity $\sigma_1(\omega)$ of CeAl$_3$ obtained from the Kramers–Kronig analysis of optical data and directly from the surface impedance measurements (after [Awa93]) at two different temperatures. The inset shows the frequency dependence of the scattering rate $\gamma^* = \Gamma_1^*/(2\pi c)$ and mass m^* at $T = 1.2$ K.

coherence temperature, and the heavy fermion state arises only at temperatures well below T_{coh}. This temperature is, as a rule, smaller the more important the correlations are. It can be viewed as an effective Fermi temperature, reduced from the Fermi temperature T_{F} which should be observed in the absence of correlations, and handwaving arguments suggest that this reduction is proportional to the mass enhancement $T_{\mathrm{F}}^*/T_{\mathrm{F}} \propto m/m^*$. With $T_{\mathrm{F}} \approx 10^4$ K, for $m^*/m \approx 1000$ the coherence temperature is of the order of 10 K. This temperature then corresponds via $k_{\mathrm{B}}T_{\mathrm{coh}} \approx \hbar\omega_{\mathrm{c}}^*$ to a crossover frequency ω_{c}, which separates the high frequency region, where correlations are not important, from the low frequency, strongly renormalized Drude response regime.

In Fig. 12.13 the optical conductivity of CeAl$_3$, a well studied model compound for heavy fermion behavior [Awa93], is displayed at two different temperatures. The temperature below which electron–electron interactions are effective and lead to a coherent, heavy fermion state is approximately $T_{coh} \approx 3$ K in this material. At $T = 10$ K, above the coherence temperature, $\sigma_1(\omega)$ is that of a simple metal with a relaxation rate $\gamma = 1/(2\pi c\tau) = 1000$ cm^{-1} and $\nu_p = \omega_p/(2\pi c) = 30\,000$ cm^{-1}. This value of the plasma frequency is in agreement with the calculated ω_p as inferred from the total spectral weight with $\omega_p = \left(4\pi Ne^2/m_b\right)^{1/2}$ with m_b the unrenormalized bandmass only slightly larger than the free-electron mass m. Therefore at this temperature CeAl$_3$ behaves as a simple metal, with no interaction induced renormalization. In contrast, at $T = 1.2$ K, well in the coherent regime, one observes a narrow resonance in $\sigma_1(\omega)$ and the data joins the 10 K data at approximately 3 cm^{-1}, i.e. at a frequency ω_c^* corresponding to T_{coh}. In order to analyze this behavior, and to extract quantities which are related to the enhanced effective mass m^*, we can assume that the narrow feature is that of a renormalized Drude response, with a renormalized (and likely frequency dependent, but at first approximation assumed to be constant) relaxation rate and mass. Spectral weight arguments are utilized to evaluate this renormalized mass. In order to do so we can evaluate the following integrals:

$$A^* = \int_0^{\omega_c^*} \sigma_1(\omega)\, d\omega = \frac{\pi Ne^2}{2m^*} \tag{12.2.9a}$$

and compare it with the sum rule

$$A_0 = \int_0^{\omega_c} \sigma_1(\omega)\, d\omega = \frac{\omega_p^2}{8} = \frac{\pi Ne^2}{2m_b} \quad, \tag{12.2.9b}$$

which is directly related to the unrenormalized bandmass; ω_c is the usual cutoff frequency just below interband transitions (see Section 3.2.2). In Eq. (12.2.9a), $\omega_c^*/2\pi c = 3$ cm^{-1} is the frequency above which the data at 10 K and at 1.2 K are indistinguishable.[5] This is also the frequency where the renormalization becomes ineffective as discussed above. From the resulting ratio A_0/A^*, we obtain $m^*/m_b = 450 \pm 50$, a value comparable with the one estimated from the thermodynamic quantities [Awa93].

Consequently, at low temperatures, where electron–electron interactions are effective, the optical properties can be described in terms of a renormalized Drude response

$$\hat{\sigma}(\omega) = \frac{Ne^2}{m^*} \frac{1}{1 - i\omega\tau^*} \tag{12.2.10}$$

with the renormalization simply incorporated into a renormalized mass. Note that

---

[5] Such division of the spectral response is somewhat arbitrary, and a fit to an extended Drude response with a frequency dependent scattering rate and mass is certainly more appropriate.

the dc conductivity – while larger at 1.2 K than at 10 K due to the usual temperature dependence observed in any metal – is not enhanced; what is enhanced is the relaxation time τ^* and mass m^*. The ratio τ^*/m^*, however, is not modified by the electron–electron interactions – at least in the limit where τ^* is determined by impurity scattering.

The analysis described above has been performed for a wide range of materials with differing strength of electron–electron interactions – and thus with different enhancements of the effective mass. This enhancement has been evaluated using the thermodynamic quantities, such as specific heat and magnetic susceptibility, with both given by Eqs (12.2.6) and (12.2.7). The mass has also been evaluated from the reduced spectral weight, and in Fig. 12.14 the two types – thermodynamic and electrodynamic – of masses are compared. Here the specific heat coefficient, γ, is plotted; as neither N nor k_F vary much from material to material, from Eq. (12.2.6) it follows that γ is approximately proportional to the mass enhancement. We find that thermodynamic and optical mass enhancements go hand in hand, leading to the notion that in these materials electron–electron interactions lead to renormalized Fermi liquids, to first approximation.

Of course, the above description is an oversimplification, in light of what was said before. The resistivity is strongly temperature dependent at low temperatures, reflecting electron–electron interactions. As approximately $k_B T/\mathcal{E}_F^*$ electrons are scattered on $k_B T/\mathcal{E}_F^*$ electrons at the temperature T (where $\mathcal{E}_F^*$ is a renormalized Fermi energy), the dc resistivity is $\rho_{dc}(T) \propto \left(k_B T/\mathcal{E}_F^*\right)^2$, and such behavior is seen by experiments. Fermi-liquid theory also predicts a renormalized frequency dependent scattering rate [Ash76]

$$\frac{1}{\tau^*(T,\omega)} \propto \left[(2\pi k_B T)^2 + (\hbar\omega)^2\right] \quad , \tag{12.2.11}$$

whereas the effective mass is not frequency dependent in lowest order. At low temperatures the dc resistivity

$$\rho_{dc} = \left(\frac{Ne^2}{m^*}\tau^*\right)^{-1} \approx \frac{m^*}{Ne^2} 2\pi k_B^2 T^2$$

is proportional to the square of the temperature (see also Eq. (12.2.8)), as expected for a (renormalized) Fermi liquid. Using both components of the conductivity as evaluated by experiments over a broad spectral range, the frequency dependence of both $\gamma^* = 1/(2\pi c\tau^*)$ and m^* can be extracted by utilizing Eqs (12.2.5); these are shown in the insert of Fig. 12.13. The leading frequency dependence of $1/\tau^*$ is indeed ω^2 as expected for a Fermi liquid, and this is followed at higher frequencies by a crossover for both $1/\tau^*$ and m^* to a high frequency region where these parameters assume their unrenormalized value. Thus, just like the temperature

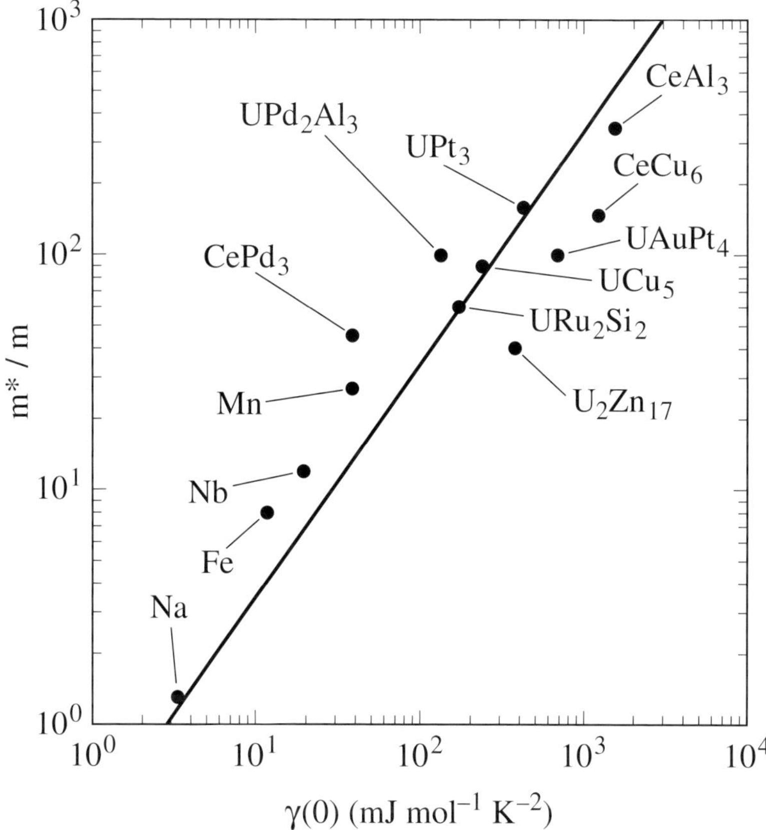

Fig. 12.14. Enhanced mass m^* relative to the free-electron mass m plotted versus the Sommerfeld coefficient γ obtained from the low temperature specific heat C/T for several heavy fermion compounds. The solid line represents $m^*/m_0 \propto \gamma(0)$. Also included are the parameters for a simple metal, sodium, and for several transition metals (after [Deg99]).

driven destruction of the correlated heavy fermion state, renormalization effects are ineffective at high frequencies.

Electron–phonon interactions also lead to an enhancement of the thermodynamic mass, as evidenced by the moderately enhanced low temperature specific heat C/T. The influence of these interactions on the optical properties of metals has been considered at length. The leading term is the emission of phonons in a metal; this so-called Holstein process involves the absorption of photons by simultaneously emitting a phonon and scattering an electron. The process obviously depends on the phonon spectrum, which we describe as the density of states $F(\omega_P)$ and by a weight factor α_P^2 which is the electron–phonon matrix element. With these parameters, an effective coupling constant

$$\lambda_{\mathbf{P}} = 2 \int \frac{d\omega_{\mathbf{P}}}{\omega_{\mathbf{P}}} \alpha_{\mathbf{P}}^2 F(\omega_{\mathbf{P}}) \tag{12.2.12}$$

can be defined; needless to say, $\alpha_{\mathbf{P}}$ can also be frequency dependent. The initial state is a photon of energy $\hbar\omega$, and the final state involves an electron and hole with energies $\mathcal{E}_1$ and $\mathcal{E}_2$, respectively, and a phonon with energy $\hbar\omega_{\mathbf{P}}$. If we assume that the matrix element is independent of frequency (except for the factor $\omega^{-1/2}$ representing the photon state), the transition probability, and thus the absorptivity A, is given as

$$A(\omega) \propto \frac{1}{\omega} \int d\mathcal{E}_1 d\mathcal{E}_2 d\omega_{\mathbf{P}} \, D_{\mathrm{e}}(\mathcal{E}_1) D_{\mathrm{h}}(\mathcal{E}_2) \, \alpha_{\mathbf{P}}^2(\omega_{\mathbf{P}}) \, F(\omega_{\mathbf{P}}) \, \delta\{\mathcal{E}_1 + \mathcal{E}_2 + \hbar\omega_{\mathbf{P}} - \hbar\omega\} \quad , \tag{12.2.13}$$

where $D_{\mathrm{e}}(\mathcal{E}_1)$ and $D_{\mathrm{h}}(\mathcal{E}_2)$ are the density of electrons and holes, respectively, $A(\omega)$ can be written as

$$A(\omega) \propto \frac{1}{\omega} \int_0^\omega (\omega - \omega_{\mathbf{P}}) \alpha_{\mathbf{P}}^2 F(\omega_{\mathbf{P}}) \, d\omega_{\mathbf{P}} \tag{12.2.14}$$

at zero temperature. For a narrow density of phonon states centered around $\omega_{\mathbf{P}}^0$, $A(\omega)$ has a structure at $\omega = \omega_{\mathbf{P}}^0$; in general, the frequency dependence of the absorption reflects the phonon density of states.

This absorptivity can be cast into a form we used earlier, by assuming a frequency dependent relaxation rate $1/\tau(\omega)$ and renormalized plasma frequency [All71]. In the relaxation regime $\omega\tau \gg 1$ and $\omega \ll \omega_{\mathrm{p}}$ – in the infrared spectral range for typical metals – the absorptivity is, according to Eq. (5.1.22),

$$A = \frac{2}{\omega_{\mathrm{p}}\tau} \quad ;$$

and therefore we can define a frequency dependent relaxation rate

$$\frac{1}{\tau(\omega)} \approx \frac{2\pi}{\omega} \int_0^\omega (\omega - \omega_{\mathbf{P}}) \, \alpha_{\mathbf{P}}^2 \, F(\omega_{\mathbf{P}}) \, d\omega_{\mathbf{P}} \quad . \tag{12.2.15}$$

This scattering time depends on the structure factor $\alpha_{\mathbf{P}}^2 F(\omega_{\mathbf{P}})$; the phonon limited scattering rate $1/\tau$ approaches zero at low frequencies, while at high frequencies it saturates at

$$\frac{1}{\tau} = \pi \lambda_{\mathbf{P}} \langle \omega_{\mathbf{P}} \rangle \quad , \tag{12.2.16}$$

where $\langle \omega_{\mathbf{P}} \rangle$ is the average phonon frequency weighted by the effective coupling constant $\lambda_{\mathbf{P}}$ we have defined in Eq. (12.2.12). Of course, having just a frequency dependent scattering rate violates the Kramers–Kronig relations; it has to be accompanied by a frequency dependent mass, which can be established along similar lines to the introduction of the frequency parameter into the Drude model earlier (Eqs (12.2.1)–(12.2.5)).

From the absorptivity $A(\omega)$ the optical parameters $1/\tau(\omega)$ and $\lambda_P(\omega)$ can be extracted, and eventually the weight factor $\alpha^2 F(\omega_P)$ can be derived. Although this works in principle, in practice $F(\omega_P)$ is modeled by, for instance, harmonic oscillators. The structure thus obtained should be related to $F(\omega_P)$ as evaluated from inelastic neutron scattering, and in cases where the analysis has been performed, this is indeed the case [Joy70, McK79].

Because of the Bose statistics of the relevant phonon modes, this renormalization is also temperature dependent. As before for electron–electron interactions, the frequency and temperature dependent relaxation rates go hand in hand (but because of the Bose–Einstein population factor the functional forms are somewhat different), and the zero temperature scattering rate, as given by Eq. (12.2.15), has its temperature dependent counterpart. One finds that

$$\frac{1}{\tau(T, \omega = 0)} = \pi \int_0^\infty d\omega_P \, \frac{\hbar\omega_P}{k_B T} \, \sin^2\left\{\frac{\hbar\omega_P}{2k_B T}\right\} \alpha^2(\omega_P) \, F(\omega_P) \quad , \qquad (12.2.17)$$

and this parameter can be extracted in principle from the temperature dependent dc conductivity.

12.2.3 Strongly disordered metals

Throughout the previous discussion of the experiments, we have assumed that the electron states are delocalized, and furthermore that they can be represented as plane waves or Bloch functions; this clearly holds for pure metals in the absence of disorder. For a small amount of disorder, or for a small amount of impurities in the sample, we expect that the effects of disorder and impurities are absorbed into the relaxation rate, with impurity scattering assumed to lead to an extra contribution to $1/\tau$. This is, furthermore, usually assumed to be independent of other scattering mechanisms, leading to the (experimentally well confirmed) Matthiessen rule in the case of dc conduction. This picture then also implies that the Drude model still applies, with a modified relaxation rate. Consequently we expect a reduced dc conductivity and a Drude roll-off which now occurs at higher frequencies. The total spectral weight associated with the conductivity and thus the plasma frequency is, however, determined only by the number of carriers and by their bandmass; consequently it remains little changed. The situation is different for strong disorder effects. It was shown by Anderson in a classic paper [And58] that for a sufficient amount of disorder all states are localized, and therefore by increasing disorder there is a transition from delocalized electron states to states with the absence of diffusion, hence to a state where all electrons are localized. In three dimensions, the condition for this is given by V/W, where V is the overall, average strength of the random potential and W is the single-particle bandwidth. In a one-dimensional electron gas, in contrast, all states are localized for any small disorder [Blo72, Mot61]. The consequences of localization are profound: the dc conductivity is zero at $T = 0$, and we have an insulator; at finite temperatures, conduction proceeds by mechanisms which are fundamentally different from those which determine the resistivity of conventional metals. We cannot do justice to the field, which goes by the names of Anderson transition or localization driven metal–insulator transition, within the limitations of this book, and therefore we simply recall the various expressions, first for the temperature dependence of the dc conductivity, which have been proposed and found, and later for the frequency dependent transport.

Right at the metal–insulator transition, scaling arguments prevail [Bel94], and they can be used to establish the temperature dependence of the conductivity. Such arguments give a power law dependence:

$$\sigma_{\mathrm{dc}}(T) = AT^{\alpha} \quad ; \tag{12.2.18}$$

for a non-interacting electron gas, $\alpha = 1/3$, for example. On the metallic side, and also on the insulating side except at rather low temperatures (or frequencies, as we will see later), the conductivity is finite at zero temperature, and the magnitude goes smoothly to zero upon approaching the transition. The temperature dependent dc conductivity is then described using the form

$$\sigma_{\mathrm{dc}}(T) = AT^{\alpha} + B \tag{12.2.19}$$

where the factor A is independent and the factor B is dependent on the parameter x which controls the transition (usually concentration of scattering centers, the pressure, or the magnetic field). The residual conductivity B goes to zero as the transition is approached from both sides:

$$B(x) \propto (x - x_{\mathrm{c}})^{\mu} \quad , \tag{12.2.20}$$

where x_{c} is the critical parameter where the metal–insulator transition occurs and μ is the characteristic exponent. In the insulating state, Eq. (12.2.19) still describes the conductivity if B is negative. At low temperatures, well in the localized regime (called the Fermi glass regime), such scaling arguments do not apply and a different picture emerges. Here thermally assisted hopping between localized states which lie within the energy range $k_{\mathrm{B}}T$ at temperature T determines the dc transport properties. The spatial proximity of such states depends on the temperature, hence the hopping is of variable range, and the evaluation of the most probable hop leads to [Mot68, Mot69]

$$\sigma_{\mathrm{dc}}(T) \propto \exp\left\{-\frac{T_0}{T}\right\}^{\beta} , \tag{12.2.21}$$

where the exponent $\beta = 1/(d + 1)$ is a function of the dimensionality d; for three dimensions $\beta = 1/4$, and it is 1/2 for the one-dimensional variable range hopping regime [Mot68]. The characteristic temperature T_0 depends on the spatial extension of the localized states, the localization length λ_0, and also on the density of states at the Fermi level $D(\mathcal{E}_{\mathrm{F}})$; the localization length diverges as we approach the transition from the localized side. As the temperature rises, a crossover from this variable range hopping regime to the scaling regime occurs; the crossover temperature where this happens increases as the disorder becomes stronger, and we move away from the transition. Electron–electron interactions, in concert with disorder, modify this picture since such interactions depress the density of states (leading to a so-called Coulomb gap) at the Fermi energy, and consequently the power laws which determine the behavior at the transition. At the strongly localized regime, which is often referred to as a Coulomb glass, both exponents are modified; in Eq. (12.2.18) $\alpha = 1/3$ for non-interacting electrons, $\alpha = 1/2$ for strong electron–electron interactions [Efr75], and β in Eq. (12.2.21) changes from 1/4 to 1/2 in three dimensions [Efr75]. All this is well confirmed by temperature dependent transport measurements [All93, Her83] on various model systems, such as $\mathrm{Nb}_x\mathrm{Si}_{1-x}$ where the metal–insulator transition occurs with decreasing x, or for phosphorus doped silicon Si:P [Loh98, Ros81] for which the impurity states, introduced by the doping, become delocalized by increasing their concentration.

Next we turn to the frequency dependent response, at temperatures low enough such that the condition $\hbar\omega \gg k_{\mathrm{B}}T$ applies; not surprisingly, we find a close correspondence between the temperature and frequency dependences predicted, and in some cases also observed. Because of localization, the conductivity vanishes at zero frequency (at zero temperature) on the insulating side, and the frequency dependence is determined by transitions between localized states induced by the applied electromagnetic field. At low frequencies the conductivity therefore increases as ω rises, but at sufficiently high frequencies a Drude roll-off is still expected as inertia effects become important there. The well known sum rule

$$\int \sigma_1(\omega)\,\mathrm{d}\omega = \frac{\omega_{\mathrm{p}}^2}{8}$$

still applies, as excitations from localized states are also possible and contribute to the spectral weight; interband transitions, of course, can influence this argument. The frequency dependence of the conductivity is shown in Fig. 12.15, for materials right at and at both sides of the metal–insulator transition. Starting from a weakly

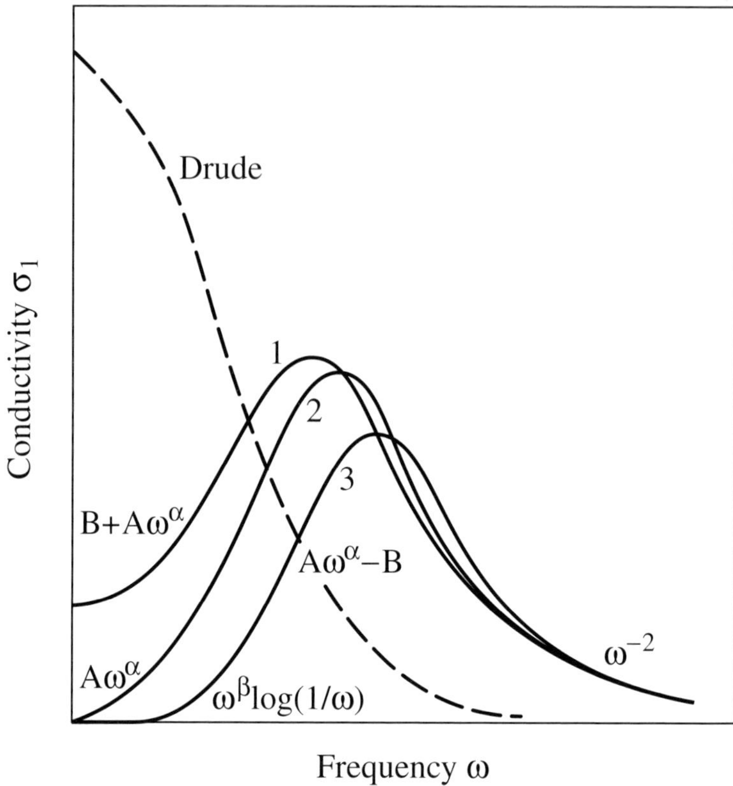

Fig. 12.15. Frequency dependent conductivity $\sigma_1(\omega)$ of weakly and strongly disordered metals, together with an insulator brought about by substantial disorder. The dashed line corresponds to a Drude behavior of a simple metal. Curve 1 is expected for weak disorder; curve 2 is right at the metal–insulator transition; and curve 3 corresponds to a strongly disordered sample on the insulating side of the transition. The parameters α and β depend on the strength of the interaction. The expression $\sigma_1(\omega) = A'\omega^{\alpha'} \pm B$ is appropriate close to the transition, while $\sigma_1(\omega) \propto \omega^\beta \log\{1/\omega\}$ is the form expected on the insulating side for strongly localized electrons.

disordered metal, the Drude response gradually gives way [Alt85, Lee85] to a conductivity which is small at low frequencies, increases with frequency, and after a maximum reverts to a Drude type roll-off $\sigma_1(\omega) \propto \omega^{-2}$. This is demonstrated [All75] in a silicon inversion layer where the density of electrons (and thus the bandwidth W) is progressively decreased, increasing the ratio of V/W as V – the random potential induced by impurities – remains constant. The frequency dependence of the low temperature conductivity is displayed in Fig. 12.16 for different carrier concentrations N_S. With increasing disorder, a pseudogap develops at low frequencies, in clear qualitative agreement with the expectations. Such experiments, however, cannot examine the detailed frequency dependence at

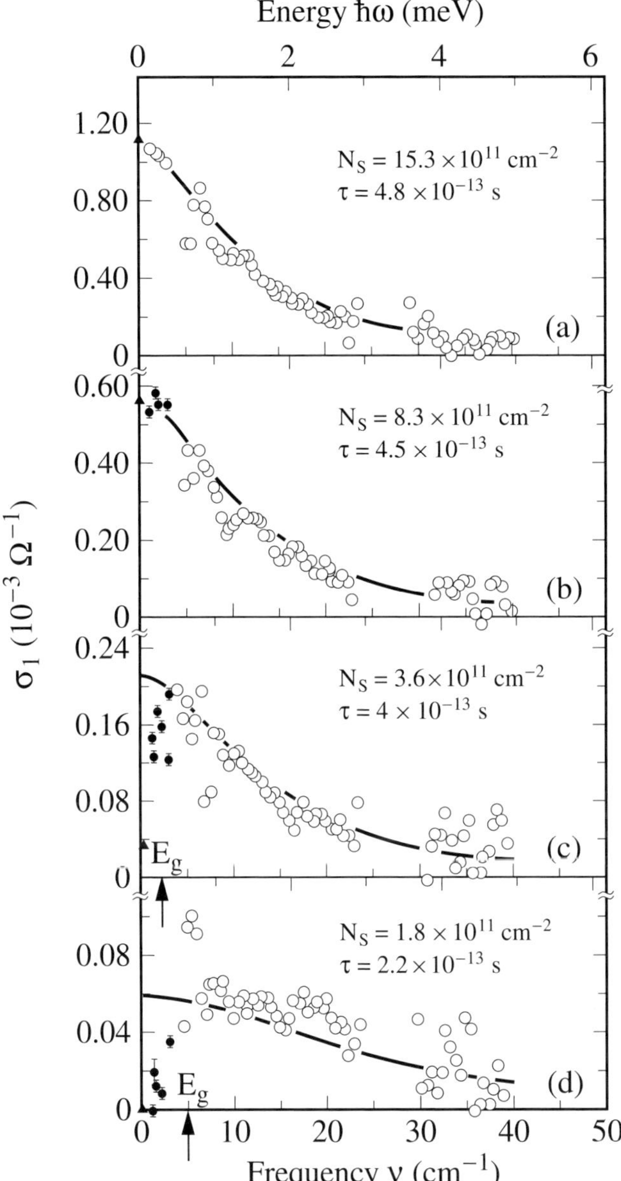

Fig. 12.16. Conductivity $\sigma_1(\omega)$ of a two-dimensional silicon inversion layer measured versus frequency at 1.2 K. The open circles are data taken by an infrared spectrometer, and the closed circles are from microwave measurements. The lines correspond to the Drude behavior predicted from the dc conductivity extrapolated to $T \to 0$ at the indicated value of the carrier concentration N_S for two dimensions and of the scattering time τ (after [All75]). $\mathcal{E}_g$ indicates the bandgap. There is a progressive crossover to a non-conducting state at low frequencies with decreasing carrier density, and thus with increasing disorder, relative to the bandwidth.

low frequencies because the low spectral range of interest is difficult to explore
at low temperature. There is an abundance of theoretical predictions on how the
conductivity should depend on frequency, along the lines of what is predicted for
the temperature dependence of the dc conductivity. Close to the transition (and also
on the localized side of the transition at rather low frequencies) the conductivity at
a temperature $T \to 0$ should follow the power law

$$\sigma_1(\omega) = A'\omega^{\alpha'} \pm B \tag{12.2.22}$$

with the positive sign appropriate for the metallic side and the negative sign for the
insulating side (obviously not at too low temperatures). The parameters depend
upon the importance of electron–electron interactions [Lee85, Vol82] just like
in Eq. (12.2.19) for $\sigma_{\mathrm{dc}}(T)$. If electron–electron interactions are important, the
exponent α' is the same as that for the temperature dependent dc conductivity.
Deep in the localized regime, arguments similar to those which were developed for
variable range hopping lead to the conductivity

$$\sigma_1(\omega) \propto \frac{D(\mathcal{E}_{\mathrm{F}})}{\lambda_0^5}\omega^2\left(\ln\left\{\frac{I_0}{\hbar\omega}\right\}\right)^4 \tag{12.2.23}$$

in three dimensions. Again, λ_0 is the localization length, $D(\mathcal{E}_{\mathrm{F}})$ is the density of
states at the Fermi level, and I_0 denotes a parameter which determines the overlap
integral of localized states and depends exponentially on λ_0. The above expression
is close to an ω^2 functional form. Just as in the case of variable range dc hopping,
electron–electron interactions modify this equation, and the leading term of the
frequency dependence is

$$\sigma_1(\omega) \propto \omega\left(\ln\left\{\frac{I_0}{\hbar\omega}\right\}\right)^4 \quad . \tag{12.2.24}$$

Similar to the temperature dependence, there is a smooth crossover to a power
law frequency dependence, described by Eq. (12.2.22) with increasing frequencies;
we can define a crossover frequency which is related to the crossover temperature
mentioned above. These frequency dependences are indicated in Fig. 12.15.

From the above a close correspondence is expected between the temperature
dependence (for $kT \gg \hbar\omega$) and the frequency dependence (for $kT \ll \hbar\omega$) of the
conductivity at low energy scales, well below the frequencies where a crossover
to the Drude roll-off occurs, or temperatures where other scattering processes
or thermal excitations play a role. In Fig. 12.17 we display the temperature (at
$\omega \to 0$) and frequency dependence (at low temperature for which $kT \ll \hbar\omega$
applies) for amorphous $\mathrm{Nb}_x\mathrm{Si}_{1-x}$ in the vicinity of the metal–insulator transition.
There is a close correspondence between the temperature and frequency dependent
conductivity in all three cases; the temperature and frequency dependencies can be

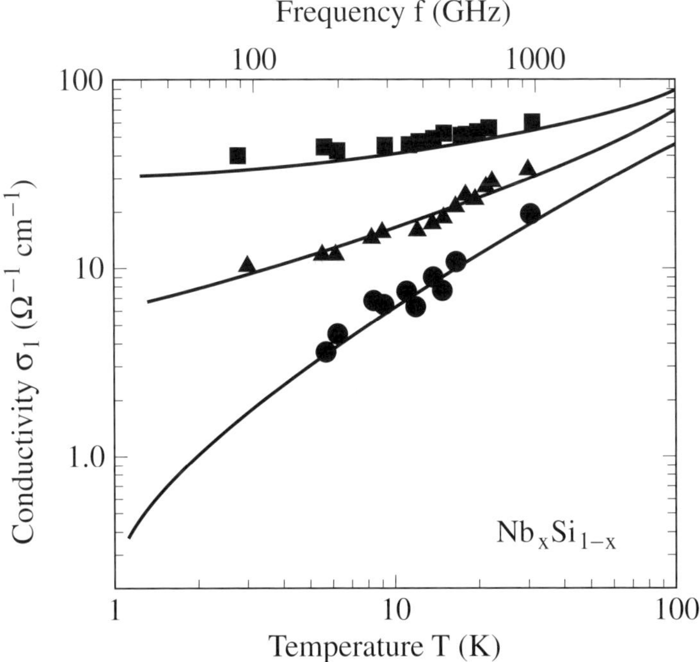

Fig. 12.17. Conductivity versus temperature (lower scale) and low temperature conductivity versus frequency (upper scale) for a metallic, a critical, and an insulating Nb_xSi_{1-x} alloy (after [Lee00]). $\sigma_1(\omega)$ for $\hbar\omega \gg k_BT$ and σ_{dc} (and in general $\sigma_1(T)$ for $\hbar\omega \ll k_BT$) have similar functional dependences, and $\sigma_{dc}(T) \approx \sigma_1(T)$ if the $k_BT = \hbar\omega$ correspondence is made.

scaled if we choose $1.3k_BT = \hbar\omega$. The numerical factor is expected to depend on the underlying electronic structure and may vary from material to material. For the so-called critical sample, both the frequency and the temperature dependence can be well accounted for by $\sigma(\omega) \propto \omega^{1/2}$ or $\sigma(T) \propto T^{1/2}$, respectively, as expected for disorder in the presence of strong electron–electron interactions. For the metallic sample, the frequency (and the temperature) dependence of the conductivity is again in agreement with the theory. For the insulating sample a power law is found at high temperatures and frequencies; this is followed by the temperature dependence as given by Eq. (12.2.21), with a smooth crossover. No experimental results are available to verify the frequency dependence of the conductivity as predicted by the equations given above.[6] We also expect that the zero temperature dielectric constant, measured at low enough frequencies, diverges as the transition is approached from the metallic side; the behavior reflects the

---

[6] Note added in proof. First experiments on doped Si indicate the crossover to a Coulomb glass in the microwave range [Hel01, Lee01].

increase of the correlation length as the transition is approached. Such behavior has been observed in Si:P [Ros81, Gay95].

References

[Abe66] F. Abelès, ed., *Optical Properties and Electronic Structure of Metals and Alloys* (North-Holland, Amsterdam, 1966)

[All71] P.B. Allen, *Phys. Rev. B* **3**, 305 (1971)

[All75] S.I. Allen, D.C. Tsui, and F.D. Rosa, *Phys. Rev. Lett.* **35**, 1359 (1975)

[All93] S.I. Allen, M.H. Paalanen, and R.N. Bhatt, *Europhys. Lett.* **21**, 927 (1993)

[Alt85] B.L. Altschuler and A.G. Aronov, in: *Electron–Electron Interaction in Disordered Systems*, edited by A.L. Efros and M. Pollak (North-Holland, Amsterdam, 1985)

[And58] P.W. Anderson, *Phys. Rev.* **109**, 1492 (1958)

[And61] P.W. Anderson, *Phys. Rev.* **124**, 41 (1961)

[Ash76] N.W. Ashcroft and N.D. Mermin, *Solid State Physics* (Holt, Rinehart and Winston, New York, 1976)

[Awa93] A.M. Awasthi, L. Degiorgi, G. Grüner, Y. Dalichaouch, and M.B. Maple, *Phys. Rev. B* **48**, 10 692 (1993)

[Bel94] D. Belik and T.R. Kirkpatrick, *Rev. Mod. Phys.* **66**, 261 (1994)

[Ben66] H.E. Bennett and J.M. Bennett, *Validity of the Drude Theory for Silver, Gold and Aluminum in the Infrared*, in: *Optical Properties and Electronic Structure of Metals and Alloys*, edited by F. Abelès (North-Holland, Amsterdam, 1966)

[Blo72] A.N. Bloch, R.B. Weismon, and C.M. Varma, *Phys. Rev. Lett.* **28**, 753 (1972)

[Cha52] R.G. Chambers, *Proc. Phys. Soc. (London) A* **65**, 458 (1952); *Proc. Roy. Soc. London A* **215**, 481 (1952)

[Deg99] L. Degiorgi, *Rev. Mod. Phys.* **71**, 687 (1999)

[Dre96] M. Dressel, O. Klein, S. Donovan, and G. Grüner, *Ferroelectrics* **176**, 285 (1996)

[Dre97] M. Dressel, G. Grüner, J.E. Eldridge, and J.M. Williams, *Synth. Met.* **85**, 1503 (1997)

[Efr75] A.L. Efros and B.I. Shklovski, *J. Phys. C: Solid State Phys.* **8**, L49 (1975)

[Ehr63] H. Ehrenreich, H.R. Philipp, and B. Segall, *Phys. Rev.* **132**, 1918 (1963)

[Ehr66] H. Ehrenreich, *Electromagnetic Transport in Solids: Optical Properties and Plasma Effects*, in: *The Optical Properties of Solids*, Proceedings of the International School of Physics 'Enrico Fermi' **34**, edited by J. Tauc (Academic Press, New York, 1966)

[Fri56] J. Friedel, *Can. J. Phys.* **34**, 1190 (1956)

[Gay95] A. Gaymann, H.P. Geserich, and H.v. Löhneysen, *Phys. Rev. B* **52**, 16 486 (1995)

[Gre91] N. Grewe and F. Steglich, *Heavy Fermions*, in: *Handbook on the Physics and Chemistry of Rare Earths* **14**, edited by K.A. Gschneider and L. Eyring (Elsevier, Amsterdam, 1991)

[Gru74a] G. Grüner, *Adv. Phys.* **23**, 941 (1974)

[Gru74b] G. Grüner and A. Zawadowski, *Rep. Prog. Phys.* **37**, 1497 (1974)

[Hel01] E. Helgren, M. Scheffler, M. Dressel, and G. Grüner, unpublished

[Her83] G. Hertel, D.J. Bishop, E.G. Spencer, J.M. Rowell, and R.C. Dynes, *Phys. Rev. Lett.* **50**, 743 (1983)

[Joy70] R.R. Joyce and P.L. Richards, *Phys. Rev. Lett.* **24**, 1007 (1970)

[Kit63] C. Kittel, *Quantum Theory of Solids* (John Wiley & Sons, New York, 1963)

[Kun62] C. Kunz, *Z. Phys.* **167**, 53 (1962); ibid. **196**, 311 (1966)

[Lee85] P.A. Lee and T.V. Rahmakrishna, *Rev. Mod. Phys.* **57**, 287 (1985)

[Lee00] H.-L. Lee, J.P. Carini, D.V. Baxter, W. Henderson, and G. Grüner, *Science* **287**, 5453 (2000)

[Lee01] M. Lee and M.L. Stutzmann, *Phys. Rev. Lett.* **87**, 056402 (2001)

[Loh98] H.v. Löhneysen, *Phil. Trans. Roy. Soc. (London) A* **356**, 139 (1998)

[Lyn85] D.W. Lynch and W.R. Hunter, *Comments on the Optical Constants of Metals and an Introduction to the Data for Several Metals*, in: *Handbook of Optical Constants of Solids*, Vol. I, edited by E.D. Palik (Academic Press, Orlando, FL, 1985), p. 275

[McK79] S.W. McKnight, S. Perkowitz, D.B. Tanner, and L.R. Testardi, *Phys. Rev. B* **19**, 5689 (1979)

[Mar62] L. Marton, J.A. Simpson, H.A. Fowler, and N. Swanson, *Phys. Rev.* **126**, 182 (1962)

[Miy89] K. Miyaka, T. Masuura, and C.M. Varma, *Solid State Commun.* **71**, 1149 (1989)

[Mot61] N.F. Mott and W.D. Twose, *Adv. Phys.* **10**, 107 (1961)

[Mot68] N.F. Mott, *J. Non Cryst. Solids* **1**, 1 (1968)

[Mot69] N.F. Mott, *Phil. Mag.* **19**, 835 (1969)

[Mye68] H.P. Myers, L. Walldén, and Å. Karlsson, *Phil. Mag.* **68**, 725 (1968)

[Nor71] C. Norris and H.P. Myers, *J. Phys.* **1**, 62 (1971)

[Ott87] H.R. Ott and Z. Fisk, in: *Handbook on the Physics and Chemistry of the Acinides* **5**, edited by A.J. Freeman and G.H. Lander (North-Holland, Amsterdam, 1987), p. 85

[Pip50] A.B. Pippard, *Proc. Roy. Soc. (London) A* **203**, 98 (1950)

[Pip57] A.B. Pippard, *Phil. Trans. Roy. Soc. (London)* **A 250**, 325 (1957)

[Pip60] A.B. Pippard, *Rep. Prog. Phys.* **33**, 176 (1960)

[Rae80] H. Raether, *Excitation of Plasmons and Interband Transitions by Electrons*, Springer Tracts in Modern Physics **88** (Springer-Verlag, Berlin, 1980)

[Reu48] G.E.H. Reuter and E.H. Sondheimer, *Proc. Roy. Soc. A* **195**, 336 (1948)

[Ros81] T.F. Rosenbaum, K. Andres, G.A. Thomas, and P.A. Lee, *Phys. Rev. Lett.* **46**, 568 (1981)

[Ste84] G. Stewart, *Rev. Mod. Phys.* **56**, 755 (1984)

[Ves98] V. Vescoli, L. Degiorgi, W. Henderson, G. Grüner, K.P Starkey, and L.K. Montgomery, *Science* **281**, 1181 (1998)

[Vol82] D. Vollhardt and P. Wölfe, *Phys. Rev. Lett.* **48**, 699 (1982)

Further reading

[Cha90] R.G. Chambers, *Electrons in Metals and Semiconductors* (Chapman and Hall, London, 1990)

[Dyr00] J.C. Dyre and T.B. Schrøder, *Rev. Mod. Phys.* **72**, 873 (2000)

[Efr85] A.L. Efros and M. Pollak, eds, *Electron–Electron Interaction in Disordered Systems*, Modern Problems in Condensed Matter Sciences **10** (North-Holland, Amsterdam, 1985)

[Fin89] J. Fink, *Recent Developments in Energy-loss Spectroscopy*, in: *Advances in Electronics and Electron Physics* **75** (Academic Press, Boston, MA, 1989)

[Fis88] Z. Fisk, D.W. Hess, C.J. Pethick, D. Pines, J.L. Smith, J.D. Thompson, and J.O. Willis, *Science* **239**, 33 (1988)

[Gol89] A.I. Golovashkin, ed., *Metal Optics and Superconductivity* (Nova Science Publishers, New York, 1989)

[Kau98] H.J. Kaufmann, E.G. Maksimov, and E.K.H. Salje, *J. Superconduct.* **11**, 755 (1998)

[Mit79] S.S. Mitra and S. Nudelman, eds, *Far Infrared Properties of Solids* (Plenum Press, New York, 1970)

[Mot79] N.F. Mott and E.H. Davis, *Electronic Processes in Non-Crystalline Materials*, 2nd edition (Clarendon Press, Oxford, 1979)

[Mot90] N.F. Mott, *Metal–Insulator Transition*, 2nd edition (Taylor & Francis, London, 1990)

[Ord85] M.A. Ordal, R.J. Bell, R.W. Alexander, L.L. Long, and M.R. Querry, *Appl. Optics* **24**, 4493 (1985)

[Shk84] B.I. Shklovskii and A.L. Efros, *Electronic Properties of Doped Semiconductors*, Springer Series in Solid-State Sciences **45** (Springer-Verlag, Berlin, 1984)

13

Semiconductors

Optical experiments on semiconductors have led to some of the most powerful confirmations of the one-electron theory of solids; these experiments provide ample evidence for direct and indirect gaps, and in addition for excitonic states. Optical studies have also contributed much to our current understanding of doping semiconductors, including the existence and properties of impurity states and the nature of metal–insulator transitions which occur by increasing the dopant concentration. Experiments on amorphous semiconductors highlight the essential differences between the crystalline and the amorphous solid state, and the effects associated with the loss of lattice periodicity. We first focus on experiments performed on pure band semiconductors for which the one-electron theory applies, where direct and indirect transitions and also forbidden transitions are observed; in these materials the subtleties of band structure have also been explored by experimentation. This is followed by examples of the optical effects associated with exciton and impurity states. Subsequently we consider the effects of electron–electron and electron–lattice interactions, and finally we discuss optical experiments on amorphous semiconductors, i.e. on materials for which band theory obviously does not apply.

13.1 Band semiconductors

The term band semiconductor refers to materials where the non-conducting state is brought about by the interaction of electrons with the periodic underlying lattice. Single-particle effects – accounted for by band structure calculations – are responsible for the optical properties under such circumstances, these properties reflecting interband transitions.

13.1.1 Single-particle direct transitions

Optical processes associated with the semiconducting state are fundamentally different from the processes which determine the response of conduction electrons. For semiconductors optical transitions between bands are responsible for the absorption of electromagnetic radiation, and thus for the optical properties. These transitions have been treated in Chapter 6 under simplified assumptions about both the transition rates and the density of states of the relevant bands.

Let us first comment on the notation vertical transitions, which reflects the observation that the speed of light is significantly larger than the relevant electron (and also phonon) velocities in crystals. We can estimate the difference in the wavevectors $\mathbf{k}_{l'} - \mathbf{k}_l = \Delta\mathbf{k}$ involved in interband transitions as the momentum of the photon $|\mathbf{q}| = \omega/c$ – assuming that the refractive index $n = 1$ and $\hbar\omega$ is the energy difference between the two states. The bands are separated by the single-particle gap $\mathcal{E}_g$, and $\mathbf{k}_{max} = \pi/a$ is the wavevector at the Brillouin zone; therefore

$$\frac{\Delta\mathbf{k}}{\mathbf{k}_{max}} = \frac{\omega/c}{\pi/a} = \frac{a\mathcal{E}_g}{\pi\hbar c} \quad . \tag{13.1.1}$$

With $a \approx 3$ Å and $\mathcal{E}_g = 1$ eV the ratio is about 5×10^{-4}, and, as assumed, the change in the momentum during the course of the optical absorption can indeed be neglected.

In Fig. 13.1 we display the optical parameters, the reflectivity $R(\omega)$, and the components of the complex dielectric constant $\epsilon_1(\omega)$ and $\epsilon_2(\omega)$, as well as the loss function $\mathrm{Im}\{1/\hat{\epsilon}(\omega)\}$ for the intrinsic semiconductor germanium measured up to very high energies [Phi63]. There are several features of interest: first there is a broad qualitative agreement between the findings and what is predicted by the Lorentz model, which has been explored in Section 6.1. The frequency dependent response of the various parameters which follow from the model are displayed in Figs 6.3–6.7: the reflectivity is finite – but smaller than unity – as we approach zero frequency, and rises with increasing frequency, reaching a plateau before rolling off at high frequencies – features prominently observed in germanium and also in other semiconductors. Both the real and the imaginary parts of the dielectric constant of germanium are also close to those of a harmonic oscillator (if we neglect the sharp changes in the reflectivity and also in the dielectric constant), as the comparison with Fig. 6.3 clearly indicates. Finally, the broad peak of the loss function at around 15 eV – just as in the case for metals – indicates a plasma resonance in the spectral range which is similar to that observed for simple metals. There are however important differences: we observe considerable structure, shoulders, and peaks in the optical properties, reflecting band structure effects (as we will discuss). Also, the frequency which we would associate with the characteristic frequency of

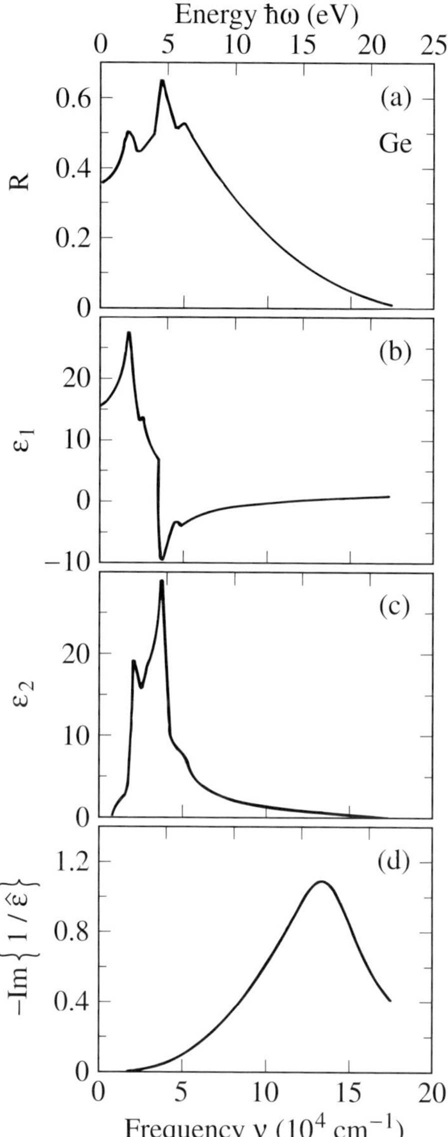

Fig. 13.1. Optical parameters of germanium as determined by reflection measurements at room temperature. (a) Frequency dependence of the reflectivity $R(\omega)$. (b) The real part of the dielectric constant $\epsilon_1(\omega)$, obtained by Kramers–Kronig analysis of the reflectivity data. (c) The imaginary part of the dielectric constant $\epsilon_2(\omega)$. (d) The energy loss function $\text{Im}\{-1/\hat{\epsilon}(\omega)\}$. (After [Phi63].)

the harmonic oscillator ω_0 – and thus related to the restoring force – is significantly larger than the single-particle gap which we obtain from the temperature dependent dc conductivity. For an intrinsic semiconductor the dc conductivity reads

$$\sigma_{\text{dc}}(T) = \sigma_0 \exp\left\{-\frac{\mathcal{E}_g}{2k_B T}\right\} \quad , \tag{13.1.2}$$

where $\mathcal{E}_g$ is the so-called thermal gap, the energy required to create an electron in the conduction band by thermal excitation, leaving a hole in the valence band. The pre-factor σ_0 contains the number of carriers and their mobility, and we assume for the moment that they are only weakly temperature dependent. Dc transport experiments on germanium give a thermal gap of $\mathcal{E}_g = 0.70$ eV, significantly smaller than the energy associated with the frequency ω_0. In fact, as we can see in Fig. 13.1a, the onset of absorption is not as smooth as the Lorentz model predicts. A closer inspection of $\epsilon_2(\omega)$ reveals a sudden onset of the absorption processes – note that the relation (2.3.26) between the imaginary part of the dielectric constant and the absorption coefficient is

$$\epsilon_2(\omega) = \frac{nc}{\omega}\alpha(\omega) \quad ,$$

where n is the real part of the refractive index (which is only weakly frequency dependent in the relevant frequency range). This onset of absorption can be associated with the single-particle gap, and we obtain for the optical gap $\mathcal{E}_g = 0.74$ eV – a value close to the thermal gap given above. The behavior close to the gap is shown in more detail in Fig. 13.2 for another intrinsic semiconductor PbS, for which detailed optical experiments on epitaxial films have been conducted in the gap region [Car68, Zem65]. It is found that $\alpha(\omega)$ increases as the square root of the frequency

$$\alpha(\omega) \propto (\hbar\omega - \mathcal{E}_g)^{1/2} \quad , \tag{13.1.3}$$

since the frequency dependence described by Eq. (6.3.11) dominates the absorption process. The gap here is 0.45 eV, in excellent agreement with what is measured by the dc transport; both of these observations provide clear evidence that this material has a direct bandgap. The dielectric constant ϵ_1 – also shown in Fig. 13.2 – is in full agreement with such an interpretation; the full line corresponds to a fit of the data by Eq. (6.3.12) with the gap $\mathcal{E}_g = 0.47$ eV.

While the sharp onset of absorption is found in a large number of materials, confirming the well defined bandgap, often the functional dependence is somewhat different from that expected for simple direct transitions. The absorptivity of InSb displayed in Fig. 13.3 illustrates this point: near the gap of 0.23 eV the absorption has the characteristic square root dependence as expected for allowed direct transitions, see Eq. (6.3.11), but a fit over a broad spectral range requires an additional term which can be described as

$$\alpha(\omega) \propto (\hbar\omega - \mathcal{E}_g)^{3/2} \quad . \tag{13.1.4}$$

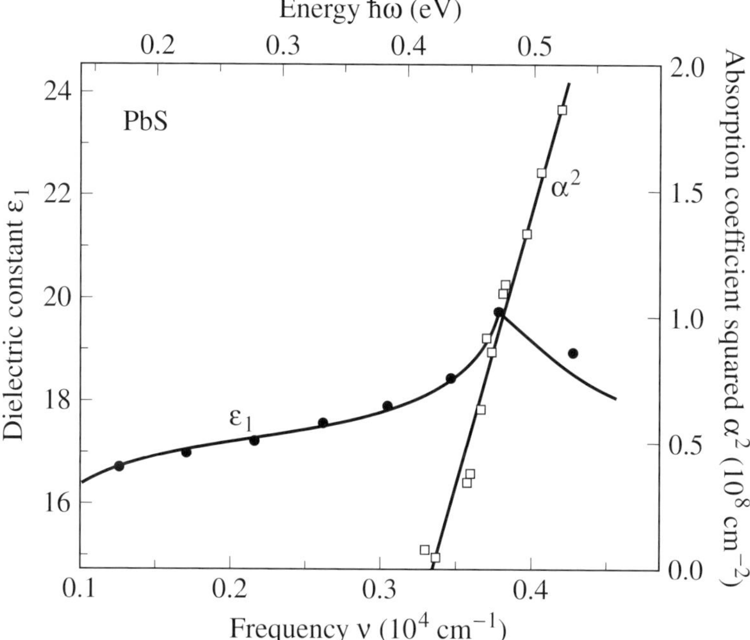

Fig. 13.2. Frequency dependent dielectric constant $\epsilon_1(\omega)$ and absorption coefficient of PbS measured at $T = 373$ K. The full lines are fits to the data according to Eqs (6.3.11) and (6.3.12); the single-particle gap is 0.47 eV (after [Car68, Zem65]).

Comparison of this behavior with Eq. (6.4.6) indicates that forbidden transitions also contribute to the absorption process; we expect these increasingly to play a role as we move away from the band edge. Alternatively a frequency dependence of the matrix element involved with the transition can lead to the behavior observed; indeed, both interpretations have been offered [Mac55].

The reason for this ambiguity is clear and emphasizes a general problem. Theory can describe the features of the band structure with sufficient accuracy, and these features enter into the various expressions for the frequency dependence of the optical parameters, as discussed in Section 6.3. However, evaluating the transition matrix elements, such as those given by the interaction Hamiltonian (6.2.10), is a significantly more difficult task; consequently the magnitude of the absorption, together with the possible additional frequency dependencies due to the matrix elements, cannot be in general evaluated. In fact often the reverse procedure, the evaluation of the transition matrix elements by comparing the theoretical expressions with the experimental findings, is the preferred route. Sum rule arguments as presented in Section 3.2.2 can also be utilized in order to estimate the transition matrix elements – such a procedure of course would require that the optical

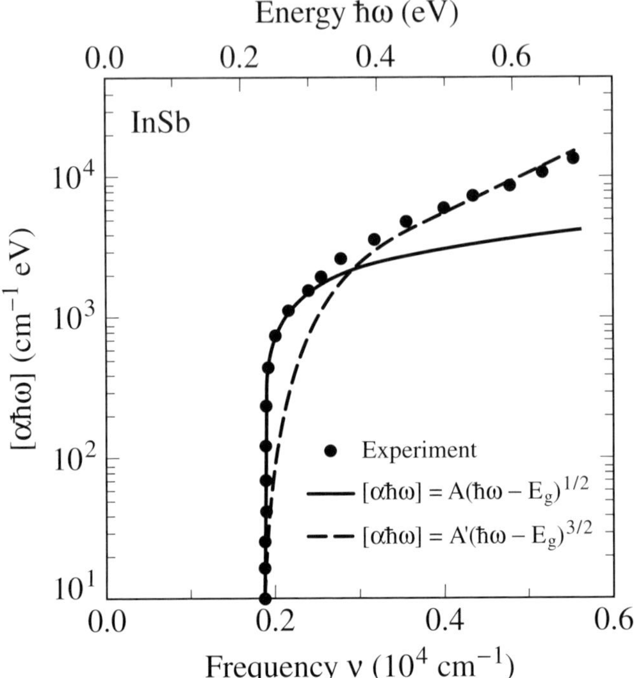

Fig. 13.3. Frequency dependence of the absorption coefficient α times the photon energy $\hbar\omega$ of InSb [Gob56]. For frequencies close to the bandgap the experimental data can be well described by Eq. (6.3.11), giving evidence for direct transitions; a $(\hbar\omega - \mathcal{E}_g)^{3/2}$ frequency dependence is observed at higher frequencies and may be due to forbidden transitions (after [Joh67]).

constants are available over a broad spectral range.

Next we turn to the peaks and shoulders which are so evident in the optical properties of semiconductors and are seen in Fig. 13.1, for example. These provide perhaps the clearest and cleanest evidences for band structure effects – but also for the power of modern band structure calculations. In order to understand these observations, we first have to describe some essential features and consequences of such band structure calculations. The starting point is the dispersion relations obtained by a band structure calculation of a particular choice. Once the band structure is established, one can evaluate the single-particle density of states given by the expression

$$D_l(\mathcal{E}) = \frac{1}{(2\pi)^3} \int \frac{dS_\mathcal{E}}{\nabla_\mathbf{k}\mathcal{E}_l(\mathbf{k})} \quad ,$$

as discussed in Section 6.3. Such dispersion curves, calculated by the so-called empirical pseudopotential method, are shown in Fig. 13.4a; Fig. 13.4b shows the

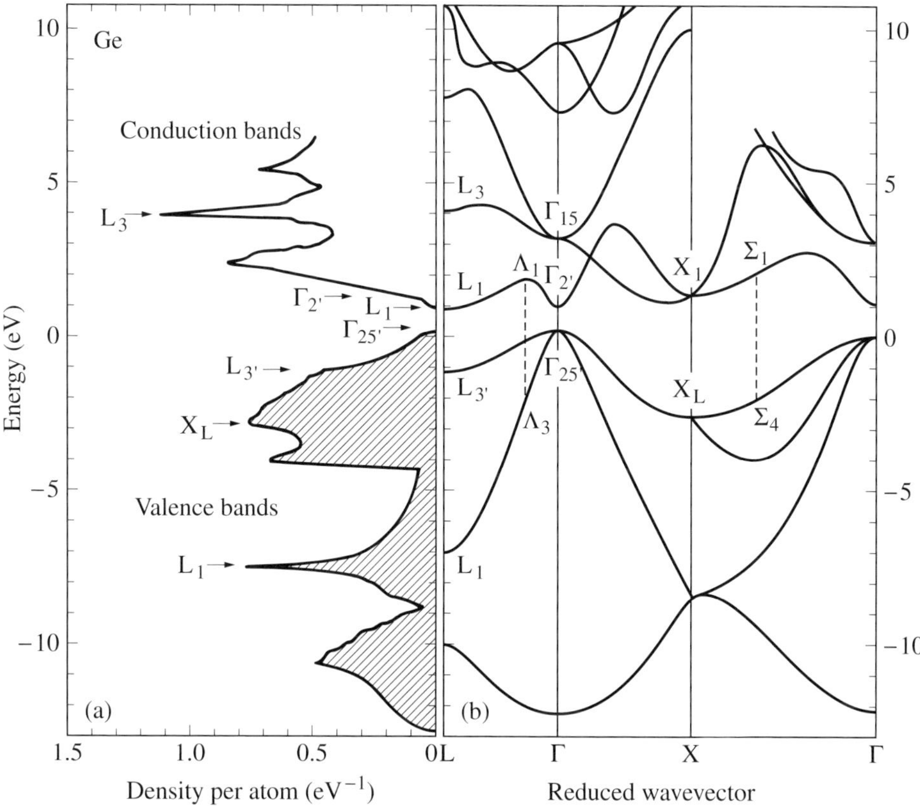

Fig. 13.4. (a) Electronic band structure of germanium in various crystallographic directions of high symmetry, and (b) the corresponding density of states $D(\mathcal{E})$, calculated by the empirical pseudopotential method. Some critical points are indicated, which correspond to a flat dispersion of $\mathcal{E}(\mathbf{k})$ (after [Her67]).

corresponding density of states obtained by the preceeding equation. Particular symmetry points are labeled corresponding to the common notation found in various books on semiconductors [Coh89, Gre68, Pan71, Yu96], and will not be discussed here. What appears in the formulas for the optical properties in the case of direct transitions is the joint density of states calculated by Eq. (6.3.2)

$$D_{l'l}(\hbar\omega) = \frac{2}{(2\pi)^3} \int \frac{\mathrm{d}S_\mathcal{E}}{\nabla_\mathbf{k}\left[\mathcal{E}_l(\mathbf{k}) - \mathcal{E}_{l'}(\mathbf{k})\right]}$$

using the dispersion relations as derived from band structure calculations. If we assume that the transition matrix element is independent of wavevector, the frequency dependence of the absorption is, through Eq. (6.3.4), determined by the frequency dependence of the joint density of states. For particular $\mathbf{k}$ values, called critical points, $\nabla_\mathbf{k}\left[\mathcal{E}_l(\mathbf{k}) - \mathcal{E}_{l'}(\mathbf{k})\right]$ has zeros, and thus the density of states displays the so-called van Hove singularities at these singularities. In the vicinity of these critical points the energy difference $\delta\mathcal{E}(\mathbf{k}) = \mathcal{E}_l(\mathbf{k}) - \mathcal{E}_{l'}(\mathbf{k})$ can be expanded as

$$\delta\mathcal{E}(\mathbf{k}) = \hbar\omega_c + \alpha_1 k_1^2 + \alpha_2 k_2^2 + \alpha_3 k_3^2 \tag{13.1.5}$$

in three dimensions. Utilizing such an expansion, the joint density of states and, through Eqs (6.3.11) and (6.3.12), the components of the complex dielectric constant can be evaluated; the behavior depends on the sign of three coefficients α_1, α_2, and α_3. In three dimensions, four distinct possibilities exist, which are labeled M_j according to the number j of negative coefficients. For M_0, for example, all coefficients are positive and

$$D(\hbar\omega) \propto \begin{cases} 0 & \text{for } \omega < \omega_c \\ (\hbar\omega - \hbar\omega_c)^{1/2} & \text{for } \omega > \omega_c \end{cases} , \tag{13.1.6}$$

with the resulting $\epsilon_2(\omega)$ shown in Fig. 13.5; the other possibilities are also displayed. Not surprisingly, the real part of the dielectric constant also displays sharp changes at these critical points. In one dimension, the situation is entirely different. First of all

$$\delta\mathcal{E}(\mathbf{k}) = \hbar\omega_c + \alpha_1 k_1^2 \tag{13.1.7}$$

and we have only two possibilities, with α either positive or negative. At the onset of an interband transition, the joint density of states diverges as

$$D(\hbar\omega) \propto \begin{cases} 0 & \text{for } \omega < \omega_c \\ (\hbar\omega - \hbar\omega_c)^{-1/2} & \text{for } \omega > \omega_c \end{cases} \tag{13.1.8}$$

at the critical points, for $\alpha_1 > 0$ for example; the behavior of $\epsilon_2(\omega)$ for the two solutions is displayed in the lower part of Fig. 13.5.

After these preliminaries we can understand the features observed in the optical experiments as consequences of critical points and relevant dimensionalities near these critical points. In Fig. 13.6 the measured imaginary part of the dielectric constant $\epsilon_2(\omega)$ of germanium is displayed together with results of calculations [Phi63]. Without discussing the various transitions and critical points in detail, it suffices to note that modern band structure calculations such as the empirical pseudopotential method give an excellent account of the optical experiments and provide evidence for the power and accuracy of such calculations. At high frequencies the behavior shown in Fig. 13.6 gives way to a smooth ω^{-2} decrease of the conductivity, the Drude roll-off associated with inertia effects. For $1/\tau$ and $\mathcal{E}_g$ much smaller than the plasma frequency, the real part of the dielectric constant

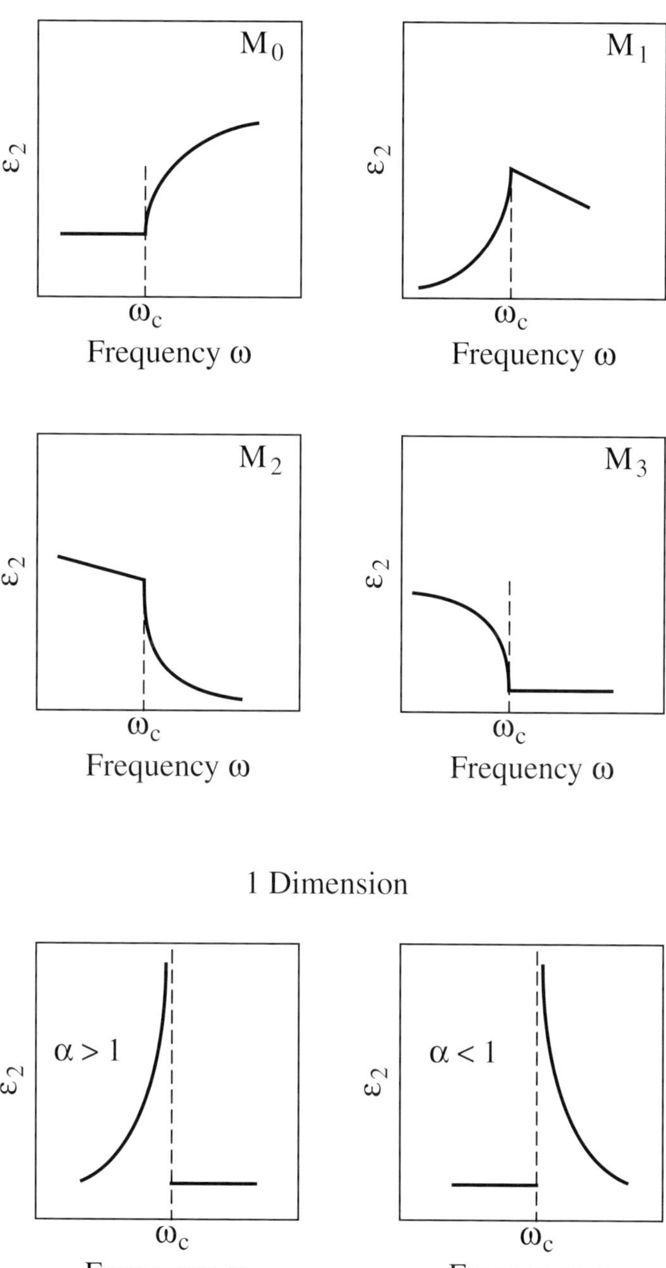

Fig. 13.5. Behavior of the dielectric loss $\epsilon_2(\omega)$ near van Hove singularities in three dimensions and in one dimension.

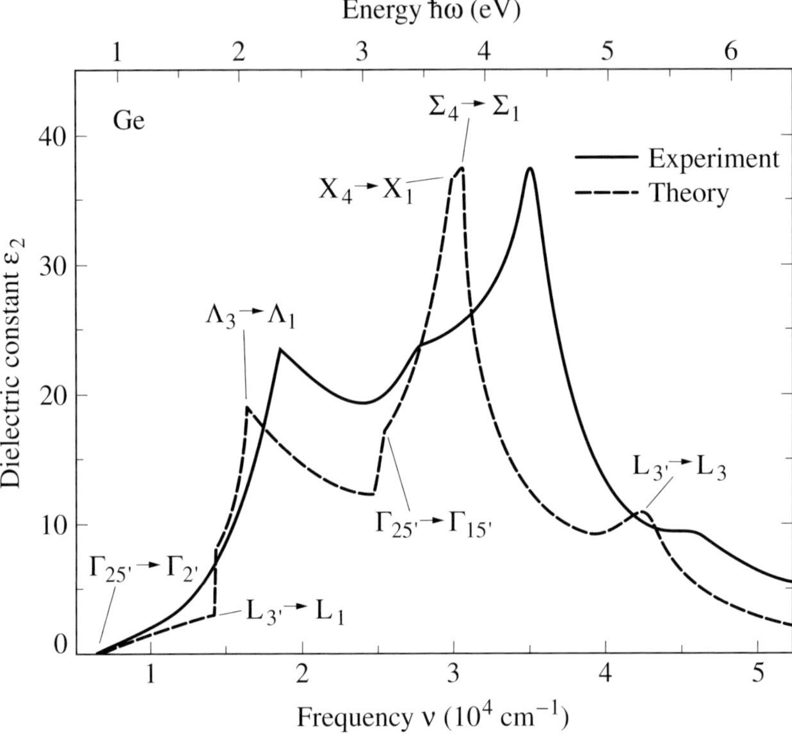

Fig. 13.6. Calculated spectral structure of $\epsilon_2(\omega)$ of germanium compared with experimental results obtained from transmission and reflection measurements (after [Phi63]). The peaks and shoulders identify the van Hove singularities; the transitions which are involved in the particular joint density of states $D_{cv}(\mathcal{E})$ are also indicated. Important critical points in the band structure are marked according to Fig. 13.4.

$\epsilon_1(\omega)$ shows a zero-crossing at

$$\omega_{\mathrm{p}}^+ = \left(\frac{4\pi n e^2}{\epsilon_\infty m_{\mathrm{b}}} \right)^{1/2} \, ,$$

the same expression which applies for metals (Eq. (12.1.8)), where interband transitions are accounted for by a high frequency dielectric constant ϵ_∞. Just as for metals, plasma oscillations can also be examined by electron energy loss spectroscopy; both methods can be used to evaluate the loss function $\mathrm{Im}\{-1/\hat{\epsilon}(\omega)\}$. This has been done for various semiconductors, and results are displayed in Fig. 13.7 for our example of germanium. The peak identifies the plasma frequency, and we observe an underdamped plasmon at energy of approximately 16 eV. The fundamental optical parameters of some semiconductors are collated in Table 13.1. For most band semiconductors, the bandgap is significantly smaller than the plasma frequency, leading also to a relatively large dielectric constant, which broadly

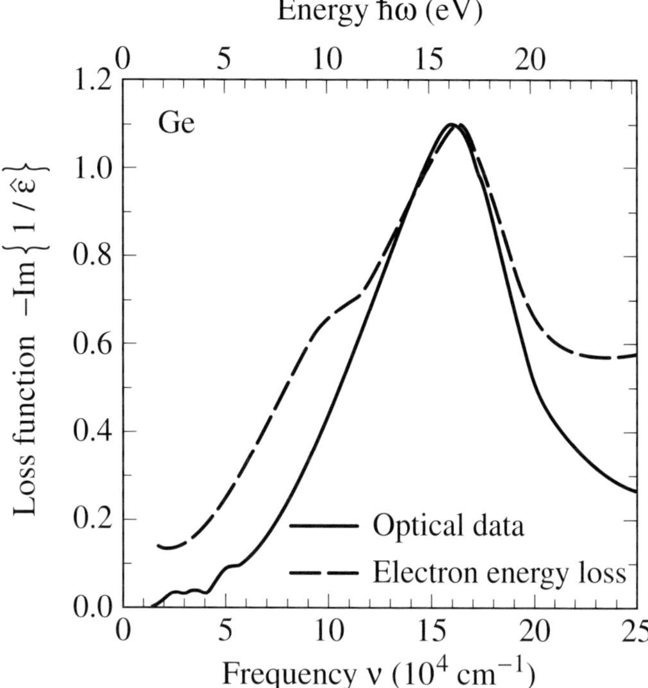

Fig. 13.7. Electron energy loss spectra $-\mathrm{Im}\{1/\hat{\epsilon}(\omega)\}$ of germanium as obtained from the optical data shown in Fig. 13.1 and from electron energy loss spectroscopy (after [Phi63, Pow60]). The data are normalized to the maximum of the measured loss.

Table 13.1. *Optical gaps* $\mathcal{E}_\mathrm{g} = \hbar\omega_\mathrm{g}$, *plasma frequencies* ω_p, *and static dielectric constant* $\epsilon_1(\omega = 0)$ *of various semiconductors as obtained at low temperatures.* '*i*' *indicates an indirect gap,* '*d*' *a direct gap. After [Kit96, Mad96, Wea90].*

Material	Gap	$\omega_\mathrm{g}/2\pi c$ (cm^{-1})	$\mathcal{E}_\mathrm{g}$ (eV)	$\omega_\mathrm{p}/2\pi c$ (cm^{-1})	$\hbar\omega_\mathrm{p}$ (eV)	$\epsilon_1(\omega = 0)$
Ge	i	5950	0.74	13.4×10^4	16.6	15.8
Si	i	9450	1.17	13.0×10^4	16.2	11.7
InSb	d	1850	0.24	9.7×10^4	12	17.9
GaAs	d	1.2×10^4	1.52	12.5×10^4	15.5	13.1
diamond	i	4.4×10^4	5.4			5.5

correlates with the ratio $\left(\hbar\omega_\mathrm{p}/\mathcal{E}_\mathrm{g}\right)^2$ as derived in Eq. (6.3.5). The above equations apply at zero temperature; at finite temperatures contributions to the optical properties coming from thermally excited electrons and holes have to be included. The number of these excitations increases exponentially with increasing temperature, and is small even at room temperature in most typical (undoped) semiconductors. If the relaxation time associated with these excitations is long, absorption due to these excitations – often called the free-electron absorption – occurs well below the gap frequency, and, as a rule, a simple Drude model accounts for the optical properties [Ext90]. In Fig. 13.8 the optical conductivity $\sigma_1(\omega)$ and $\sigma_2(\omega)$ of silicon is displayed for two different doping levels: for an acceptor concentration of 1.1×10^{15} cm^{-3} leading to a resistivity of 9.0 Ω cm, and a donor concentration of 4.2×10^{14} cm^{-3} which gives $\rho_\mathrm{dc} = 8.1$ Ω cm. The frequency dependence measured by time domain spectroscopy at room temperature can be well described for both cases by a simple Drude model with somewhat different scattering rates: $1/(2\pi\tau c) = 40$ cm^{-1} and 20 cm^{-1}, respectively.

The onset of the absorption is different for a highly anisotropic band structure. Of course, strictly one-dimensional materials are difficult to make, but in several so-called linear chain compounds (the name indicates the atomic or molecular arrangements in the crystal) the band structure is anisotropic enough such that it can be regarded as nearly or quasi-one-dimensional. Near to the gap, the optical conductivity of such materials can be described by the one-dimensional form of semiconductors we have discussed in Chapter 6. Expression (6.3.16) assumes that the transition matrix element between the valence and conduction states is independent of frequency; thus the functional dependence of σ_1 on ω is determined solely by the joint density of states. In general this is not the case, as can be seen from the following simple arguments. The periodic potential leads to a strong modification of the electronic wavefunctions only for states close to the gap; for wavevectors away from the gap (and therefore for transition energies significantly different from the gap energy $\mathcal{E}_\mathrm{g}$) such perturbation is less significant, and the electron states are close to the original states which correspond to a single band. As transitions between states labeled by different $\mathbf{k}$ values in a single band are not possible (note that optical transitions correspond to $\mathbf{q} = 0$), the transition probability is presumed to be strongly reduced as the energy of such transitions is increased from the gap energy $\mathcal{E}_\mathrm{g}$. Thus the conductivity is expected to display a stronger frequency dependence than that given by Eq. (6.3.16). All this can be made quantitative by considering the transitions in one dimension [Lee74]. One finds that

$$\sigma_1(\omega) = \frac{\pi N e^2 \mathcal{E}_\mathrm{g}^2}{2\hbar^2 m} \frac{1}{\omega^3 \left[1 - (\mathcal{E}_\mathrm{g}/\hbar\omega)^2\right]^{1/2}} \;, \qquad (13.1.9)$$

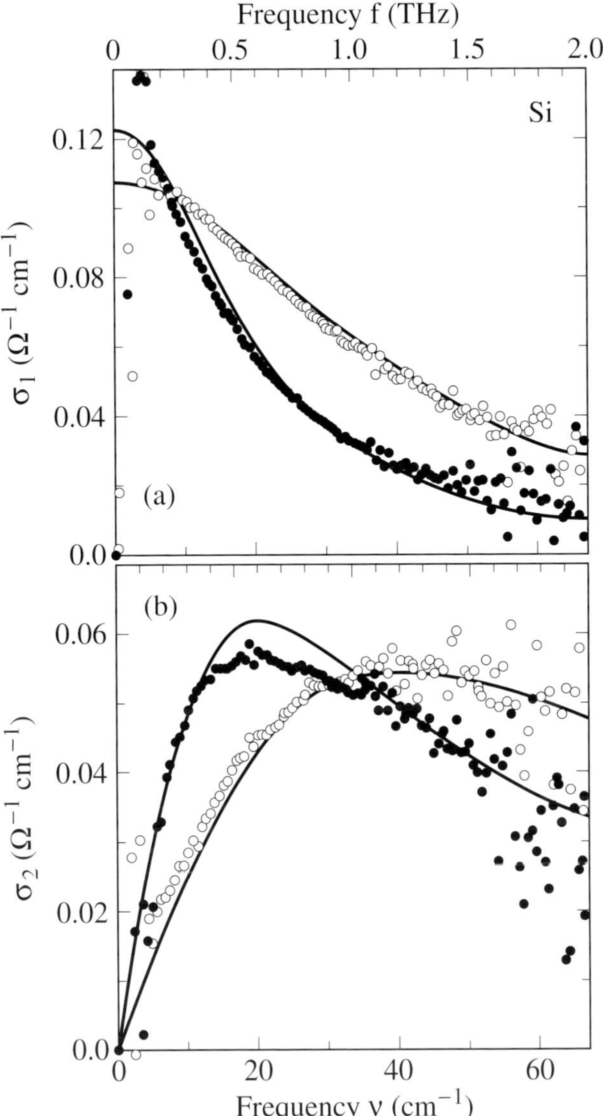

Fig. 13.8. Frequency dependence of the (a) real and (b) imaginary parts of the optical conductivity of silicon weakly doped by holes (1.1×10^{15} cm^{-3}, open circles) or electrons (4.2×10^{14} cm^{-3}, solid circles). The experiments were performed at $T = 300$ K in the time domain; the full lines correspond to $\sigma_1(\omega)$ and $\sigma_2(\omega)$ as obtained by the Drude model with the parameters given in the text (after [Ext90]).

an expression which has a leading ω^{-3} frequency dependence for frequencies well exceeding $\mathcal{E}_g/\hbar$; near to the gap the familiar square root frequency dependence is recovered. There is no effective mass μ here, as in the perturbation treatment of

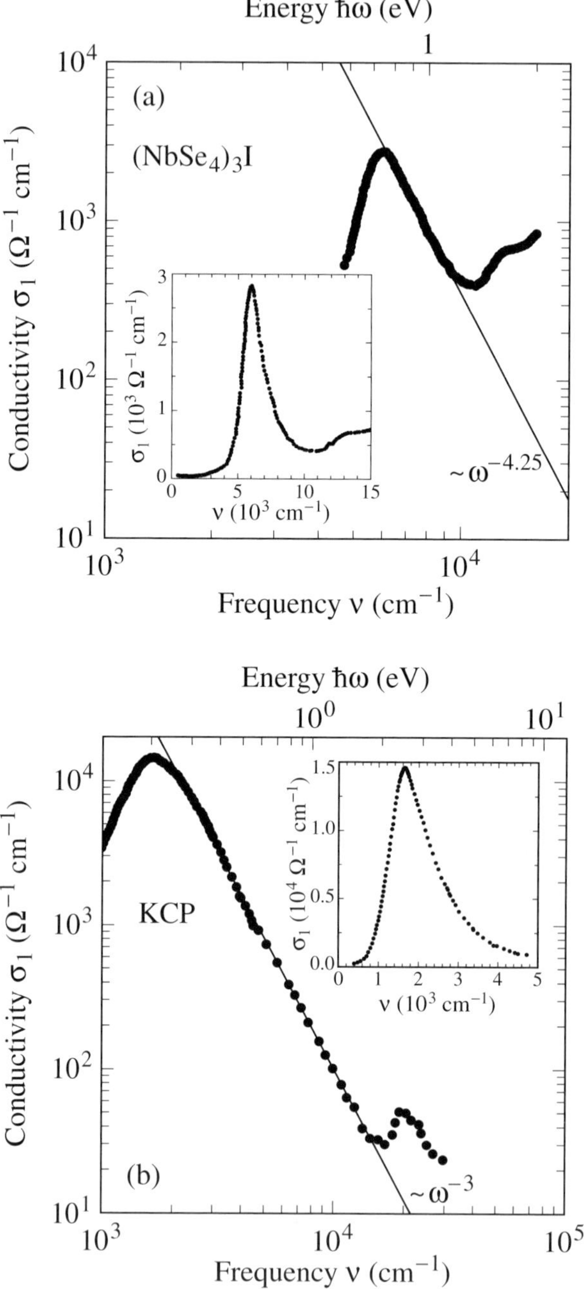

Fig. 13.9. Frequency dependence of the conductivity in highly anisotropic semiconductors $K_2[Pt(CN)_4]Br_{0.3}\cdot 3H_2O$ (KCP) at 40 K and $(NbSe_4)_3I$ at room temperature (data taken from [Bru75, Ves00]). The lines correspond to a power law behavior with exponents as indicated. The insets show the same data on linear scales.

the periodic potential the valence and conduction bands have the same curvature and the same mass. Both can be expressed in terms of the gap energy $\mathcal{E}_g$.

A strongly frequency dependent optical conductivity has indeed been found in several one-dimensional or nearly one-dimensional semiconductors, such as $K_2[Pt(CN)_4]Br_{0.3}\cdot 3H_2O$ (better known by the abbreviation KCP) [Bru75] and $(NbSe_4)_3I$ [Ves00]. The conductivity of these much studied compounds is displayed in Fig. 13.9. The full line plotted in Fig. 13.9a is the expected ω^{-3} frequency dependence; for $(NbSe_4)_3I$, however, we find a stronger dependence, presumably due to the narrow band, with the bandwidth comparable to the optical frequencies where the measurements were made.

Electron–electron interactions may modify this picture. Such interactions lead, in strictly one dimension, to an electron gas which is distinctively different from a Fermi liquid. In not strictly but only nearly one dimension such interactions lead to a gap, and the features predicted by theory [Gia97] are expected to be observed only at high energies. These features include a power law dependence of the conductivity, with the exponents on ω different from -3 (see above). Experiments on highly anisotropic materials, such as $(TMTSF)_2PF_6$ [Sch98], have been interpreted as evidence for such novel, so-called Luttinger liquids.

A strict singularity at the single-particle gap, of course, cannot be expected for real materials; deviations from one dimensionality, impurities, and the frequency dependence of the transition matrix element involved, all tend to broaden the optical transition. In materials where the band structure is strongly anisotropic, another effect is also of importance. For a highly anisotropic system where the bands do not depend on the wavevector $\mathbf{k}$, but with decreasing anisotropy, the parallel sheets of the band edges become warped, and the gap assumes a momentum dependence. Consequently the gap – defined as the minimum energy separation between both bands – becomes indirect. This can be seen most clearly by contrasting the thermal and optical gaps in materials of varying degrees of band anisotropy with the thermal measurements (i.e. dc resistivity) resulting in smaller gaps as expected and the difference increasing with decreasing anisotropy [Mih97].

13.1.2 Forbidden and indirect transitions

Forbidden transitions, as a rule, have low transition probabilities and therefore are difficult to observe experimentally. A possible example is presented in Fig. 13.2 for PbS, where the characteristic $(\omega - \omega_g)^{3/2}$ dependence of the absorption is seen, together with the $(\omega - \omega_g)^{1/2}$ dependence close to the bandgap; this latter gives evidence for direct allowed interband transitions as discussed above.

Less controversial is the observation of indirect transitions; the reason for this is that – in addition to the different frequency dependence – these transitions are

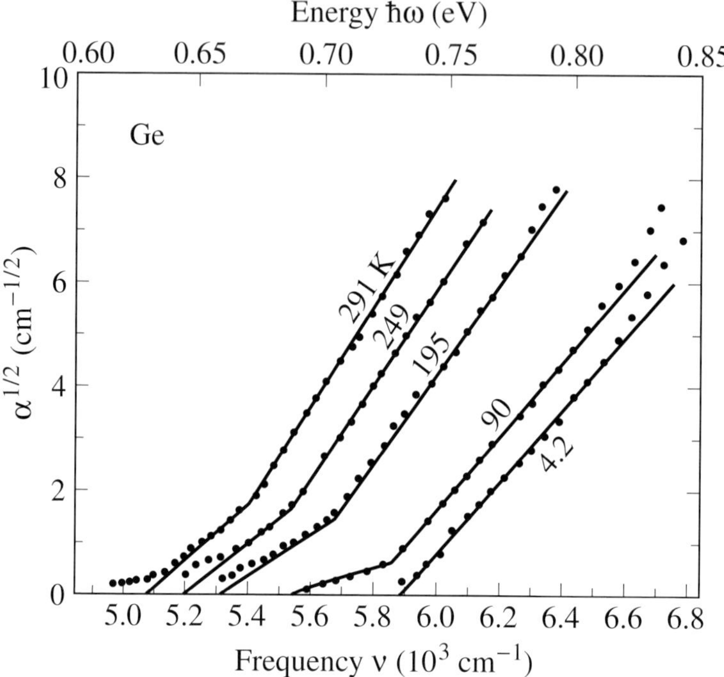

Fig. 13.10. Absorption spectra of germanium in the vicinity of the indirect bandgap at various temperatures as indicated (after [Mac55, Yu96]). The two onsets of absorption identify the two indirect processes; for both the absorption coefficient $\alpha(\omega)$ is proportional to the square of the frequency. The temperature dependence – analyzed in terms of the one-phonon model – provides evidence for the indirect transition.

also strongly temperature dependent. Since the two possible processes discussed in Section 6.4.1 involve different phonon states, two different frequencies where the onset of the absorption occurs are expected. This is nicely demonstrated in Fig. 13.10, in which the data on germanium are taken at different temperatures. The functional dependence of the absorption $(\hbar\omega - \mathcal{E}_g)^2$ is also in accordance with that calculated for indirect transitions in Eq. (6.4.4). In addition to the two absorption processes we have discussed earlier, the low temperature gap, determined by the onset of absorption at 20 K, can be estimated to approximately 0.74 eV.

13.1.3 Excitons

In addition to the transitions involving single-particle states, optical absorption associated with the creation of excitons is observed in various semiconductors. Let us first estimate the energies and spatial extension of excitons, following our discussion in Section 6.5. For a typical semiconductor the static dielectric constant

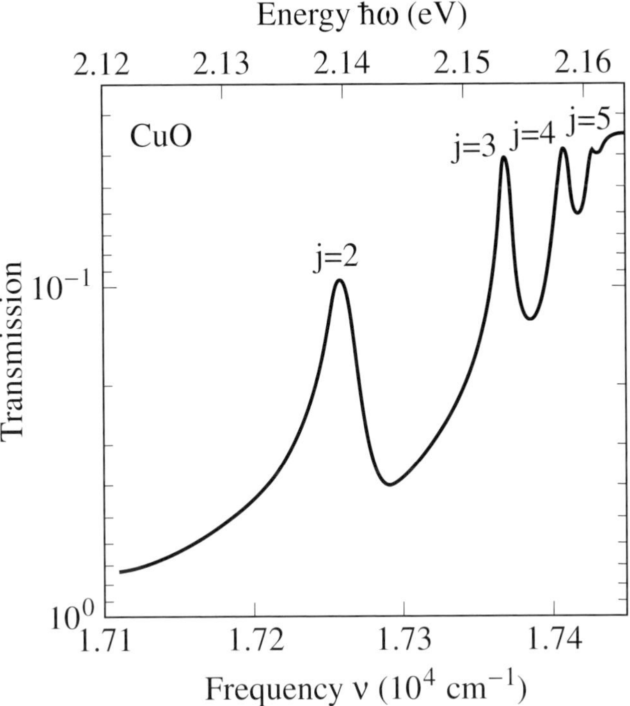

Fig. 13.11. Optical transmission through a 0.1 mm slab of cuprous oxide (CuO) in the frequency region below the single-particle gap measured at $T = 77$ K (after [Bau61]). Note the inverted transmission scale; the peaks correspond to absorption of electrodynamic radiation by the transitions to the various excitation states j.

$\epsilon_1(\omega \to 0) \approx 10$. If we assume that both m_c and m_v are $1/3$ of the free-electron mass, $\mu^{-1} = m_v^{-1} + m_c^{-1}$ is approximately $6/m$, and we arrive at a binding energy – the energy required to break up the exciton – of approximately

$$\mathcal{E}_1^{\text{exc}} = \mathcal{E}_g - \frac{e^4 \mu}{2\hbar^2 \epsilon_1^2} \approx 15 \text{ meV} \quad ;$$

while the spatial extension r^{exc} corresponding to the $j = 1$ state is about 20 Å, significantly larger than the typical lattice constant. The binding energy $\mathcal{E}_j^{\text{exc}}$ decreases significantly for semiconductors with smaller gaps. As $\epsilon_1 \approx 1 + \left(\hbar\omega_p/\mathcal{E}_g\right)^2$, it follows from Eq. (6.3.6) that the exciton energy $\mathcal{E}_j^{\text{exc}}$ depends on the fourth power of the single-particle gap $\mathcal{E}_g$. Well defined and unambiguously resolved exciton lines can hence be observed only in wide bandgap semiconductors, and in Fig. 13.11 the absorption spectrum of CuO is displayed as an example. The quantum numbers $j = 2, 3, \ldots$ are arrived at on fitting the observed energies to the Rydberg series, Eq. (6.5.1).

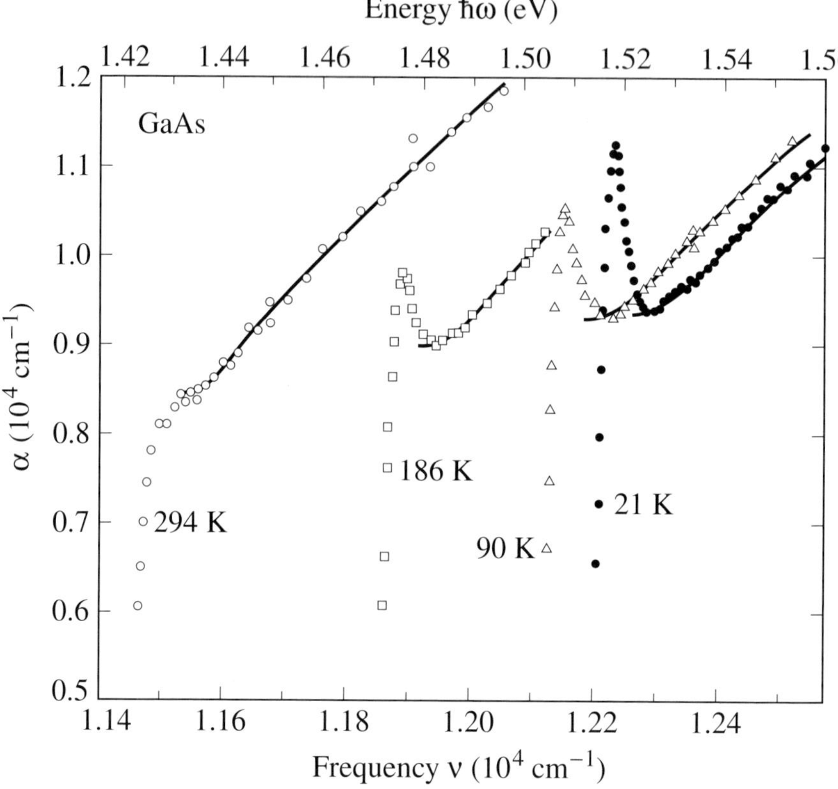

Fig. 13.12. Exciton absorption in gallium arsenide (GaAs) at different temperatures. The full line is a fit to theory [Ell57] (after [Stu62]).

For small semiconductor gaps, the binding energies are also small, and individual transitions from the top of the valence band to the various exciton energy levels (corresponding to different j values) cannot be resolved. The levels, by definition, form a continuum near to the single-particle gap as $j \to \infty$; this continuum then leads to a broad absorption and thus modifies the characteristic onset near $\mathcal{E}_g$. The intensity of this absorption is straightforward to estimate, using results available for the hydrogen atom. Combining the expressions available for the absorption associated with the various Rydberg states with the density of states as $j \to \infty$, we obtain [Yu96]

$$\epsilon_2(\omega) \approx \frac{32\pi^2 e^2 |\mathbf{p}_{cv}|^2 \mu^2}{3m^3 \mathcal{E}_g^2} \quad , \tag{13.1.10}$$

where $|\mathbf{p}_{cv}|$ is the transition matrix element of exciting one state for $j \to \infty$. The above expression – valid at the band edge – corresponds to a frequency independent absorption at $\mathcal{E}_g$, in contrast to the absorption process which goes to zero for singe-particle transitions as the bandgap is approached from above. Calculations can be performed for transitions with energies near to the gap, and in general one finds a broad peak at $\mathcal{E}_g$ due to excitonic absorption which merges with the single-particle absorption as the energy increases [Ell57]. This has been seen in a number of small bandgap semiconductors, and the frequency dependence of the absorption coefficient of GaAs measured at different temperatures is displayed in Fig. 13.12 as an example. The full lines are fits according to an expression derived for the energy dependence of the process [Stu62]; note also the strong temperature dependence and the thermal smearing of the excitonic absorption peak – both natural consequences of the small energy scales involved.

13.2 Effects of interactions and disorder

Optical transitions discussed previously occur for pure and crystalline semicon-ductors where the nature of transition and its dependence on the underlying band structure depends essentially on the existence of well defined **k** states. Now we turn to situations where this strict periodicity is broken, either by the introduction of impurities or by the preparation conditions which prevent the lattice forming. In the former case we talk about impure semiconductors; and in the latter case about amorphous semiconductors.

13.2.1 Optical response of impurity states of semiconductors

Impurity states have been extensively studied by a wide range of optical methods, mainly because of the enormous role these electronic states play in the semiconduc-tor industry. Extrinsic conduction associated with such impurity states is a standard issue for solid state physics, and transport effects which depend on the impurity concentration are also well studied. For small concentrations these impurity states are localized to the underlying lattice, but an insulator–metal transition occurs at zero temperature [Cas89] as the impurity concentration increases.

While there are numerous examples for these impurity effects phenomena, they have been studied in detail for phosphorus doped silicon Si:P. From transport and various spectroscopic studies it is found, for example, that the ionization energy, i.e. the energy needed to promote an electron from the $j = 1$ level of the donor phosphorus to the conduction band, is 44 meV; via the description (6.5.3) in terms of Bohr's model this corresponds to the spatial extension of the impurity states of approximately 25 Å, significantly larger than the lattice periodicity. It is also known that due to valley–orbital interaction the various s and p impurity levels are slightly split; this splitting is about one order of magnitude smaller than the ionization energy. Localized impurity states can be described by the Rydberg formula (6.5.3) for isolated atoms; the energy levels are close to the top of the valence band for p-type impurities or below the bottom of the conduction band for n-doped material, as in the case of Si:P. Because of valley–orbit splitting, the optical absorption shows a complicated set of narrow lines located in the spectral range of a few millielectronvolts, in good accord with our estimations in Section 6.5.2. The results [Tho81] are displayed in Fig. 13.13b. For small impurity concentrations, the sharp lines correspond to transitions between atomic levels; the broad maxima above 400 cm^{-1} correspond to transitions to the conduction band. With increasing impurity density the lines become broadened due to the interaction between neighboring donor states; because of the random donor positions this is most likely an inhomogeneous broadening. Further increase in the phosphorus concentration leads to further broadening; and the appropriate model is that of donor clusters, with progressively more and more impurity wavefunctions overlapping as the average distance between the donors becomes smaller. Because of the large spatial extension of the phosphorus impurity states, all this occurs at low impurity concentrations. A detailed account of these effects for Si:P is given by Thomas and coworkers [Tho81], but studies on other dopants or other materials can also be interpreted using this description.

For increasing impurity concentration N, the dc conductivity and also various thermodynamic and transport measurements provide evidence for a sharp zero temperature transition to a conducting state. In contrast to classical phase transitions with thermal fluctuations, this is regarded as a quantum phase transition. The delocalization of the donor states increases, and at a critical concentration of $N_\mathrm{c} = 3.7 \times 10^{18}$ cm^{-3} in the case of Si:P an impurity band is formed. The optical conductivity [Gay93] for samples which span this transition is displayed in Fig. 13.13a. On the insulating side the conductivity is typical of that of localized states: zero conductivity at low frequencies with a smooth increase to a maximum above which the conductivity rolls off at high frequency according to ω^{-2} (the frequency of this Drude roll-off is nevertheless small because of the small donor concentration). The observed peaks are associated with the transitions already seen for lower donor concentrations (Fig. 13.13b). On the metallic side, the conductivity is of Drude type, with all its ramifications, such as the conductivity which extrapolates to a finite dc value, and absorption which is proportional to the square of the frequency (both also seen in a variety of highly doped semiconductors). Transitions between the various donor states lead to additional complications which can nevertheless be accounted for. While the insulator–metal transition is clearly seen by the sudden change of the conductivity extrapolated to zero frequency $\sigma_1(\omega \to 0)$, the optical sum rule is obeyed both below and above the transition; the integral

$$\int_\mathrm{imp} \sigma_1(\omega) = \frac{\pi N e^2}{2m} = \frac{(\omega_\mathrm{p}^\mathrm{imp})^2}{8} \tag{13.2.1}$$

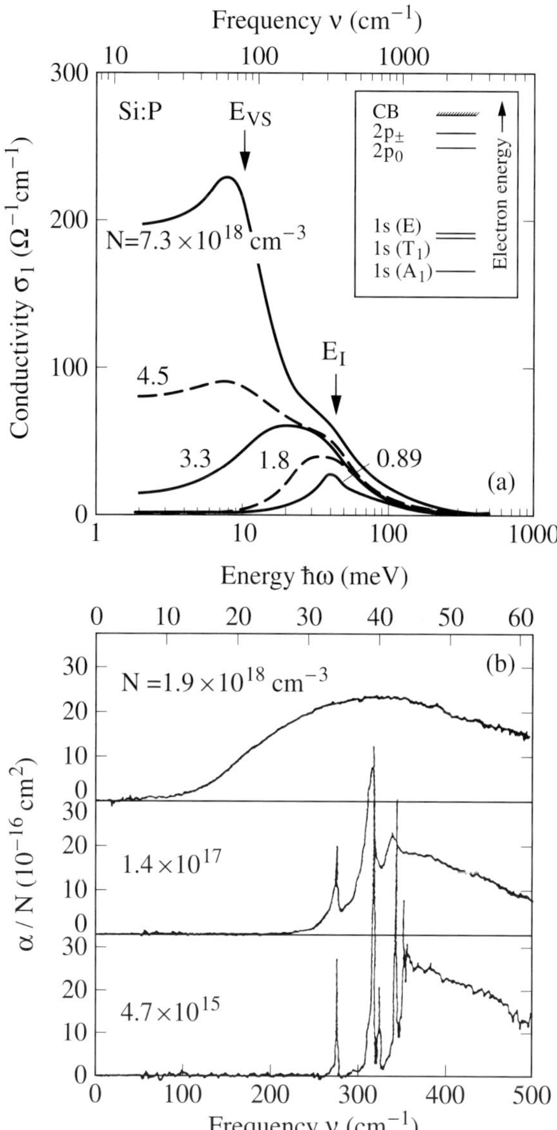

Fig. 13.13. (a) Conductivity $\sigma_1(\omega)$ of different samples of phosphorus doped silicon obtained by Kramers–Kronig analysis of reflection measurements ($T = 10$ K). The inset shows the level scheme of Si:P states in the dilute limit including valley–orbit splitting of $1s$ states: CB = conduction band; E_{I} = ionization energy of the impurity atom; E_{VS} = valley–orbit splitting between the $1s(A_1)$ and the closely spaced $1s(T_1)$ and $1s(E)$ levels (after [Gay93]). (b) Absorption coefficient (normalized to the carrier concentration N) as a function of frequency ω for three donor densities N in samples of Si:P measured at $T \approx 2$ K (after [Tho81]). The curves illustrate regimes of broadening ($N = 1.4 \times 10^{17}$ cm^{-3}) and larger cluster absorption (1.9×10^{18} cm^{-3}).

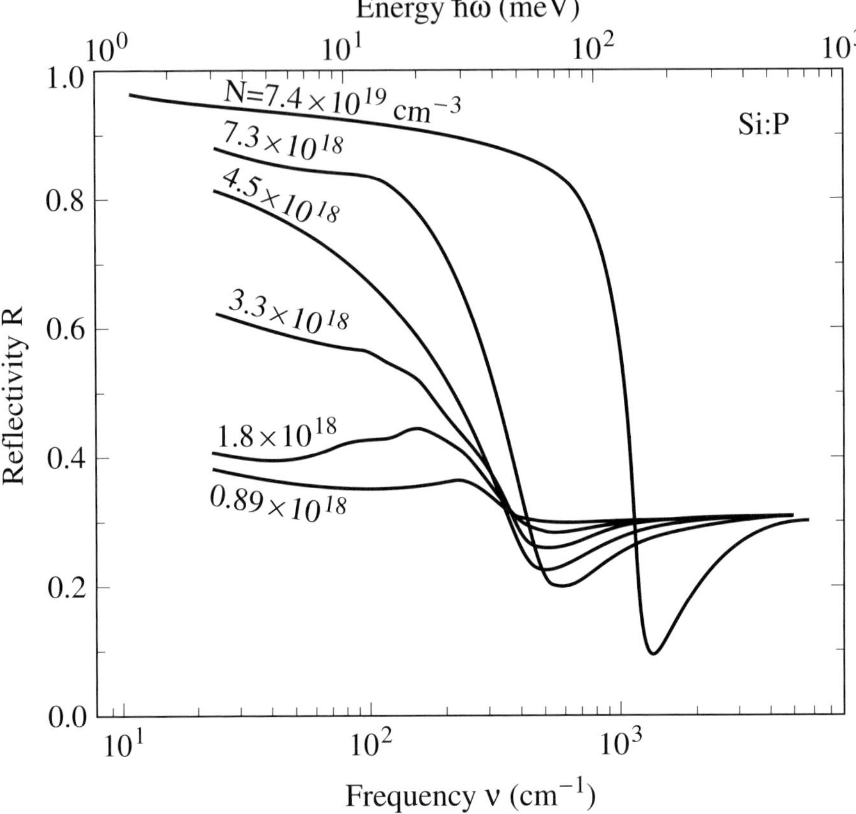

Fig. 13.14. Reflectance spectra $R(\omega)$ of insulating and metallic Si:P samples determined at $T = 10$ K (after [Gay93]). The transition from the insulating state to a metallic conduction state occurs at the impurity concentration of $N_c = 3.7 \times 10^{18}$ cm^{-3}.

is proportional to the donor concentration N at both sides of the critical density N_c. The reflectivity $R(\omega)$ of samples near to the critical concentration, displayed in Fig. 13.14, provides evidence for the metal–insulator transition. While there is no plasma edge for samples below N_c [Gay93], in the conducting state a well defined plasma edge appears in the optical reflectivity; the impurity concentration where this appears is in full agreement with the sum rule given above. Increasing the impurity concentration further, we find that the impurity states merge with the conduction band. This is evidenced, for instance, by the plasma edge which can be analyzed to give the full number of states in the silicon band in contrast to the number of impurity states. Optical studies of the so-called critical region, very close to the critical concentration (studies such as those conducted on the system Nb_xSi_{1-x} and discussed in Section 12.2.3) have not been performed to date.

13.2.2 Electron–phonon and electron–electron interactions

Interactions between electrons and also electron–lattice interactions may lead to a non-conducting state even for a partially filled electron band which, in the absence of these interactions, would be a metal. Several routes to such non-conducting states – which we call a semiconducting state – have been explored; and the emergence of these states depends, broadly speaking, on the relative importance of the kinetic energy of the electron gas, and the interaction energy.

The former is usually cast in the form of

$$\mathcal{H} = \sum_{i,j} t_{ij} \left(c^{+}_{i,\sigma} c_{j,\sigma} + c^{+}_{j,\sigma} c_{i,\sigma} \right) \quad , \tag{13.2.2}$$

where t_{ij} is the transition matrix element between electron states i and j, and the terms in the parentheses describe electron transitions between sites i and j; the spin is indicated by σ. In general this Hamiltonian is treated in the tight binding approximation. Often only nearest neighbor interactions are included, and in this case the transfer integral is t_0. Electron–electron interactions are described, as a rule, by the Hamiltonian

$$\mathcal{H} = U \sum_{i} n_{i,\sigma} n_{i,-\sigma} \quad , \tag{13.2.3}$$

where U represents the Coulomb interaction between electrons residing at the same site. This term would favor electron localization by virtue of the tendency for electrons to avoid each other. When electron–lattice interactions are thought to be important, they are accounted for by the Hamiltonian

$$\mathcal{H} = \alpha \sum_{i,j} \left(u_i - u_j \right) \left(c^{+}_i c_j + c^{+}_j c_i \right) \quad , \tag{13.2.4}$$

where u refers to the lattice position. Here the spin of the electrons can, at first sight, be neglected; α is the electron–lattice coupling constant. The consequences of these interactions have been explored in detail for one-dimensional lattices, with nearest neighbor electron transfer included. In this case the Hamiltonian reads

$$\mathcal{H} = \alpha \sum_{j} \left(u_{j+1} - u_j \right) \left(c^{+}_{j+1} c_j + c^{+}_j c_{j+1} \right) \quad . \tag{13.2.5}$$

For a half filled band both interactions lead – if the interactions are of sufficient strength – to a non-conducting state, the essential features of which are indicated in Fig. 13.15. Coulomb interactions lead – if $U > W$ (the bandwidth given by Eq. (13.2.2)) – to a state with one electron localized at each lattice site, with an antiferromagnetic ground state in the presence of spin interactions. The state is usually referred to as a Mott–Hubbard insulator. The lattice period, which is defined as the separation of spins with identical orientation, is doubled. The

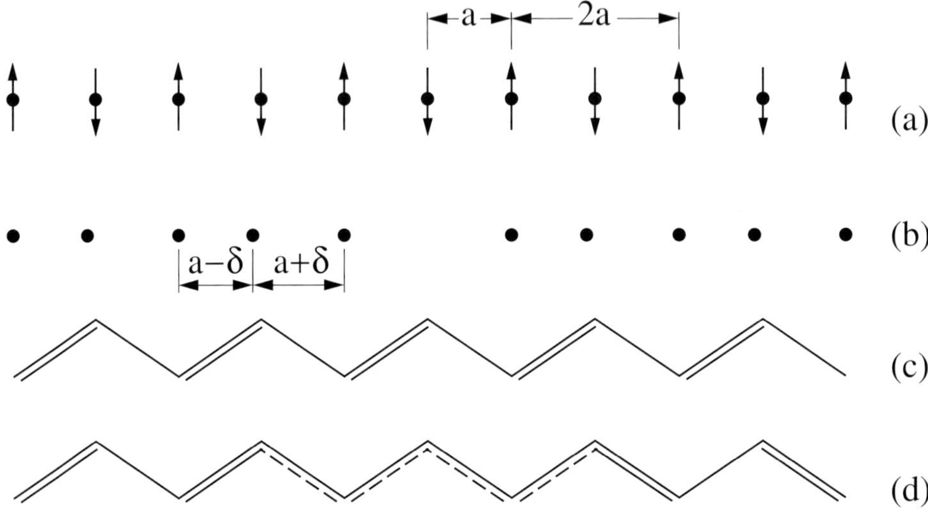

Fig. 13.15. Simple representation of correlation driven semiconducting states for a half filled band. (a) Coulomb correlations lead to the magnetic ground state and (b) electron–lattice interactions to the state with a bond alternation, or (c) to a lattice with a period doubling. In (d) a soliton state is shown, in the usual representation appropriate for the polymer trans-$(CH)_x$.

period of this broken symmetry ground state is $2a$; thus it is commensurate with the lattice period a. The broken symmetry state which arises as a consequence of electron–lattice interactions represents either a displacement of the ionic positions or an alternating band structure, such as shown in Fig. 13.15; such states are referred to as Peierls insulators. Again the structure is commensurate with the underlying lattice period. In contrast to incommensurate density waves which develop in a partially filled band (see Chapter 7), here the order parameter is real. Consequently phase oscillations of the ground state do not occur; instead, non-linear excitations of the broken symmetry states – domain walls or solitons – are of importance.

The single particle gap Δ in the insulating state – for strong interactions – is

$$\Delta \propto U - W \quad ,$$

where U is the strength of the electron–electron or electron–lattice interaction and W is the bandwidth; for weaker interactions Δ is a non-analytic function of the parameters U and W. The optical properties of these states are – in the absence of non-linear excitations – those of a semiconductor with $\sigma_1(\omega)$ depending on the dimensionality of the electron structure, as discussed before. This is fairly straightforward and also comes out of calculations.

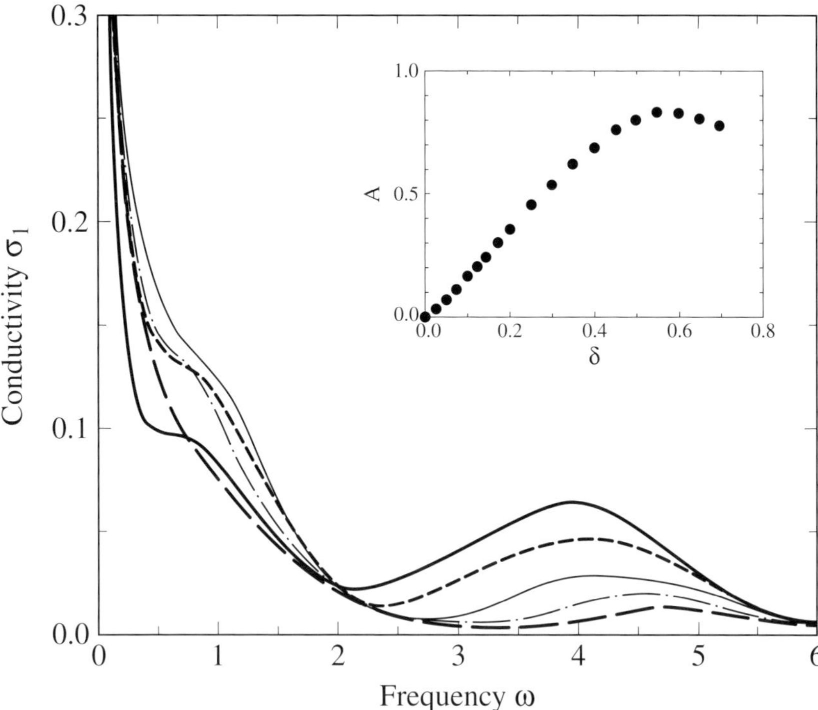

Fig. 13.16. Doping dependence of the optical conductivity $\sigma_1(\omega)$ from quantum Monte Carlo calculations with $U = 4$ on the hypercubic lattice with $t_{ij} = (4d)^{-1/2}$ (after [Jar95]). Going from the thick solid line to the long-dash line the curves correspond to increasing doping levels ($\delta = 0.068$, 0.1358, 0.2455, 0.35, and 0.45); and they lead to an increased conductivity at low, and decreased conductivity at high, frequencies – showing a gradual shift of the spectral weight to the zero energy mode with increasing dopant concentration. The inset shows the evolution of the Drude weight A as a function of doping. The bandwidth is $W = 1$.

More interesting is the question of what spectral features are recovered just before interactions drive the metal into an insulator: does the metallic state look like an uncorrelated Drude metal, or is there a precursor gap feature in the metallic state together with a zero frequency Drude component? Both theory and experiment on materials where Coulomb interactions are important point to the second scenario. The optical conductivity, as derived using the so-called quantum Monte Carlo (QMC) technique in infinite dimensions [Jar95], is shown in Fig. 13.16. Here the different curves correspond to different doping levels: deviations from a strictly half filled band. There is a gap feature with the smooth onset around $\omega = 2$, and in addition a narrow Drude peak. The spectral weight of the peak A shown in the inset is zero for a half filling, and the material is clearly an insulator. Upon doping, the Drude peak assumes a finite intensity, increasing, for small deviations from the half

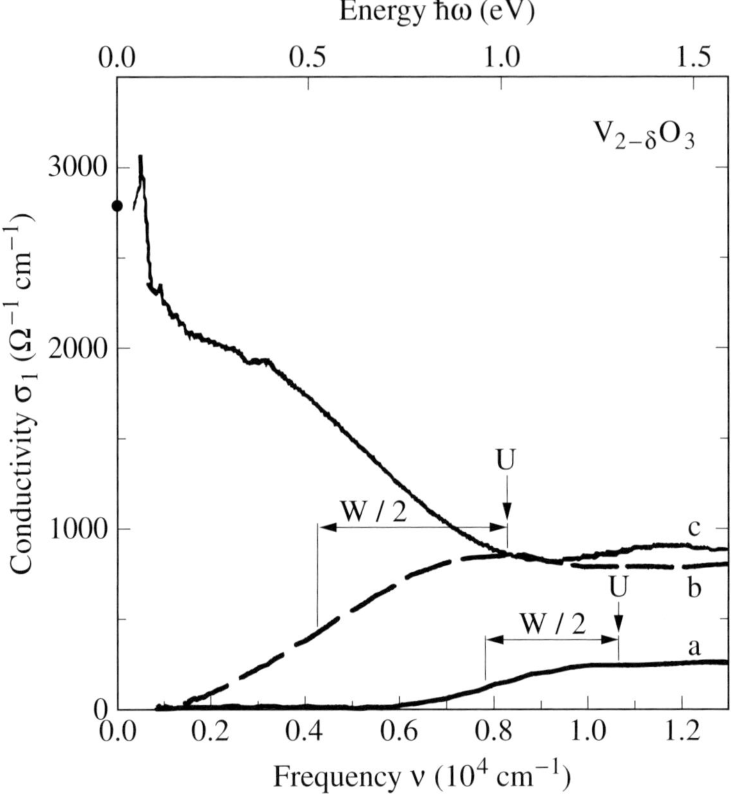

Fig. 13.17. Frequency dependent conductivity of metallic and insulating $V_{2-\delta}O_3$ samples. The full circle indicates the measured dc conductivity. Curve a corresponds to an undoped, $\delta = 0$ sample at $T = 70$ K; curve b is slightly doped, $\delta = 0.013$ at $T = 10$ K; and c corresponds to undoped V_2O_3 in the metallic state at $T = 170$ K (after [Geo96, Tho94]). The parameters of the interaction U and the bandwidth W refer to values obtained by fitting the experimental curves to calculated spectra from the Hubbard model.

filled case, linearly with concentration. The material V_2O_3 is a classic example of the Mott–Hubbard insulating state. Optical experiments, displayed in Fig. 13.17, clearly reveal a gap in the insulating state, which occurs at low temperatures both in the undoped and for the slightly doped material $V_{2-\delta}O_3$ (in the latter case disorder localizes the small number of carriers which, in the absence of disorder, would lead to a Drude response with a small spectral weight A). The parameters W and U are derived through comparison with $\sigma_1(\omega)$ calculated by theory [Geo96, Tho94]. Upon heating, the material undergoes an insulator–metal transition at $T = 150$ K [Wha73], and just on the metallic side (curve c in Fig. 13.17) it is apparent that the measured conductivity is similar to that calculated when a comparison with Fig. 13.16 is made.

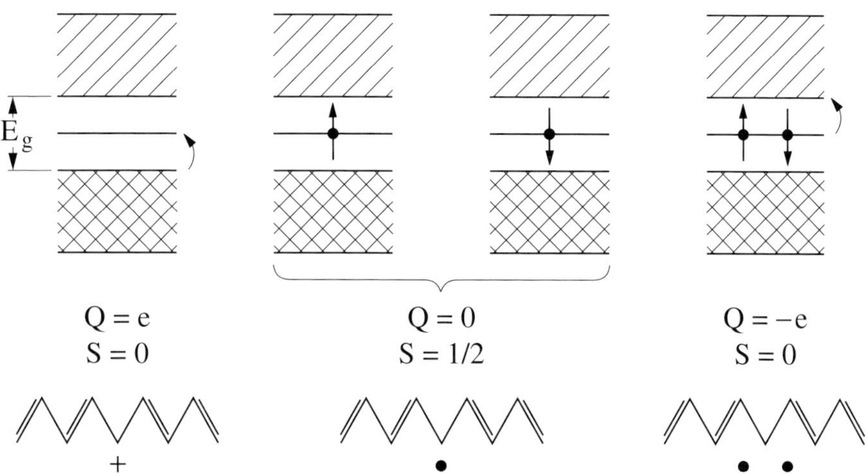

$$Q = e \qquad Q = 0 \qquad Q = -e$$
$$S = 0 \qquad S = 1/2 \qquad S = 0$$

Fig. 13.18. Various charge and spin states of a soliton, showing the localized chemical shorthand description for these delocalized structures. Only interband optical transitions occur for a neutral $Q = 0$ soliton; for charged solitons transitions occur from the valence to the soliton states (for $Q = e$) and from the soliton to the conduction band (for $Q = -e$). Note that in this notation e is positive (after [Hee88]).

Non-linear excitations are most prominent for one-dimensional lattices. If the lattice is fixed at the positions indicated in Fig. 13.15b and c no such non-linear configuration would arise. However, such excitations might be created by rearranging the bond order, for example in a way shown in Fig. 13.15d. The bond arrangement on the right and left is different, corresponding to the different broken symmetry configurations, and the state separating the two is called a soliton. The topological excitation can extend over several lattice constants, and the spatial extension ξ depends on the strength of the electron–lattice interaction with respect to the single-particle bandwidth. The states can be induced thermally or by doping, and have strange spin charge relations; depending on the dopant atoms or molecules [Hee88], these states are summarized in Fig. 13.18. The various soliton states occur in the gap region; calculations [Fel82] for parameters which are appropriate for the polymer trans-$(CH)_x$ – known as polyacetylene – give a soliton energy of approximately $0.6\mathcal{E}_g$. Optical transitions between the soliton and single-particle states can occur if the soliton has a charge e or $-e$, and these transitions are indicated by the arrows. Results of optical experiments on trans-$(CH)_x$ have all the optical signatures of a one-dimensional semiconductor with a gap $\mathcal{E}_g \approx 1.5$ eV, and a prominent midgap excitation upon doping. This can be associated with a soliton state induced by doping – a value in agreement with band structure calculations [Gra83]. The spectral weight of this state is significantly larger than the spectral weight which would be associated with an electron of mass m. The reason for

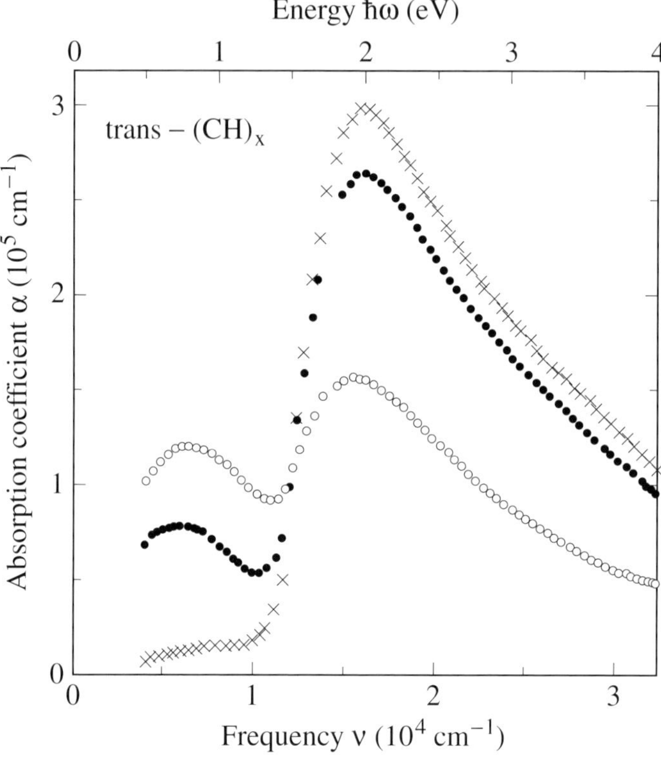

Fig. 13.19. Absorption coefficient $\alpha(\omega)$ of neutral ($\times$) and doped polyacetylene trans-$(CH)_x$ (after [Rot95, Suz80]).

this is that – due to the relatively large spatial extension of the soliton ξ – the spectral weight is enhanced; this enhancement factor is approximately given by ξ/a [Kiv82, Su79]. Of course the total spectral weight is conserved, and upon doping the increased contribution of soliton states is at the expense of the decreased optical intensity associated with electron–hole excitations across the single-particle gap; this can be clearly seen in Fig. 13.19.

13.2.3 Amorphous semiconductors

In amorphous semiconductors the loss of lattice periodicity removes all the effects, which we have associated with long range order. Consequently, signatures of band structure effects such as the van Hove singularities, cannot be observed. Short range interactions, however, prevail, and these set the relevant overall energy scale, such as the width of the bands, and also the magnitude of the (smeared) gap. Of course, the mere notions of a band and a bandgap are not well defined under such

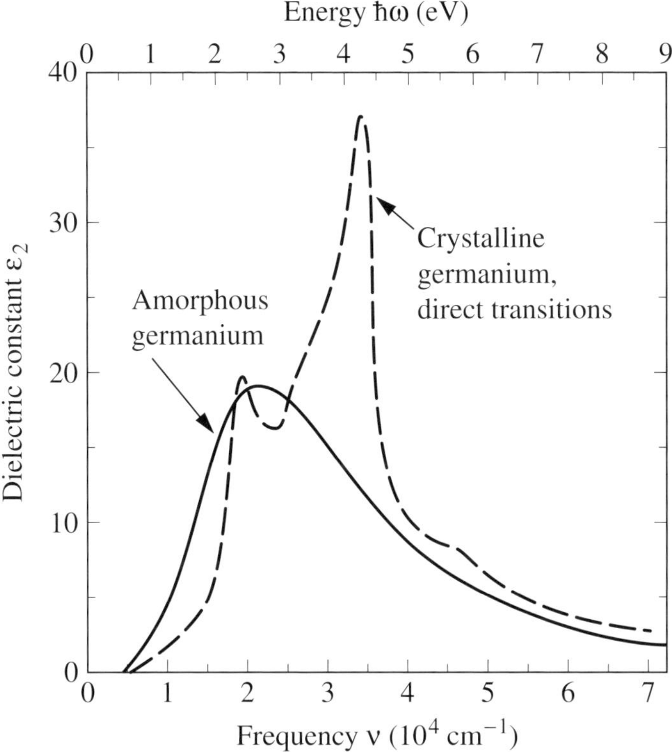

Fig. 13.20. Frequency dependence of the imaginary part of the dielectric constant $\epsilon_2(\omega)$ of crystalline and amorphous germanium (after [Spi70]).

circumstances; they merely refer to energy ranges with large density of energy levels, separated by regions where the density of states is small.

Momentum conservation is, of course, also removed, and thus in the optical absorption the product of the two density of states in the valence and in the conduction band is observed, in contrast to the joint density of states which occurs in Eq. (6.3.4). The transition probability and thus the absorption coefficient is simply given by the density of states in the valence and the conduction bands, and

$$\alpha(\omega) \propto \int D_{\mathrm{v}}(\mathcal{E}) D_{\mathrm{c}}(\mathcal{E} + \hbar\omega)\, \mathrm{d}\mathcal{E} \quad , \tag{13.2.6}$$

where we have assumed that the transition probability is independent of energy. In Fig. 13.20 we compare the imaginary part of the dielectric constant $\epsilon_2(\omega)$ of crystalline germanium – shown in Fig. 13.1c – with $\epsilon_2(\omega)$, which was measured on an amorphous specimen. For amorphous germanium a broad infrared peak is seen which can be associated with a band-to-band transition, with the overall energy scale similar to $\mathcal{E}_{\mathrm{g}}$, which characterizes the crystalline modification. The suppression of the absorption at low frequencies is due to the development of a pseudogap instead of a real gap. There are also arguments [Mot79] which suggest that the electronic states are localized only in the pseudogap region, but not at energies away from the pseudogap where the density of states is large. Measurements of the dc conductivity indeed show a well defined thermal gap; such a transition from localized to delocalized states as a function of frequency, however, is not obvious from the optical data. While there is a more or less well defined energy for the onset of appreciable absorption, called the absorption edge $\mathcal{E}_c$, this edge may or may not correspond to the thermal gap. The subject is further complicated by the fact that a variety of different absorption features are found in different amorphous semiconductors and glasses. Often the absorption displays an exponential behavior above $\mathcal{E}_c$, but often also a power law dependence

$$\alpha(\omega) \propto (\hbar\omega - \mathcal{E}_c)^n \tag{13.2.7}$$

with n ranging from 1 to 3 is found. The reason for these different behaviors is not clear and is the subject of current research, summarized by Mott and Davis [Mot79].

References

[Bau61] P.W. Baumeister, *Phys. Rev.* **121**, 359 (1961)

[Bru75] P. Brüesch, *Optical Properties of the One-Dimensional Pt Complex Compounds*, in: *One-Dimensional Conductors*, edited by H.G. Schuster, Lecture Notes in Physics **34** (Spinger-Verlag, Berlin, 1975), p. 194

[Car68] M. Cardona, *Electronic Properties of Solids*, in: *Solid State Physics, Nuclear Physics and Particle Physics*, edited by I. Saavedia (Benjamin, New York, 1968)

[Cas89] T.G Castner and G. Thomas, *Comments Solid State Phys.* **9**, 235 (1989)

[Coh89] M.L. Cohen and I. Chelikowsky, *Electronic Structure and Optical Properties of Semiconductors*, 2nd edition (Springer-Verlag, Berlin, 1989)

[Ell57] R.J. Elliot, *Phys. Rev.* **108**, 1384 (1957)

[Ext90] M. van Exter and D. Grischkowsky, *Phys. Rev. B* **41**, 12 140 (1990)

[Fel82] A. Feldblum, J.H. Kaufmann, S. Etmad, A.J. Heeger, T.C. Chung, and A.G.MacDiarmid, *Phys. Rev. B* **26**, 815 (1982)

[Gay93] A. Gaymann, H.P. Geserich, and H.V. Löhneysen, *Phys. Rev. Lett.* **71**, 3681 (1993); *Phys. Rev. B* **52**, 16 486 (1995)

[Geo96] A. Georges, G. Kotliar, W. Krauth, and M.J. Rozenberg, *Rev. Mod. Phys.* **68**, 13 (1996)

[Gia97] T. Giamarchi, *Phys. Rev. B* **44**, 2905 (1991); *Phys. Rev. B* **46**, 9325 (1992); Physica B **230–2**, 975 (1997)

[Gob56] G.W. Gobeli and H.Y. Fan, *Semiconductor Research*, Second Quarterly Report (Purdue University, Lafayette, 1956)

[Gra83] P.M. Grant and I. Batra, *J. Phys. (Paris) Colloq.* **44**, C3-437 (1983)

[Gre68] D.L. Greenaway and G. Harbeke, *Optical Properties and Band Structure of Semiconductors* (Pergamon Press, New York, 1968)

[Hee88] A.J. Heeger, S. Kivelson, J.R. Schrieffer, and W.P. Su, *Rev. Mod. Phys.* **60**, 781 (1988)

[Her67] F. Herman, R.L. Kortum, C.D. Kuglin, and J.L Shay, *Energy Band Structure and Optical Spectrum of Several II–VI Compounds*, in: *II–VI Semiconducting Compounds*, edited by D.G. Thomas (Benjamin, New York, 1967), p. 503

[Jar95] A. Jarrell, J.K. Freericks, and T. Pruschke, *Phys. Rev. B* **51**, 11 704 (1995)

[Joh67] E.J. Johnson, *Absorption near the Fundamental Edge. Optical Properties of III–V Compounds*, in: *Semiconductors and Semimetals* **3**, edited by R.K. Willardson and A.C. Beer (Academic Press, New York, 1967)

[Kit96] C. Kittel, *Introduction to Solid State Physics*, 7th edition (John Wiley & Sons, New York, 1996)

[Kiv82] S. Kivelson, T.-K. Lee, Y.R. Lin-Liu, I. Peschel, and L. Yu, *Phys. Rev. B* **25**, 4173 (1982)

[Lee74] P.A. Lee, T.M. Rice, and P.W. Anderson, *Solid State Commun.* **14**, 703 (1974)

[Mac55] G.G. MacFarlane and V. Roberts, *Phys. Rev.* **97**, 1714 (1955); *ibid.* **98**, 1865 (1955)

[Mad96] O. Madelung, ed., *Semiconductors: Basic Data*, 2nd edition (Springer-Verlag, Berlin, 1996)

[Mih97] G. Mihaly, A.Virosztek, and G. Grüner, *Phys. Rev. B* **55**, R13456 (1997)

[Mot79] N.F. Mott and E.H. Davis, *Electronic Processes in Non-Crystalline Materials*, 2nd edition (Clarendon Press, Oxford, 1979)

[Pan71] J.I. Pankove, *Optical Processes in Semiconductors* (Prentice-Hall, Englewood Cliffs, NJ, 1971)

[Phi63] H.R. Philipp and H. Ehrenreich, *Phys. Rev.* **129**, 1550 (1963); *Ultraviolet Optical Properties. Optical Properties of III–V Compounds*, in: *Semiconductors and Semimetals* **3**, edited by R.K. Willardson and A.C. Beer (Academic Press, New York, 1967)

[Pow60] C.J. Powell, *Proc. Phys. Soc.* **76**, 593 (1960)

[Rot95] S. Roth, *One-Dimensional Metals* (VCH, Weinheim, 1995)

[Sch98] A. Schwartz, M. Dressel, G. Grüner, V. Vescoli, L. Degiorgi, and T. Giamarchi, *Phys. Rev. B* **58**, 1261 (1998)

[Spi70] W.E. Spicer and T.M. Donoven, *J. Non Cryst. Solids* **2**, 66 (1970)

[Stu62] M.D. Sturge, *Phys. Rev.* **127**, 768 (1962)

[Su79] W.-P. Su, J.R. Schrieffer, and A.J.Heeger, *Phys. Rev. Lett.* **42**, 1698 (1979)

[Suz80] N. Suzuki, M. Ozaki, S. Etemad, A.J. Heeger, and A.G. MacDiamid, *Phys. Rev. Lett.* **45**, 1209 (1980)

[Tho81] G.A. Thomas, M. Capizzi, F. DeRosa, R.N. Bhatt, and T.M. Rice, *Phys. Rev. B* **23**, 5472 (1981)

[Tho94] G.A. Thomas, D. H. Rapkine, S.A. Carter, A.J. Millis, T.F. Rosenbaum, P. Metcalf, and J.M. Honig, *Phys. Rev. Lett.* **73**, 1529 (1994); G.A. Thomas, D. H. Rapkine, S.A. Carter, T.F. Rosenbaum, P. Metcalf, and J.M. Honig, *J. Low Temp. Phys.* **95**, 33 (1994)

[Ves00] V. Vescoli, F. Zwick, J. Voit, H. Berger, M. Zacchigna, L. Degiorgi, M. Grioni, and G. Grüner, *Phys. Rev. Lett.* **84**, 1272 (2000)

[Wea90] R.C. Weast, ed., *CRC Handbook of Chemistry and Physics*, 70th edition (CRC

Press, Boca Raton, FL, 1990)

[Wha73] D.B. Whan, A. Menth, J.P. Remeika, W.F. Brinkman, and T.M. Rice, *Phys. Rev. B* **7**, 1920 (1973)

[Yu96] P.Y. Yu and M. Cardona, *Fundamentals of Semiconductors* (Springer-Verlag, Berlin, 1996)

[Zem65] Y.N. Zemel, J.D. Jensen, and R.B. Schoolar, *Phys. Rev.* **140A**, 330 (1965)

Further reading

[Kli95] C.F. Klingshirn, *Semiconductor Optics* (Springer-Verlag, Berlin, 1995)

[Mit70] S.S. Mitra and S. Nudelman, eds, *Far Infrared Properties of Solids* (Plenum Press, New York, 1970)

[Mot87] N.F. Mott, *Conduction in Non-Crystalline Materials* (Oxford University Press, Oxford, 1987)

[Mot90] N.F. Mott, *Metal–Insulator Transition*, 2nd edition (Taylor & Francis, London, 1990)

[Nud69] S. Nudelman and S.S. Mitra, eds, *Optical Properties of Solids* (Plenum Press, New York, 1969)

[Par81] G.R. Parkins, W.E. Lawrence, and R.W. Christy, *Phys. Rev. B* **23**, 6408 (1981)

[Pei55] R.E. Peierls, *Quantum Theory of Solids* (Clarendon, Oxford, 1955)

[Phi66] J.C Phillips in: *Solid State Physics* **18**, edited by F. Seitz and D. Thurnbull (Academic Press, New York, 1966)

[Shk84] B.I. Shklovskii and A.L. Efros, *Electronic Properties of Doped Semiconductors*, Springer Series in Solid-State Sciences **45** (Springer-Verlag, Berlin, 1984)

[Tau66] J. Tauc, ed., *The Optical Properties of Solids*, Proceedings of the International School of Physics 'Enrico Fermi' **34** (Academic Press, New York, 1966)

14

Broken symmetry states of metals

The exploration of the electrodynamic response has played an important role in establishing the fundamental properties of both the superconducting state and the density wave states. The implications of the BCS theory (and related theories for density waves) – the gap in the single-particle excitation spectrum, the phase coherence in the ground state built up of electron–electron (or electron–hole) pairs, and the pairing correlations – have fundamental implications which have been examined by theory and by experiment, the two progressing hand in hand. The ground state couples directly to the electromagnetic fields with the phase of the order parameter being of crucial importance, while single-particle excitations lead to absorption of electromagnetic radiation – both features are thoroughly documented in the various broken symmetry states. Such experiments have also provided important early evidence supporting the BCS theory of superconductivity.

There is, by now, a considerable number of superconductors for which the weak coupling theory or the assumption of the gap having an s-wave symmetry do not apply. In several materials the superconducting state is accounted for by assuming strong electron–phonon coupling, and in this case the spectral characteristics of the coupling can be extracted from experiments. Strong electron–electron interactions also have important consequences on superconductivity, not merely through renormalization effects but also leading possibly to new types of broken symmetry. In another group of materials, such as the so-called high temperature superconductors, the symmetry of the ground state is predominantly d-wave, as established by a variety of studies. All these aspects have important consequences for the electrodynamics of the superconducting state.

14.1 Superconductors

Experiments on the electrodynamic properties of the superconducting state include the use of a variety of methods, ranging from dc magnetization measurements of the penetration depth, through the measurements of the radio frequency and microwave losses at frequencies below the superconductivity gap, to the evaluation of the single-particle absorption – and thus the gap – by optical studies. Experiments on the superconducting state of simple metals are a, more or less, closed chapter of this field, with the attention being focused on novel properties of the superconducting state found in a variety of new materials, which – in the absence of a better name – are called unconventional or non-BCS superconductors.

14.1.1 BCS superconductors

We first discuss the experimental results on superconductors for which the pairing is s-wave and the energy gap opens along the entire Fermi surface, with the gap anisotropy reflecting merely the subtleties of the band structure. Furthermore the weak coupling approximation applies, which leads to the BCS expressions for the various parameters, such as Eq. (7.1.15) for the ratio of the gap to the transition temperature. Most of the so-called conventional superconductors, i.e. simple metals with low transition temperatures, fall into this category.

We start with the penetration depth, one of the spectacular attributes of the superconducting state. Experiments are too numerous to survey here and are reviewed in several books [Pip62, Tin96, Wal64]. The London penetration depth derived in Section 7.2.1 can be written as

$$\lambda_\mathrm{L}(T) = \left(\frac{m_\mathrm{b} c^2}{4\pi N_\mathrm{s}(T) e^2} \right)^{1/2} \tag{14.1.1}$$

at finite temperatures, where $N_\mathrm{s}(T)$ is the temperature dependent condensate density and m_b is the bandmass of electrons; the expectation is that m_b is the same as the bandmass determined via the plasma frequency in the normal state. The above expression holds in the limit where local electrodynamics applies and also in the clean limit where the mean free path ℓ is significantly larger than the penetration depth λ_L; in this limit $\lambda_\mathrm{L} = c/\omega_\mathrm{p}$ at zero temperature. As discussed in Section 7.4, corrections to the above expressions are in order if the assumed conditions are not obeyed; we will return to this point later. One common method of measuring the penetration depth $\lambda(T)$ is to monitor the frequency of a resonant structure, part of which contains the material which becomes superconducting. In the majority of cases enclosed cavities are used which operate in the microwave spectral range (see Sections 9.3 and 11.3.3). The resonant frequency is (crudely speaking) proportional to the effective dimensions of the cavity, and this includes the surface layer over which the electromagnetic field penetrates into the material, i.e. the penetration depth. The canonical quantity which is evaluated is the surface reactance X_S, related to the penetration depth by Eq. (7.4.24). The other quantity which is measured is the surface resistance R_S, i.e. the loss associated with the absorption of the electromagnetic field within the surface layer; this loss decreases the quality of the resonance. Both components of the surface impedance $\hat{Z}_S = R_S + iX_S$ can be evaluated, although often only one of the components is accessible in practice.

The temperature dependence of the real and imaginary parts of the surface impedance, measured in niobium at frequencies well below the gap frequency, is displayed in Fig. 14.1a. Both parameters approach their normal state values as the transition temperature $T_c = 9$ K is reached from below. At sufficiently low temperatures both parameters display an exponential temperature dependence, establishing the existence of a well defined superconducting gap. If the gap opens up along the entire Fermi surface, $\exp\{-\Delta/k_B T\}$ is the leading term in the temperature dependence of $R_S(T)$ and $X_S(T)$; the correct expressions as derived from the BCS theory are given by Eqs (7.4.23),

$$R_S(T) \propto \frac{(\hbar\omega)^2}{k_B T} \ln\left\{\frac{4k_B T}{\hbar\omega}\right\} \exp\left\{-\frac{\Delta}{k_B T}\right\} \quad , \tag{14.1.2a}$$

and from the relative difference in the penetration depth (Eq. (7.4.25))

$$X_S(T) \propto 1 + \left(\frac{\pi\Delta}{2k_B T}\right)^{1/2} \exp\left\{-\frac{\Delta}{k_B T}\right\} \tag{14.1.2b}$$

in the regime where both $\hbar\omega$ and $k_B T$ are much smaller than the superconducting gap Δ. A fit of the experimental data to the low temperature part of $R_S(T)$ and $X_S(T)$ gives $2\Delta \approx 3.7 k_B T_c$, suggesting that the material is close to the weak coupling limit, for which $2\Delta/k_B T_c = 3.53$. Often, only the ratio of the superconducting to the normal state impedance is evaluated, and this parameter is compared with the prediction of the BCS theory; the calculations by Mattis and Bardeen [Mat58] presented in Section 7.4.3 provide the theoretical basis. Note that the losses are approximately proportional to the square of the measuring frequency; this is in contrast to the normal state where, according to Eq. (5.1.18), in the Hagen–Rubens regime $R_S(\omega)$ increases as the square root of the frequency. With simple (but somewhat misleading) arguments, we can understand the above expression for the surface resistance as follows: let us assume that the electrons which form the condensate and the electrons thermally excited across the single-particle gap form two quantum liquids which both respond to the applied electromagnetic field. At low temperatures the number of thermally excited electrons is small, and we also assume that for these electrons the Hagen–Rubens limit applies, i.e. $\sigma_1 \gg \sigma_2$, the latter subsequently being neglected. For the condensate, on the other hand, we can safely disregard losses ($\sigma_1 = 0$), and keep only the contribution to σ_2, given by

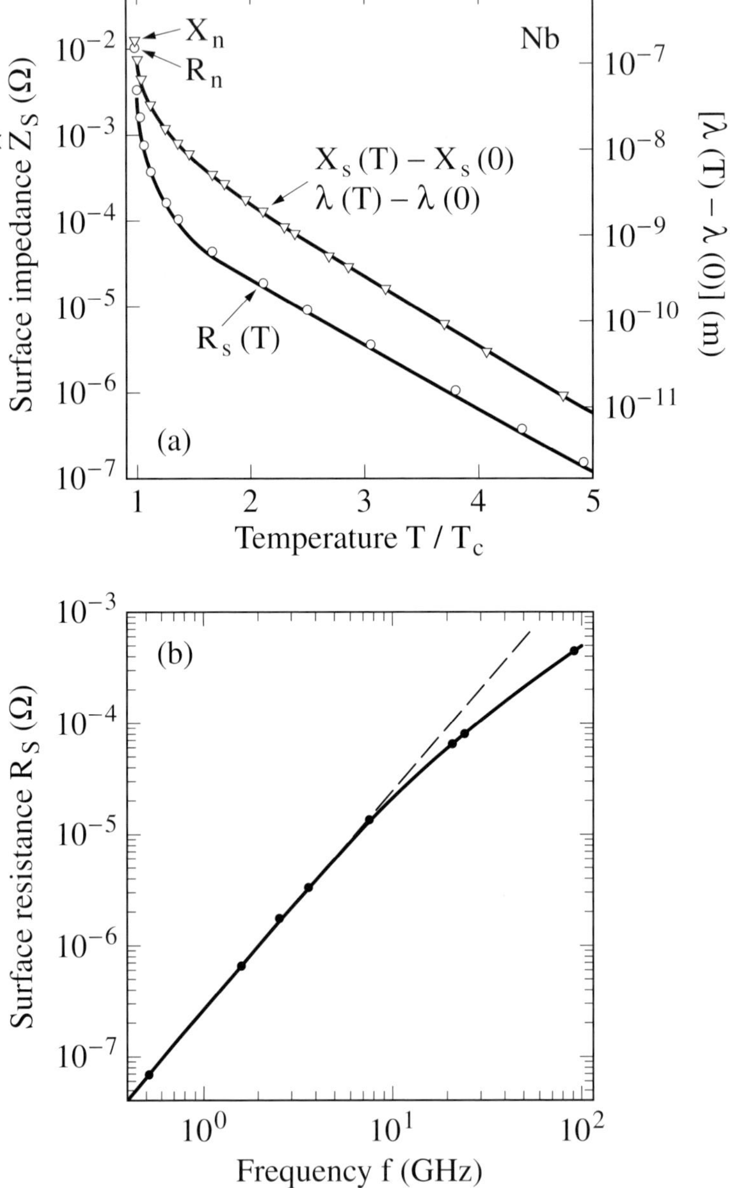

Fig. 14.1. (a) Temperature dependence of the surface reactance $X_S(T)$ and surface resistance $R_S(T)$ in superconducting niobium ($T_c = 9$ K) measured at 6.8 GHz. R_n and X_n refer to the normal state values. The change in surface reactance $X_S(T) - X_S(T = 0)$ (left axis) is proportional to the change in penetration depth $\lambda(T) - \lambda(T = 0)$ (right axis). The fit by the Mattis–Bardeen theory (full lines) leads to a superconducting gap of $2\Delta = 3.7 k_B T_c$. (b) Frequency dependence of the surface resistance $R_S(\omega)$ of niobium measured at $T = 4.2$ K. The dashed line indicates an ω^2 frequency dependence, and the full line is calculated by taking a gap anisotropy into account (after [Tur91]).

Eq. (7.2.11a). Then the conductivity is given by

$$\hat{\sigma}(\omega, T) = \sigma_1(T) + i\sigma_2(\omega) = \sigma_1(T) + \frac{iN_s e^2}{m_b \omega} \quad ,$$

where we have assumed that the concentration of superconducting carriers N_s is close to its $T = 0$ value and therefore the imaginary component is only weakly temperature dependent. Inserting this expression into Eq. (2.3.32a) for the surface resistance leads to $R_S \propto \omega^{3/2}$. Assuming further that the number of thermally excited electrons determines the conductivity (by assuming that the mobility is independent of temperature), we find that the temperature and frequency dependencies can be absorbed into an expression

$$R_S \propto \omega^{3/2} \exp\left\{-\frac{\Delta}{k_B T}\right\} \quad .$$

The assumption of the two independent quantum fluids is not entirely correct, as we have to take into account mutual screening effects; this leads to a somewhat different frequency dependence, which is then derived on the basis of the correct Mattis–Bardeen expression [Hal71]. The temperature dependence of the surface resistance was first measured on aluminum by Biondi and Garfunkel [Bio59] at frequencies both below and above the gap frequency. A strong increase in R_S with increasing frequency was found in the superconducting state. The temperature dependence as well as the frequency dependence observed can be accounted for by the Mattis–Bardeen theory; the experiments conducted at low frequencies and at low temperatures are in good semiquantitative agreement with Eq. (14.1.2a), and as such also provide clear evidence for a superconducting gap. More detailed data on niobium are displayed in Fig. 14.1b and show an $R_S \propto \omega^{1.8}$ behavior, close to that predicted by the previous argument. Note that the frequency dependence as given above follows from general arguments about the superconducting state, and is expected to be valid as long as the response of the condensate is inductive and the two-fluid description is approximately correct.

Other important consequences of the BCS theory are the so-called coherence factors introduced in Section 7.3.1. The case 2 coherence factor leads to the enhancement of certain parameters just below the superconducting transition; and the observation that the nuclear spin–lattice relaxation rate is enhanced [Heb57] provided early confirmation of the BCS theory. The real part of the conductivity σ_1, which is proportional to the absorption, is also governed by the case 2 coherence factors; this then leads to the enhancement of $\sigma_1(T)$ just below the superconducting transition [Kle94]. The temperature dependence of this absorption, calculated for frequencies significantly less than $\Delta/\hbar$, is displayed in Fig. 7.2. When evalu-

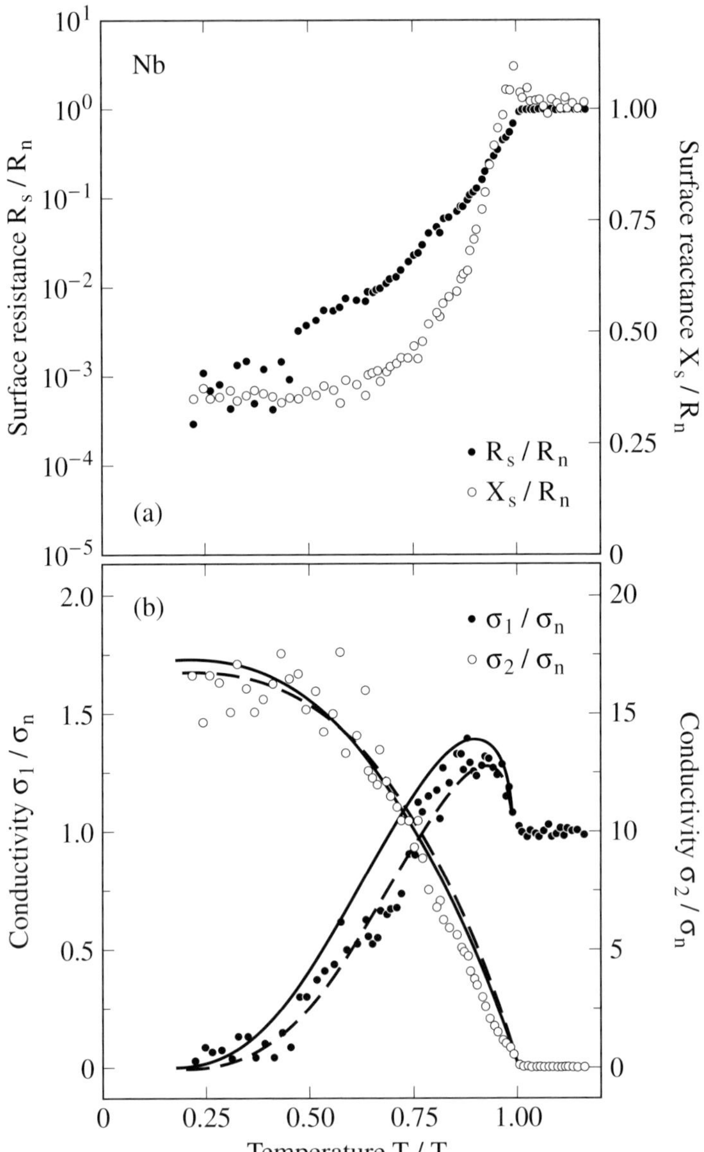

Fig. 14.2. (a) Real component $R_S(T)$ (referring to the logarithmic scale on the left axis) and imaginary component $X_S(T)$ (right axis) of the surface impedance of niobium, normalized to the normal state surface resistance R_n measured at 60 GHz as a function of temperature. (b) Temperature dependence of the components of the conductivity $\sigma_1(T)$ and $\sigma_2(T)$ in niobium calculated from the results of the surface impedance shown in (a). Note the enhancement of $\sigma_1(T)$ just below $T_c = 9.2$ K. The full lines are calculated using the Mattis–Bardeen formalism (7.4.20), and the dashed lines follow from the Eliashberg theory of a strong coupling superconductor (after [Kle94]).

ated from the measured surface resistance and reactance[1] such coherence factors become evident, as shown by the data for niobium, displayed in Fig. 14.2. In the figure the full lines are the expressions (7.4.22) based on the Mattis–Bardeen theory (see Section 7.4.3); the dashed line is calculated by assuming that niobium is a strong coupling superconductor – a notion which will be discussed later.

Another important observation by Biondi and Garfunkel was that for the highest frequencies measured the surface resistance does not approach zero as $T \to 0$, but saturates at finite values, providing evidence that at these frequencies carrier excitations induced by the applied electromagnetic field across the gap are possible. The reflectivity $R(\omega)$ for a superconductor below the gap energy 2Δ approaches 100% as the temperature decreases towards $T = 0$, and this is shown in Fig. 14.3 for the example of niobium nitrate. Using a bolometric technique the absorptivity $A(\omega)$ was directly measured and clearly shows a drop below the gap frequency; this also becomes sharper as the temperature is lowered [Kor91]. The fringes in both data are due to multireflection within the silicon substrate. There is an excellent agreement with the prediction by the Mattis–Bardeen formalism (7.4.20), the consequences of which are shown in Fig. 7.5. The full frequency dependence of the electrodynamic response in the gap region has been mapped out in detail for various superconductors by Tinkham and coworkers [Gin60, Ric60], and in Fig. 14.4 we display the results for lead, conducted at temperatures well below T_c. The data, expressed in terms of the frequency dependent conductivity $\sigma_1(\omega)$ and normalized to the (frequency independent) normal state value σ_n, have been obtained by measuring both the reflectivity from and transmission through thin films. There is a well defined threshold for the onset of absorption which defines the BCS gap; the conductivity smoothly increases with increasing frequencies for $\omega > 2\Delta/\hbar$, again giving evidence for the case 2 coherence factor as the comparison with Fig. 7.2 clearly demonstrates. The frequency for the onset of conductivity leads to a gap $2\Delta/\hbar$ of approximately 22 cm^{-1} in broad agreement with weak coupling BCS theory, and the full line follows from the calculations of Mattis and Bardeen – as before, the agreement between theory and experiment is excellent. The data also provide evidence that the gap is well defined and has no significant anisotropy; if this were the case, the average overall orientations for the polycrystalline sample (such as the lead film which was investigated) would yield a gradual onset of absorption. With increasing temperature $T < T_c$ the normal carriers excited across the gap become progressively important, causing an enhanced low frequency response. This is shown in Fig. 14.5, where $\sigma_1(\omega)$ and $\sigma_2(\omega)$, measured directly on a thin niobium film at various temperatures using a Mach–Zehnder interferometer, are displayed [Pro98]. Similar results have been

---

[1] Note that both parameters $R_S(T)$ and $X_S(T)$ have to be measured precisely in order to evaluate the conductivity $\sigma_1(T)$ using Eqs (2.3.32).

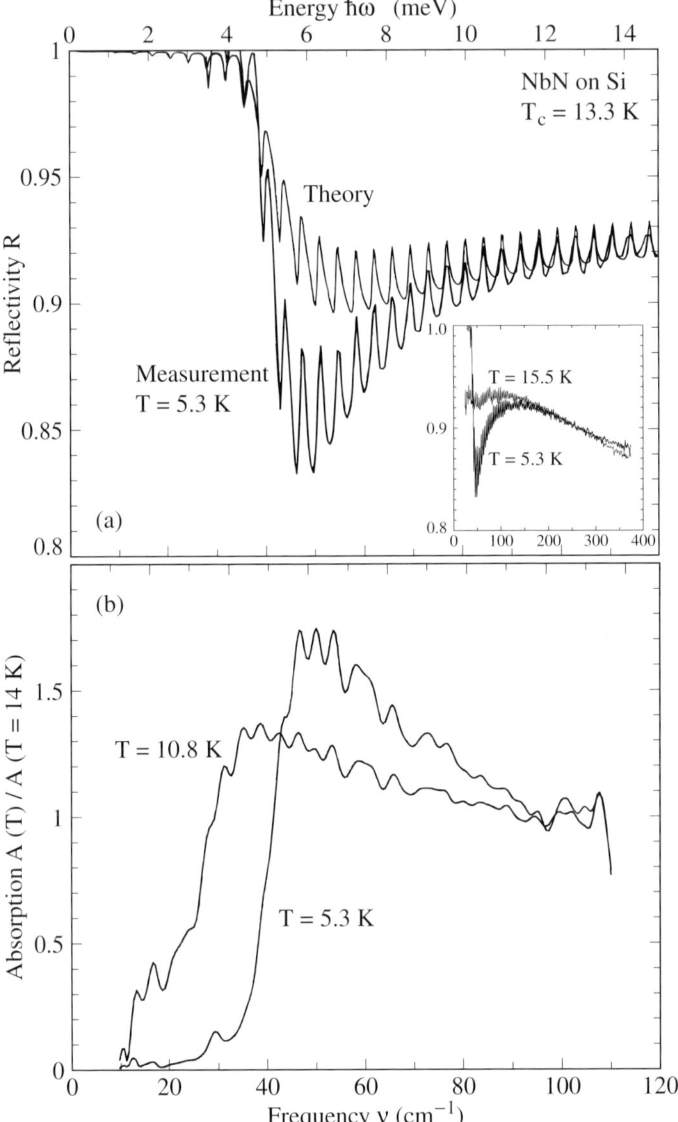

Fig. 14.3. (a) Frequency dependence of the reflectivity $R(\omega)$ of a 250 nm NbN film. The interferences are caused by the 391 μm Si substrate. Note that the reflectivity is (within experimental error) 100% for frequencies below 40 cm^{-1}; compare with Fig. 7.5. The calculation is done using the Mattis–Bardeen theory, Eqs (7.4.20). The inset shows the frequency dependence of the reflectivity above and below T_c over a wider frequency range. (b) Frequency dependent absorptivity $A(\omega)$ of the same sample for two different temperatures $T < T_c$ normalized to the absorptivity just above the transition temperature $T_c = 13.3$ K (after [Kor91]).

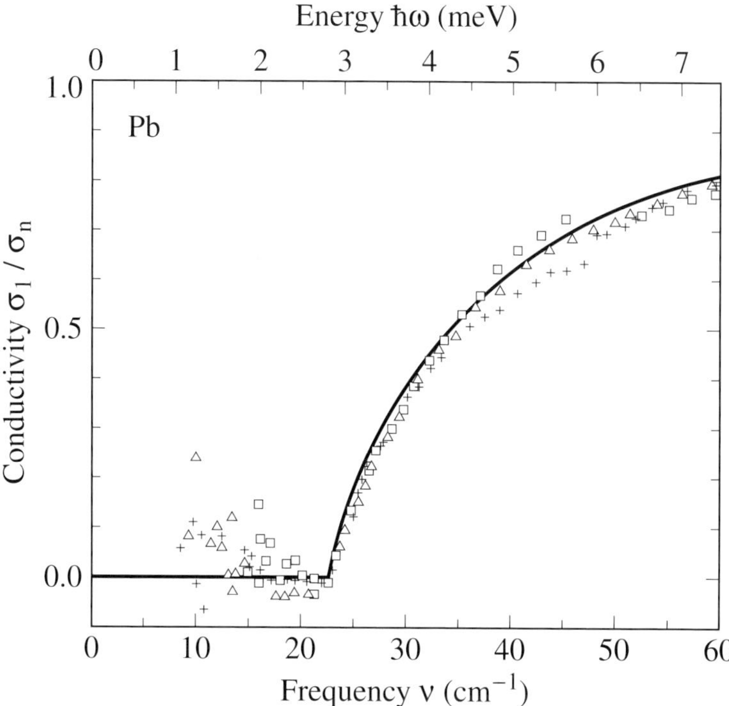

Fig. 14.4. Frequency dependence of the conductivity $\sigma_1(\omega)$ in different lead films, normalized to the normal state conductivity σ_n. The results are obtained from reflection and transmission measurements at $T = 2$ K. The full line is calculated using the Mattis–Bardeen theory with a gap of $2\Delta/\hbar = 22.5$ cm^{-1} (after [Pal68]).

obtained by time domain spectroscopy (cf. Section 10.2), as shown in Fig. 10.5. The data are fitted by the Mattis–Bardeen equations (7.4.20) to extract the temperature dependence of the superconducting gap.

As discussed in Section 7.4, various length scales and their relative magnitude determine the electrodynamic properties of the superconducting state. Let us first estimate these length scales; this can be done for each material in question by using the optical properties of the normal state together with the Fermi velocity and the superconducting gap. For the plasma frequency of a typical metal of $\hbar\omega_p = 10$ eV, the London penetration depth $\lambda_L = c/\omega_p$ is of the order of 100 Å at zero temperature. The second length scale, the correlation length $\xi_0 = \hbar v_F/\pi\Delta$, is at $T = 0$ for a typical Fermi velocity of 5×10^7 cm s^{-1} and for a superconducting energy gap of 1 meV, about 1000 Å. Finally the mean free path $\ell = v_F\tau$, where the relaxation time τ (the parameter which depends on the purity of the specimen, and as such can be easily varied for a metal) is assumed to be the same in both the normal and the superconducting states, is approximately 500 Å with the previous

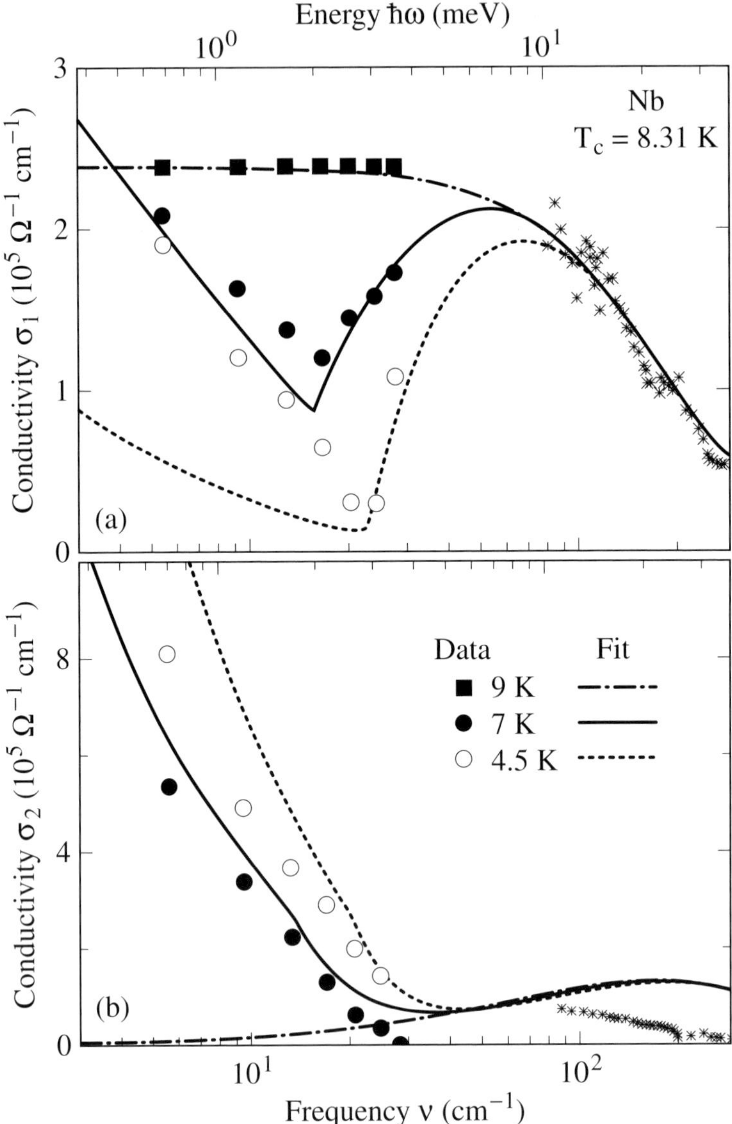

Fig. 14.5. Frequency dependence of the real and imaginary parts of the conductivity, $\sigma_1(\omega)$ and $\sigma_2(\omega)$, in niobium at various temperatures. The transmission through a 150 Å thick film on sapphire was measured by a Mach–Zehnder interferometer; the stars were obtained by reflection measurements. The full lines are calculated using the theory of Mattis and Bardeen (after [Pro98]).

value of v_F and for a relaxation time of 10^{-13} s, typical of good metals at low temperatures. The three length scales for clean metals are therefore of the same order of magnitude.

The most obvious parameter, and therefore also the quantity experiments have the easiest access to, is the mean free path and correlation length dependent penetration depth which we call λ – in contrast to the London penetration depth λ_L, which is measured only in the clean limit and for $\xi_0 \ll \lambda_L$ and also for $\xi_0 \ll \ell$. Let us start by considering the effects related to the finite mean free path ℓ. If this length scale is smaller than the coherence length ξ_0 (i.e. the spatial extension of the Cooper pairs), the effectiveness of pair formation is reduced, and thus the penetration depth is increased since λ is inversely related to the number of condensed pairs. In Section 7.4.1 we derived an appropriate expression based on spectral weight arguments (the Tinkham–Ferrell sum rule) in the limit when ℓ is short; Eq. (7.4.8) can be extended to a long mean free path (for which we should recover the London penetration depth) by the interpolation expression (7.4.18).

The above applies when $\xi_0 > \lambda$. As discussed in Section 7.4, the penetration depth plays a similar role to the skin depth in the case of normal metals, and – through the ineffectiveness concept – the coherence length takes the role of the mean free path. Therefore the inequality above corresponds to what is called the normal skin effect in metals. In the opposite, i.e. the anomalous, limit Eq. (7.4.16) applies, in full analogy to the expression of the skin depth in metals. Again we can postulate an interpolation of the form

$$\lambda_{\text{eff}} = \lambda_L \left(1 + B \frac{\xi_0}{\lambda_L} \right)^{1/3} \quad ,$$

where the constant B is of the order of unity. The validity of these considerations can be explored by measuring the penetration depth in various materials of different purity. In Table 14.1 we have collected the values of λ obtained for some typical metals, together with the parameters, such as the London penetration depth, the coherence length, and the mean free path, all estimated using the procedures discussed before. The values calculated for the penetration depth λ are in excellent agreement with the measured data, both for the corrections due to short mean free path (7.4.18) and large coherence length (7.4.16).

The experimental results, displayed in Figs 14.4 and 14.5 have been obtained on lead and niobium films with thicknesses less than the penetration depth λ_L and less than the coherence length ξ_0. In this case the electric field within the film is independent of position and thus non-local effects can be neglected. These effects, however, become important for experiments on bulk samples; and here the full $\mathbf{q}$ dependence of the electromagnetic response has to be included. This was done by using expressions for the $\mathbf{q}$ dependent surface impedance as derived by Reuter and Sondheimer [Reu48] and by Dingle [Din53], and the corresponding calculations for σ_1/σ_n were performed by Miller [Mil60]. Of course these issues are also related to the behavior of the superconducting state in the presence of a magnetic field;

Table 14.1. *Values of the actual penetration depth λ as calculated and as measured, the London penetration depth λ_{L}, the mean free path ℓ, the coherence length ξ_0, the transition temperature T_{c}, and the gap ratio $2\Delta/k_{\mathrm{B}}T_{\mathrm{c}}$ of several metals. Data taken from [Bar61, Kle94, Tur91].*

Material	T_{c} (K)	$2\Delta/k_{\mathrm{B}}T_{\mathrm{c}}$	λ_{L} (Å)	ξ_0 (Å)	ℓ (Å)	λ (exp.) (Å)	λ (theory) (Å)
Al	1.18	3.4	157	16 000		500	530
Sn	3.7	3.5	355	2 300		510	560
Pb	7.2	4.1	370	830	2000	390	480
Nb	9.0	3.7	330	380	200	440	450

studies under such conditions confirm the fundamental differences between type I ($\lambda_{\mathrm{L}} < \xi_0$) and type II ($\lambda_{\mathrm{L}} > \xi_0$) superconductors.

14.1.2 Non-BCS superconductors

By now a wide range of superconductors have been discovered where the superconducting state cannot be described by the simplest BCS calculations, i.e. the solution which relies on two assumptions: weak electron–phonon coupling and s wave symmetry of the condensate wavefunction.

Materials with relatively high transition temperatures clearly fall into the regime of what is called strong coupling; for a small coupling constant the transition temperature is (exponentially) small. In other superconductors, which are mainly based on CuO and go by the name high temperature superconductors, the order parameter is dominantly d wave, while in the so-called heavy fermion superconductors, the unusual normal state properties and also the coexistence or progressive development of several superconducting phases signal the possibility of novel superconducting states. There are indications that electron–phonon coupling is not solely responsible for superconductivity.

In metals where the electron–phonon interactions are not particularly weak, the BCS theory has to be modified and details of the electron–phonon interaction have to be included. The appropriate theory has been worked out by Eliashberg [Eli60] and is reviewed among others in [Sca69]; strong electron–phonon interaction results in a renormalized (and complex) gap. The real part corresponds to the mass enhancement

$$\frac{m^*}{m} = 1 + \lambda_{\mathbf{P}} \quad , \tag{14.1.3}$$

with the value given by

$$\lambda_{\mathbf{P}} = 2 \int_0^{\infty} \frac{\alpha^2(\omega_{\mathbf{P}}) F(\omega_{\mathbf{P}})}{\omega_{\mathbf{P}}} \, d\omega_{\mathbf{P}} \quad ;$$

i.e. Eq. (12.2.12), which we encountered before when the electron–phonon interaction in metals was discussed. There are several consequences: first, the magnitude of the gap is enhanced over what is predicted by the BCS theory $2\Delta/k_{\mathrm{B}}T_{\mathrm{c}} = 3.53$; approximate expressions which relate the gap to the enhancement [Mit84] are in good agreement with the experimental findings. Second, the selfconsistent gap equation is modified, and consequently the temperature dependence of the various quantities, such as the penetration depth, surface impedance, and the coherence effects are also modified; this was experimentally verified. Also seen is a modified frequency dependence of the absorption which can be related to the phonon spectrum through the Holstein process discussed in Section 12.2.2. Because of the existence of a single-particle gap and the particular behavior of the density of states near to the gap, the Holstein process of phonon emission is modified. Instead of Eq. (12.2.15), we find that in the superconducting state the absorptivity is

$$A(\omega) \propto \frac{1}{\omega} \int_0^{\omega-2\Delta} d\omega_{\mathbf{P}} \, (\omega - \omega_{\mathbf{P}}) \left(1 - \frac{4\Delta^2}{(\hbar\omega - \hbar\omega_{\mathbf{P}})^2} \right) \alpha^2 F(\omega_{\mathbf{P}}) \quad . \qquad (14.1.4)$$

This result can be arrived at by following the same procedure as for the normal state [All71].

The frequency dependent absorption can be measured both above and below T_{c} and thus the difference between the absorption in the normal state and superconducting state can be evaluated with high precision. The advantage then is that many of the unknown parameters drop out by such normalization, and consequently the above expression can be analyzed in detail. In addition, the same information is provided by tunneling experiments [McM69] conducted in the superconducting state. The factors $\alpha^2 F(\omega)$ as evaluated from tunneling and from optical measurements of lead are displayed in Fig. 14.6. The two are in good agreement and give a reliable description of this so-called Eliashberg factor. This factor can be inserted into the relevant expressions of the transition temperature

$$T_{\mathrm{c}} \approx \frac{\theta_{\mathrm{D}}}{1.45} \exp \left\{ \frac{1.04(1 + \lambda_{\mathbf{P}})}{\lambda_{\mathbf{P}} - \mu^*(1 + 0.62\lambda_{\mathbf{P}})} \right\} \qquad (14.1.5)$$

to establish (if it agrees with experiment) the mechanism of Cooper pair formation [McM68]. Here θ_{D} is the Debye temperature and μ^* refers to the effective screened Coulomb interaction. Needless to say, the procedure works equally well for other types of bosonic fluctuations which may mediate superconducting pairing; optical experiments can potentially be extremely useful in clarifying such coupling mechanisms.

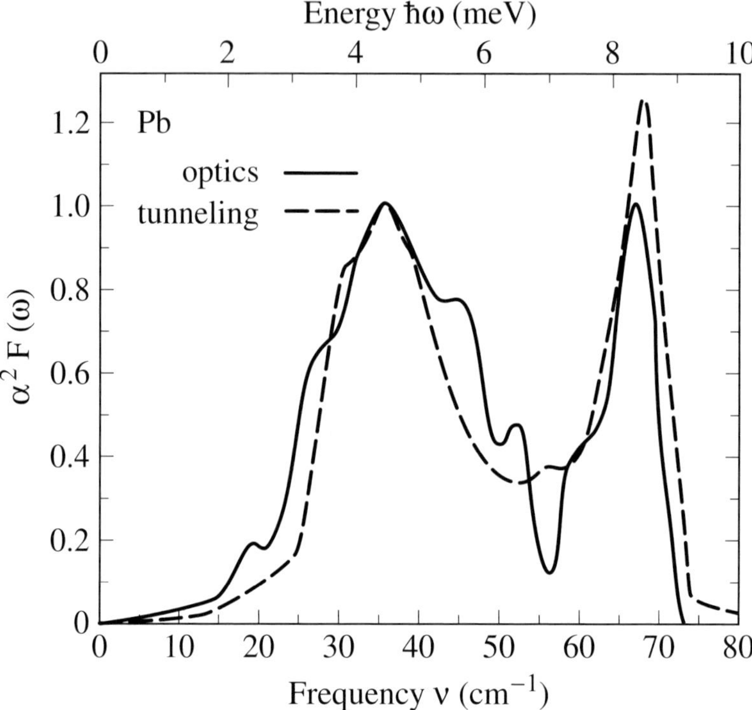

Fig. 14.6. Weighted phonon density of states of superconducting lead obtained from far-infrared reflectivity measurements at 1.2 K (solid curve) using an inversion routine based on Allen's theory [All71]. The dashed curve is the tunneling result [McM69]; both results are normalized to unity at the peak near 35 cm^{-1} (after [Far76]).

The normal state properties of materials with strong electron–electron interactions have been discussed in Section 12.2.2. In the limit of low frequencies these interactions lead to a strongly enhanced effective mass m^* and relaxation time τ^*, also the plasma frequency $\omega_\mathrm{p}^* = (4\pi N e^2/m^*)^{1/2}$ is renormalized; since at high frequencies (and high temperatures $T > T_\mathrm{coh}$) these parameters are not influenced by electron–electron interactions, the unrenormalized values are recovered. Whether such interactions play an important role for the parameters which characterize the superconducting state depends on whether the energy scale, which determines the (smooth) crossover to the normalized low frequency response, is larger or smaller than the superconducting gap; the former is given by the so-called coherence temperature T_coh that we have encountered in Section 12.2.2. In heavy fermion superconductors the transition temperature is small, and $2\Delta < k_\mathrm{B} T_\mathrm{coh}$; consequently, the Cooper pairs are formed by the strongly renormalized electron states. Thus in the clean limit we obtain

$$\lambda_{\mathrm{L}} = \frac{c}{\omega_{\mathrm{p}}^*} = \left(\frac{m^* c^2}{4\pi N e^2} \right)^{1/2} \quad , \tag{14.1.6}$$

and the penetration depth is significantly enhanced by electron–electron interactions. This enhancement can also be derived using a two-component Fermi-liquid theory [Var86]. The above equation – with m^* values evaluated from the thermodynamic or electrodynamic response of the normal state; see Fig. 12.14 – accounts well for the extremely large penetration depth values found [Hef96] in various heavy fermion superconductors. The above analysis assumes that the materials are in the clean limit – a by no means obvious assumption. If this is not the case, the Tinkham–Ferrell sum rule arguments (7.4.6) should be used with the frequency dependent mass and relaxation rate included.

Mechanisms different from the well known phonon mediated superconductivity are also possible, and they may lead to pairing symmetry different from the s wave symmetry we have considered before. Higher momentum pairing implies momentum dependent gaps (in analogy to the atomic wavefunctions corresponding to different quantum numbers) with zeros expected for certain $\mathbf{k}$ values in momentum space and a possible change in phase. If this occurs, the temperature dependencies of various quantities are modified, which can be tested by experiment; also the influence of impurities is modified, for example. This scenario happens for the so-called high temperature superconductors, such as $YBa_2Cu_3O_{7-\delta}$ (here δ refers to the varying oxygen content which corresponds to a doping of charge carriers) where the pairing has d wave symmetry. The momentum dependence of the gap is drawn in Fig. 14.7, together with the momentum dependence of the superconducting gap as expected for isotropic and anisotropic s wave pairing where the gap does not have spherical symmetry, but its momentum dependence reflects the underlying crystal symmetry; in both cases there is a minimum gap value $\Delta_{\min}$ (which might be zero).

If the gap is anisotropic with a finite minimum $\Delta_{\min}$, the temperature dependence of the penetration depth is led by an exponential term

$$\frac{\lambda(T) - \lambda(0)}{\lambda(0)} \propto \exp\left\{ -\frac{\Delta_{\min}}{k_{\mathrm{B}} T} \right\} \quad . \tag{14.1.7}$$

Higher momentum pairing, such as a pairing with d wave symmetry, leads to several novel features. First, because the modes in the gap extend to zero energy, the exponential temperature dependence turns into a power law behavior, and

$$\frac{\lambda(T) - \lambda(0)}{\lambda(0)} \propto \left(\frac{T}{T_0} \right)^n \quad , \tag{14.1.8}$$

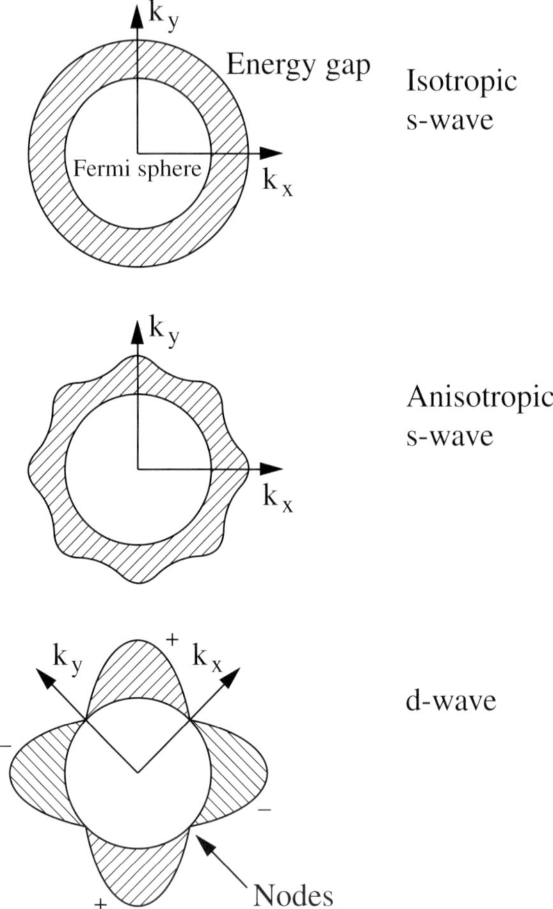

Fig. 14.7. Momentum dependence of the superconducting gaps for isotropic and anisotropic *s* wave and for *d* wave pairing. Note that for the last case the gap function changes sign by going across the nodes.

where the exponent n depends upon whether the nodes are points or lines in **k** space; for line nodes, $n = 1$. Second, impurities introduce states at very small energies and thus allow scattering for the quasi-particles; the electromagnetic response is therefore significantly modified by impurities – much more than for *s* wave pairing. Penetration depth measurements on the model compound of the high temperature superconductor $YBa_2Cu_3O_{7-\delta}$ give a linear temperature dependence and provide evidence that lines of nodes exist in this type of superconductor. Impurity states are supposed to turn the above behavior into a quadratic temperature dependence, again in agreement with experiments; the current status of the field is reviewed by [Bon96].

Needless to say, the rapid advances in the area of unusual superconductors may lead to other pairing symmetries or superconducting phases with competing order parameters. The exploration of the electrodynamic response is likely to play an important role.

14.2 Density waves

The electrodynamic properties of density waves have many features common to those that describe superconducting state. In both types of condensates, the collective mode couples to the electromagnetic field, and there is also a single-particle gap with single-particle excitations occurring for photon energies $\hbar\omega > 2\Delta$. There are, however, several important differences. As discussed in Section 7.3, the coherence factors appear in different combinations in the two types of condensates, leading to significant differences as far as the conductivities and spectral weights are concerned. Also, the effective mass of the charge density wave is large, and the sum rule arguments for superconductors have to be modified. In addition, due to lattice imperfections, the collective mode is pinned to the underlying lattice, shifting the mode to finite frequencies.

14.2.1 The collective mode

For a perfect crystal and for an incommensurate density wave, the collective mode contribution occurs at $\omega = 0$ due to the translational invariance of the ground state. In the presence of impurities, however, this translational invariance is broken, and the collective modes are tied to the underlying lattice due to interactions with impurities [Gru88, Gru94a]. To first order this can be described by an average restoring force $K = \omega_0^2 m^*$. The interaction between the collective mode and the lattice imperfections, impurities etc. may also lead to a finite relaxation time τ^*. With these effects the equation of motion for the phase becomes

$$\frac{\mathrm{d}^2\phi}{\mathrm{d}t^2} + \frac{1}{\tau^*}\frac{\mathrm{d}\phi}{\mathrm{d}t} + \omega_0^2\phi = -\frac{Ne}{m^*}E(t) \quad . \tag{14.2.1}$$

This equation is somewhat different from that derived in Eq. (7.2.12). Here we have neglected the term which is associated with the spatial deformation of the mode, and we have included a damping term – which leads to dissipation – together with a restoring force term – which shifts the response from zero to finite frequencies. Also, $\mathbf{k}_F$ has been replaced by $k_F = \pi/N$, a relation valid in one dimension. This differential equation is solved with a response $\hat{\sigma}^{\mathrm{coll}}(\omega)$ similar to that of the Lorentz model (6.1.14):

$$\sigma_1^{\text{coll}}(\omega) \;=\; \frac{Ne^2}{m^*}\,\frac{\omega^2/\tau^*}{\left(\omega_0^2-\omega^2\right)^2+(\omega/\tau^*)^2} \tag{14.2.2a}$$

$$\sigma_2^{\text{coll}}(\omega) \;=\; -\frac{Ne^2}{m^*}\,\frac{\omega\left(\omega_0^2-\omega^2\right)}{\left(\omega_0^2-\omega^2\right)^2+(\omega/\tau^*)^2}\;, \tag{14.2.2b}$$

and the collective mode contribution to $\hat{\sigma}(\omega)$ now appears as a harmonic oscillator at finite frequencies – with the same oscillator strength as given in Eq. (7.2.15a).

The mass m^* of the condensate is large in the case of charge density waves, for which the condensate develops as the consequence of electron–phonon interactions. As given by Eq. (7.2.13), the effective mass m^*/m is large,

$$\frac{m^*}{m_{\text{b}}} = 1 + \frac{4\Delta^2}{\lambda_{\mathbf{P}}\hbar^2\omega_{\mathbf{P}}^2}\;, \tag{14.2.3}$$

if the gap is larger than the corresponding phonon frequency $\omega_{\mathbf{P}}$. Because of the larger mass, the oscillator strength associated with the collective mode is small, and thus the Tinkham–Ferrell sum rule is modified, as we have discussed in Section 7.5.3. For an effective mass $m^*/m_{\text{b}} \gg 1$, nearly all of the contributions to the total spectral weight come, even in the clean limit, from single-particle excitations.

This behavior can be clearly seen by experiment if the optical conductivity is measured over a broad spectral range. In Fig. 14.8 the frequency dependent conductivity $\sigma_1(\omega)$ is displayed for a number of materials in their charge and spin density wave states [Gru88, Gru94a]. In all cases two absorption features are seen: one typically in the microwave and one in the infrared spectral range. The former corresponds to the response of the collective mode at finite frequency ω_0, and the latter is due to single-particle excitations across the charge density wave gap. A few remarks are in order. First, it has been shown that impurities are responsible for pinning the mode to a well defined position in the crystal, thus ω_0 is impurity concentration dependent. Second, the spectral weight of the mode is small; fitting the observed resonance to a harmonic oscillator as described in Eq. (14.2.1) leads to a large effective mass m^*. Third, there is a well defined onset for the single-particle excitations at 2Δ (typically in the infrared spectral range). The gap is in good overall agreement with the gap obtained from the temperature dependent conductivity in the density wave state. At temperatures well below the transition temperature, where $\Delta(T)$ is close to its $T = 0$ value Δ, the dc conductivity reads

$$\sigma_{\text{dc}}(T) = \sigma_0 \exp\left\{-\frac{\Delta}{k_{\text{B}}T}\right\}\;, \tag{14.2.4}$$

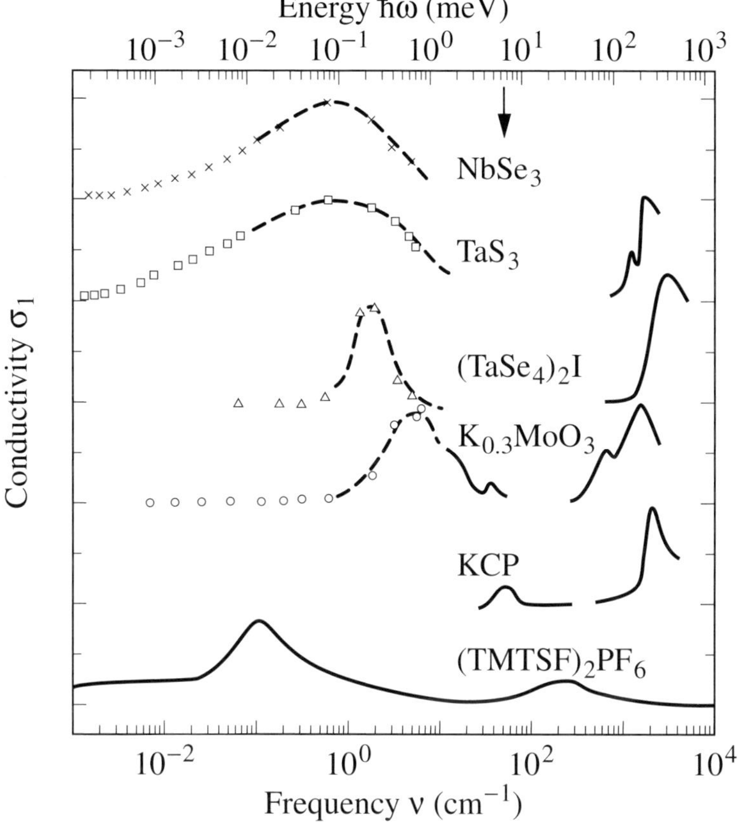

Fig. 14.8. Frequency dependent conductivity $\sigma_1(\omega)$ measured in several compounds in their charge and spin density states (after [Don94, Gru88]). The data are shown for the direction parallel to the highly conducting axis. The arrow indicates the gap measured by tunneling for NbSe$_3$. The dashed lines are fits to the data by Eq. (14.2.2a).

and the gap can also be found from dc transport measurements. Fourth, the behavior of $\sigma_1(\omega)$ above the gap is distinctively different from the $\sigma_1(\omega)$ we observe for superconductors. Instead of the smooth increase of $\sigma_1(\omega)$ as $\hbar\omega$ exceeds the gap energy, here one finds a maximum above 2Δ, much like what one would expect for the case 1 coherence factor displayed in Fig. 7.2.

One can use optical data, such as displayed in Fig. 14.8, to establish the relationship between the mass enhancement m^*/m and the single-particle gap 2Δ. In Fig. 14.9 the effective mass obtained from the fit of the low frequency resonance by Eqs (14.2.2) is plotted versus the gap energy Δ. The full line is the result of Eq. (14.2.3), with $\lambda_{\mathrm{P}} = 0.5$ and $\hbar\omega_{\mathrm{P}} = 34$ meV – both reasonable values for the materials summarized in the figure.

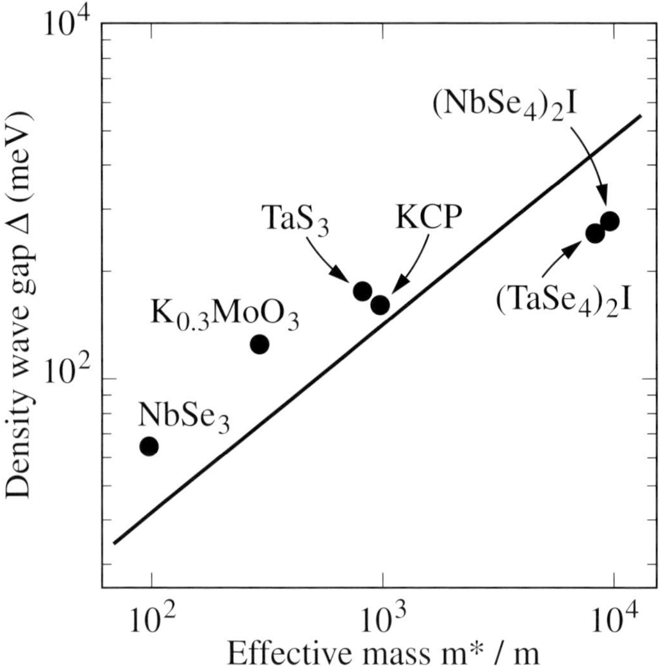

Fig. 14.9. Single-particle gaps Δ of various materials with a charge density wave ground state versus the effective mass values m^*/m of the condensate; the data are obtained by analyzing optical experiments. The full line is Eq. (7.2.13) with $\lambda_\mathbf{P} = 0.5$ and $\omega_\mathbf{P}/2\pi c = 35$ cm^{-1} (after [Gru94b]).

In contrast to superconductors, at zero frequency the response is capacitive because of the restoring force acting on the condensate. In the presence of impurities, the force K – and consequently ω_0 – are small, and the static dielectric constant is therefore enormous. Including also the contributions from the single-particle excitations, the zero frequency dielectric constant is written as

$$\epsilon_1(\omega \to 0) = 1 + \frac{4\pi Ne}{m^*\omega_0^2} + \frac{4\pi Ne^2\hbar^2}{6m_\mathrm{b}\Delta^2} = 1 + \epsilon_1^{\mathrm{coll}} + \epsilon_1^{\mathrm{sp}} \quad , \qquad (14.2.5)$$

where the second and third terms on the right hand side represent the collective and single-particle contributions to the dielectric constant. In the above equation we have used the tight binding model of a one-dimensional semiconductor to account for the contribution of single-particle excitations to the dielectric constant. With $\omega_0/2\pi = 10^{10}$ s^{-1}, and for a mass $m^*/m = 10^3$, the zero frequency dielectric constant is expected to be of the order of $\epsilon_1(\omega \to 0) \approx 10^6$; indeed such enormous values have been observed in materials with charge density wave ground states [Gru88].

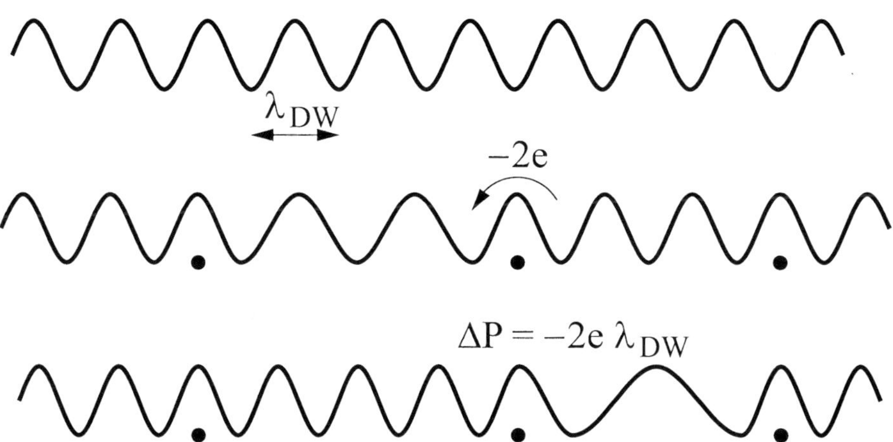

Fig. 14.10. Dynamics of the internal deformations of density waves. The top part of the figure is the undistorted density wave; the middle part shows the mode distorted due to interaction with impurities (full circles); and the bottom part displays the rearrangement of the internal distortion by displacing the density wave period over the impurity as indicated by the arrow. The process leads to an internal polarization $\Delta \mathbf{P} = -2e\lambda_{\mathrm{DW}}$, where λ_{DW} is the wavelength of the density wave.

To describe the effect of impurities by an average restoring force K is a gross oversimplification since it neglects the dynamics of the local deformations of the collective modes. The types of processes which have been neglected are shown in Fig. 14.10. The top part of the figure displays an undistorted density wave, with a period $\lambda_{\mathrm{DW}} = \pi/k_{\mathrm{F}}$ and a constant phase ϕ. In the presence of impurities, the density wave is pinned as shown in the middle section of the figure. A low lying excitation, which involves the dynamics of the internal deformations, is indicated at the bottom; here a density wave segment has been displaced by λ_{DW}, leading to a stretched density wave to the left and to a compressed part to the right side of the impurity. The local deformation leads to an internal polarization of the mode by virtue of the displaced charge which accompanies the stretched or compressed density wave. This polarization $\Delta \mathbf{P}$ is given by the spatial derivative of the phase

$$\mathbf{P}(\mathbf{r}) = -4\pi e \frac{\partial \phi(\mathbf{r})}{\partial \mathbf{r}} \quad .$$

For randomly positioned impurities we expect a broad distribution of the time and energy scales for the processes which reflect the dynamics of such interband deformations. Such effects (a response typical to a glass) are described by a broad superposition of Debye type relaxation processes, and various phenomenological expressions have been proposed to account for the low frequency and long-time behavior of the electrical response. Among these, the so-called Cole–Cole expression

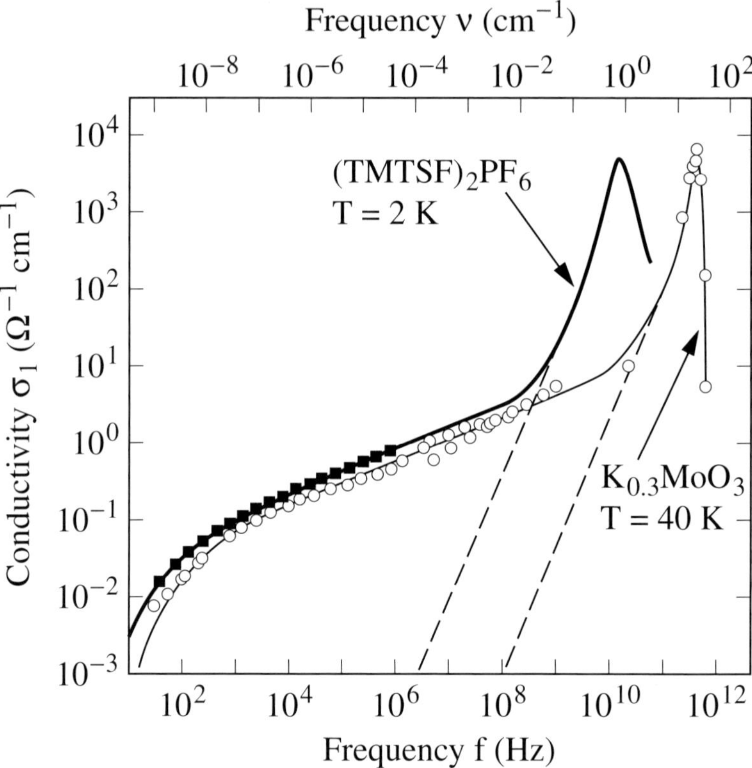

Fig. 14.11. Low frequency conductivity of $K_{0.3}MoO_3$ and $(TMTSF)_2PF_6$ in their density wave states (after [Don94]). The well defined peaks correspond to the pinning frequencies ω_0, and the dashed lines indicate the response of harmonic oscillators as expected for the collective mode with no internal deformation. The full lines model the low frequency response by Eq. (14.2.6).

(see for example [Nga79]),

$$\hat{\epsilon}(\omega) = \frac{\hat{\epsilon}(\omega \to 0)}{1 + (i\omega\tau_0)^{1-\alpha}} \quad , \tag{14.2.6}$$

where τ_0 is an average relaxation time and $\alpha < 1$, is frequently used to describe the so-called glassy behavior of a variety of random systems. Such low frequency relaxation effects lead to an enhanced ac conductivity at low frequencies, such as that shown in Fig. 14.11 for two materials with density wave ground states: $(TMTSF)_2PF_6$, which undergoes a spin density wave transition at $T_{SDW} = 12$ K, and $K_{0.3}MoO_3$, which enters the charge density wave ground state below $T_{CDW} = 180$ K. The dashed lines indicate the description in terms of a harmonic oscillator, with pinning frequencies of 6×10^9 Hz and 10^{11} Hz, respectively, and the full lines are fits to Eq. (14.2.6). While this description offers little insight into the

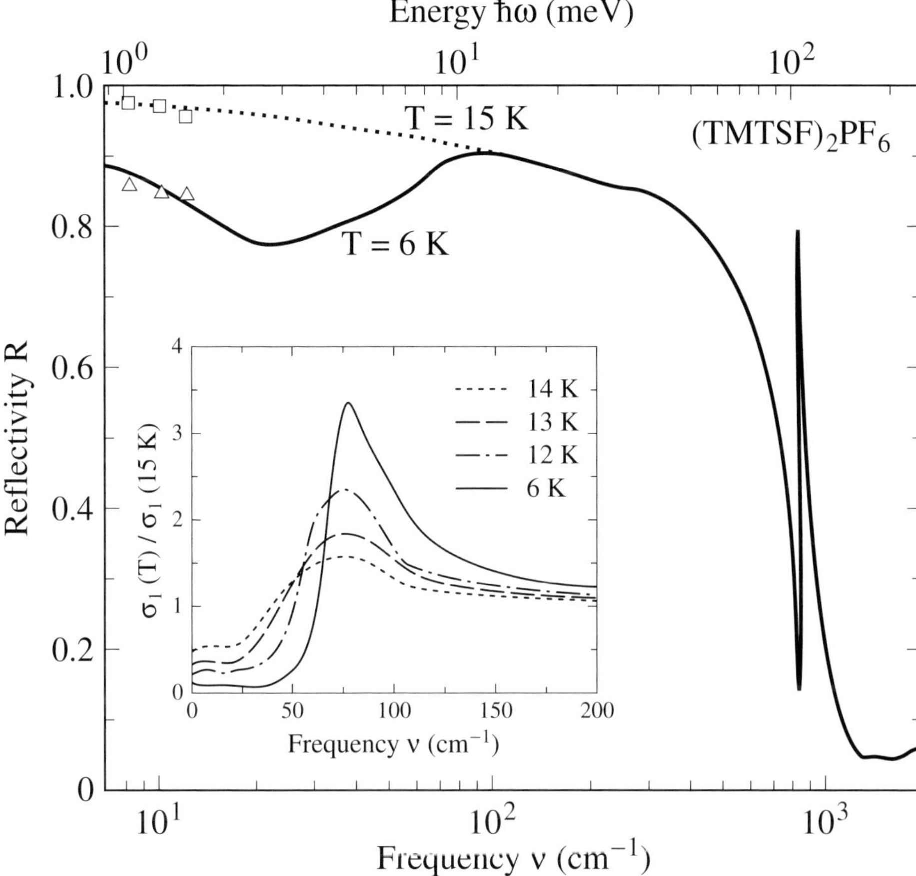

Fig. 14.12. Frequency dependent reflectivity $R(\omega)$ of $(TMTSF)_2PF_6$ measured perpendicular to the highly conducting axis at temperatures above and below $T_{SDW} = 12$ K. At $T = 15$ K the low frequency response is that of a Drude metal; below T_{SDW} the SDW gap opens as the temperature is lowered (after [Ves99]). The inset shows the normalized conductivity spectra at various temperatures. The optical properties are those of a semiconductor with a gap of $\Delta = 70$ cm^{-1}, but can also be well described by the Mattis–Bardeen formalism, with case 1 coherence factors [Dre99].

microscopic details of the density wave dynamics, it is useful to establish that the response is due to a broad distribution of relaxation times, a feature which often occurs in glasses and amorphous structures – here we have a density wave glass, the properties of which have not yet been fully explored.

14.2.2 Single-particle excitations

Single-particle excitations of density waves have a character which is different from single-particle excitations of superconductors due to the different coherence factors which occur in the two types of condensates. Case 1 coherence factors lead to a peak in the optical conductivity at the gap frequency as displayed in Fig. 7.2, in contrast to the gradual rise of the conductivity above the gap in the case of superconductors for which case 2 coherence factors apply.

This has also been most clearly observed in materials which undergo transitions to an incommensurate spin density wave state, with the best example the organic linear chain compound $(TMTSF)_2PF_6$. The normal state properties of this material when measured along the highly conducting direction – the direction along which the incommensurate density structure develops – cannot be described by a straightforward Drude response, and therefore a simple analysis of the optical properties of the density state is not possible. When measured with electric fields perpendicular to the highly conducting axis, such complications do not arise; also, along this direction the density wave is commensurate with the underlying lattice and thus the collective mode contribution to the conductivity is absent, due to this so-called commensurability pinning. Below the transition to the spin density wave state, a well defined gap develops, as evidenced by the drop in reflectivity at frequencies around 70 cm^{-1}; the data are displayed in Fig. 14.12. What we observe is similar to what can be calculated for case 1 coherence factors and what is displayed in Fig. 7.8. The reflectivity can also be analyzed to lead to the frequency dependent conductivity $\sigma_1(\omega)$, which for several different temperatures is displayed in the inset of the figure. The singularity at the gap of 70 cm^{-1}, at temperatures much lower than the transition temperature, is characteristic to a one-dimensional semiconductor, and this value, together with the transition temperature $T_{SDW} = 12$ K, places this material in the strong coupling spin density wave limit. The gap feature progressively broadens, and also moves to lower frequencies, and an appropriate analysis can be performed. Such studies conducted at different temperatures can also be used to evaluate the temperature dependence of the single gap [Ves99]; there is an excellent agreement with results of other methods [Dre99].

14.2.3 Frequency and electric field dependent transport

A few comments on the non-linear response are in order here. Because of the weak restoring force acting on density wave condensates, moderate electric fields may depin the collective mode, leading to a dc, non-linear conduction process. Of course, the smaller the restoring force – and thus the larger the low frequency dielectric constant ϵ_1 – the smaller the threshold field $\mathbf{E}_T$ which is required for depinning. The arguments lead to a particularly simple relation between the dielectric constant and threshold field:

$$\epsilon_1(\omega = 0)E_T = 4\pi e N_{DW} \quad , \tag{14.2.7}$$

where N_{DW} is the number of atoms in the area perpendicular to the direction along which the density wave develops. This relation has indeed been confirmed in a wide range of materials with charge density wave ground states [Gru89]. The intimate relationship between the dielectric constant and fields which characterize the non-linear response is, however, more general; a relation similar to that above can be derived, for example, for Zener tunneling of semiconductors. The topic of non-linear and frequency dependent response with all of its ramifications is, however, beyond the scope of this book.

References

[All71]   P.B. Allen, *Phys. Rev. B* **3**, 305 (1971)

[Bar61]   J. Bardeen and J.R. Schrieffer, *Recent Developments in Superconductivity*, in: *Progess in Low Temperature Physics* **3**, edited by C.J. Gorter (North-Holland, Amsterdam, 1961), p. 170

[Bio59]   M.A. Biondi and M.P. Garfunkel, *Phys. Rev.* **116**, 853 (1959)

[Bon96]   D.A. Bonn and W.N. Hardy, *Microwave Surface Impedance of High Temperature Superconductors*, in: *Physical Properties of High Temperature Superconductors*, Vol. 5, edited by D.M. Ginsberg (World Scientific, Singapore, 1996), p. 7

[Din53]   R.B. Dingle, *Physica* **19**, 348 (1953)

[Don94]   S. Donovan, Y. Kim, L. Degiorgi, M. Dressel, G. Grüner, and W. Wonneberger, *Phys. Rev. B* **49**, 3363 (1994)

[Dre96]   M. Dressel, A. Schwartz, G. Grüner, and L. Degiorgi, *Phys. Rev. Lett.* **77**, 398 (1996)

[Dre99]   M. Dressel, *Physica C* **317-318**, 89 (1999)

[Eli60]   G.M. Eliashberg, *Soviet Phys. JETP* **11**, 696 (1960)

[Far76]   B. Farnworth and T. Timusk, *Phys. Rev. B* **14**, 5119 (1976)

[Gin60]   D.M. Ginsberg and M. Tinkham, *Phys. Rev.* **118**, 990 (1960)

[Gru88]   G. Grüner, *Rev. Mod. Phys.* **60**, 1129 (1988)

[Gru89]   G. Grüner and P. Monceau, *Dynamical Properties of Charge Density Waves*, in: *Charge Density Waves in Solids*, Modern Problems in Condensed Matter Sciences **25**, edited by L. P. Gor'kov and G. Grüner (North-Holland, Amsterdam, 1989), p. 137

[Gru94a]   G. Grüner, *Rev. Mod. Phys.* **66**, 1 (1994)

[Gru94b]   G. Grüner, *Density Waves in Solids* (Addison-Wesley, Reading, MA, 1994)

[Hal71]   J. Halbritter, *Z. Phys.* **243**, 201 (1971)

[Heb57]   L.C. Hebel and C.P. Slichter, *Phys. Rev.* **107**, 901 (1957)

[Hef96]   R.K. Heffner and A.R. Norman, *Comments Mod. Phys. B* **17**, 325 (1996)

[Kle94]   O. Klein, E.J. Nicol, K. Holczer, and G. Grüner, *Phys. Rev. B* **50**, 6307 (1994)

[Kor91]   K. Kornelsen, M. Dressel, J.E. Eldridge, M.J. Brett, and K.L. Westra, *Phys. Rev. B* **44**, 11 882 (1991)

[Mat58]   D.C. Mattis and J. Bardeen, *Phys. Rev.* **111**, 412 (1958)

[McM68]   W.L. McMillan, *Phys. Rev.* **167**, 331 (1968)

[McM69]   W.L. McMillan and J.M. Rowell, *Tunneling and Strong-Coupling*

Superconductivity in: *Superconductivity*, edited by R.D. Parks (Marcel Dekker, New York, 1969), p. 561

[Mil60]　P.B. Miller, *Phys. Rev.* **118**, 928 (1960)

[Mit84]　B. Mitrovic, H.G. Zarate, and J.P. Cabrotte, *Phys. Rev. B* **29**, 184 (1984)

[Nga79]　K.L. Ngai, *Comments Solid State Phys.* **9**, 127 (1979)

[Pal68]　L.H. Palmer and M. Tinkham, *Phys. Rev.* **165**, 588 (1968)

[Pip62]　A.B. Pippard, *The Dynamics of Conduction Electrons*, in: *Low Temperature Physics*, edited by C. DeWitt, B. Dreyfus, and P.G. deGennes (Gordon and Breach, New York, 1962)

[Pro98]　A.V. Pronin, M. Dressel, A. Pimenov, A. Loidl, I. Roshchin, and L.H. Greene, *Phys. Rev. B* **57**, 14 416 (1998)

[Reu48]　G.E.H. Reuter and E.H. Sondheimer, *Proc. Roy. Soc. A* **195**, 336 (1948)

[Ric60]　P.L. Richards and M. Tinkham, *Phys. Rev.* **119**, 575 (1960)

[Sca69]　D.J. Scalopino, *The Electron-Phonon Interaction and Strong Coupling Superconductors*, in: *Superconductivity*, edited by R.D. Parks (Marcel Dekker, New York, 1969), p. 449

[Tin96]　M. Tinkham, *Introduction to Superconductivity*, 2nd edition (McGraw-Hill, New York, 1996)

[Tur91]　J.P. Turneaure, J. Halbritter, and K.A.Schwettman, *J. Supercond.* **4**, 341 (1991)

[Var86]　C.M. Varma, K. Miyake, and S. Schmitt-Rink, *Phys. Rev. Lett.* **57**, 626 (1986)

[Ves99]　V. Vescoli, L. Degiorgi, M. Dressel, A. Schwartz, W. Henderson, B. Alavi, G. Grüner, J. Brinckmann, and A. Virosztek, *Phys. Rev. B* **60**, 8019 (1999)

[Wal64]　J.R. Waldram, *Adv. Phys.* **13**, 1 (1964)

Further reading

[Bio58]　M.A. Biondi *et al.*, *Rev. Mod. Phys.* **30**, 1109 (1958)

[Cox95]　D.E. Cox and B. Maple, *Phys. Today* **48**, 32 (Feb. 1995)

[Gin69]　D.M. Ginsberg and L.C. Hebel, *Nonequilibrium Properties: Comparision of Experimental Results with Predictions of the BCS Theory*, in: *Superconductivity*, edited by R.D. Parks (Marcel Dekker, New York, 1969)

[Gol89]　A.I. Golovashkin, ed., *Metal Optics and Superconductivity* (Nova Science Publishers, New York, 1989)

[Gor89]　L. P. Gor'kov and G. Grüner, eds, *Charge Density Waves in Solids*, Modern Problems in Condensed Matter Sciences **25** (North-Holland, Amsterdam, 1989)

[Mon85]　P. Monceau, ed., *Electronic Properties of Inorganic Quasi-One Dimensional Compounds* (Riedel, Dordrecht, 1985)

[Par69]　R.D. Parks, ed., *Superconductivity* (Marcel Dekker, New York, 1969)

[Sig91]　M. Sigrist and K. Ueda, *Rev. Mod. Phys.* **63**, 239 (1991)

Part four

Appendices

Appendix A
Fourier and Laplace transformations

In various chapters of the book we made intensive use of the Fourier transformation and the Laplace transformation. Although an essential part of any mathematical course for physicists, we want to summarize the main relations, in particular those important for our task.

A.1 Fourier transformation

The Fourier transformation describes the relationship between a time dependent function and its spectral components, for example, or between a spatial dependent function and its wavevectors. Basically any waveform can be generated by adding up harmonic waves with the proper weight factor. The general relations between a function $f(t)$ and its Fourier transform $F(\omega)$ are

$$F(\omega) \;=\; \int_{-\infty}^{\infty} f(t) \exp\{-i\omega t\} \, dt \qquad \text{(A.1a)}$$

$$f(t) \;=\; \frac{1}{2\pi} \int_{-\infty}^{\infty} F(\omega) \exp\{i\omega t\} \, d\omega \quad . \qquad \text{(A.1b)}$$

In some texts, the Fourier transform and retransform are defined symmetrically with $1/\sqrt{2\pi}$ as pre-factors. No pre-factors occur in the definition of the Fourier transform and its inverse if $f = \omega/2\pi$ is used as frequency; however, the exponent then becomes $\{2\pi i f t\}$. According to the applications of the Fourier transformation in this book, we consider ω as an angular frequency and t as the time; but all the expressions hold for wavevector $\mathbf{q}$ and spatial coordinate $\mathbf{r}$ as well, separately for each of the three vector components. Also of interest is the convolution of two functions f and g

$$h(t) = [f(t') * g(t')](t) = \int_{-\infty}^{\infty} f(t')g(t - t') \, dt'$$

Table A.1. *Some important functions $f(t)$ and their Fourier transforms $F(\omega)$.*
$F(\omega) = \int_{-\infty}^{\infty} f(t) \exp\{-i\omega t\}\, dt$ *for the transformation and*
$f(t) = \frac{1}{2\pi} \int_{-\infty}^{\infty} F(\omega) \exp\{i\omega t\}\, d\omega$ *for the retransformation.*

$f(t)$	$F(\omega)$
1	$2\pi\,\delta\{\omega\}$
$\delta\{t\}$	1
$\cos\{\omega t\}$	$\pi\,(\delta\{\omega - \omega_0\} + \delta\{\omega + \omega_0\})$
$\sin\{\omega t\}$	$-i\pi\,(\delta\{\omega - \omega_0\} - \delta\{\omega + \omega_0\})$
$\exp\{i\omega t\}$	$2\pi\,\delta\{\omega - \omega_0\}$
$\exp\left\{-\dfrac{t^2}{2(\Delta t)^2}\right\}$	$\dfrac{2\pi}{\Delta\omega}\,\exp\left\{-\dfrac{\omega^2}{2(\Delta\omega)^2}\right\}$
$\exp\{-\Delta\omega|t|\}$	$\dfrac{2\Delta\omega}{(\Delta\omega)^2 + \omega^2}$

since the convolution theorem states that the Fourier transform of a product of two functions f and g equals the convolution product of the individual spectra, and vice versa:

$$H(\omega) = \int_{-\infty}^{\infty} [f^* * g](t)\exp\{-i\omega t\}\, dt = F(\omega)G(\omega) \tag{A.2a}$$

$$h(t) = [f * g^*](t) = \frac{1}{(2\pi)^2} \int_{-\infty}^{\infty} F(\omega)G^*(\omega)\exp\{i\omega t\}\, d\omega \ . \tag{A.2b}$$

Applied to the autocorrelation we arrive at Parseval's identity, which expresses the fact that the total energy of a time dependent function, measured as the integral over $|f(t)|^2$, is equal to the total energy of its spectrum $|F(\omega)|^2$ (the so-called Wiener–Khinchine theorem). From these formulas we can easily derive the Nyquist criterion, which states that any waveform (which can be composed by harmonic functions) can be sampled unambiguously and without any loss of information using a sampling frequency greater than or equal to twice the bandwidth of the system (Shannon's sampling theorem). In a Fourier transform spectrometer the data points have to be taken at a distance of mirror displacement shorter than $\lambda/2$ of the maximum frequency $f = c/\lambda$ which should be obtained. These considerations also limit the resolution of a Fourier transform spectrometer due to the maximum length of the path difference. According to Rayleigh's criterion the interferogram has to be measured up to a path length of at least δ_{max} in order to resolve two spectral lines separated by a frequency c/δ_{max}. The scaling of the function in t in the form $f(at)$ leads to the inverse scaling of the Fourier transform: $\frac{1}{|a|}F\left(\frac{\omega}{a}\right)$. The differentiation $\frac{d}{dt}f(t)$ becomes a multiplication in the Fourier transform: $i\omega F(\omega)$.

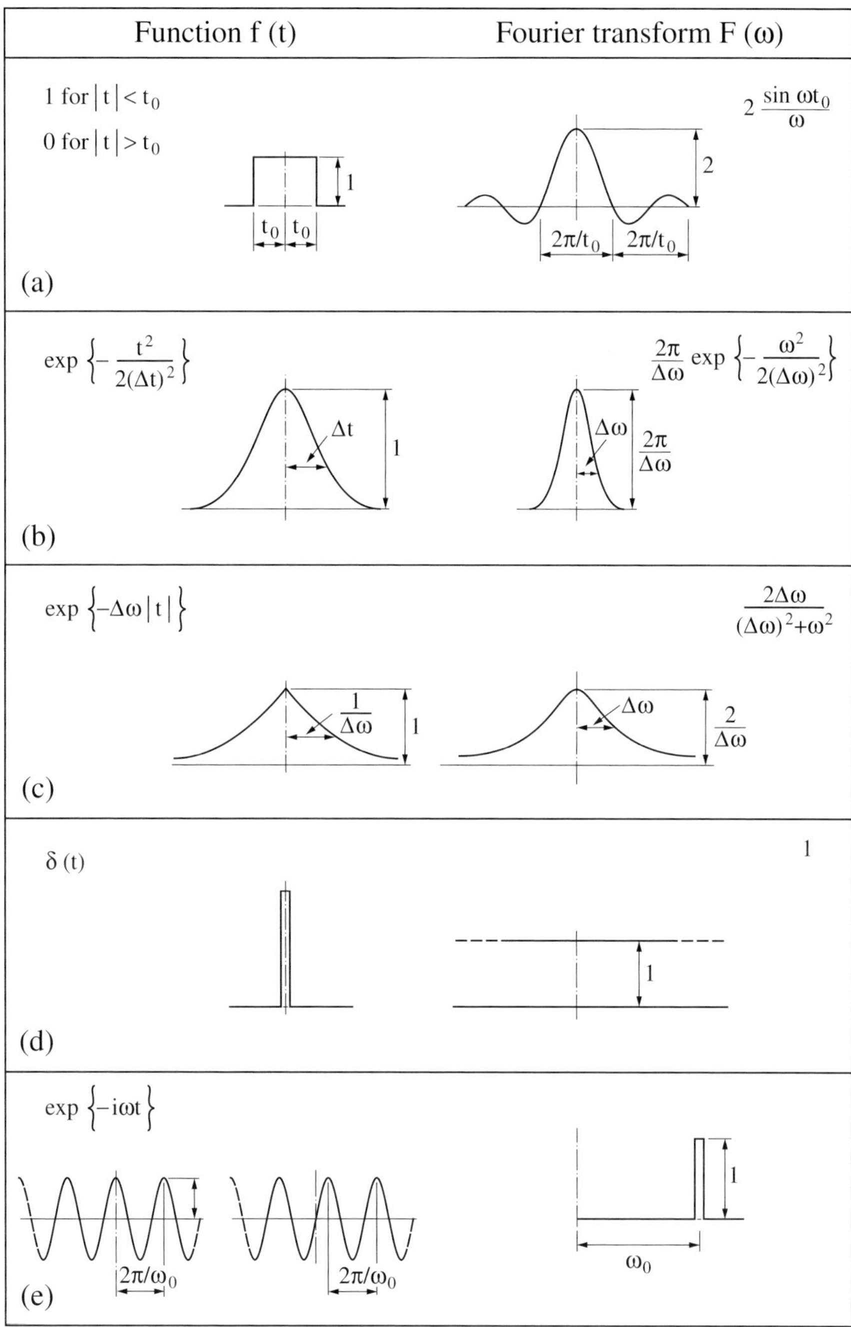

Fig. A.1. Graphs of different functions $f(t)$ and their Fourier transforms $F(\omega) = 1/(2\pi) \int_{-\infty}^{\infty} f(t)\, \exp\{i\omega t\}\, dt$: (a) box function; (b) Gaussian curve; (c) Lorentzian curve; (d) δ-function at $t = 0$; (e) harmonic wave.

In the following we want to give some examples of the Fourier transformations depicted in Fig. A.1. A box function, for instance,

$$f(t) = \begin{cases} 1 & |t| < t_0 \\ 0 & |t| > t_0 \end{cases} \tag{A.3a}$$

leads to a sinc function (Fig. A.1a)

$$F(\omega) = \frac{2\sin\{\omega t_0\}}{\omega} \quad , \tag{A.3b}$$

well known from the diffraction pattern of a slit. The function $f(t) = t^{-1/2}$ remains unchanged during the Fourier transformation: $F(\omega) = (\omega/2\pi)^{-1/2}$. In the same way, the Gaussian curve $f(t) = \exp\left\{-\frac{t^2}{2(\Delta t)^2}\right\}$ of the width Δt is transformed to

$$F(\omega) = \frac{2\pi}{\Delta\omega} \exp\left\{-\frac{\omega^2}{2(\Delta\omega)^2}\right\} \quad , \tag{A.4}$$

as displayed in Fig. A.1b. The width of the spectrum $\Delta\omega$ is related to the width of its Fourier transform by $\Delta t = 1/\Delta\omega$. A Lorentzian curve (Fig. A.1c) is obtained by transforming the exponential decay $f(t) = \exp\{-\Delta\omega|t|\}$:

$$F(\omega) = \frac{2\Delta\omega}{(\Delta\omega)^2 + \omega^2} \quad , \tag{A.5}$$

the line shape of atomic transitions, for instance. A delta function $f(t) = \delta\{t\}$ leads to a flat response $F(\omega) = 1$ as depicted in Fig. A.1d; for the converse transformation of $f(\omega) = \delta\{\omega\}$ we obtain $F(t) = 1/2\pi$ since $\delta\{\omega/2\pi\} = 2\pi\delta\{\omega\}$. Fig. A.1e shows that from a harmonic wave $f(t) = \exp\{i\omega t\}$ we get a δ-function at the frequency ω_0 as the Fourier transform:

$$F(\omega) = 2\pi\delta\{\omega - \omega_0\} \quad . \tag{A.6}$$

Table A.1 summarizes the most important examples of the Fourier transformation.

The Fourier transformation is a powerful tool which, besides fast signal analysis, can lead to deep insight into the properties of time or space dependent phenomena. Although we have only discussed the one-dimensional case, the Fourier transformation can be extended to two and three dimensions, which can be useful for the description of problems on surfaces or in crystals, for example.

A.2 Laplace transformation

The attempt to apply the Fourier transformation in a straightforward manner in order to obtain the Fourier transform of a step function

$$f(t) = \begin{cases} -\frac{1}{2} & t < 0 \\ 0 & t = 0 \\ \frac{1}{2} & t > 0 \end{cases} \tag{A.7}$$

fails because the integral in Eq. (A.1a) does not converge. The problem can be avoided by multiplying $f(t)$ by a convergency factor $\exp\{-\eta t\}$, which then allows the integral to be solved by finally taking the $\lim_{\eta \to 0}$. Since the function is odd, we can express it in terms of the sinc function

$$f(t) = \lim_{\eta \to 0} \frac{1}{\pi} \int_0^\infty \frac{\omega}{\eta^2 + \omega^2} \sin\{\omega t\} \, d\omega \tag{A.8}$$

$$= \frac{1}{\pi} \int_0^\infty \frac{\sin\{\omega t\}}{\omega} \, d\omega \quad, \tag{A.9}$$

and thus $F(\omega) = \frac{1}{\sqrt{\pi}} \frac{1}{\omega}$. If

$$f(t) = \begin{cases} 0 & t < 0 \\ 1 & t > 0 \end{cases} \quad, \tag{A.10a}$$

the Fourier representation has the form

$$f(t) = \frac{1}{2} + \frac{1}{\pi} \int_0^\infty \frac{\sin\{\omega t\}}{\omega} \, d\omega \quad. \tag{A.10b}$$

This can be expressed more elegantly by the Laplace transformation. The definition of the Laplace transform and its retransform is

$$P(\omega) = \int_0^\infty f(t) \exp\{-\omega t\} \, dt \tag{A.11a}$$

$$f(t) = \frac{1}{2\pi i} \int_{c-i\infty}^{c+i\infty} P(\omega) \exp\{\omega t\} \, d\omega \quad, \tag{A.11b}$$

where ω is complex (Fig. A.2). The Fourier transform is the degenerate form of the Laplace transform if the latter has purely imaginary arguments ($c \to 0$).

The transformation is linear and the convolution becomes a multiplication

$$P(\omega) = \int_0^\infty \left[\int_0^t f_1(t - t') f_2(t') dt' \right] \exp\{-\omega t\} \, dt$$

$$= \int_0^\infty f_1(t) \exp\{-\omega t\} \, d \cdot \int_0^\infty f_2(t) \exp\{-\omega t\} \, dt \quad. \tag{A.12}$$

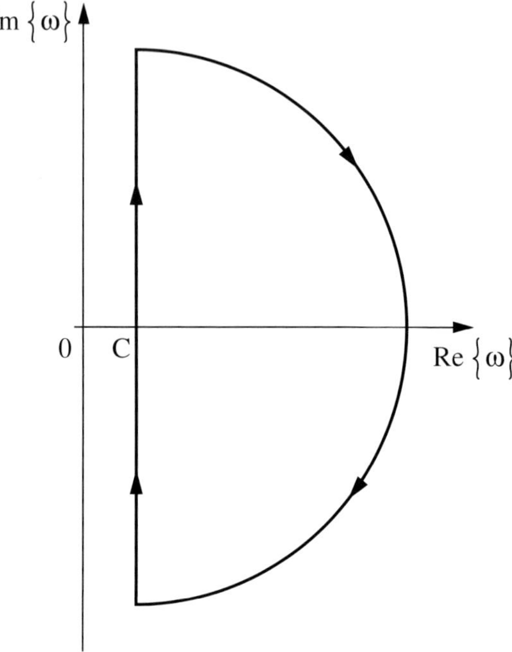

Fig. A.2. Contour for inversion integration used by Laplace transformation; the integral is calculated for the limit that the radius of the semicircle goes to infinity.

The properties can best be seen by a few examples, as summarized in Table A.2.

Table A.2. *Some important functions $f(t)$ and their Laplace transforms $P(\omega)$.*
$$P(\omega) = \int_0^\infty f(t)\exp\{-\omega t\}dt \text{ for the transformation and}$$
$$f(t) = \frac{1}{2\pi i}\int_{c-i\infty}^{c+i\infty} P(\omega)\exp\{\omega t\}\,d\omega \text{ for the retransformation.}$$

$f(t)$	$P(\omega)$
1	$\frac{1}{\omega}$
$\exp\{-at\}$	$\frac{1}{\omega+a}$
$\cos\{\omega t\}$	$\pi\,(\delta\{\omega-\omega_0\}+\delta\{\omega+\omega_0\})$
$\sin\{at\}$	$\frac{a}{\omega^2+a^2}$
$\frac{1}{t^2+a^2}$	$\frac{1}{a}\sin a\omega$
$\delta\{t-a\}$	$\exp\{-a\omega\}$

References

[Cha73] D.C. Champeney, *Fourier Transforms and Their Physical Applications* (Academic Press, London, 1973)

[Doe74] G. Doetsch, *Introduction to the Theory and Application of the Laplace Transformation* (Springer-Verlag, Berlin, 1974)

[Duf83] P.M. Duffieux, *The Fourier Transform and Its Applications to Optics* (John Wiley & Sons, New York, 1983)

[Fra49] P. Franklin, *An Introduction to Fourier Methods and the Laplace Transformation* (Dover, New York, 1949)

[Mer65] L. Mertz, *Transformations in Optics* (John Wiley & Sons, New York, 1965)

[Ste83] E.G. Steward, *Fourier Optics: An Introduction* (Halsted Press, New York, 1983)

Appendix B

Medium of finite thickness

In the expressions (2.4.15) and (2.4.21) we arrived at the power ratio reflected by or transmitted through the surface of an infinitely thick medium, which is characterized by the optical constants n and k. For a material of finite thickness d, the situation becomes more complicated because the electromagnetic radiation which is transmitted through the first interface does not entirely pass through the second interface; part of it is reflected from the back of the material. This portion eventually hits the surface, where again part of it is transmitted and contributes to the backgoing signal, while the remaining portion is reflected again and stays inside the material. This multireflection continues infinitely with decreasing intensity as depicted in Fig. B.1.

In this appendix we discuss some of the optical effects related to multireflection which becomes particularly important in media with a thickness smaller than the skin depth but (significantly) larger than half the wavelength. Note, the skin depth does not define a sharp boundary but serves as a characteristic length scale which indicates that, for materials which are considerably thicker than δ_0, most of the radiation is absorbed before it reaches the rear side. First we introduce the notion of film impedance before the concept of impedance mismatch is applied to a multilayer system. We finally derive expressions for the reflection and transmission factors of various multilayer systems.

B.1 Film impedance

First we define the impedance of a film with thickness d which is smaller than the skin depth δ_0. In this case Eq. (2.4.24) is not appropriate because for its derivation we used the assumption that the medium is an infinite half plane. For very thin

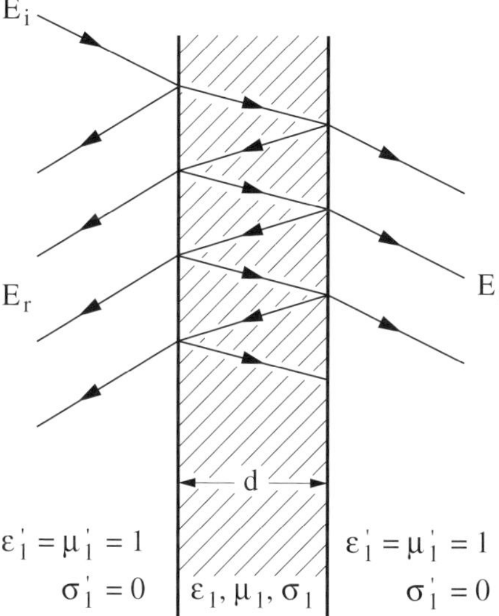

Fig. B.1. Reflection off and transmission through a dielectric slab with thickness d and optical parameters ϵ_1, σ_1, and μ_1. The multireflections cause interference. $\mathbf{E}_i$, $\mathbf{E}_t$, and $\mathbf{E}_r$ indicate the incident, transmitted, and reflected electric fields, respectively. The optical properties of vacuum are given by $\epsilon'_1 = \mu'_1 = 1$ and $\sigma'_1 = 0$.

films ($d \ll \delta_0$) the film impedance is given by

$$\hat{Z}_F = \frac{\hat{E}}{\hat{H}_1 - \hat{H}_2} \approx \frac{\hat{E}}{\hat{J}} = \frac{1}{\hat{\sigma} d} \tag{B.1}$$

if we assume that the current density $\hat{J}$ is uniform throughout the film. $\hat{H}_1$ and $\hat{H}_2$ are the magnetic fields at the two sides of the film; $\hat{E}$ denotes the electric field. For intermediate thickness $d \approx \delta_0$, we have to integrate over the actual field distribution in the film, leading to [Sch94]:

$$E(z) = E_0 \frac{\cosh\{i\hat{q}z\}}{\cosh\{i\hat{q}d/2\}} \quad , \tag{B.2}$$

where the wavevector is given by $\hat{q} = \frac{\omega}{c}\sqrt{\hat{\epsilon}}$. In the general case of a thin film of thickness d, width b, and length l, the film impedance is given by [Sch75]

$$\hat{Z}_F = \frac{l}{2b}\left(\frac{4\pi i\omega}{c^2\hat{\sigma}}\right)^2 \coth\left\{\frac{d(i\pi\omega\hat{\sigma})^{1/2}}{c}\right\} \quad . \tag{B.3}$$

These finite thickness corrections are especially important in the radiofrequency range since for these frequencies the skin depth δ_0 is of the order of the typical film thickness $d \approx 1$ μm.

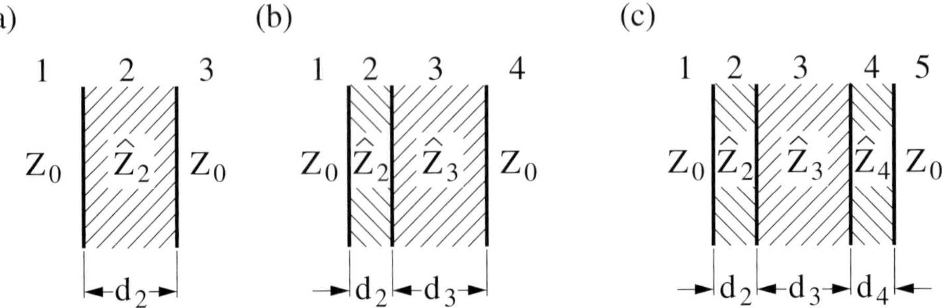

Fig. B.2. (a) Three-layer system: free standing film with impedance $\hat{Z}_2$ and thickness d_2 surrounded by vacuum (Z_0). (b) Four-layer arrangement: film ($\hat{Z}_2$, d_2) on a substrate with impedance $\hat{Z}_3$ and thickness d_3. (c) Substrate ($\hat{Z}_3$, d_3) covered on both sides by films with impedances $\hat{Z}_2$ and $\hat{Z}_4$ and thicknesses d_2 and d_4, respectively.

B.2 Impedance mismatch

A very powerful approach to the problems of multilayer reflectivity utilizes the surface impedance of an interface [Hea65, Ram94]. It is useful to start looking at the situation where there are three different regions. Let us assume that a wave traveling in medium 1 hits the boundary to medium 2 and finally enters medium 3 (Fig. B.2a). The reflectivity of the entire system is not completely determined by the impedance of the first and second material but also by the properties of the third material, because part of the light is reflected at the interface between the second and third layers. This effect is taken into account by considering the effective impedance of the second material to be changed by the third one. For normal incidence the light with frequency ω sees the so-called load impedance $\hat{Z}_{L2}$ at the first interface between medium 1 and 2; it is given by

$$\hat{Z}_{L2} = \hat{Z}_2 \frac{\hat{Z}_3 \cosh\{-i\hat{q}_2 d_2\} + \hat{Z}_2 \sinh\{-i\hat{q}_2 d_2\}}{\hat{Z}_2 \cosh\{-i\hat{q}_2 d_2\} + \hat{Z}_3 \sinh\{-i\hat{q}_2 d_2\}} \quad , \tag{B.4}$$

where $\hat{q}_2 = \frac{\omega}{c}\sqrt{(\hat{\epsilon})_2}$ is the value of the wavevector in medium 2, and d_2 is the thickness of the second material. The reflectivity at this interface is easily calculated from Eq. (2.4.29) by substituting for $\hat{Z}_2$ by this load impedance:

$$\hat{r} = \frac{\hat{Z}_1 - \hat{Z}_{L2}}{\hat{Z}_1 + \hat{Z}_{L2}} \quad . \tag{B.5}$$

These formulas are applied repeatedly in order to analyze any series of dielectric slabs, no matter what thickness or surface impedance. For example, if another dielectric layer is added at the back as shown in Fig. B.2b, the load impedance $\hat{Z}_{L3}$ seen by region 2 is calculated using an equation similar to Eq. (B.4). Then

$\hat{Z}_{L3}$ is used in place of $\hat{Z}_3$ in the calculation of $\hat{Z}_{L2}$. Following this procedure the load impedance of any multilayer system is evaluated by starting from the rear and adding layer by layer to the front.

In the limiting case of a thin metal film (with impedance $\hat{Z}_2$ and with d_2 very small) placed in air ($\hat{Z}_3 = Z_0$), Eq. (B.4) becomes

$$\hat{Z}_{L2} = \frac{Z_0 - i\hat{q}_2\hat{Z}_2 d_2}{1 - Z_0(i\hat{q}_2/\hat{Z}_2)d_2} \ . \tag{B.6}$$

The second term in the numerator is negligible for small d_2 because the complex dielectric constant $(\hat{\epsilon})_2$ of the film cancels. In the denominator, however, the term proportional to d_2 must not be neglected as the factor $\hat{q}_2/\hat{Z}_2$ is proportional to $(\hat{\epsilon})_2$. In the limit where d_2 is small, we then arrive at

$$\hat{Z}_{L2} = \frac{Z_0}{1 - Z_0(\frac{i\omega}{4\pi}(\hat{\epsilon})_2)d_2} \approx \frac{Z_0}{1 + Z_0(\hat{\sigma})_2 d_2} \tag{B.7}$$

for the equation of the load impedance of a thin film.

For a two-layer system consisting of a conducting film on a dielectric substrate, Z_0 is replaced by the effective impedance of the substrate $\hat{Z}_{L3}$:

$$\hat{Z}_{L2} = \frac{\hat{Z}_{L3}}{1 + \hat{Z}_{L3}(\hat{\sigma})_2 d_2} \ ,$$

where the load impedance seen is

$$\hat{Z}_{L3} = \hat{Z}_D \frac{Z_0 \cosh\{-i\hat{q}_3 d_3\} + \hat{Z}_3 \sinh\{-i\hat{q}_3 d_3\}}{\hat{Z}_3 \cosh\{-i\hat{q}_3 d_3\} + Z_0 \sinh\{-i\hat{q}_3 d_3\}} \ , \tag{B.8}$$

where $\hat{Z}_3$ is the substrate impedance and d_3 is its thickness. In the case where the backing region is another dielectric layer, as shown in Fig. B.2c, the value of Z_0 in Eq. (B.8) is replaced by an effective surface impedance which is calculated using the same equations. This procedure can be repeated for any series of layers with different optical properties (n and k) and film thickness d.

B.3 Multilayer reflection and transmission

In the following we discuss the overall reflectivity and transmission of various multilayer systems, starting from one thin layer, going to double layer compounds, etc.

B.3.1 Dielectric slab

The evaluation of the amplitude and phase of the electromagnetic wave is complicated but straightforward (for example by using matrices), and here we merely give the end result, valid for an isotropic and homogeneous medium [Bor99, Hea65]. By calculating the multiple reflections and transmissions at the two symmetrical boundaries (given by $z = 0$ and $z = d$), the final expression for the reflectivity R_F of a material with finite thickness d is

$$R_F = R \frac{(1 - \exp\{-\alpha d\})^2 + 4 \exp\{-\alpha d\} \sin^2 \beta}{(1 - R \exp\{-\alpha d\})^2 + 4R \exp\{-\alpha d\} \sin^2\{\beta + \phi_r\}} \quad , \tag{B.9}$$

with the bulk reflectivity

$$R = \left| \frac{1 - \hat{N}}{1 + \hat{N}} \right|^2 = \frac{(1 - n)^2 + k^2}{(1 + n)^2 + k^2} \tag{B.10a}$$

as in Eq. (2.4.15), obtained in the limiting case for $d \to \infty$; the phase change upon reflection is

$$\phi_r = \arctan \left\{ \frac{-2k}{1 - n^2 - k^2} \right\} \tag{B.10b}$$

from Eq. (2.4.14). The power absorption coefficient $\alpha = 4\pi k / \lambda_0$ was defined in Eq. (2.3.18); it describes the attenuation of the wave. The angle

$$\beta = \frac{2\pi n d}{\lambda_0} \tag{B.11}$$

indicates the phase change on once passing through the medium of thickness d and refractive index n. Here λ_0 is the wavelength in a vacuum, and hence β describes the ratio of film thickness and wavelength in the medium. Sometimes it is convenient to combine both in a complex angle $\delta = \beta + i\alpha d/2$. We obtain for the transmission

$$T_F = \frac{[(1 - R)^2 + 4R \sin^2 \phi_r] \exp\{-\alpha d\}}{(1 - R \exp\{-\alpha d\})^2 + 4R \exp\{-\alpha d\} \sin^2\{\beta + \phi_r\}} \quad , \tag{B.12a}$$

$$\phi_t = \frac{2\pi n d}{\lambda_0} - \arctan \left\{ \frac{k(n^2 + k^2 - 1)}{(k^2 + n^2)(2 + n)n} \right\}$$

$$+ \arctan \left\{ \frac{R \exp\{-\alpha d\} \sin^2\{\beta + \phi_r\}}{1 - R \exp\{-\alpha d\} \cos^2\{\beta + \phi_r\}} \right\} \quad . \tag{B.12b}$$

The second term in the denominator of Eq. (B.12a) describes the interference; for strong absorption it might be neglected, and the equation then reduces to

$$T_F = \frac{(1 - R)^2 (1 + \frac{k^2}{n^2}) \exp\{-\alpha d\}}{1 - R^2 \exp\{-2\alpha d\}} \quad . \tag{B.13}$$

This relation also describes the case where the wavelength λ is larger than the film thickness d, and for that reason no interference is present. In the limit of infinite thickness ($d \rightarrow \infty$), obviously there is no transmission through the material and $T_F = 0$; the radiation is either reflected or fully absorbed. For optically very thin plates ($nd \ll \lambda$), both components of the complex angle αd and β are very small, and therefore the reflection R_F is also close to zero. The transmission through the material is then given by

$$T_F \approx (1-R)^2 \exp\left\{-\frac{4\pi kd}{\lambda_0}\right\} = (1-R)^2 \exp\left\{-\frac{2\omega kd}{c}\right\} = (1-R)^2 \exp\{-\alpha d\} \quad ; \tag{B.14}$$

the rest of the signal is absorbed. At intermediate optical thickness one obtains a series of maxima and minima in the transmitted power as the frequency is varied (Fig. B.3). The interference extrema are separated by

$$\Delta f = \frac{c}{2nd} \quad , \tag{B.15}$$

and similar oscillations occur when the thickness is varied keeping the frequency fixed: $\Delta d = \frac{c}{2nf}$. This phenomenon is utilized to determine the dielectric properties of thin films or transparent media.

It is interesting to consider a very thin metallic film in air as we have done with the impedance approach; in this case Eqs (B.9) and (B.12a) simplify to

$$R_F \approx \frac{(\epsilon_1^2 + \epsilon_2^2)4\pi^2 d^2/\lambda_0^2}{4 + 8\epsilon_2\pi d/\lambda_0 + (\epsilon_1^2 + \epsilon_2^2)4\pi^2 d^2/\lambda_0^2} \approx \left(\frac{\sigma_{dc} Z_0 d}{\sigma_{dc} Z_0 d + 2}\right)^2$$

$$= \left(1 + \frac{c}{2\pi\sigma_{dc}d}\right)^{-2} \tag{B.16}$$

$$T_F \approx \frac{4}{4 + 8\epsilon_2\pi d/\lambda_0 + (\epsilon_1^2 + \epsilon_2^2)4\pi^2 d^2/\lambda_0^2} \approx \left(\frac{2}{\sigma_{dc} Z_0 d + 2}\right)^2 \quad ,$$

$$= \left(1 + \frac{2\pi\sigma_{dc}d}{c}\right)^{-2} \tag{B.17}$$

where $Z_0 = 377\ \Omega$ is the wave impedance of a vacuum, and σ_{dc} denotes the dc conductivity of the film. The second approximation used in both relations holds if $\epsilon_1^2 \ll \epsilon_2^2$, and the optical properties then are frequency independent, as it becomes obvious after substituting $2\pi\epsilon_2 d/\lambda = \sigma_{dc} Z_0 d$. The absorptivity of the film $A_F = 1 - R_F - T_F$ is given by

$$A_F \approx \frac{8\epsilon_2\pi d/\lambda_0}{4 + 8\epsilon_2\pi/\lambda_0 + (\epsilon_1^2 + \epsilon_2^2)4\pi^2 d^2/\lambda_0^2}$$

$$\approx \frac{4\sigma_{dc} Z_0 d}{(\sigma_{dc} Z_0 d + 2)^2} = \frac{c}{\pi\sigma_{dc}d}\left(1 + \frac{c}{2\pi\sigma_{dc}d}\right)^{-2} \tag{B.18}$$

and is displayed in Fig. B.4. Since the reflection and transmission depend on d,

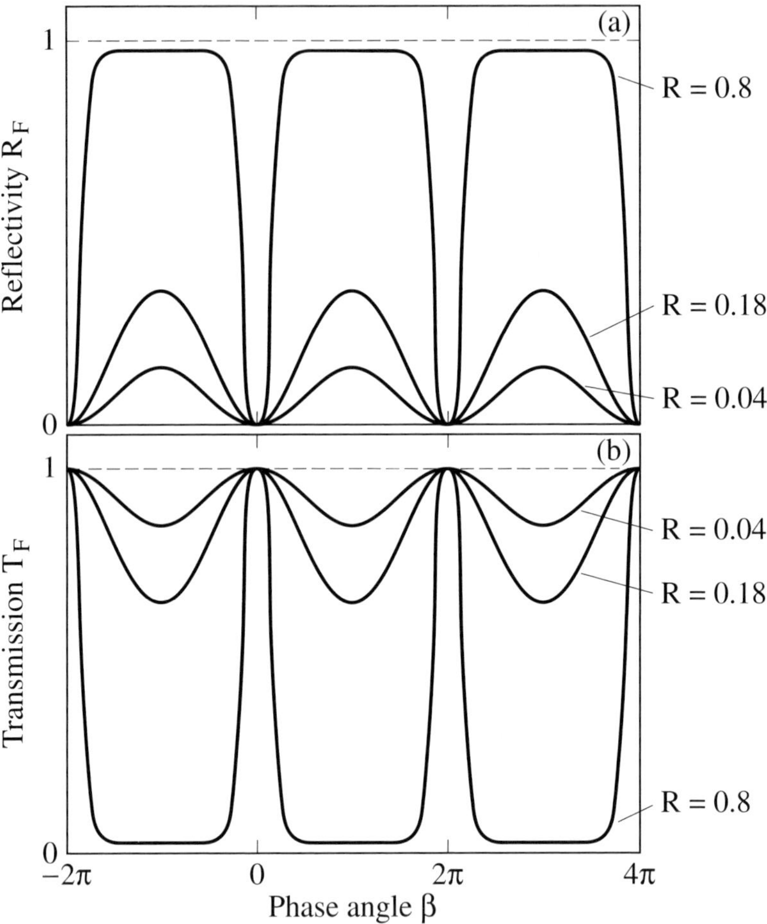

Fig. B.3. Frequency dependence of the (a) reflection off and (b) transmission through an insulating material ($\alpha = 0$) with thickness d calculated by Eqs (B.19) and (B.21). The curves indicated by the bulk reflectivity $R = 0.04$, 0.18, and 0.8 correspond to different refractive indices: $n = 1.5$, 2.5, and 18.

the absorption is also thickness dependent. A_{F} exhibits a maximum of 0.5 at the so-called Woltersdorff thickness $d_{\mathrm{W}} = 2/(\sigma_{\mathrm{dc}} Z_0)$; in addition, $R_{\mathrm{F}} = T_{\mathrm{F}} = 0.25$ for this thickness [Wol34]. For $\sigma_{\mathrm{dc}} = 100 \ \Omega^{-1} \ \mathrm{cm}^{-1}$ we obtain $d_{\mathrm{W}} = 2/\sigma_{\mathrm{dc}} Z_0 = 0.53 \ \mu\mathrm{m}$, for instance. For $d < d_{\mathrm{W}}$ most of the light is transmitted; for $d > d_{\mathrm{W}}$ most of it is reflected. This behavior has to be considered if the absorptivity of metallic films is important for applications.

As a first example, let us consider the optical properties of an insulating material ($\alpha = 0$ and therefore $k = 0$) with finite thickness d. The film reflectivity derived

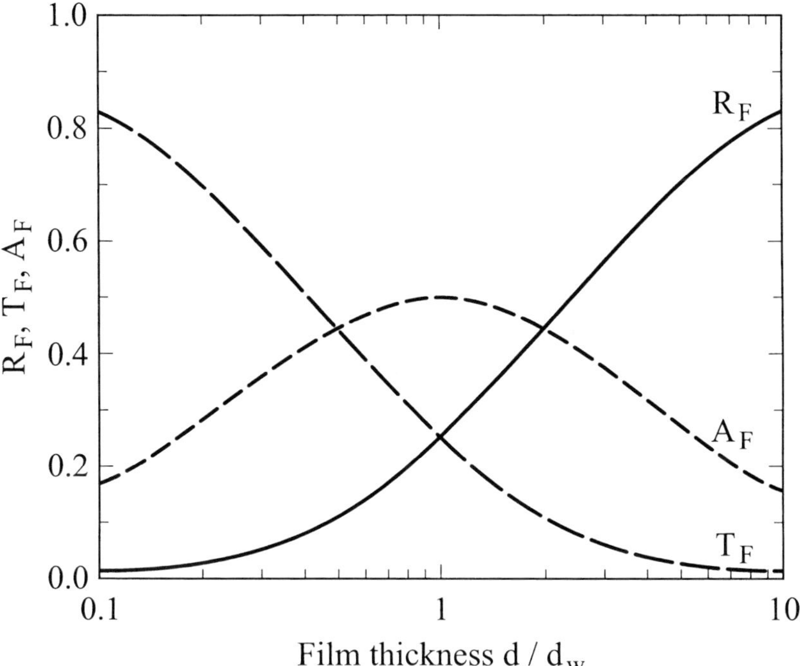

Fig. B.4. Reflection R_F, transmission T_F, and absorption A_F of thin metal films as a function of thickness d normalized to the Woltersdorff thickness d_W which is given by $d_\mathrm{W} = 2/\sigma_\mathrm{dc} Z_0$.

in Eq. (B.9) then simplifies to the Airy function

$$R_\mathrm{F} = \frac{4R \sin^2\{\beta\}}{(1-R)^2 + 4R \sin^2\{\beta\}} = \frac{2R - 2R \cos\{2\beta\}}{1 + R^2 - 2R \cos\{2\beta\}} \quad , \qquad (B.19)$$

as plotted in Fig. B.3a. The interference leads to extrema in the reflectivity given by

$$R_\mathrm{F}^{\text{extrema}} = R \left(\frac{1 \pm \exp\{-\alpha d\}}{1 \pm R \exp\{-\alpha d\}} \right)^2 \quad ; \qquad (B.20)$$

for $\alpha = 4\pi \nu k = 0$ the minima drop to zero and the maxima reach $R_\mathrm{F}^{\max} = 4R/(1+R)^2$, and are evenly spaced by a frequency of $\Delta f = c/(2nd)$. The transmission T_F shows similar fringes (Fig. B.3b):

$$T_\mathrm{F} = \frac{(1-R)^2}{(1-R)^2 + 4R \sin^2\{\beta\}} = \frac{(1-R)^2}{1 + R^2 - 2R \cos\{2\beta\}} \quad , \qquad (B.21)$$

where the extrema are given by

$$T_\mathrm{F}^\mathrm{extrema} = \frac{(1-R)^2 \exp\{-\alpha d\}}{(1 \pm R \exp\{-\alpha d\})^2} \quad , \tag{B.22}$$

which for $\alpha = 0$ reaches $T_\mathrm{F}^\mathrm{max} = 1$ and $T_\mathrm{F}^\mathrm{min} = (1-R)^2/(1+R)^2$. Hence for a perfect match of wavelength and film thickness all the radiation passes through the film, even for finite refractive index n (e.g. Fabry–Perot filter, etalon).

Following the discussion of the bulk properties in Section 5.1.2, the case of a conductor requires us to look at different frequency ranges separately. The frequency dependence of the reflectivity of a typical metal film is displayed in Fig. B.5a, where we assume a dc conductivity $\sigma_\mathrm{dc} = 10^5\ \Omega^{-1}\ \mathrm{cm}^{-1}$ and a plasma frequency $\nu_\mathrm{p} = \omega_\mathrm{p}/(2\pi c) = 10^4\ \mathrm{cm}^{-1}$. It is basically 1 for $\omega < 1/\tau$, and decreases significantly only when approaching ω_p. Only for very thin films does the plasma edge become broader and the low frequency value of the reflectivity decrease. Note that we do not consider surface scattering or other deviations of the thin film material properties from the bulk behavior. If the layer is much thicker than the skin depth given by Eq. (2.3.15b), there is no interference and the intensity dies off following Beer's law (2.3.14). This implies that the amplitude of the electric field $\mathbf{E}$ decreases to $\exp\{-2\pi\} \approx 1/536$ per wavelength propagating into the metal. In the low frequency limit $\lambda > d$, Eq. (B.13) applies, which leads to a frequency independent transmission. Fig. B.5b shows that the transmission has a constant value up to a point where the skin depth δ_0 becomes comparable to the sample thickness d, above which it decreases exponentially. With the frequency increasing further, we exceed the Hagen–Rubens limit ($\omega \ll 1/\tau$) and the reflection off the surface decreases, causing a dramatic increase in the transmitted radiation (ultraviolet transparency). For very thin films (e.g. $d = 0.1\ \mu\mathrm{m}$ in Fig. B.5b), the transmission does not change. The absorptivity $A_\mathrm{F} = 1 - (R_\mathrm{F} + T_\mathrm{F})$ shows a similar behavior (Fig. B.5c). For $\delta_0 \ll d$ the absorptivity is frequency independent; its value is inversely proportional to the thickness of the film. In the range $\omega \ll 1/\tau$ it follows approximately the Hagen–Rubens relation (5.1.17)

$$A_\mathrm{F} \approx 1 - R_\mathrm{F} = \left(\frac{2\omega}{\pi\sigma_\mathrm{dc}}\right)^{1/2} \quad , \tag{B.23}$$

since T_F is many orders of magnitude smaller. Above the scattering rate the absorptivity stays constant up to the plasma frequency.

Finally the optical properties of a thin semiconducting film show distinct differences from those of the bulk material. While the reflectivity R_F of an infinitely thick semiconductor slab drops off according to the Hagen–Rubens relation (B.23), the film with a finite thickness has clear modulations due to the multireflections on

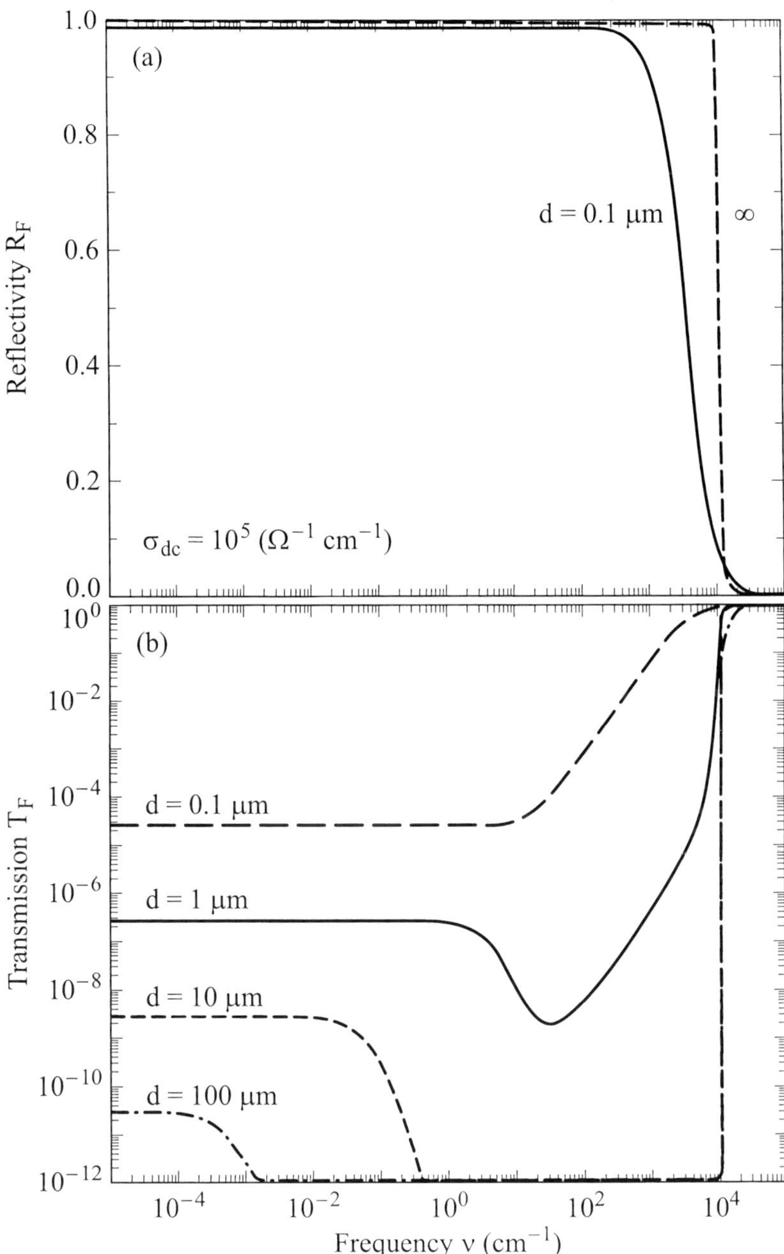

Fig. B.5. Optical properties of a thin metal film versus frequency with a dc conductivity of $\sigma_{dc} = 10^5 \ \Omega^{-1} \ cm^{-1}$ and a plasma frequency $\nu_p = 10^4 \ cm^{-1}$. (a) Reflectivity R_F of films with thickness $d = 0.1$ mm compared to the bulk properties. (b) Transmission T_F of a thin metal film for different thicknesses d of the film.

top of this behavior (Eq. (B.9)). The median reflectivity is given by

$$R_{\mathrm{F}}^{\mathrm{median}} = R \frac{1 + \exp\{-2\alpha d\}}{1 + R^2 \exp\{-2\alpha d\}} \quad ,$$

while the values of the minima and maxima can be calculated by Eq. (B.20). Comparison with the insulating case shows that the spacing $\Delta f = c/(2nd)$ does not change with increasing losses in the material. The transmission T_{F} and the absorptivity $A_{\mathrm{F}} = 1 - R_{\mathrm{F}} - T_{\mathrm{F}}$ show an interference pattern. The extrema of the transmission are given above by Eq. (B.22); the average transmission can be written as

$$T_{\mathrm{F}}^{\mathrm{median}} = \frac{(1 - R)^2 \exp\{-\alpha d\}}{1 + R^2 \exp\{-2\alpha d\}} \quad ,$$

which in our example is only slightly frequency dependent.

B.3.2 Multilayers

The analysis of the optical properties of a sandwich structure consisting of many layers has to start from the complex Fresnel formulas as derived in Section 2.4

$$\hat{r}_{12} = \frac{\hat{N}_1 - \hat{N}_2}{\hat{N}_1 + \hat{N}_2} \qquad \text{and} \qquad \hat{t}_{12} = \frac{2\hat{N}_2}{\hat{N}_1 + \hat{N}_2} \tag{B.24}$$

for light traveling from medium 1 to medium 2. For a one-layer system (i.e. a material of thickness d, refractive index n, and extinction coefficient k which is situated between materials labeled by the subscripts 1 and 3), the total reflection and transmission coefficients are

$$\hat{r}_{123} = \frac{\hat{r}_{12} + \hat{r}_{23} \exp\{2i\delta\}}{1 + \hat{r}_{12}\hat{r}_{23} \exp\{2i\delta\}} \qquad \text{and} \qquad \hat{t}_{123} = \frac{\hat{t}_{12}\hat{t}_{23} \exp\{i\delta\}}{1 + \hat{r}_{12}\hat{r}_{23} \exp\{2i\delta\}} \quad , \tag{B.25}$$

where the reflection and transmission coefficients of each interface are calculated according to Eqs (B.24) and the complex angle

$$\delta = \beta + i\alpha d/2 = 2\pi d(n + ik)/\lambda_0$$

as defined above. On calculating the power reflection coefficient, we see that these results are identical to the Eqs (B.9) and (B.12a) obtained in the previous subsection.

Obviously this procedure can be repeated for multilayer systems by applying the same method developed in the impedance mismatch approach (Section B.2). The complex reflection coefficient $\hat{r}_{23}$ of the last interface has to be replaced by the total reflection coefficient of two subsequent interfaces according to Eq. (B.25). Thus we obtain after some rearrangement

$$\hat{r}_{1234} = \frac{\hat{r}_{12} + \hat{r}_{23}\exp\{2i\delta_2\} + \hat{r}_{34}\exp\{2i(\delta_2 + \delta_3)\} + \hat{r}_{12}\hat{r}_{23}\hat{r}_{34}\exp\{2i\delta_3\}}{1 + \hat{r}_{12}\hat{r}_{23}\exp\{2i\delta_2\} + \hat{r}_{23}\hat{r}_{34}\exp\{2i\delta_3\} + \hat{r}_{12}\hat{r}_{34}\exp\{2i(\delta_2 + \delta_3)\}} \ . \tag{B.26}$$

The corresponding formula describing the transmission of layer 2 and layer 3 between media 1 and media 4 has the form

$$\hat{t}_{1234} = \frac{\hat{t}_{12}\hat{t}_{23}\hat{t}_{34}\exp\{i(\delta_2 + \delta_3)\}}{1 + \hat{r}_{12}\hat{r}_{23}\exp\{2i\delta_2\} + \hat{r}_{23}\hat{r}_{34}\exp\{2i\delta_3\} + \hat{r}_{12}\hat{r}_{34}\exp\{2i(\delta_2 + \delta_3)\}} \ , \tag{B.27}$$

where $\hat{t}_{pq}$ and $\hat{r}_{pq}$ are the complex Fresnel transmission and reflection coefficients at the boundaries between media p and media q (with $p, q = 1, 2, 3, 4$), and the complex angle $\delta_p = \beta_p + i\alpha_p d_p/2 = 2\pi d_p(n_p + ik_p)/\lambda_0$.

By iteration we arrive at the equations for a three-layer system:

$$\hat{r}_{12345} = \frac{A}{B} \ , \tag{B.28}$$

where

$$\begin{aligned}
A \ = \ & \hat{r}_{12} + \hat{r}_{23}\exp\{2i\delta_2\} + \hat{r}_{34}\exp\{2i(\delta_2 + \delta_3)\} + \hat{r}_{45}\exp\{2i(\delta_2 + \delta_3 + \delta_4)\} \\
& + \hat{r}_{12}\hat{r}_{23}\hat{r}_{34}\exp\{2i\delta_3\} + \hat{r}_{23}\hat{r}_{34}\hat{r}_{45}\exp\{2i(\delta_2 + \delta_4)\} \\
& + \hat{r}_{12}\hat{r}_{34}\hat{r}_{45}\exp\{2i\delta_4\} + \hat{r}_{12}\hat{r}_{23}\hat{r}_{45}\exp\{2i(\delta_3 + \delta_4)\}
\end{aligned}$$

and

$$\begin{aligned}
B \ = \ & 1 + \hat{r}_{12}\hat{r}_{23}\exp\{2i\delta_2\} + \hat{r}_{23}\hat{r}_{34}\exp\{2i\delta_3\} + \hat{r}_{34}\hat{r}_{45}\exp\{2i\delta_4\} \\
& + \hat{r}_{12}\hat{r}_{34}\exp\{2i(\delta_2 + \delta_3)\} + \hat{r}_{23}\hat{r}_{45}\exp\{2i(\delta_3 + \delta_4)\} \\
& + \hat{r}_{12}\hat{r}_{45}\exp\{2i(\delta_2 + \delta_3 + \delta_4)\} + \hat{r}_{12}\hat{r}_{23}\hat{r}_{34}\hat{r}_{45}\exp\{2i(\delta_2 + \delta_4)\} \ .
\end{aligned}$$

The transmission coefficient of the three-layer system is given by (layers 2, 3, and 4 between media 1 and 5)

$$\hat{t}_{12345} = \frac{\hat{t}_{12}\hat{t}_{23}\hat{t}_{34}\hat{t}_{45}\exp\{i(\delta_2 + \delta_3 + \delta_4)\}}{B} \ , \tag{B.29}$$

where the notations are the same as used before.

In practice a transparent known substrate covered on one side with the material of interest represents a typical two-layer system. The values n and k of the bare substrate are known or measured beforehand. In Fig. B.6 the influence of the thickness of a conducting film on the interference pattern is demonstrated. One side of a sapphire substrate ($d_3 = 0.41$ mm) is covered with a metallic film of NbN with $d_2 = 20$ Å, 140 Å, and 300 Å (curves 2, 3, and 4, respectively). The

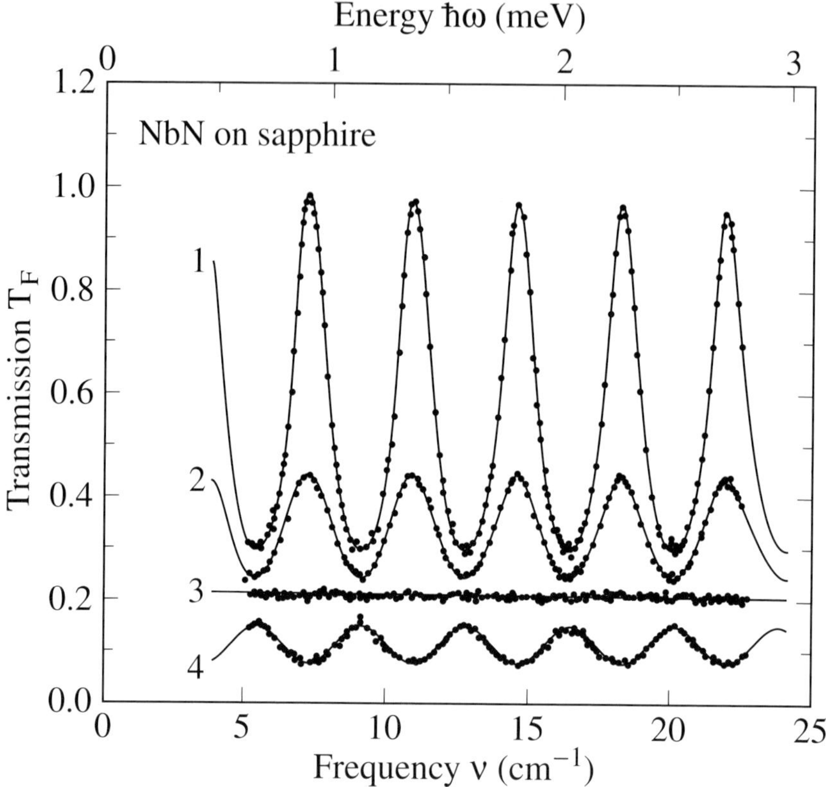

Fig. B.6. Transmission through a plane parallel sapphire plate (thickness $d_3 = 0.41$ mm) with a metallic NbN film of different thicknesses: bare substrate (curve 1); $d_2 = 20$ Å (curve 2); $d_2 = 140$ Å (curve 3); and $d_2 = 300$ Å (curve 4). The spectra are taken at room temperature. The lines are fits according to Eq. (B.27) (from [Gor93a]).

interference pattern can be well described by Eq. (B.27) using $n_3 = 3.39$ and $k_3 = 0.0003$ as the refractive index of the substrate [Gor93a]. It is interesting to compare these results with those displayed in Fig. 11.17, where not the thickness of the film but the conductivity of the material changes; the transmission and the phase shift depend on the surface impedance of the film $\hat{Z}_F = (\hat{\sigma}d)^{-1}$ evaluated by Eq. (B.1).

In Fig. B.7a the transmission spectra of a 115 Å thick NbN film at two different temperatures above and below the superconducting transition are shown. The solid lines in the upper panel represent the theoretical fit of the multireflection in this two-layer system. In the normal state transmission spectra vary only slightly while the temperature decreases from 300 K down to transition temperature $T_c \approx 12$ K. Below T_c we see that at small frequencies the transmission decreases by about

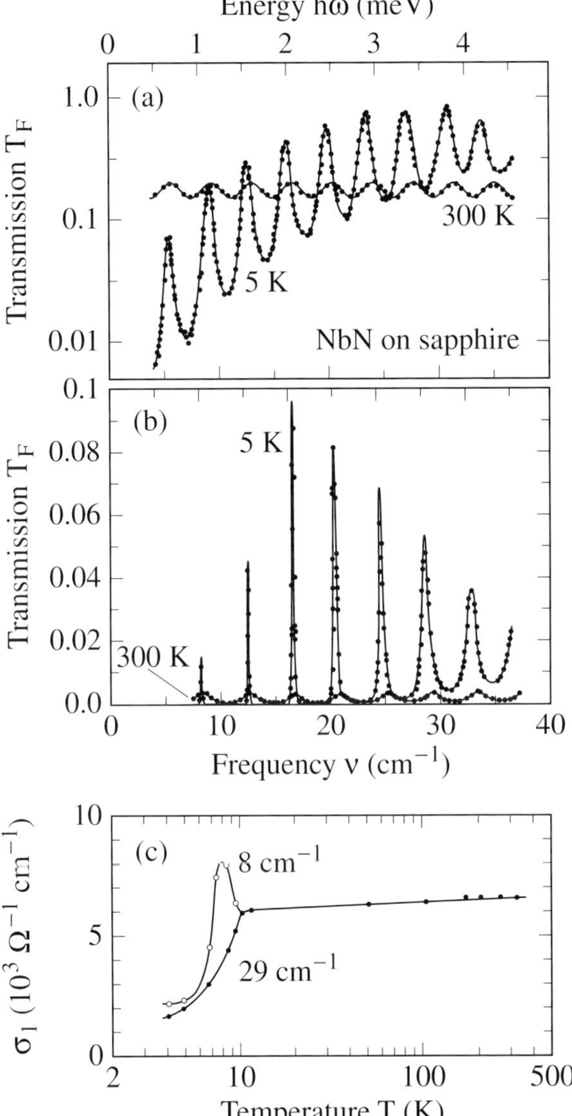

Fig. B.7. (a) Transmission spectra $T_F(\nu)$ of an asymmetrical Fabry–Perot resonator, formed by a NbN film ($d_2 = 115$ Å) on a sapphire substrate (0.43 mm), measured at two temperatures above and below critical temperature $T_c \approx 12$ K. (b) Transmission spectra $T_F(\nu)$ of a plane parallel sapphire plate (0.41 mm) covered with NbN films on both sides ($d_2 = 250$ Å and $d_4 = 60$ Å), again measured at 300 K and 5 K. The solid lines represent the theoretical fit. (c) For the latter case the temperature dependence of the optical conductivity $\sigma_1(T)$ of NbN film is calculated for two frequencies, 8 cm^{-1} and 29 cm^{-1} (after [Gor93a, Gor93b]). Note the good agreement with the theoretical predictions shown in Fig. 7.4.

an order of magnitude, while at higher frequencies $\omega > 2\Delta/\hbar$ it increases. If the substrate is covered on both sides, a three-layer system is obtained where the internal reflections in the substrate are significantly enhanced. A system with two metallic films on a dielectric substrate is essentially a symmetrical Fabry–Perot resonator with a much higher quality factor Q (or finesse $\mathcal{F}$ according to Eq. (11.3.1)) compared to the asymmetrical one. The sensitivity of the measurement is increased because the interaction of the wave with the material is multiplied. The transmission coefficients of such a Fabry–Perot resonator with NbN mirrors is shown in Fig. B.7b. The Q factors of the resonances increase strongly when the films become superconducting, which is connected with decrease of losses and increase of reflectivities of the mirrors. The material parameters of the substrate do not change significantly. Fig. B.7c shows temperature behavior of optical conductivity of the NbN film, obtained by least mean square fitting treatment of the transmission spectra. It is clearly seen that at the superconducting transition the conductivity drops for $T \to 0$, but for low frequencies ($\nu = 8$ cm^{-1}) a peak develops right below T_{c}. The results are in good agreement with the theoretical predictions by the BCS theory as calculated by the Mattis–Bardeen formula (7.4.20) and displayed in Fig. 7.4.

References

[Bor99] M. Born and E. Wolf, *Principles of Optics*, 6th edition (Cambridge University Press, Cambridge, 1999)

[Gor93a] B.P. Gorshunov, G.V. Kozlov, A.A. Volkov, S.P. Lebedev, I.V. Fedorov, A.M. Prokhorov, V.I. Makhov, J. Schützmann, and K.F. Renk, *Int. J. Infrared Millimeter Waves* **14**, 683 (1993)

[Gor93b] B.P. Gorshunov, I.V. Fedorov, G.V. Kozlov, A.A. Volkov, and A.D. Semenov, *Solid State Commun.* **87**, 17 (1993)

[Hea65] O.S. Heavens, *Optical Properties of Thin Solid Films* (Dover, New York, 1963)

[Hec98] E. Hecht, *Optics*, 3rd edition (Addison Wesley, Reading, MA, 1998)

[Ram94] S. Ramo, J.R. Whinnery, and T.v. Duzer, *Fields and Waves in Communication Electronics*, 3rd edition (John Wiley & Sons, New York, 1994)

[Sch75] H. Schunk, *Stromverdrängung* (Hüthig Verlag, Heidelberg, 1975)

[Sch94] G. Schaumburg and H.W. Helberg, *J. Phys. III (France)* **4**, 917 (1994)

[Wol34] W. Woltersdorff, *Z. Phys.* **91**, 230 (1934)

Further reading

[Kle86] M.V. Klein, *Optics*, 2nd edition (John Wiley & Sons, New York, 1986)

Appendix C

$\mathbf{k} \cdot \mathbf{p}$ perturbation theory

The method of $\mathbf{k} \cdot \mathbf{p}$ perturbation theory was originally introduced by J. Bardeen and F. Seitz as a means of determining effective masses and crystal wavefunctions near high symmetry points of $\mathbf{k}$ [Sei40]. Considering small $\mathbf{q}$ as a perturbation, the $\mathbf{k} \cdot \mathbf{p}$ perturbation theory can be utilized to evaluate the transition between $|\mathbf{k}\rangle$ and $|\mathbf{k}'\rangle$ states. The matrix element which appears in Eq. (4.3.20) can be evaluated using Bloch functions (Eq. (4.3.5)). Using the wavefunction $\psi_{\mathbf{k}l} = \Omega^{-1/2} u_{\mathbf{k}l} \exp\{i\mathbf{k}\cdot\mathbf{r}\} = |\mathbf{k}l\rangle$ the Schrödinger equation $\mathcal{H}_{\mathbf{k}} \psi_{\mathbf{k}l} = \mathcal{E}_{\mathbf{k}l} \psi_{\mathbf{k}l}$ becomes

$$\mathcal{H}_{\mathbf{k}} u_{\mathbf{k}l} = \mathcal{E}_{\mathbf{k}l} u_{\mathbf{k}l} \quad ,$$

with the Hamilton operator

$$\mathcal{H}_{\mathbf{k}} = \frac{\mathbf{p}^2}{2m} + \frac{\hbar}{m} \mathbf{k} \cdot \mathbf{p} + \frac{\hbar^2 k^2}{2m} + V(\mathbf{r}) \quad . \tag{C.1}$$

In order to find $u_{\mathbf{k}+\mathbf{q},l'}$ in the Schrödinger equation $\mathcal{H}_{\mathbf{k}+\mathbf{q}} u_{\mathbf{k}+\mathbf{q},l'} = \mathcal{E}_{\mathbf{k}+\mathbf{q},l'} u_{\mathbf{k}+\mathbf{q},l'}$, we assume that $\mathbf{q}$ is small, and consequently the Hamiltonian becomes

$$\mathcal{H}_{\mathbf{k}+\mathbf{q}} = \mathcal{H}_{\mathbf{k}} + \frac{\hbar \mathbf{q} \cdot (\mathbf{p} + \hbar\mathbf{k})}{m} + \frac{\hbar^2 q^2}{2m} \quad , \tag{C.2}$$

where the second term is considered a perturbation. The state $u_{\mathbf{k}+\mathbf{q},l'}$ can be found by performing a perturbation around $u_{\mathbf{k}l'}$

$$
\begin{aligned}
|\mathbf{k} + \mathbf{q}l'\rangle_* &= |\mathbf{k}l'\rangle_* + \sum_{l''} |\mathbf{k}l''\rangle_* \frac{\langle \mathbf{k}l''|\mathcal{H}_{\mathbf{k}+\mathbf{q}} - \mathcal{H}_{\mathbf{k}}|\mathbf{k}l',\rangle_*}{\mathcal{E}_{\mathbf{k}l'} - \mathcal{E}_{\mathbf{k}l''}} \\
&= |\mathbf{k}l'\rangle_* + \frac{\hbar}{m} \sum_{l''} |\mathbf{k}l''\rangle_* \frac{\langle \mathbf{k}l''|\mathbf{q} \cdot \mathbf{p}|\mathbf{k}l'\rangle_*}{\mathcal{E}_{\mathbf{k}l'} - \mathcal{E}_{\mathbf{k}l''}} \quad ,
\end{aligned} \tag{C.3}
$$

where the integration over the squared terms ($\mathbf{q} \cdot \mathbf{k}$ and q^2) in Eq. (C.2) is zero since they are not operators and the Bloch functions are orthogonal. Here the $\langle\ \rangle_*$

indicates that we consider a unit cell (Eq. (4.3.12)). Thus

$$\langle \mathbf{k}l|\mathbf{k}+\mathbf{q},l'\rangle_* = \langle \mathbf{k}l|\mathbf{k}l'\rangle_* + \frac{\hbar}{m}\sum_{l''}\frac{\langle \mathbf{k}l|\mathbf{k}l''\rangle_*\langle \mathbf{k}l''|\mathbf{q}\cdot\mathbf{p}|\mathbf{k}l'\rangle_*}{\mathcal{E}_{\mathbf{k}l'}-\mathcal{E}_{\mathbf{k}l''}}$$

$$= \delta_{ll'} + \frac{\hbar}{m}\sum_{l'}\frac{\langle \mathbf{k}l|\mathbf{q}\cdot\mathbf{p}|\mathbf{k}l'\rangle_*}{\mathcal{E}_{\mathbf{k}l'}-\mathcal{E}_{\mathbf{k}l}} \quad , \tag{C.4}$$

since $\langle \mathbf{k}l|\mathbf{k}l''\rangle_* = 0$ for $l \neq l''$ because the periodic part is orthogonal $\langle u_{\mathbf{k}l}u_{\mathbf{k}l'}\rangle = \delta_{ll'}$. By defining the momentum operator $P_{ll'}$ in the direction of $\mathbf{q}$ and the energy difference $\hbar\omega_{l'l}$ as

$$P_{ll'} = \Delta^{-1}\int d\mathbf{r}\, u_{\mathbf{k}l}\, p\, u_{\mathbf{k}l'}$$

$$\hbar\omega_{l'l} = \mathcal{E}_{\mathbf{k}l'}-\mathcal{E}_{\mathbf{k}l}$$

with Δ the volume of the unit cell, we obtain

$$\langle \mathbf{k}l|\mathbf{k}+\mathbf{q},l'\rangle_* = \delta_{ll'} + \frac{\hbar q}{m}\sum_{l'}\frac{P_{ll'}}{\hbar\omega_{l'l}} \quad . \tag{C.5}$$

Since the summation is reduced to the exclusion of the term $l = l'$, we finally obtain for the expression in Eq. (4.3.20)

$$|\langle \mathbf{k}+\mathbf{q},l'|\exp\{i\mathbf{q}\cdot\mathbf{r}\}|\mathbf{k}l\rangle_*|^2 = \delta_{ll'} + (1-\delta_{ll'})\left(\frac{q}{m\omega_{l'l}}\right)^2|P_{l'l}|^2 \quad , \tag{C.6}$$

which is equal to unity for $l = l'$, but has to be taken into account in the case of interband transitions.

Reference

[Sei40] F. Seitz, *The Modern Theory of Solids* (McGraw-Hill, New York, 1940)

Further reading

[Jon73] W. Jones and N.H. March, *Theoretical Solid State Physics* (John Wiley and Sons, New York, 1973)

[Mer70] E. Merzbacher, *Quantum Mechanics*, 2nd edition (John Wiley & Sons, New York, 1970)

[Woo72] F. Wooten, *Optical Properties of Solids* (Academic Press, San Diego, CA, 1972)

Appendix D

Sum rules

The interaction of an electronic system with the electromagnetic field lead to an absorption process. The total absorption is finite, and this is expressed by sum rules. These can be derived by examining the absorption process, but often the Kramers–Kronig relations are used as a starting point (together with some general assumptions about the absorption itself).

D.1 Atomic transitions

The interaction of a free atom with the electromagnetic field is treated in numerous textbooks (see for example [Mer70]), and therefore only the main results will be recalled here. For an atom initially in the ground state, the response to an electromagnetic field of the harmonic form $\mathbf{E} = E_0 \cos\{\omega t\}$ leads to transitions between the ground and excited states. The relevant matrix element is related to the dipole moment

$$r_{l0} = \int \Psi_l^* \, e\mathbf{r} \, \Psi_0 \, d\mathbf{r} \quad ,$$

where Ψ_0 and Ψ_l are the wavefunctions of the ground state and excited states. The transition probability between the ground state and state l is described by the so-called oscillator strength (Eq. (6.1.7)):

$$f_{l0} = \frac{2m}{\hbar^2} \hbar\omega_{l0} \, |r_{l0}|^2 \quad , \tag{D.1}$$

where m is the electron mass and $\hbar\omega_{l0} = \mathcal{E}_l - \mathcal{E}_0$ is the energy difference between the ground state and the excited state in question; $\hbar\omega_{l0}$ is positive if the energy of the final state l is higher (upward transition) and negative for downward transitions. Here the transition matrix element is given in terms of the electric dipole moment, but we can equally well use the momentum matrix element

$$\mathbf{p}_{l0} = \int \Psi_l^* \, (i\hbar\nabla) \, \Psi_0 \, d\mathbf{r} \quad ,$$

and in this form f_{l0} becomes

$$f_{l0} = \frac{2\,|\mathbf{p}_{l0}|^2}{m\hbar\omega_{l0}} \quad . \tag{D.2}$$

The oscillator strength f_{l0} measures the relative probability of a quantum mechanical transition between two atomic levels. The transition obeys the so-called Thomas–Reiche–Kuhn sum rule

$$\sum_l f_{l0} = 1 \quad , \tag{D.3}$$

which can be proven by considering some general commutation rules.

The absorbed power per unit time can also be calculated in a straightforward manner, and one finds that

$$P = \frac{\pi e^2 E_0^2 \, |r_{l0}|^2 \, \omega_{l0}}{\hbar} \quad , \tag{D.4}$$

which can be cast into the form of

$$P = \frac{2m\hbar\omega_{l0}\,|r_{l0}|^2}{\hbar^2} \, \frac{\pi e^2 E_0^2}{2m} = \frac{\pi e^2 E_0^2}{2m} \sum_l f_{l0} \quad . \tag{D.5}$$

The expression in front of the sum is just the power absorption by a classical oscillator, which itself must be equal to P, and thus $\sum_l f_{l0} = 1$ as stated above. If these are a collection of atoms, as in a solid, which might also have more than a single electron, the sum rule is modified to become [Ste63]

$$\sum_l f_{l0} = N \quad , \tag{D.6}$$

where N is the number of electrons per unit volume. Note that the sum rule does not contain information, and thus is independent of the wavefunctions and energies of the individual energy levels. Of course the oscillator strength associated with the transitions is sensitive to these; however, the sum of the oscillator strengths must be fixed, and must be independent from the details of the system in question.

D.2 Conductivity sum rules

Here the absorbed power has beeen expressed in terms of the oscillator strength associated with the various transitions between the energy levels of the atoms, but the sum rule can also be expressed in terms of the optical conductivity $\sigma_1(\omega)$. In order to see this, let us calculate the rate of energy absorption per unit time for a time-varying field as given before. This rate is given by Eq. (3.2.34) as

$$\frac{d\mathcal{E}}{dt} = \text{Re}\left\{\frac{\partial \mathbf{D}}{\partial t} \cdot \mathbf{E}\right\} = \text{Re}\left\{(\epsilon_1 + i\epsilon_2)\frac{\partial \mathbf{E}}{\partial t} \cdot \mathbf{E}\right\} = \omega\epsilon_2 \frac{E_0^2}{2} = \frac{\sigma_1 E_0^2}{8\pi} \quad . \tag{D.7}$$

The total absorption per unit time is

$$P = \int \frac{E_0^2}{4\pi}\omega\epsilon_2(\omega)\,d\omega = E_0^2 \int \sigma_1(\omega)\,d\omega \quad , \tag{D.8}$$

which, in view of our previous expression of the absorbed power, becomes

$$\int \sigma_1(\omega)\,d\omega = \frac{\pi N e^2}{2m} = \frac{\omega_p^2}{8} \quad , \tag{D.9}$$

with $\omega_p = \left(4\pi N e^2/m\right)^{1/2}$ the so-called plasma frequency. It is clear how valuable such a sum rule argument is: while the frequency dependence of the conductivity can be influenced by many features, including band structure and interaction effects, the sum rule is independent of these and must be obeyed.

D.3 Sum rule from Kramers–Kronig relations

The same relation can be derived by utilizing the Kramers–Kronig relations, and from some general arguments about the conductivity or dielectric constant. In order to do this, first note that the complex dielectric function has, at high frequencies, the following limiting form:

$$\epsilon_1(\omega) = 1 - \frac{\omega_p^2}{\omega^2} \quad , \tag{D.10}$$

already derived in Eq. (3.2.25). This in general follows from the fact that inertial effects dominate the response at high frequencies. Next let us look at the Kramers–Kronig relation (3.2.12a)

$$\epsilon_1(\omega) - 1 = \frac{2}{\pi}\mathcal{P}\int_0^\infty \frac{\omega'\epsilon_2(\omega')}{\omega'^2 - \omega^2}\,d\omega' \quad .$$

This, in the high frequency limit – at the frequency above which there is no loss, and consequently $\epsilon_2(\omega) = 0$ – then reads

$$\epsilon_1(\omega) - 1 \approx -\frac{2}{\pi\omega^2}\int_0^\infty \omega'\epsilon_2(\omega')\,d\omega' \quad .$$

Equating this with expression (D.10), we find that

$$\int_0^\infty \omega\epsilon_2(\omega)\,d\omega = \frac{\pi}{2}\omega_p^2 \quad , \tag{D.11}$$

or alternatively

$$\int_0^\infty \sigma_1(\omega)\,\mathrm{d}\omega = \frac{\pi N e^2}{2m} = \frac{\omega_{\mathrm{p}}^2}{8} \quad , \tag{D.12}$$

the same as found in Eq. (D.9).

D.4 Sum rules in a crystal

The sum rule given above holds for a collection of atoms in a solid; and in a crystal it can be directly related to the electron structure in certain limits. Using Bloch wavefunctions with the periodic part $u_{\mathbf{k}l}$ as defined in Eq. (4.3.5) to describe the electronic states in a crystal, the Schrödinger equation reads $\mathcal{H}_{\mathbf{k}}u_{\mathbf{k}l} = \mathcal{E}_{\mathbf{k}l}u_{\mathbf{k}l}$, where the Hamilton operator is given by

$$\mathcal{H}_{\mathbf{k}} = \mathcal{H}_0 + \mathcal{H}_1 + \mathcal{H}_2 = \left[\frac{\mathbf{p}^2}{2m} + V(\mathbf{r}) \right] + \frac{\hbar}{m}\mathbf{k}\cdot\mathbf{p} + \frac{\hbar^2 \mathbf{k}^2}{2m} \quad . \tag{D.13}$$

$V(\mathbf{r})$ denotes the periodic potential. The term in the squared brackets describes the energy levels at $\mathbf{k} = 0$. $\mathcal{H}_1$ is a first order perturbation and $\mathcal{H}_2$ is a second order perturbation. With the zero order expressions $u_{\mathbf{k}l} = u_{0l}$ and $\mathcal{E}_{\mathbf{k}l} = \mathcal{E}_{0l}$, we obtain for the first order terms

$$\mathcal{E}_{\mathbf{k}l} = \mathcal{E}_{0l} + \frac{\hbar}{m}\mathbf{k}\cdot\langle 0l|\mathbf{p}|0l\rangle_* \quad ,$$

where the correction to the energy cancels for crystals with inversion symmetry. Therefore, the second order correction to the energy also becomes important, and

$$\mathcal{E}_{\mathbf{k}l} = \mathcal{E}_{0l} + \frac{\hbar^2 \mathbf{k}^2}{2m} + \frac{\hbar^2}{m^2}\sum_{l'\neq l}\frac{|\mathbf{k}\cdot\langle 0l'|\mathbf{p}|0l\rangle_*|^2}{\mathcal{E}_{0l} - \mathcal{E}_{0l'}} \quad , \tag{D.14}$$

where the labels l and l' refer to different electron states in the crystal.

A Taylor series expansion for the energy of an electron in the neighborhood of $\mathbf{k} = 0$ gives:

$$\mathcal{E}_{\mathbf{k}l} = \mathcal{E}_{0l} + \frac{\partial \mathcal{E}_{0l}}{\partial k_i}k_i + \frac{1}{2}\frac{\partial^2 \mathcal{E}_{0l}}{\partial k_i \partial k_j}k_i k_j = \mathcal{E}_{0l} + \frac{\partial \mathcal{E}_{0l}}{\partial k_i}k_i + \frac{\hbar^2}{2m_{ij}^*}k_i k_j \quad , \tag{D.15}$$

where we have substituted the effective mass tensor m_{ij}^* defined in analogy to Eq. (12.1.17):

$$\left(\frac{1}{m^*}\right)_{ij} = \frac{1}{\hbar^2}\frac{\partial^2 \mathcal{E}(\mathbf{k})}{\partial k_i \partial k_j} \quad .$$

We can now compare the third term of this expansion with the corresponding term of Eq. (D.14) obtained by second order perturbation, and we find that

$$\frac{m}{m^*_{ij}} = \delta_{ij} + \frac{2}{m} \sum_{l' \neq l} \frac{\langle 0l | p_i | 0l' \rangle_* \langle 0l' | p_j | 0l \rangle_*}{\mathcal{E}_{0l} - \mathcal{E}_{0l'}} \quad . \tag{D.16}$$

With the expression of the oscillator strength in terms of the momentum matrix element, given by Eq. (D.2), where again $\hbar \omega_{l'l} = \mathcal{E}_{l'} - \mathcal{E}_l$, we immediately obtain for a cubic crystal

$$\frac{m}{m^*} = 1 - \sum_{l \neq l'} f_{l'l} \quad . \tag{D.17}$$

Thus the bandmass is related to the interband transition. For free electrons $m^* = m$, and thus $\sum_{l \neq l'} f_{ll'} = 0$; there is no absorption by a free-electron gas. This is not surprising, as – in the absence of collisions which absorb momentum – the different electronic states correspond to different momenta, and thus the electromagnetic radiation for $\mathbf{q} = 0$ cannot induce transitions between these states. This also follows from the Drude model, in the collisionless, $\tau \to \infty$ limit, for which

$$\hat{\sigma}(\omega) = i \frac{Ne^2}{m\omega}$$

and there is no dissipation at any finite frequency. For τ finite, however, momentum and energy are absorbed during collisions, and these processes lead to the absorption of electromagnetic radiation at finite frequencies, consequently the sum rule is restored, and

$$\int \sigma_1^{\text{Drude}}(\omega) \, d\omega = \frac{\pi Ne^2}{2m} \quad ,$$

as can be verified by direct integration of the simple Drude expression (5.1.8).

D.5 Electron gas with scattering

Of course, the Drude model, with a frequency independent relaxation rate and mass, is a crude approximation, and even in the absence of a lattice electron–electron and electron–phonon interactions result in frequency dependences of these parameters. In the absence of interband transitions, the sum rule for the conduction band must be conserved – i.e. the plasma frequency is independent of the interactions. This then sets conditions for $1/\tau(\omega)$ and $m(\omega)$, and it is expected that the proper Kramers–Kronig relations with $1/\tau(\omega)$ and $m(\omega)$ will leave the total spectral weight associated with $\sigma_1(\omega)$ unchanged.

This also holds when these interactions lead to broken symmetry ground states, such as superconductivity or density waves: for these states the sum rule including both collective and single-particle excitations must be the same as the sum rule which is valid above the transition, in the metallic state. The so-called Tinkham–Ferrell sum rule for superconductors – discussed in Section 7.4.1 – has its origin in this condition, and analogous sum rule arguments can be developed for the density wave states (Section 7.5.3).

References

[Mer70] E. Merzbacher, *Quantum Mechanics*, 2nd edition (John Wiley & Sons, New York, 1970)

[Ste63] F. Stern, *Elementary Theory of the Optical Properties of Solids*, in: *Solid State Physics* **15**, edited by F. Seitz and D. Turnbull (Academic Press, New York, 1963), p. 299

Further reading

[Jon73] W. Jones and N.H. March, *Theoretical Solid State Physics* (John Wiley and Sons, New York, 1973)

[Mah90] G.D. Mahan, *Many-Particle Physics*, 2nd edition (Plenum Press, New York, 1990)

[Rid93] B.K. Ridley, *Quantum Processes in Semiconductors*, 3rd edition (Clarendon Press, Oxford, 1993)

[Smi85] D.Y. Smith, *Dispersion Theory, Sum Rules, and Their Application to the Analysis of Optical Data*, in: *Handbook of Optical Constants of Solids*, Vol. 1, edited by E.D. Palik (Academic Press, Orlando, FL, 1985), p. 35

[Woo72] F. Wooten, *Optical Properties of Solids* (Academic Press, San Diego, CA, 1972)

Appendix E
Non-local response

Local electrodynamics assumes that the response at one point in the material only depends on the electric field at this point. This assumption breaks down in the case of metals with a long mean free path ℓ and in the case of superconductors with a long coherence length ξ_0. The anomalous skin effects become important if $\delta < \ell$. For superconductors the London limit is exceeded if $\lambda_L < \xi_0$. In both cases the influence of electrons, which feel a different electric field at some distant point, becomes important.

E.1 Anomalous skin effect

A brief discussion of the anomalous skin effect was given in Section 5.2.5. Here we want to go beyond the ineffectiveness concept of Pippard and discuss the surface impedance in particular. For a full treatment of the anomalous skin effect [Abr88, Cha90, Pip62, Sok67, Son54] we must solve the Maxwell equations for this special boundary problem. We will not consider the effect of an external magnetic field, which leads to a quite different behavior, such as helicons or Alfvén waves [Abr88, Kan68]. A detailed discussion of the anomalous skin effect in the presence of an external magnetic field can be found in [Kar86]. If the surface of a metal ($\mu_1 = 1$) is in the xy plane, the wave equation for the electric field with harmonic time dependence (2.2.16a) has the form

$$\frac{\mathrm{d}^2 E(r)}{\mathrm{d}z^2} + \frac{\omega^2 \epsilon_1}{c^2} E(r) = -\frac{4\pi \mathrm{i}\omega}{c^2} J(r) \quad . \tag{E.1}$$

Assuming the electrons are spatially reflected at the surface of the metal (for diffusive reflection only a numerical factor changes slightly [Kit63, Reu48, Sok67]), the boundary condition can be written as

$$\left(\frac{\partial E}{\partial z}\right)_{z \to 0} = -\left(\frac{\partial E}{\partial z}\right)_{0 \leftarrow z} \quad .$$

The discontinuity of its first derivative modifies the wave equation in the following way

$$\frac{\mathrm{d}^2 E(r)}{\mathrm{d}z^2} + \frac{\omega^2 \epsilon_1}{c^2} E(r) = -\frac{4\pi \mathrm{i}\omega}{c^2} J(r) + 2 \left(\frac{\mathrm{d}E}{\mathrm{d}z}\right)_{z=0} \delta(z) \quad , \tag{E.2}$$

which, when transformed in Fourier space (see Section A.1), yields

$$-q^2 E(q) + \frac{\omega^2 \epsilon_1}{c^2} E(q) = -\frac{4\pi \mathrm{i}\omega}{c^2} J(q) + 2 \left(\frac{\mathrm{d}E}{\mathrm{d}z}\right)_{z=0} \quad . \tag{E.3}$$

We obtain the general solution from the Chambers formula

$$\mathbf{J} = \frac{2e^2}{(2\pi)^3} \int \mathrm{d}\mathbf{k} \left(-\frac{\partial f^0}{\partial \mathcal{E}}\right) \frac{\mathbf{v}(\mathbf{k})}{v} \int_{\mathbf{r}} \mathbf{E}(r') \cdot \mathbf{v}(\mathbf{k}, r') \exp\left\{-\frac{r'}{\ell}\right\} \mathrm{d}r' \quad . \tag{E.4}$$

Assuming that the scattering rate is independent of $\mathbf{q}$, we can also utilize Boltzmann's transport theory, and for the current density in the low temperature limit we get an expression similar to Eq. (5.2.13). In an even simpler approximation, we can apply the generalized Ohm's law which connects the current and the electric field by the conductivity $\mathbf{J}(\mathbf{q}, \omega) = \sigma_1(\mathbf{q}, \omega)\mathbf{E}(\mathbf{q}, \omega)$. By neglecting the displacement term in Eq. (E.3) containing ω^2/c^2, we obtain for the electric field

$$E(q) = \frac{2}{\sqrt{2\pi}} \left(\frac{\mathrm{d}E}{\mathrm{d}z}\right)_{z=0} \left[\frac{4\pi \mathrm{i}\omega}{c^2} \sigma_1(q, \omega) - q^2\right]^{-1} \quad .$$

Substituting the first order approximation ($\ell \to \infty$) of the conductivity from Eq. (5.2.24a) yields

$$E(q) = \sqrt{\frac{2}{\pi}} \left(\frac{\mathrm{d}E}{\mathrm{d}z}\right)_{z=0} \left[\frac{3\pi^2 \omega}{c^2} \frac{\sigma_{\mathrm{dc}}}{\ell} \frac{\mathrm{i}}{q} - q^2\right]^{-1} \quad . \tag{E.5}$$

The decay function is obtained by the inverse Fourier transform

$$E(z) = \frac{1}{\pi} \left(\frac{\mathrm{d}E}{\mathrm{d}z}\right)_{z=0} \int_{-\infty}^{\infty} \mathrm{d}q \, \frac{\exp\{-\mathrm{i}qz\}}{\frac{3\pi^2 \omega}{c^2} \frac{\sigma_{\mathrm{dc}}}{\ell} \frac{\mathrm{i}}{q} - q^2} \quad , \tag{E.6}$$

and by substituting the integration variable $\zeta = q \left(\frac{c^2}{3\pi^2\omega} \frac{\ell}{\sigma_{\mathrm{dc}}}\right)^{1/3}$ we get at the surface

$$\begin{aligned}
E(z = 0) &= \frac{1}{\pi} \left(\frac{\mathrm{d}E}{\mathrm{d}z}\right)_{z=0} \frac{-2\mathrm{i}c^2 \ell}{3\pi^2 \omega \sigma_{\mathrm{dc}}} \int_0^{\infty} \frac{\mathrm{d}\zeta}{1 + \mathrm{i}\zeta^3} \\
&= -\frac{1}{\pi} \left(\frac{\mathrm{d}E}{\mathrm{d}z}\right)_{z=0} \frac{2\pi}{3} \left(\frac{c^2}{3\pi^2 \omega} \frac{\ell}{\sigma_{\mathrm{dc}}}\right)^{1/3} \left(1 + \frac{\mathrm{i}}{\sqrt{3}}\right) \quad . \tag{E.7}
\end{aligned}$$

As for the normal skin effect, also in the anomalous regime, the penetration of the electric field into the metal has approximately an exponential form, and we can define a characteristic length of how far the field penetrates the metal,

$$\delta = \left(\frac{c^2}{3\pi^2 \omega} \frac{\ell}{\sigma_{dc}} \right)^{1/3} \quad , \tag{E.8}$$

which has the same functional form as obtained from our phenomenological approach, Eq. (5.2.30), with $\gamma \approx 3\pi/2 \approx 4.71$. The electric field falls quickly with distance z so that the major part of the field is confined to a depth much less than ℓ. This rapid decrease is not maintained, however, and eventually, far from the surface, the field is approximately $(z/\ell)^{-2} \exp\{-z/\ell\}$, so there is a long tail of small amplitude extending into the metal to a distance of the order of the mean free path ℓ.

From Eqs (2.4.23) and (2.4.25), the surface impedance of a material is given by

$$\hat{Z}_S = \frac{4\pi}{c^2} i\omega \frac{E(z=0)}{(dE/dz)_{z=0}} \quad . \tag{E.9}$$

For an anomalous conductor the result can be obtained by inserting Eq. (E.7) into Eq. (E.9):

$$\hat{Z}_S(\omega) = \frac{4\pi}{c^2} \frac{2i\omega}{3} \left(\frac{c^2}{3\pi^2\omega} \frac{\ell}{\sigma_{dc}} \right)^{1/3} \left(1 + \frac{i}{\sqrt{3}} \right) = \frac{8}{9} \left(\frac{\sqrt{3}\pi\omega^2}{c^4} \frac{\ell}{\sigma_{dc}} \right)^{1/3} (1 + \sqrt{3}i) \tag{E.10}$$

in the limit $qv_F \gg |1 - i\omega\tau|$; the factor $8/9$ drops for diffusive scattering at the surface [Reu48]. The most important feature of the anomalous conductors is that the surface resistance and surface reactance are not equal, but $X_S = -\sqrt{3}R_S$ as displayed in Fig. E.1. Also the frequency dependence $\omega^{2/3}$ is different from that relating to the classical skin effect regime, where $\hat{Z}_S \propto \omega^{1/2}$ was found (Eq. (2.3.30)). A more rigorous derivation of the anomalous skin effect in conducting and superconducting metals given by Mattis and Bardeen [Mat58] leads to the same functional relations as obtained in the semiclassical approach.

Up to this point in the discussion of the non-local conductivity and anomalous skin effect, we have not taken into account effects of the band structure and non-spherical Fermi surface. In order to do so, we have to consider the range of integration in Chambers' expression of $\mathbf{J}$ as introduced in Section 5.2.4 and in Eq. (E.4). Electrons passing through the volume element dV with wavevectors $d\mathbf{k}$ at the time t_0 have followed some trajectory since their last collision. The distribution function f is obtained by adding the contributions from all electrons scattered into the trajectory at a time t prior to t_0. The probability that scattering has not occurred in this period between t and t_0 is given by the expression $\exp\left\{ -\int_t^{t_0} [\tau(t')]^{-1} dt' \right\}$,

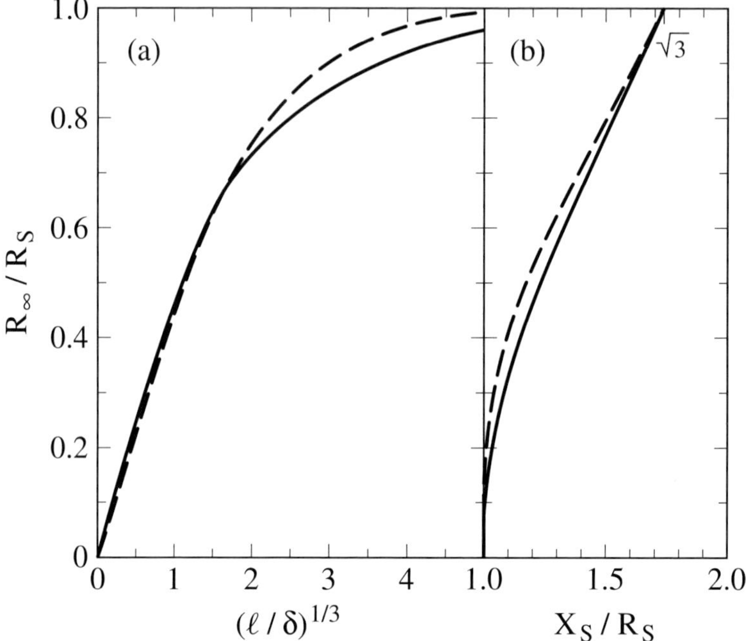

Fig. E.1. (a) Dependence of the surface resistance R_S on the mean free path with $R_\infty = R_S(\ell \to \infty)$. For large ℓ/δ (good conductor) the surface resistance goes to zero. (b) Ratio of imaginary and real parts of the surface impedance as a function of the inverse surface resistance. For a perfect conductor ($\ell = \infty$) the ratio approaches $X_S/R_S = \sqrt{3}$. The dashed lines correspond to the case of diffusive scattering and the solid line shows the specular scattering (after [Reu48]).

where the relaxation time in general depends on t' through the dependence of the velocity $\mathbf{v}$ on the wavevector $\mathbf{k}$. Thus, we obtain

$$
\begin{aligned}
f^1(\mathbf{r}, \mathbf{v}) &= \frac{e}{v} \left(-\frac{\partial f^0}{\partial \mathcal{E}} \right) \int_{\mathbf{r}} \mathbf{E}(r') \cdot \mathbf{v} \, \exp\{-r'/\tau v\} \, \mathrm{d}r' \\
&= e \left(-\frac{\partial f^0}{\partial \mathcal{E}} \right) \int_{-\infty}^0 \mathbf{E}(t) \cdot \mathbf{v} \, \exp\left\{ -\int_t^{t_0} \frac{\mathrm{d}t'}{\tau(\mathbf{k}(t'))} \right\} \mathrm{d}t \quad . \text{(E.11)}
\end{aligned}
$$

It can be shown [Cha52, Kit63] that this expression is a solution of the Boltzmann equation (5.2.7) within the relaxation time approximation. Using Eq. (5.2.12), we can rewrite Eq. (E.4) as

$$
\mathbf{J} = \frac{2e^2}{(2\pi)^3} \int \mathrm{d}\mathbf{k} \left(-\frac{\partial f^0}{\partial \mathcal{E}} \right) \mathbf{v}(\mathbf{k}) \int_{-\infty}^0 \mathbf{E}(t) \cdot \mathbf{v}(\mathbf{k}, t) \exp\left\{ -\int_t^{t_0} \frac{\mathrm{d}t'}{\tau(\mathbf{k}(t'))} \right\} \mathrm{d}t \quad .
$$

$$\text{(E.12)}$$

For a detailed discussion see [Pip54b, Pip62] and other textbooks.

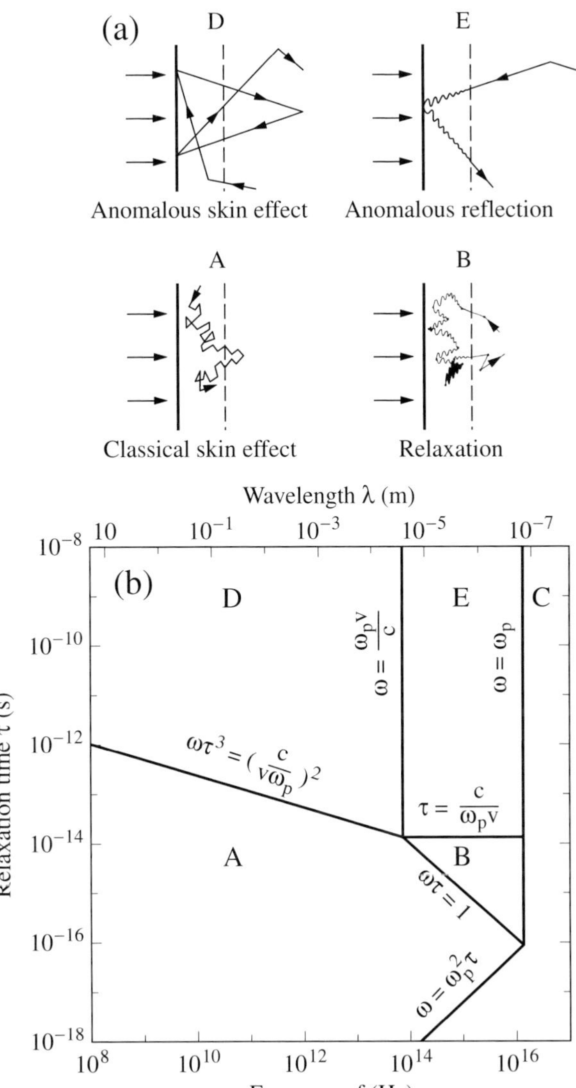

Fig. E.2. (a) Electron trajectories in the skin layer illustrating the nature of the interaction between photons and electrons in metals. The thick line represents the surface of the metal. The arrows indicate the incident light waves, the dashed line marks the skin layer, the zigzag line represents a path of the electron for successive scattering events, and the wavy line represents the oscillatory motion of the electron due to the alternating electric field. The optical characteristics in the four regions illustrated depend upon the relative value of the mean free path ℓ, the skin depth δ, and the mean distance traveled by an electron in a time corresponding to the inverse frequency of the light wave. (A) Classical skin effect, $\ell \ll \delta$ and $\ell \ll v_F/\omega$. (B) Relaxation regime, $v_F/\omega \ll \ell \ll \delta$. (D) Anomalous skin effect, $\delta \ll \ell$ and $\delta \ll v_F/\omega$. (E) Extreme anomalous skin effect, $v_F/\omega \ll \delta \ll \ell$. (b) Logarithmic plot of the scattering time τ versus frequency ω showing the regions just described. Region C characterizes the transparent regime (after [Cas67]).

E.2 Extreme anomalous regime

The concept of the anomalous skin effect no longer holds if the distance the charge carriers travel per period of the alternating field $\mathbf{E}(\omega)$ becomes comparable to the mean free path, i.e. the field direction alternates before the electrons have scattered. The absorption falls back to the classical value, and this observation is called the extreme anomalous skin effect. In order to understand this limit, we have to consider a third characteristic length scale of the problem, in addition to the mean free path ℓ and the skin depth δ: the distance v_F/ω the electrons move during one period of the electromagnetic wave. The classical skin effect only holds if two conditions are met: the skin depth δ is larger than the mean free path ℓ, and the frequency is lower than the scattering rate. In this regime, the electrons suffer many collisions during the time they spend in the skin layer and during one period of the electromagnetic wave (Fig. E.2, region A). Thus the region is well described by a local, instantaneous relationship between the current and the total electric field. If the frequency is larger than the plasma frequency, the metal becomes transparent (region C), and in the intermediate spectral range between the scattering rate $1/\tau$ and the plasma frequency ω_p the absorption is frequency independent (Eq. (5.1.22)). In this so-called relaxation regime, many periods of the radiation fall between two scattering events; however, the electrons still experience a large number of collisions while they travel within the skin layer. The collisions become less important and the light basically experiences a layer of free electrons responding to the rapidly oscillating electric field (region B).

Similar considerations hold for the anomalous regime. The anomalous skin effect becomes important when the mean free path ℓ exceeds the skin depth δ (Fig. E.2, region D). The electrons leave the skin layer before they are scattered, and the collisions are of little importance. If in this case the frequency increases, we do not see any change if $\omega\tau < 1$, but a new regime is reached when $v_F/\omega \ll \delta \ll \ell$. During the time the electron spends in the skin layer, it experiences an increasing number of oscillations of the electric field (region E): $\delta/v_F > 1/\omega$. The region is called the extreme anomalous skin effect or anomalous reflection. Since the electrons in the skin layer respond to the electric field as essentially free electrons, region E differs only slightly from region B. Most of the collisions happen at the surface. In the $\omega\tau$ diagram of Fig. E.2b the five regimes are shown and the borders between them indicated [Cas67]. Fig. E.3 illustrates the relationship between the three different length scales.

E.3 The kernel

The non-local conductivity and the effects of a finite mean free path can be elegantly treated with the help of the non-local kernel $K_{\mu\nu}$. The kernel is a response

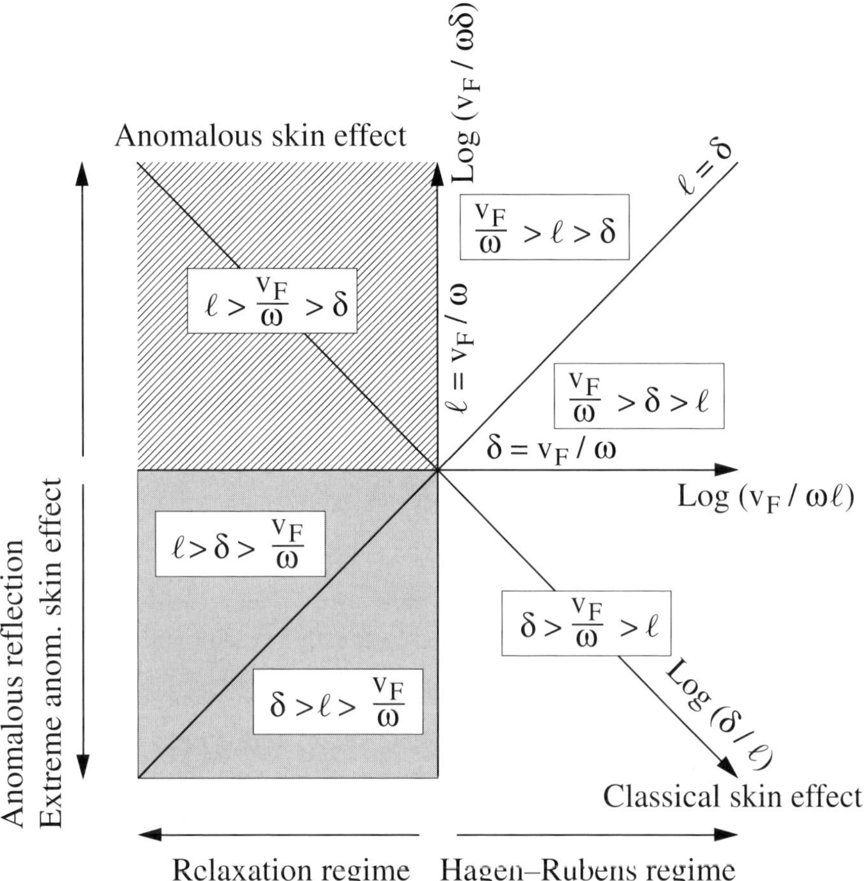

Fig. E.3. Schematic representation of the various skin depth regimes in the parameter space given by the three different length scales, the skin depth δ, the mean free path ℓ, and the distance the charge carriers travel during a period of the light wave v_F/ω.

function which relates the current density $\mathbf{J}$ to the vector potential $\mathbf{A}$; in general, it is a second order tensor. This approach is particularly useful for the discussion of impurity effects in superconductors.

In Eq. (4.1.16) we arrived at the Kubo formula for the conductivity as the most general description of the electrodynamic response. Here we want to reformulate this expression by defining a non-local kernel $K_{\mu\nu}$. We write the position of time dependent current as

$$J_\mu(\mathbf{r}, t) = -\frac{c}{4\pi} \int_{-\infty}^{\infty} \int_{\Omega} \sum_{\nu=1}^{3} K_{\mu\nu}(\mathbf{r}, t, \mathbf{r}', t') A_\nu(\mathbf{r}', t')\, \mathrm{d}\mathbf{r}'\mathrm{d}t' \quad , \qquad (E.13)$$

with the vector potential $\mathbf{A}(\mathbf{r}, t)$ and the kernel

$$K_{\mu\nu}(\mathbf{r}, t, \mathbf{r}', t') = -\frac{4\pi i}{c^2} \langle 0 | [J_{\mu}^{\mathrm{p}}(\mathbf{r}, t), J_{\nu}^{\mathrm{p}}(\mathbf{r}', t')] | 0 \rangle \Theta(t - t')$$

$$-\frac{4\pi e}{mc^2} \langle 0 | \rho_0(\mathbf{r}, t) | 0 \rangle \delta(\mathbf{r} - \mathbf{r}') \delta(t - t') \delta_{\mu\nu} (1 - \delta_{\nu,0}) \quad . \quad \text{(E.14)}$$

$J_i^{\mathrm{p}}(\mathbf{r}, t)$ indicates the ith component of the paramagnetic current density introduced in Eq. (4.1.6), ρ_0 is the charge density, and Θ refers to the step function

$$\Theta(t - t') = \begin{cases} 1 & \text{if } t > t' \\ 0 & \text{if } t < t' \; ; \end{cases}$$

the details are discussed in [Sch83] and similar textbooks. The expectation value of the current density in an external field can be calculated from the fluctuation-dissipation theorem by first order perturbation

$$J_{\mu}(\mathbf{r}, t) = \frac{e}{mc} \langle 0 | \rho_0(\mathbf{r}, t) | 0 \rangle A_{\mu}(\mathbf{r}, t) [1 - \delta_{\mu,0}] - i \langle 0 | \left[J_{\mu}^{\mathrm{p}}(\mathbf{r}, t), \int_{-\infty}^{t} \mathcal{H}_{\mathrm{int}}(t') \, \mathrm{d}t' \right] | 0 \rangle \quad .$$

$$\text{(E.15)}$$

Since $\hat{\sigma}(\mathbf{q}, \omega) = -\frac{\mathrm{i}c^2}{4\pi\omega} K(\mathbf{q}, \omega)$, the kernel is directly related to the conductivity.

E.4 Surface impedance of superconductors

In the superconducting case, the kernel was studied intensively by [Abr63]. For the Pippard case $\lambda \ll \xi(0)$ we obtain for frequencies below the gap $\omega < 2\Delta/\hbar$

$$K(\mathbf{q}, \omega) = \frac{4\pi}{c^2} \frac{Ne^2}{m} \frac{3\pi}{4q v_{\mathrm{F}}}$$

$$\times \left[\int_{\Delta}^{\Delta + \hbar\omega} \tanh\left\{ \frac{\mathcal{E}}{2k_{\mathrm{B}}T} \right\} \frac{\mathcal{E}(\mathcal{E} - \hbar\omega) + \Delta^2}{(\mathcal{E}^2 - \Delta^2)^{1/2}(\Delta^2 - (\mathcal{E} - \hbar\omega)^2)^{1/2}} \, \mathrm{d}\mathcal{E} \right.$$

$$+ \mathrm{i} \int_{\Delta}^{\infty} \left(\tanh\left\{ \frac{\mathcal{E}}{2k_{\mathrm{B}}T} \right\} - \tanh\left\{ \frac{\mathcal{E} + \hbar\omega}{2k_{\mathrm{B}}T} \right\} \right)$$

$$\left. \times \frac{\mathcal{E}(\mathcal{E} + \hbar\omega) + \Delta^2}{(\mathcal{E}^2 - \Delta^2)^{1/2}((\mathcal{E} + \hbar\omega)^2 - \Delta^2)^{1/2}} \, \mathrm{d}\mathcal{E} \right]$$

and for $\hbar\omega > 2\Delta$

$$K(\mathbf{q}, \omega) = \frac{4\pi}{c^2} \frac{Ne^2}{m} \frac{3\pi}{4q v_{\mathrm{F}}}$$

$$\times \left[\int_{\hbar\omega - \Delta}^{\hbar\omega + \Delta} \tanh\left\{ \frac{\mathcal{E}}{2k_{\mathrm{B}}T} \right\} \frac{\mathcal{E}(\mathcal{E} - \hbar\omega) + \Delta^2}{(\mathcal{E}^2 - \Delta^2)^{1/2}[\Delta^2 - (\mathcal{E} - \hbar\omega)^2]^{1/2}} \, \mathrm{d}\mathcal{E} \right.$$

$$+ \mathrm{i} \int_{\Delta}^{\hbar\omega - \Delta} \frac{\mathcal{E}(\mathcal{E} - \hbar\omega) + \Delta^2}{(\mathcal{E}^2 - \Delta^2)^{1/2}[(\mathcal{E} - \hbar\omega)^2 - \Delta^2]^{1/2}} \, \mathrm{d}\mathcal{E}$$

$$+ \, \mathrm{i} \int_{\Delta}^{\infty} \left(\tanh\left\{\frac{\mathcal{E}}{2k_{\mathrm{B}}T}\right\} - \tanh\left\{\frac{\mathcal{E}+\hbar\omega}{2k_{\mathrm{B}}T}\right\} \right)$$

$$\times \, \frac{\mathcal{E}(\mathcal{E}+\hbar\omega)+\Delta^2}{(\mathcal{E}^2-\Delta^2)^{1/2}[(\mathcal{E}+\hbar\omega)^2-\Delta^2]^{1/2}} \, \mathrm{d}\mathcal{E} \Bigg] \qquad . \tag{E.16}$$

Electromagnetic energy is absorbed only in the latter case. In order to relate the complex surface impedance $\hat{Z}_{\mathrm{S}}$ to K, two cases have to be distinguished. For diffusive scattering we obtain

$$\hat{Z}_{\mathrm{S}}(\omega) = 4\pi^2 \mathrm{i}\omega \left(\int_0^{\infty} \ln\{1 + K(\mathbf{q},\omega)/q^2\} \, \mathrm{d}\mathbf{q} \right)^{-1} \quad ,$$

and for specular scattering

$$\hat{Z}_{\mathrm{S}}(\omega) = 8\mathrm{i}\omega \int_0^{\infty} \frac{\mathrm{d}\mathbf{q}}{q^2 + K(\mathbf{q},\omega)} \quad .$$

In general the ratio of the surface impedance $\hat{Z}_{\mathrm{S}}(\omega)$ to the real part of the normal state surface resistance $R_{\mathrm{n}}(\omega)$ is considered. In the Pippard case we obtain

$$\frac{\hat{Z}_{\mathrm{S}}(\omega)}{R_{\mathrm{n}}(\omega)} = -2\mathrm{i} \left(\frac{3\pi^2 N e^2}{c^2 m} \, \frac{\hbar\omega}{q\,v_{\mathrm{F}} K(\omega)} \right)^{1/3} \quad . \tag{E.17}$$

At $T=0$ the integrals of the kernel can be evaluated by using the complete elliptic integrals E and K

$$K(\omega) = \begin{cases} \dfrac{3\Lambda}{2q v_{\mathrm{F}}}\dfrac{N e^2}{c^2 m}\mathrm{E}\left\{\hbar\omega/2\Delta\right\} & \omega < 2\Delta/\hbar \\[3ex] \dfrac{3\Delta}{q v_{\mathrm{F}}}\dfrac{N e^2}{c^2 m}\left[\dfrac{\hbar\omega}{\Delta}\mathrm{E}\{2\Delta/\hbar\omega\} - \left(\dfrac{\hbar\omega}{\Delta} - \dfrac{4\Delta}{\hbar\omega}\right)\mathrm{K}\{2\Delta/\hbar\omega\}\right] & \omega > 2\Delta/\hbar \\[2ex] \quad +\mathrm{i}\dfrac{3\Delta}{q v_{\mathrm{F}}}\dfrac{N e^2}{c^2 m}\Theta\{\hbar\omega/\Delta - 2\}\left[\dfrac{\hbar\omega}{\Delta}\mathrm{E}\left\{\left(1 - 4\Delta^2/\hbar^2\omega^2\right)^{1/2}\right\}\right. \\[2ex] \quad \left. -4\Delta/\hbar\omega\mathrm{K}\left\{\left(1 - 4\Delta^2/\hbar^2\omega^2\right)^{1/2}\right\}\right] \quad . \end{cases}$$

For finite temperatures no full analytical expression can be given and only the limiting cases can be evaluated [Gei74]. For low frequencies and temperatures ($\hbar\omega \ll k_{\mathrm{B}}T \ll \Delta$) we get for the normalized surface impedance

$$\frac{\hat{Z}_{\mathrm{S}}(\omega)}{R_{\mathrm{n}}(\omega)} = 2\left(\frac{\hbar\omega}{\pi\Delta}\right)^{1/3}\left\{\frac{4}{3}\sinh\left\{\frac{\hbar\omega}{2k_{\mathrm{B}}T}\right\}\mathrm{K}_0\left\{\frac{\hbar\omega}{2k_{\mathrm{B}}T}\right\}\exp\left\{\frac{-\Delta}{k_{\mathrm{B}}T}\right\}\right.$$

$$\left. -\mathrm{i}\left[1 + \frac{1}{3}\left(\frac{\hbar\omega}{4\Delta}\right)^2 + \frac{2}{3}\exp\left\{\frac{\hbar\omega}{2k_{\mathrm{B}}T}\right\}\mathrm{I}_0\left\{\frac{\hbar\omega}{2k_{\mathrm{B}}T}\right\}\exp\left\{\frac{-\Delta}{k_{\mathrm{B}}T}\right\}\right]\right\}. \tag{E.18}$$

where I_0 and K_0 are the modified zero order Bessel functions of first and second kind, respectively. In the case of higher temperatures but low frequencies ($\hbar\omega \ll \Delta \ll k_B T$) we can write

$$
\frac{\hat{Z}_S(\omega)}{R_n(\omega)} = 2\left(\frac{\hbar\omega/\pi\Delta}{\tanh\{\Delta/2k_B T\}}\right)^{1/3}\left[\frac{2}{3\pi}\frac{\hbar\omega/k_B T}{\sinh\{\Delta/k_B T\}}\ln\left\{2\,(2\Delta/\hbar\omega)^{1/2}\right\}\right.
$$

$$
+\frac{1}{3\pi}\frac{\hbar\omega}{\Delta}\left(\coth\left\{\frac{\Delta}{2k_B T}\right\} - 1\right) - \frac{7}{3\pi^3}\frac{\Delta\,\hbar\omega}{(k_B T)^2}\zeta(3)\coth\left\{\frac{\Delta}{2k_B T}\right\}
$$

$$
\left.-\mathrm{i}\left(1 + \frac{\hbar^2\omega^2}{48\Delta^2}\right)\right] \tag{E.19}
$$

if $\Delta^2 \gg \hbar\omega k_B T$; otherwise

$$
\frac{\hat{Z}_S(\omega)}{R_n(\omega)} = -2\mathrm{i}\left(-\mathrm{i} + \frac{\pi\Delta^2}{2k_B T\hbar\omega}\right)^{-1/3} \quad . \tag{E.20}
$$

Finally we want to consider the cases where the frequency exceeds the temperature. In the range $k_B T \ll \hbar\omega \approx \Delta$ and $\omega < 2\Delta(0)/\hbar$,

$$
\frac{\hat{Z}_S(\omega)}{R_n(\omega)} = 2\left(\frac{\hbar\omega/2\Delta}{E\{\hbar\omega/2\Delta\}}\right)^{1/3}\left\{\frac{\exp\{-\Delta/k_B T\}}{3E\{\hbar\omega/2\Delta\}}\left[\pi k_B T\left(\frac{1}{\hbar\omega} + \frac{1}{2\Delta}\right)\right]^{1/2}\right.
$$

$$
\left.-\mathrm{i}\left[1 + \frac{\exp\{-\Delta/k_B T\}}{3E\{\hbar\omega/2\Delta\}}\left(\frac{\pi k_B T}{\hbar\omega} - \frac{\pi k_B T}{2\Delta}\right)^{1/2}\right]\right\} \quad , \tag{E.21}
$$

and for large frequencies $\omega \gg \Delta(0)/\hbar$ but low temperatures

$$
\frac{\hat{Z}_S(\omega)}{R_n(\omega)} = 1 + \left(\frac{\Delta}{\hbar\omega}\right)^2\left[\frac{2}{3}\ln\left\{\frac{2}{\Delta(0)}\right\} + \frac{1}{3} - \frac{\pi}{\sqrt{3}}\right]
$$

$$
-\mathrm{i}\sqrt{3}\left[1 + \left(\frac{\Delta}{\hbar\omega}\right)^2\left(\frac{2}{3}\ln\left\{\frac{2}{\Delta(0)}\right\} + \frac{1}{3} - \frac{\pi}{3\sqrt{3}}\right)\right] \quad . \tag{E.22}
$$

The simple ω^2 dependence of the surface resistance derived in Eq. (7.4.23) is the limit of Eq. (E.21) for small frequencies.

E.5 Non-local response in superconductors

The effects of a finite mean free path ℓ can also be discussed in terms of the kernel. Here we follow the outline of M. Tinkham [Tin96] and G. Rickayzen [Ric65]. In the limit $\delta, \lambda \ll \ell$ the local relations between current and field are no longer valid, and we have to use the expressions derived for the anomalous regime (Section E.1), summarized in $\hat{Z}_S/R_n = 2\left(-\hat{\sigma}/\sigma_n\right)^{-1/3}$. We are interested in the frequency and temperature dependent response and how it varies for different wavevectors.

By comparing Eq. (E.13) with Eq. (4.3.30) we see that

$$K(0) = \frac{1}{\lambda_L^2} = \frac{4\pi N_s e^2}{mc^2} \quad .\tag{E.23}$$

This immediately implies that at low temperatures the kernel K as well as the penetration depth λ are frequency independent. Let us first discuss the temperature dependence of $K(\mathbf{q}, T)$ in the long wavelength ($\mathbf{q} \to 0$) limit. In the expression

$$
\begin{aligned}
\mathbf{J}(\mathbf{q}, \omega) \;=\;& \mathbf{J}^{\mathrm{p}}(\mathbf{q}, \omega) + \mathbf{J}^{\mathrm{d}}(\mathbf{q}, \omega) \\[4pt]
\;=\;& \lim_{\tau \to \infty} \frac{1}{\Omega c}\left(\frac{e\hbar}{2m}\right)^2 \sum_{\mathbf{k}} 2\mathbf{k}[\mathbf{A}(\mathbf{q}, \omega) \cdot 2\mathbf{k}] \\[4pt]
& \times \left[\left(\frac{1 - f(\mathcal{E}_{\mathbf{k}}) - f(\mathcal{E}_{\mathbf{k}+\mathbf{q}})}{\mathcal{E}_{\mathbf{k}} + \mathcal{E}_{\mathbf{k}+\mathbf{q}} - \hbar\omega - i\hbar/\tau} + \frac{1 - f(\mathcal{E}_{\mathbf{k}}) - f(\mathcal{E}_{\mathbf{k}+\mathbf{q}})}{\mathcal{E}_{\mathbf{k}} + \mathcal{E}_{\mathbf{k}+\mathbf{q}} + \hbar\omega + i\hbar/\tau}\right) \right. \\[4pt]
& \times (u_{\mathbf{k}}^* v_{\mathbf{k}+\mathbf{q}} - v_{\mathbf{k}} u_{\mathbf{k}+\mathbf{q}}^*)^2 \\[4pt]
& + \left.\left(\frac{f(\mathcal{E}_{\mathbf{k}}) - f(\mathcal{E}_{\mathbf{k}+\mathbf{q}})}{-\mathcal{E}_{\mathbf{k}} + \mathcal{E}_{\mathbf{k}+\mathbf{q}} - \hbar\omega - i\hbar/\tau} + \frac{f(\mathcal{E}_{\mathbf{k}}) - f(\mathcal{E}_{\mathbf{k}+\mathbf{q}})}{-\mathcal{E}_{\mathbf{k}} + \mathcal{E}_{\mathbf{k}+\mathbf{q}} + \hbar\omega + i\hbar/\tau}\right) \right. \\[4pt]
& \times \left. (u_{\mathbf{k}} u_{\mathbf{k}+\mathbf{q}}^* + v_{\mathbf{k}}^* v_{\mathbf{k}+\mathbf{q}})^2\right] - \frac{Ne^2}{mc}\mathbf{A}(\mathbf{q}, \omega) \quad ,
\end{aligned}\tag{E.24}
$$

the paramagnetic current density simplifies to

$$\mathbf{J}^{\mathrm{p}}(0, T) = \frac{c\hbar}{m}\sum_{\mathbf{k}} \mathbf{k}(f_{\mathbf{k}0} - f_{\mathbf{k}1}) = \frac{2e^2 h^2}{m^2 c}\sum_{\mathbf{k}}[\mathbf{A}(0) \cdot \mathbf{k}]\mathbf{k}\left(-\frac{\partial f}{\partial \mathcal{E}_{\mathbf{k}}}\right)$$

in the small field approximation. Thus we obtain for the kernel

$$K^{\mathrm{P}}(0, T) = -\frac{4\pi N e^2}{mc^2}\frac{4\mathcal{E}_{\mathrm{F}}}{3}\sum_{\mathbf{k}}\left(-\frac{\partial f}{\partial \mathcal{E}_{\mathbf{k}}}\right) = -\lambda_L^{-2}(0)\int_{-\infty}^{\infty}\left(-\frac{\partial f}{\partial \mathcal{E}_{\mathbf{k}}}\right)\,\mathrm{d}\zeta \quad ,$$

where $\zeta = \mathcal{E}_{\mathbf{k}} - \mathcal{E}_{\mathrm{F}}$. The total kernel is given by

$$K(0, T) = \lambda_L^{-2}(T) = \lambda_L^{-2}(0)\left[1 - 2\int_{\Delta}^{\infty}\left(-\frac{\partial f}{\partial \mathcal{E}_{\mathbf{k}}}\right)\frac{\mathcal{E}}{(\mathcal{E}^2 - \Delta^2)^{1/2}}\,\mathrm{d}\mathcal{E}\right] \quad .\tag{E.25}$$

For temperatures $T \geq T_{\mathrm{c}}$ the material is in the normal state and $\Delta = 0$; since the integral reduces to $f(0) = 1/2$, we obtain $K(0, T \geq T_{\mathrm{c}}) = 0$ as expected. This corresponds to the exact cancellation of the paramagnetic and diamagnetic currents and no Meissner effect is observed. At low temperatures ($T < 0.5T_{\mathrm{c}}$), the derivative of the Fermi distribution can be approximated by a linear function, and

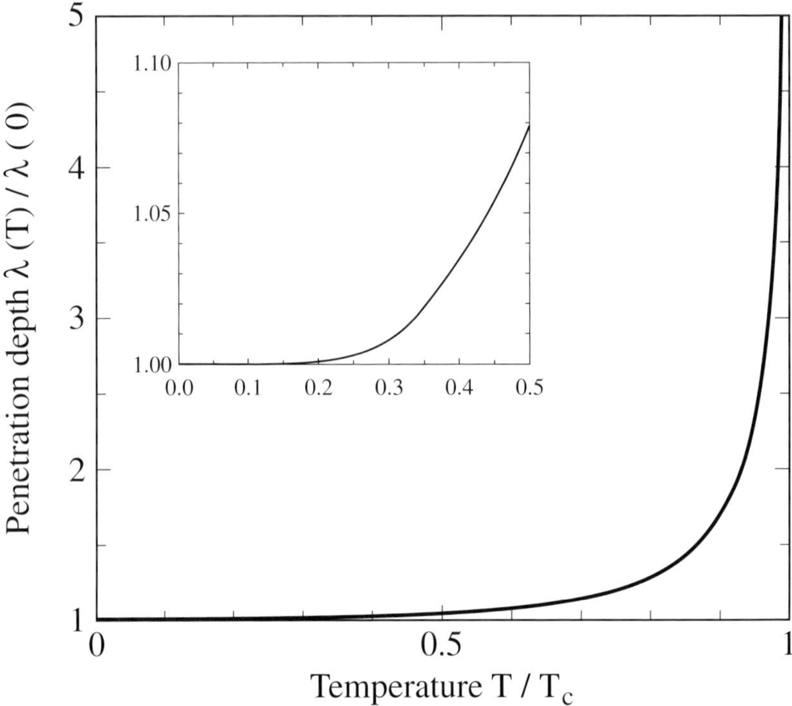

Fig. E.4. Temperature dependence of the normalized penetration depth $\lambda(T)/\lambda(0)$. The low temperature regime (inset) shows an exponential behavior according to Eq. (E.26). In the high temperature range Eq. (E.27) is plotted.

thus the previous expression reduces to

$$\frac{\lambda(T) - \lambda(0)}{\lambda(0)} = \left(\frac{\pi \Delta}{2k_{\rm B}T}\right)^{1/2} \exp\left\{-\frac{\Delta}{k_{\rm B}T}\right\} \quad ; \tag{E.26}$$

while in the range $0.8T_{\rm c} < T < T_{\rm c}$, the temperature dependence can be approximated by

$$\frac{\lambda(T)}{\lambda(0)} = \left[1 - \left(\frac{T}{T_{\rm c}}\right)^4\right]^{-1/2} . \tag{E.27}$$

In Fig. E.4 the temperature dependence of the penetration depth is shown according to Eqs (E.26) and (E.27). The temperature dependence of $K(0, T)$ is plotted in Fig. E.5a, where $K(0, T)/K(0, 0) = [\lambda_{\rm L}(0)/\lambda_{\rm L}(T)]^2$. As T decreases below $T_{\rm c}$, the superconducting gap opens, i.e. $\Delta(T)/k_{\rm B}T$ increases, causing the kernel in Eq. (E.25) to diminish, and eventually become exponentially small, and $K(0, T \to 0) = \lambda_{\rm L}^{-2}(0)$.

So far we have restricted our analysis to the $\mathbf{q} = 0$ limit; let us consider next

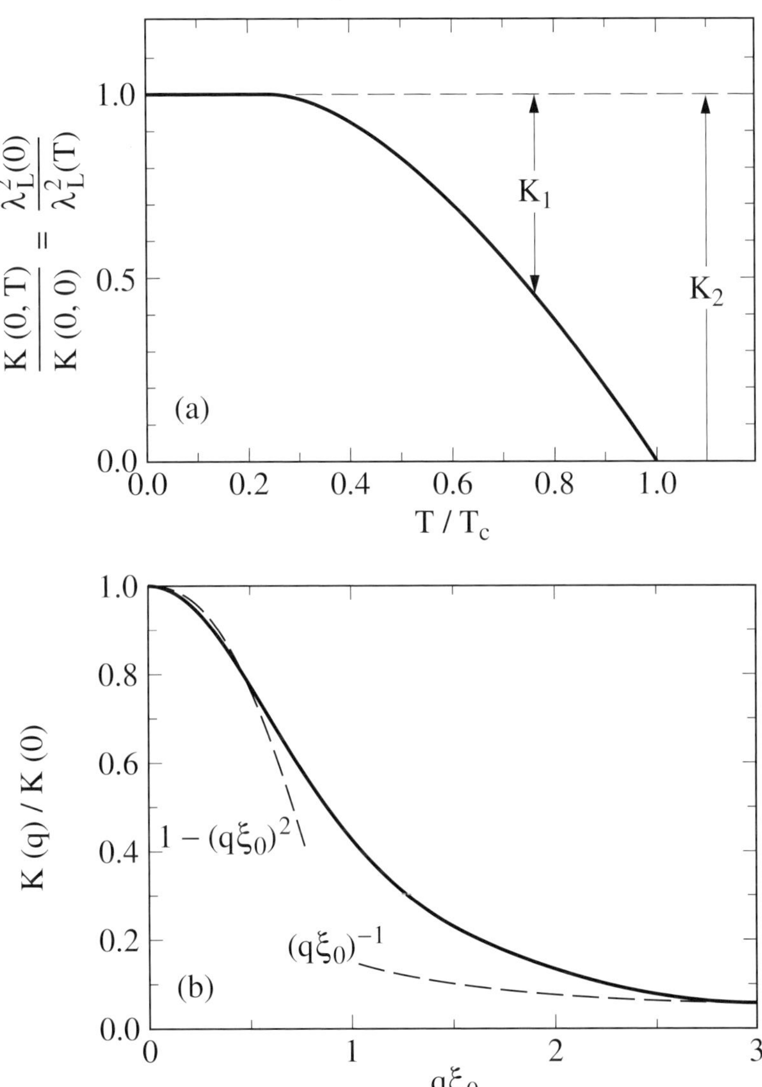

Fig. E.5. (a) Temperature dependence of the normalized kernel $K(0, T)/K(0, 0)$ (after [Tin96]). (b) Normalized kernel $K(\mathbf{q})/K(0)$ as a function of the wavevector $\mathbf{q}$.

the $\mathbf{q}$ dependence of $K(\mathbf{q}, T)$ for $T = 0$. If we calculate the experimental value of $\mathbf{J}^{\mathrm{P}}(\mathbf{q})$ following Eq. (E.24) we obtain

$$\mathbf{J}^{\mathrm{P}}(\mathbf{q}) = \frac{2e^2\hbar^2}{m^2c} \sum_{\mathbf{k}} \frac{(v_{\mathbf{k}}u^*_{\mathbf{k}+\mathbf{q}} - u^*_{\mathbf{k}}v_{\mathbf{k}+\mathbf{q}})^2}{\mathcal{E}_{\mathbf{k}} + \mathcal{E}_{\mathbf{k}+\mathbf{q}}} [\mathbf{k} \cdot \mathbf{A}(\mathbf{q})]\mathbf{k}$$

similar to the relation (4.3.31) we arrived at in the case of normal metals. For the

q dependent kernel

$$K(\mathbf{q}, 0) = \lambda_{\mathrm{L}}^{-2}(0) \left[1 - 2 \int_{\Delta}^{\infty} \frac{(v_{\mathbf{k}} u_{\mathbf{k+q}}^{*} - u_{\mathbf{k}}^{*} v_{\mathbf{k+q}})^2}{\mathcal{E}_{\mathbf{k}} + \mathcal{E}_{\mathbf{k+q}}} \mathrm{d}\zeta \right] \quad ; \tag{E.28}$$

and for small $\mathbf{q}$, we find $K(\mathbf{q}, T)\lambda_{\mathrm{L}}^{2}(0) \approx 1 - q^2 \xi_0^2$, where the coherence length ξ_0 is defined in Eq. (7.4.2). If $q \gg 1/\xi_0$ (Pippard limit), $K(\mathbf{q}, 0) = K(0, 0)\frac{3\pi}{4q\xi_0}$, as displayed in Fig. E.5b. In both the local and the London limit $\xi_0 \to 0$, and the $\mathbf{q}$ dependence of the kernel is negligible: $K(\mathbf{q}, 0) \to K(0, 0)$.

The effect of impurity scattering can be taken into account in a phenomenological way similar to Chambers' approach discussed in Section 5.2.4,

$$\frac{\lambda_{\mathrm{L}}^{2}(T)}{\lambda_{\mathrm{l}}^{2}(\ell, T)} = \frac{K(0, T, \ell)}{K(0, T, \ell \to \infty)} = \frac{1}{\xi_0} \int_{0}^{\infty} J(s', T) \exp\{-s'/\ell\} \, \mathrm{d}s' \quad , \tag{E.29}$$

where we have used the $J(s', T)$ function as defined by $\int_{0}^{\infty} J(s', T) \, \mathrm{d}s' = \xi_0 = \int_{0}^{\infty} \exp\{-s'/\xi_0\} \, \mathrm{d}s'$. An increased scattering rate makes ℓ small and the response more local. In the case $\ell \ll \xi_0$ (dirty limit) $\lambda_{\mathrm{l}}(\ell, T) = \lambda_{\mathrm{L}}(T) (\xi_0/\ell)^{1/2} [J(0, T)]^{-1/2}$. If we replace $J(s', T)$ by $\exp\{-s'/\xi_0\}$, which is valid for large ℓ, we obtain

$$\lambda_{\mathrm{l}}(\ell, T) = \lambda_{\mathrm{L}}(T) \left[1 + \frac{\xi_0'}{\ell} \right]^{1/2} \quad , \tag{E.30}$$

where Pippard's coherence length, ξ_0', and ξ' are given by

$$\frac{1}{\xi'} = \frac{1}{\xi_0'} + \frac{1}{\ell} = \frac{J(0, T)}{\xi_0} + \frac{1}{\ell} \quad , \tag{E.31}$$

with $J(0, T = 0) = 1$ and $J(0, T_{\mathrm{c}}) = 1.33$. A full discussion of non-local effects is given in [Tin96]. In the $\mathbf{q} \to \infty$ London limit ($\lambda_{\mathrm{L}} \gg \xi_0$), i.e. the extreme anomalous regime, $K(\mathbf{q}) = 3\pi/(4\lambda_{\mathrm{L}}^{2}\xi_0 q)$ leads to $\lambda_{\mathbf{q}\to\infty} = 0.58 \left(\lambda_{\mathrm{L}}^{2}\xi_0' \right)^{1/3}$ for specular scattering.

The temperature dependence of the penetration depth in the local limit is given by

$$\frac{\lambda_{\mathrm{l}}(T)}{\lambda_{\mathrm{L}}(0)} = \left[\frac{\ell}{\xi_0} \frac{\Delta(T)}{\Delta(0)} \tanh \left\{ \frac{\Delta(T)}{2k_{\mathrm{B}}T} \right\} \right]^{-1/2} \quad , \tag{E.32}$$

which resembles Eq. (7.4.21). In the anomalous regime (assuming diffusive scattering), we obtain

$$\frac{\lambda_{\mathrm{P}}(T)}{\lambda_{\mathrm{L}}(0)} = \left[\frac{3\pi^2}{4} \frac{\lambda_{\mathrm{L}}(0)}{\xi_0} \frac{\Delta(T)}{\Delta(0)} \tanh \left\{ \frac{\Delta(T)}{2k_{\mathrm{B}}T} \right\} \right]^{-1/3} \quad , \tag{E.33}$$

where $\lambda_{\mathrm{L}}(T)/\lambda_{\mathrm{L}}(0)$ is given by Eq. (E.25).

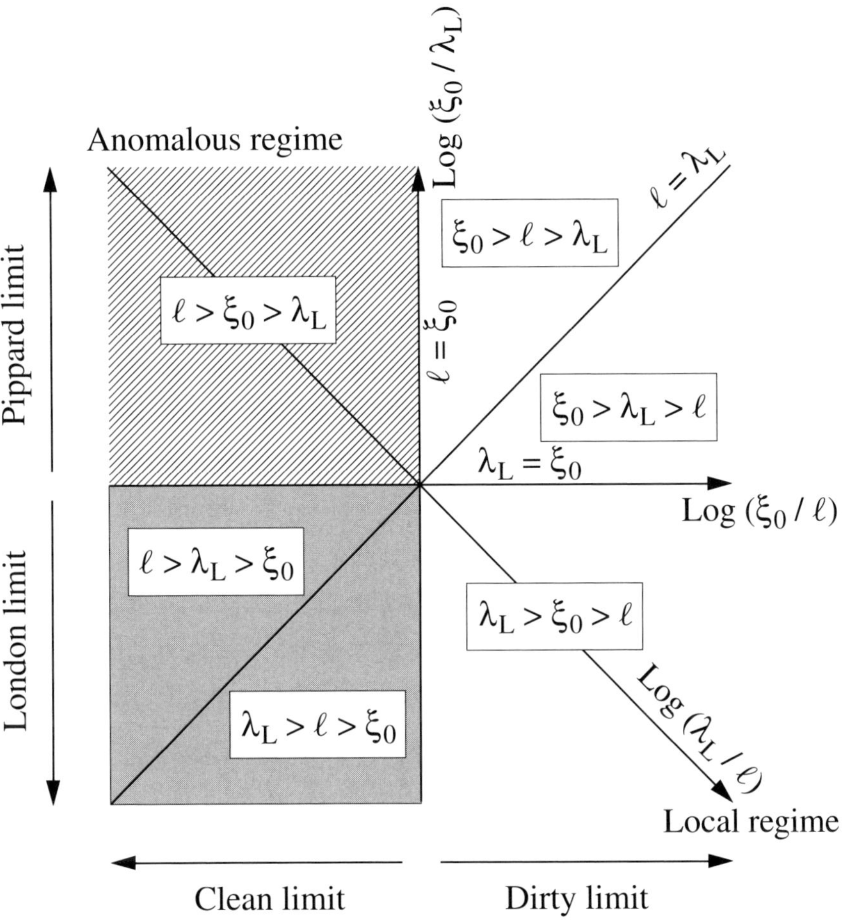

Fig. E.6. Schematic representation of the local, London, and Pippard limits in the parameter space given by the three length scales, the coherence length ξ_0, the London penetration depth λ_L, and the mean free path ℓ (after [Kle93]).

In Fig. E.6 we compare the three length scales important to the superconducting state; depending on the relative magnitude of the length scales, several limits are of importance. The first is the local regime in which ℓ is smaller than the distance over which the electric field changes, $\ell < \xi(0) < \lambda$. When $\ell/\xi(0) \to 0$ the superconductor is in the so-called dirty limit. The opposite situation in which $\ell/\xi(0) \to \infty$ is the clean limit, in which non-local effects are important and we have to consider Pippard's treatment. This regime can be subdivided if we also consider the third parameter λ. The case $\xi(0) > \lambda$ is the Pippard or anomalous regime, which is the regime of type I superconductors; and $\xi(0) < \lambda$ is the London regime, the regime of the type II superconductors.

References

[Abr63] A.A. Abrikosov, L.P. Gorkov, and I.E. Dzyaloshinski, *Methods of Quantum Field Theory in Statistical Physics* (Prentice-Hall, Englewood Cliffs, NJ, 1963)

[Abr88] A.A. Abrikosov, *Fundamental Theory of Metals* (North-Holland, Amsterdam, 1988)

[Cas67] H.B.G. Casimir and J. Ubbink, *Philips Tech. Rev.* **28**, 271, 300, and 366 (1967)

[Cha52] R.G. Chambers, *Proc. Phys. Soc. (London) A* **65**, 458 (1952); *Proc. Roy. Soc. A* **215**, 481 (1952)

[Cha90] R.G. Chambers, *Electrons in Metals and Semiconductors* (Chapman and Hall, London, 1990)

[Gei74] B.T. Geilikman and V.Z. Kresin, eds, *Kinetic and Nonsteady State Effects in Superconductors* (John Wiley & Sons, New York, 1974)

[Kan68] E.A. Kaner and V.G. Skobov, *Adv. Phys.* **17**, 605 (1968)

[Kar86] E. Kartheuser, L.R. Ram Mohan, and S. Rodrigues, *Adv. Phys.* **35**, 423 (1986)

[Kit63] C. Kittel, *Quantum Theory of Solids* (John Wiley & Sons, New York, 1963)

[Kle93] O. Klein, *PhD Thesis*, University of California, Los Angeles, 1993

[Mat58] D.C. Mattis and J. Bardeen, *Phys. Rev.* **111**, 561 (1958)

[Pip54b] A.B. Pippard, *Metallic Conduction at High Frequencies and Low Temperatures*, in: *Advances in Electronics and Electron Physics* **6**, edited by L. Marton (Academic Press, New Yok, 1954), p. 1

[Pip62] A.B. Pippard, *The Dynamics of Conduction Electrons*, in: *Low Temperature Physics*, edited by C. DeWitt, B. Dreyfus, and P.G. deGennes (Gordon and Breach, New York, 1962),

[Reu48] G.E.H. Reuter and E.H. Sondheimer, *Proc. Roy. Soc. A* **195**, 336 (1948)

[Ric65] G. Rickayzen, *Theory of Superconductivity* (John Wiley & Sons, New York, 1965)

[Sch83] J.R. Schrieffer, *Theory of Superconductivity*, 3rd edition (W.A. Benjamin, New York, 1983)

[Sok67] A.V. Sokolov, *Optical Properties of Metals* (American Elsevier, New York, 1967)

[Son54] E.H. Sondheimer, *Proc. Roy. Soc. A* **224**, 260 (1954)

[Tin96] M. Tinkham, *Introduction to Superconductivity*, 2nd edition (McGraw-Hill, New York, 1996)

Further reading

[Hal74] J. Halbritter, *Z. Phys.* **266**, 209 (1974)

[Lyn69] E.A. Lynton, *Superconductivity*, 3rd edition (Methuen & Co, London, 1969)

[Par69] R.D. Parks, *Superconductivity* (Marcel Dekker, New York, 1969)

[Pip47] A.B. Pippard, *Proc. Roy. Soc. A* **191**, 385 (1947)

[Pip50] A.B. Pippard, *Proc. Roy. Soc. A* **203**, 98 (1950)

[Pip54a] A.B. Pippard, *Proc. Roy. Soc. A* **224**, 273 (1954)

[Pip57] A.B. Pippard, *Phil. Trans. Roy. Soc. (London) A* **250**, 325 (1957)

[Pip60] A.B. Pippard, *Rep. Prog. Phys.* **33**, 176 (1960)

Appendix F

Dielectric response in reduced dimensions

With a few exceptions we have considered mainly bulk properties in the book. The physics of reduced dimensions is not only of theoretical interest, for many models can be solved analytically in one dimension only. A variety of interesting phenomena are bounded to restricted dimensions. On the other hand, fundamental models such as the theory of Fermi liquids developed for three dimensions break down in one or two dimensions. In recent years a number of possibilities have surfaced to explain how reduced dimensions can be achieved in real systems. One avenue is the study of real crystals with an extremely large anisotropy. The second approach considers artificial structures such as interfaces which might be confined further to reach the one-dimensional limit.

F.1 Dielectric response function in two dimensions

Reducing the dimension from three to two significantly changes many properties of the electron gas. If the thickness of the layer is smaller than the extension of the electronic wavefunction, the energy of the system is quantized (size quantization). We consider only the ground state to be occupied. For any practical case, just the electrons are confined to a thin sheet, while the field lines pass through the surrounding material which usually is a dielectric. A good approximation of a two-dimensional electron gas can be obtained in surfaces, semiconductor interfaces, and inversion layers; a detailed discussion which also takes the dielectric properties of the surrounding media into account can be found in [And82, Hau94]. In this section we discuss the idealized situation of a two-dimensional electron gas; however, the Coulomb interaction has a three-dimensional spatial dependence. Following the discussion of the three-dimensional case, we first consider the static limit. The full wavevector and frequency dependence of the longitudinal response in two dimensions is derived by the formalism used for the three-dimensional case.

F.1.1 Static limit

Let us assume an electric field $\mathbf{E}(\mathbf{q}, \omega) = \mathbf{E}_0 \exp\{i(\mathbf{q} \cdot \mathbf{r} - \omega t)\}$ which is purely longitudinal ($\mathbf{q} \times \mathbf{E}_0 = 0$) acting on a two-dimensional electron gas confined in the z direction which is surrounded by a vacuum. An external source produces an additional electrostatic potential Φ, which is related to the charge density ρ by Eq. (2.1.7): $\nabla \cdot (\epsilon \nabla \Phi) = -4\pi\rho$. Here ϵ is the dielectric constant of the system and $\rho = \rho_{\mathrm{ext}} + \rho_{\mathrm{ind}}$ as usual; the particle density is indicated by N. We can express the induced charge density as a function of the local potential, and linearizing it yields

$$\rho_{\mathrm{ind}}(\mathbf{r}) = -e[N(\Phi) - N(0)]\delta\{z\} = -e^2\Phi(\mathbf{r})\frac{\mathrm{d}N}{\mathrm{d}\mathcal{E}_{\mathrm{F}}}\delta\{z\} \quad .$$

This allows us to rewrite Poisson's equation

$$\nabla \cdot (\epsilon \nabla \Phi) - 2\lambda\Phi(\mathbf{r})\delta\{z\} = -4\pi\rho_{\mathrm{ext}} \quad ,$$

where we define the screening parameter in two dimensions as

$$\lambda = 2\pi e^2 \frac{\partial N}{\partial \mathcal{E}_{\mathrm{F}}} = \frac{4\pi N m e^2}{v_{\mathrm{F}}^2} \quad . \tag{F.1}$$

The classical form $\lambda_{\mathrm{DH}} = 2\pi N e^2 / (k_{\mathrm{B}}T)$ appears as the two-dimensional form of the Debye screening length. Not surprisingly, in two dimensions the screening effects are less efficient than in three dimensions. Here we utilized the fact that the Fermi surface for the two-dimensional electron system is a curve and in the simplest case becomes a circle with the radius being the Fermi wavevector $k_{\mathrm{F}} = (2\pi N)^{1/2}$. The density of states is given by

$$D(\mathcal{E}) = \frac{m}{\pi\hbar^2} = \frac{N}{\mathcal{E}_{\mathrm{F}}} \tag{F.2}$$

if only the lowest band is occupied; N is the number of electrons per unit area. Note that the density of states is independent of $\mathcal{E}$. In contrast to the three-dimensional case, the screening length $\lambda^{-1} = \hbar^2 / (2me^2)$ for two dimensions is independent of the charge density. In three dimensions we obtained Eq. (5.4.5) as an equivalent to Eq. (F.1), leading to the screened potential given by Eq. (5.4.9). In the two-dimensional case the spatial dependence of the potential is found by a

Fourier–Bessel expansion

$$\Phi(\mathbf{r}) = \int_0^\infty q\, A_q J_0(qr)\, dq \quad ,$$

where J_0 is the Bessel function of the order zero and $A_q = e(q + \lambda)^{-1}$. The statically screened potential in two dimensions is therefore

$$\Phi(q) = \frac{2\pi e}{q + \lambda} \quad . \tag{F.3}$$

Even in the absence of screening, the $1/q$ dependence of the quasi-two-dimensional Coulomb potential is different from the $1/q^2$ dependence found in three dimensions. In this approximation, the static dielectric function in two dimensions becomes

$$\epsilon_1(q) = 1 + \frac{\lambda}{q} \tag{F.4}$$

which is analogous to Eq. (5.4.10) derived in the three-dimensional case.

There are a large number of studies dealing with coupled layers since these questions became relevant in connection with the layered high temperature super-conductors; for a review see [Mar95].

F.1.2 Lindhard dielectric function

From Eq. (5.4.15) we obtain for the complex dielectric response function

$$\hat{\chi}(\mathbf{q}, \omega) = \frac{e^2}{\Omega} \lim_{1/\tau \to 0} \sum_{\mathbf{k}} \frac{f^0(\mathcal{E}_{\mathbf{k}+\mathbf{q}}) - f^0(\mathcal{E}_{\mathbf{k}})}{\mathcal{E}(\mathbf{k}+\mathbf{q}) - \mathcal{E}(\mathbf{k}) - \hbar\omega - i\hbar/\tau} \quad , \tag{F.5}$$

where the summation is taken over all one-electron states. At $T = 0$ the sum can be evaluated, and the result of the real and imaginary parts is obtained by considering the limit for $\tau \to \infty$. For the real part of the Lindhard function we obtain:

$$
\chi_1(\mathbf{q}, \omega) =
\begin{cases}
\begin{aligned}
-e^2 D(\mathcal{E}_F) &\left\{ 1 - C_- \frac{k_F}{q} \left[\left(\frac{q}{2k_F} - \frac{\omega}{q v_F} \right)^2 - 1 \right]^{1/2} \right. \\
&\left. - C_+ \frac{k_F}{q} \left[\left(\frac{q}{2k_F} + \frac{\omega}{q v_F} \right)^2 - 1 \right]^{1/2} \right\}
\end{aligned} \\
\qquad \text{for} \quad \left| \frac{q}{2k_F} - \frac{\omega}{q v_F} \right| > 1 < \left| \frac{q}{2k_F} + \frac{\omega}{q v_F} \right| \\[2ex]
-e^2 D(\mathcal{E}_F) \left\{ 1 - C_- \frac{k_F}{q} \left[\left(\frac{q}{2k_F} - \frac{\omega}{q v_F} \right)^2 - 1 \right]^{1/2} \right\} \\
\qquad \text{for} \quad \left| \frac{q}{2k_F} - \frac{\omega}{q v_F} \right| > 1 > \left| \frac{q}{2k_F} + \frac{\omega}{q v_F} \right| \\[2ex]
-e^2 D(\mathcal{E}_F) \left\{ 1 - C_+ \frac{k_F}{q} \left[\left(\frac{q}{2k_F} + \frac{\omega}{q v_F} \right)^2 - 1 \right]^{1/2} \right\} \\
\qquad \text{for} \quad \left| \frac{q}{2k_F} - \frac{\omega}{q v_F} \right| < 1 < \left| \frac{q}{2k_F} + \frac{\omega}{q v_F} \right| \\[2ex]
-e^2 D(\mathcal{E}_F) \qquad \text{for} \quad \left| \frac{q}{2k_F} - \frac{\omega}{q v_F} \right| < 1 > \left| \frac{q}{2k_F} + \frac{\omega}{q v_F} \right|
\end{cases}
\tag{F.6a}
$$

where we define $C_\pm = \mathrm{sgn} \left\{ \frac{q}{2k_F} \pm \frac{\omega}{q v_F} \right\}$, and the density of states at the Fermi energy is $D(\mathcal{E}_F) = N / \mathcal{E}_F$ in two dimensions; the imaginary part of the Lindhard function is

$$
\chi_2(\mathbf{q}, \omega) =
\begin{cases}
e^2 D(\mathcal{E}_F) \frac{k_F}{q} \left\{ \left[1 - \left(\frac{q}{2k_F} - \frac{\omega}{q v_F} \right)^2 \right]^{1/2} - \left[1 - \left(\frac{q}{2k_F} + \frac{\omega}{q v_F} \right)^2 \right]^{1/2} \right\} \\
\qquad \text{for} \quad \left| \frac{q}{2k_F} - \frac{\omega}{q v_F} \right| > 1 < \left| \frac{q}{2k_F} + \frac{\omega}{q v_F} \right| \\[2ex]
e^2 D(\mathcal{E}_F) \frac{k_F}{q} \left[1 - \left(\frac{q}{2k_F} - \frac{\omega}{q v_F} \right)^2 \right]^{1/2} \\
\qquad \text{for} \quad \left| \frac{q}{2k_F} - \frac{\omega}{q v_F} \right| > 1 > \left| \frac{q}{2k_F} + \frac{\omega}{q v_F} \right| \\[2ex]
e^2 D(\mathcal{E}_F) \frac{k_F}{q} \left[1 - \left(\frac{q}{2k_F} + \frac{\omega}{q v_F} \right)^2 \right]^{1/2} \\
\qquad \text{for} \quad \left| \frac{q}{2k_F} - \frac{\omega}{q v_F} \right| < 1 < \left| \frac{q}{2k_F} + \frac{\omega}{q v_F} \right| \\[2ex]
0 \qquad \text{for} \quad \left| \frac{q}{2k_F} - \frac{\omega}{q v_F} \right| < 1 > \left| \frac{q}{2k_F} + \frac{\omega}{q v_F} \right|
\end{cases}
\tag{F.6b}
$$

The real and imaginary parts of $\hat{\chi}(\mathbf{q}, \omega)$ are plotted in Fig. F.1 as a function of reduced frequency and wavevector. In two dimensions the Fermi surface is a circle

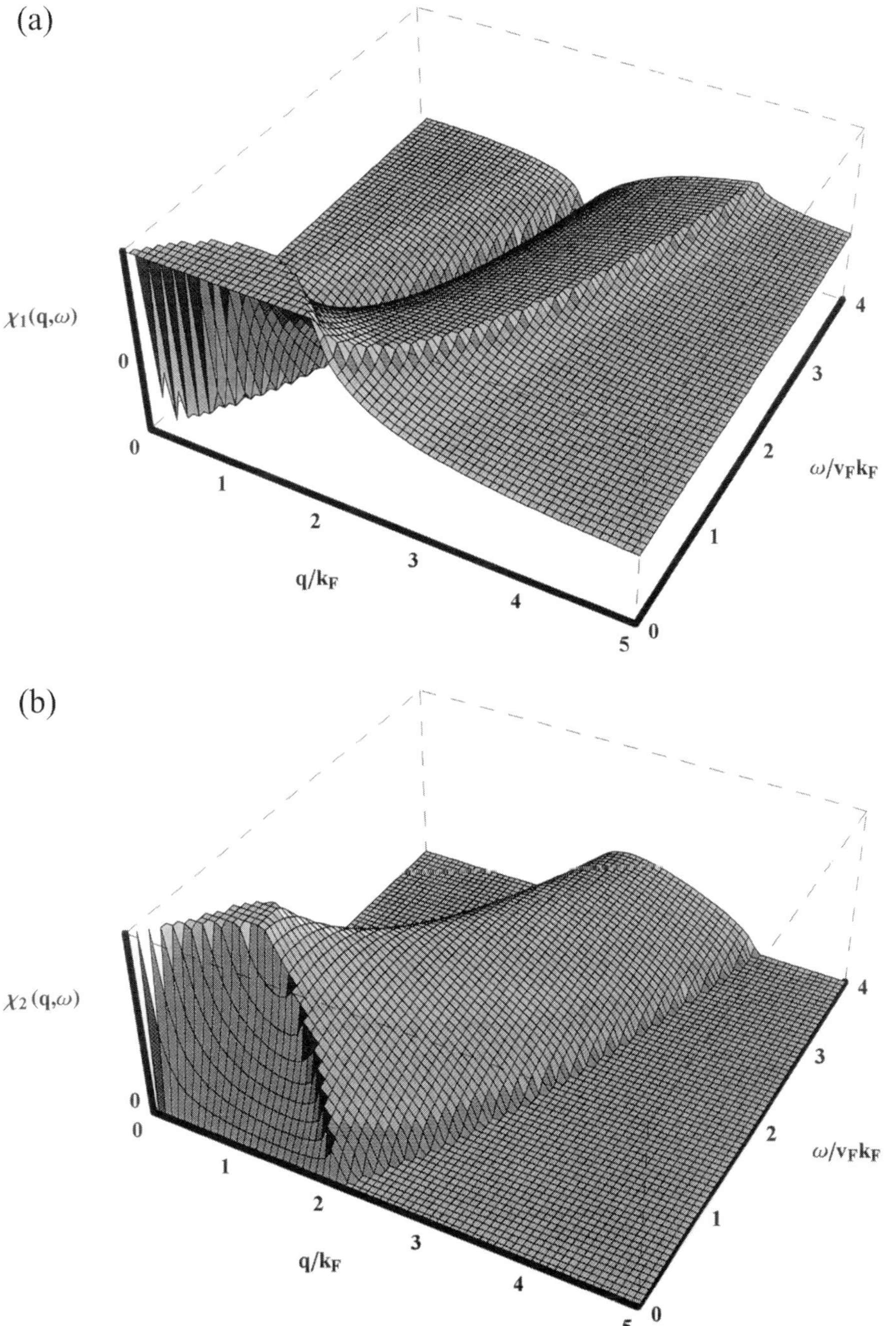

Fig. F.1. Frequency and wavevector dependence of (a) the real and (b) the imaginary parts of the Lindhard response function $\hat{\chi}(\mathbf{q}, \omega)$ of a free-electron gas at $T = 0$ in two dimensions evaluated using Eqs (F.6).

and χ_1 is constant up to $\mathbf{q} = 2\mathbf{k}_F$ over the entire semicircle. In this range there are no excitations possible and thus χ_2 vanishes. In the static limit ($\omega = 0$) the imaginary part of the response function $\chi_2(\mathbf{q}, 0) = 0$ everywhere, and the real part becomes

$$\chi_1(\mathbf{q}, 0) = \begin{cases} -e^2 D(\mathcal{E}_F) & \text{for } q < 2k_F \\ -e^2 D(\mathcal{E}_F)\left\{1 - \frac{2k_F}{q}\left[\left(\frac{q}{2k_F}\right)^2 - 1\right]^{1/2}\right\} & \text{for } q > 2k_F \end{cases},$$

(F.7)

where the long wavelength limit is given by $-e^2 D(\mathcal{E}_F) = -2Ne^2/mv_F^2 = -e^2 N/\mathcal{E}_F$.

F.1.3 Dielectric constant

The complex dielectric constant $\hat{\epsilon}(\mathbf{q}, \omega)$ in two dimensions is immediately derived from Eq. (F.5):

$$\hat{\epsilon}(\mathbf{q}, \omega) = 1 - \frac{2e^2}{\pi q^2}\int d\mathbf{k} \frac{f^0(\mathcal{E}_{\mathbf{k}+\mathbf{q}}) - f^0(\mathcal{E}_{\mathbf{k}})}{\mathcal{E}(\mathbf{k}+\mathbf{q}) - \mathcal{E}(\mathbf{k}) - \hbar(\omega + i/\tau)} \quad . \tag{F.8}$$

The real and imaginary parts of the dielectric constant $\hat{\epsilon}(\mathbf{q}, \omega)$ have the form

$$\epsilon_1(\mathbf{q}, \omega) = \begin{cases} 1 - \frac{4e^2 k_F}{\hbar v_F q^2}\left\{1 - C_-\frac{k_F}{q}\left[\left(\frac{q}{2k_F} - \frac{\omega}{q v_F}\right)^2 - 1\right]^{1/2}\right. \\ \qquad\qquad \left. -C_+\frac{k_F}{q}\left[\left(\frac{q}{2k_F} + \frac{\omega}{q v_F}\right)^2 - 1\right]^{1/2}\right\} \\ \qquad\qquad \text{for } \left|\frac{q}{2k_F} - \frac{\omega}{q v_F}\right| > 1 < \left|\frac{q}{2k_F} + \frac{\omega}{q v_F}\right| \\[4pt] 1 - \frac{4e^2 k_F}{\hbar v_F q^2}\left\{1 - C_-\frac{k_F}{q}\left[\left(\frac{q}{2k_F} - \frac{\omega}{q v_F}\right)^2 - 1\right]^{1/2}\right\} \\ \qquad\qquad \text{for } \left|\frac{q}{2k_F} - \frac{\omega}{q v_F}\right| > 1 > \left|\frac{q}{2k_F} + \frac{\omega}{q v_F}\right| \\[4pt] 1 - \frac{4e^2 k_F}{\hbar v_F q^2}\left\{1 - C_+\frac{k_F}{q}\left[\left(\frac{q}{2k_F} + \frac{\omega}{q v_F}\right)^2 - 1\right]^{1/2}\right\} \\ \qquad\qquad \text{for } \left|\frac{q}{2k_F} - \frac{\omega}{q v_F}\right| < 1 < \left|\frac{q}{2k_F} + \frac{\omega}{q v_F}\right| \\[4pt] 1 - \frac{4e^2 k_F}{\hbar v_F q^2} \qquad\qquad \text{for } \left|\frac{q}{2k_F} - \frac{\omega}{q v_F}\right| < 1 > \left|\frac{q}{2k_F} + \frac{\omega}{q v_F}\right| \end{cases}$$

(F.9a)

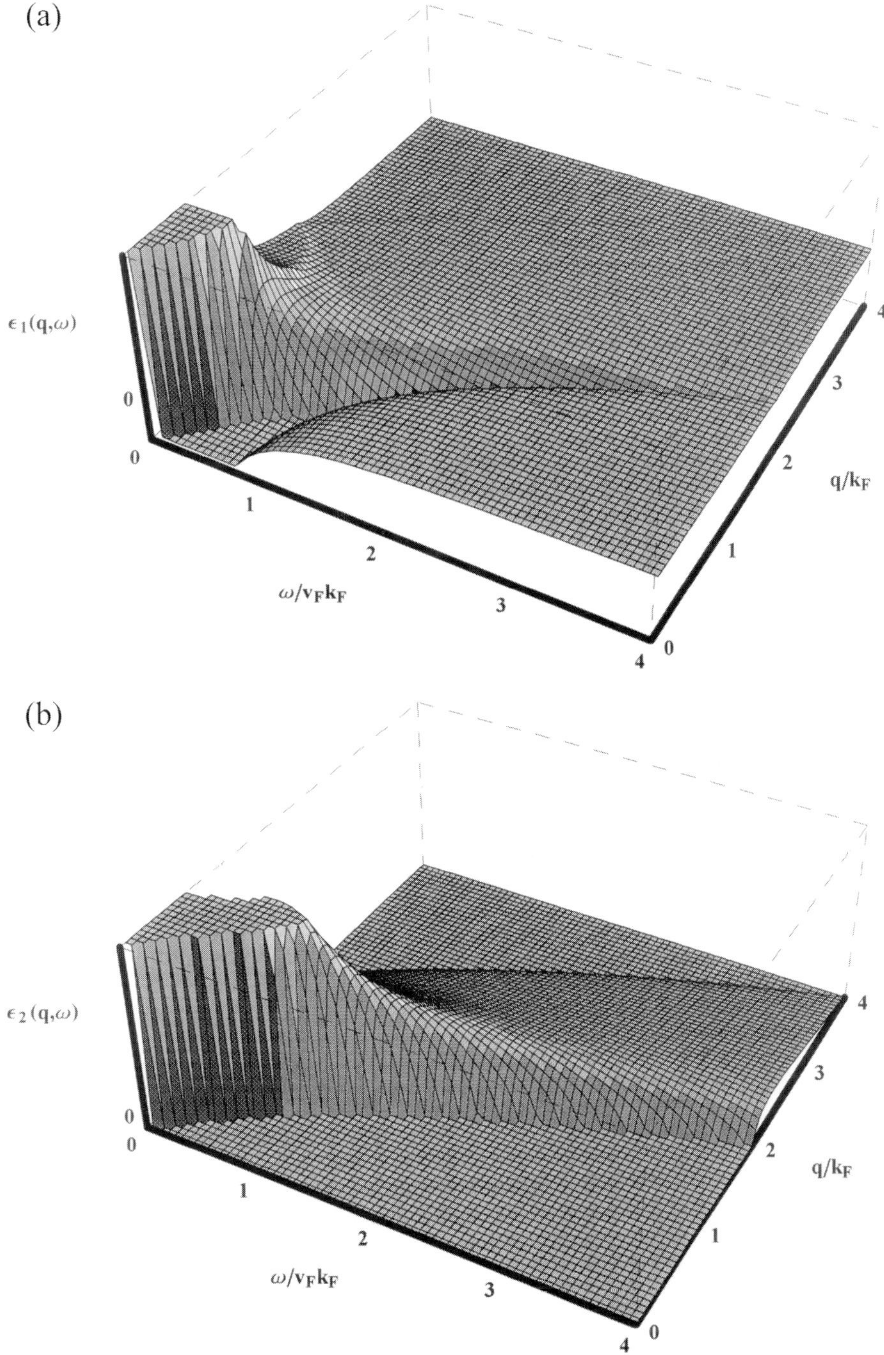

Fig. F.2. Frequency and wavevector dependence of (a) the real and (b) the imaginary parts of the dielectric constant $\hat{\epsilon}(\mathbf{q}, \omega)$ of a free-electron gas at $T = 0$ in two dimensions after Eqs (F.9). The zero-crossing of $\epsilon_1(\mathbf{q}, \omega)$ indicates the plasmon frequency.

with $C_{\pm} = \mathrm{sgn}\left\{\frac{q}{2k_F} \pm \frac{\omega}{q v_F}\right\}$, and

$$
\epsilon_2(\mathbf{q}, \omega) =
\begin{cases}
\frac{4e^2 k_F^2}{\hbar v_F q^3}\left\{\left[1 - \left(\frac{q}{2k_F} - \frac{\omega}{q v_F}\right)^2\right]^{1/2} - \left[1 - \left(\frac{q}{2k_F} + \frac{\omega}{q v_F}\right)^2\right]^{1/2}\right\} \\
\qquad\qquad \text{for} \quad \left|\frac{q}{2k_F} - \frac{\omega}{q v_F}\right| > 1 < \left|\frac{q}{2k_F} + \frac{\omega}{q v_F}\right| \\[2ex]
\frac{4e^2 k_F^2}{\hbar v_F q^3}\left[1 - \left(\frac{q}{2k_F} - \frac{\omega}{q v_F}\right)^2\right]^{1/2} \\
\qquad\qquad \text{for} \quad \left|\frac{q}{2k_F} - \frac{\omega}{q v_F}\right| > 1 > \left|\frac{q}{2k_F} + \frac{\omega}{q v_F}\right| \\[2ex]
\frac{4e^2 k_F^2}{\hbar v_F q^3}\left[1 - \left(\frac{q}{2k_F} + \frac{\omega}{q v_F}\right)^2\right]^{1/2} \\
\qquad\qquad \text{for} \quad \left|\frac{q}{2k_F} - \frac{\omega}{q v_F}\right| < 1 < \left|\frac{q}{2k_F} + \frac{\omega}{q v_F}\right| \\[2ex]
0 \qquad\qquad\quad \text{for} \quad \left|\frac{q}{2k_F} - \frac{\omega}{q v_F}\right| < 1 > \left|\frac{q}{2k_F} + \frac{\omega}{q v_F}\right| \;.
\end{cases}
\tag{F.9b}
$$

Both functions are displayed in Fig. F.2.

At $\mathbf{q} = 2\mathbf{k}_F$ we see a change of slope in the dielectric function which reflects that the screening is cut off for large wavelengths. This leads to Friedel oscillations in the response of the system to a localized perturbation, which also occur in the three-dimensional case (Eq. (5.4.20)). For large distances we obtain

$$
\Phi(r) \propto \frac{e\lambda 4 k_F^2}{(2k_F + \lambda)^2} \frac{\cos\{2k_F r\}}{(2k_F r)^2}
\tag{F.10}
$$

for the spatial dependence of the potential in a two-dimensional electron gas.

For static fields ($\omega \to 0$), Eq. (F.9a) gives the approximation:

$$
\epsilon_1(q, 0) =
\begin{cases}
1 + \frac{\lambda}{q} & \text{for} \quad q \leq 2k_F \\[2ex]
1 + \frac{\lambda}{q}\left\{1 - \left[1 - \left(\frac{2k_F}{q}\right)^2\right]^{1/2}\right\} & \text{for} \quad q > 2k_F
\end{cases}
\tag{F.11}
$$

which for small $\mathbf{q}$ leads to the result derived in Eq. (F.4) for static screening. For $q > 2k_F$ the screening effects fall off much more rapidly. It is interesting to note that via k_F the screening now depends on the charge concentration.

F.1.4 Single-particle and collective excitations

In the two-dimensional electron gas the conditions for single-particle excitations are essentially the same as discussed in Section 5.3 for three dimensions. There is a particular kind of singularity at $\Delta\mathcal{E}(\mathbf{q}) = \left|\frac{\hbar^2}{2m}(q^2 - 2q k_F)\right|$ which becomes more important for one dimension [Cza82], as we will see in the following section. Up to $\mathbf{q} = 2k_\mathrm{F}$, but in particular near $\mathbf{q} = 0$, excitations are possible, and Eqs (5.3.1)–(5.3.3) derived in three dimensions apply.

Plasmons are the longitudinal collective excitations of a two-dimensional electron gas which are sustained if $\epsilon_1(q, \omega) = 0$. For long wavelengths ($m\omega \gg \hbar q k_\mathrm{F}$), we obtain

$$q^2 - \frac{\omega^2}{c^2} = \left(\frac{m\omega^2}{2\pi N e^2}\right)^2 \quad, \tag{F.12}$$

where the right hand side can be neglected for very long wavelengths ($q < 2\pi N e^2/mc^2$). For short wavelengths Eq. (F.9a) leads to the plasma frequency in two dimensions

$$\omega_\mathrm{p}^2 \approx \frac{2\pi N e^2 q}{m} + \frac{3}{4}q^2 v_\mathrm{F}^2 \quad. \tag{F.13}$$

The well known square root plasma dispersion of the leading term $\omega_\mathrm{p} \propto \sqrt{q}$ is not affected by the finite thickness of the electron gas or by correlation effects; it is obtained for degenerate as well as non-degenerate electron gases and has been observed in various systems, such as electrons on liquid helium or inversion layers in semiconductors [And82].

F.2 Dielectric response function in one dimension

The one-dimensional response functions are of great theoretical importance although the realization can only be achieved as the limiting case of a very anisotropic material, such as quasi-one-dimensional conductors. In recent years the problem gained relevance due to the progress in confining the two-dimensional electron gas at semiconductor interfaces or arranging metallic atoms at surfaces along lines. As already pointed out in the discussion of the two-dimensional case, we consider an idealized situation of a one-dimensional electron gas, but the electric field lines extend in three-dimensional space. Concerning the size quantization effect, we again consider only the ground state of the system to be occupied.

F.2.1 Static limit

The charge density of the screening cloud around a point charge decreases very slowly at long distances. In one dimension the q dependence of the potential cannot be defined in the same way as for two and three dimensions (cf. Eqs (F.3) and (5.4.9)); the Coulomb potential can be approximated by [Hau94]

$$\Phi(r) = \frac{e}{r} \quad.$$

The case of a quantum wire with finite thickness d is discussed in [Lee83]. The one-dimensional density of charge carriers is $N = 2k_F/\pi$; the density of states diverges at the band edge with

$$D(\mathcal{E}) = \frac{2m}{\pi\hbar^2 k} = \frac{N}{2\mathcal{E}} \quad . \tag{F.14}$$

F.2.2 Lindhard dielectric function

As in three dimensions, the complex dielectric function in one dimension is obtained by solving

$$\hat{\chi}(q,\omega) = \frac{e^2}{\Omega} \lim_{1/\tau \to 0} \sum_k \frac{f^0(\mathcal{E}_{k+q}) - f^0(\mathcal{E}_k)}{\mathcal{E}(k+q) - \mathcal{E}(k) - \hbar(\omega + i/\tau)} \quad . \tag{F.15}$$

Since in one dimension the Fermi surface consists of two lines, the sum over k space is reduced to $2\sum \frac{f^0(\mathcal{E}_k)}{\mathcal{E}(k+q)-\mathcal{E}(k)-\hbar\omega}$. After some algebra, we obtain the analytical form

$$\hat{\chi}(q,\omega) = -e^2 D(\mathcal{E}_F)\frac{k_F}{2q}\left[\mathrm{Ln}\left\{ \frac{\frac{q}{2k_F} - \frac{\omega+i/\tau}{qv_F} + 1}{\frac{q}{2k_F} - \frac{\omega+i/\tau}{qv_F} - 1} \right\} + \mathrm{Ln}\left\{ \frac{\frac{q}{2k_F} + \frac{\omega+i/\tau}{qv_F} + 1}{\frac{q}{2k_F} + \frac{\omega+i/\tau}{qv_F} - 1} \right\} \right] \quad . \tag{F.16}$$

The real part and the imaginary part of the Lindhard dielectric function have the following form:

$$\chi_1(q,\omega) = -e^2 D(\mathcal{E}_F)\frac{k_F}{2q}\left[\ln\left| \frac{\frac{q}{2k_F} - \frac{\omega}{qv_F} + 1}{\frac{q}{2k_F} - \frac{\omega}{qv_F} - 1} \right| + \ln\left| \frac{\frac{q}{2k_F} + \frac{\omega}{qv_F} + 1}{\frac{q}{2k_F} + \frac{\omega}{qv_F} - 1} \right| \right] \quad , \tag{F.17a}$$

$$\chi_2(q,\omega) = \begin{cases} 0 & \frac{\omega}{qv_F} + \frac{q}{2k_F} < 1 \\[2mm] -e^2 D(\mathcal{E}_F)\frac{k_F\pi}{2q} & \left|\frac{\omega}{qv_F} - \frac{q}{2k_F}\right| < 1 < \frac{\omega}{qv_F} + \frac{q}{2k_F} \\[2mm] 0 & \left|\frac{\omega}{qv_F} - \frac{q}{2k_F}\right| > 1 \end{cases} \quad . \tag{F.17b}$$

In one dimension the density of states is $D(\mathcal{E}_F) = \frac{N}{2\mathcal{E}_F} = \frac{2m}{\pi\hbar^2 k_F}$ for both spin directions. Both parts of the complex Lindhard function are plotted in Fig. F.3.

The Fermi surface in one dimension contains two points and we have perfect nesting for $q = 2k_F$ indicated by the peak at zero energy ($\omega = 0$). The semicircle is given by $\hbar\omega = \frac{\hbar}{2m}(q^2 - 2qk_F)$. The region around the plasma frequency ω_p shows a zero-crossing similar to the three-dimensional case (Fig. 5.15). The real part of the Lindhard function is plotted in Fig. F.3a for different frequencies and wavevectors. In Fig. 5.14 the static limit of the dielectric response function $\chi_1(\mathbf{q})$ is also shown

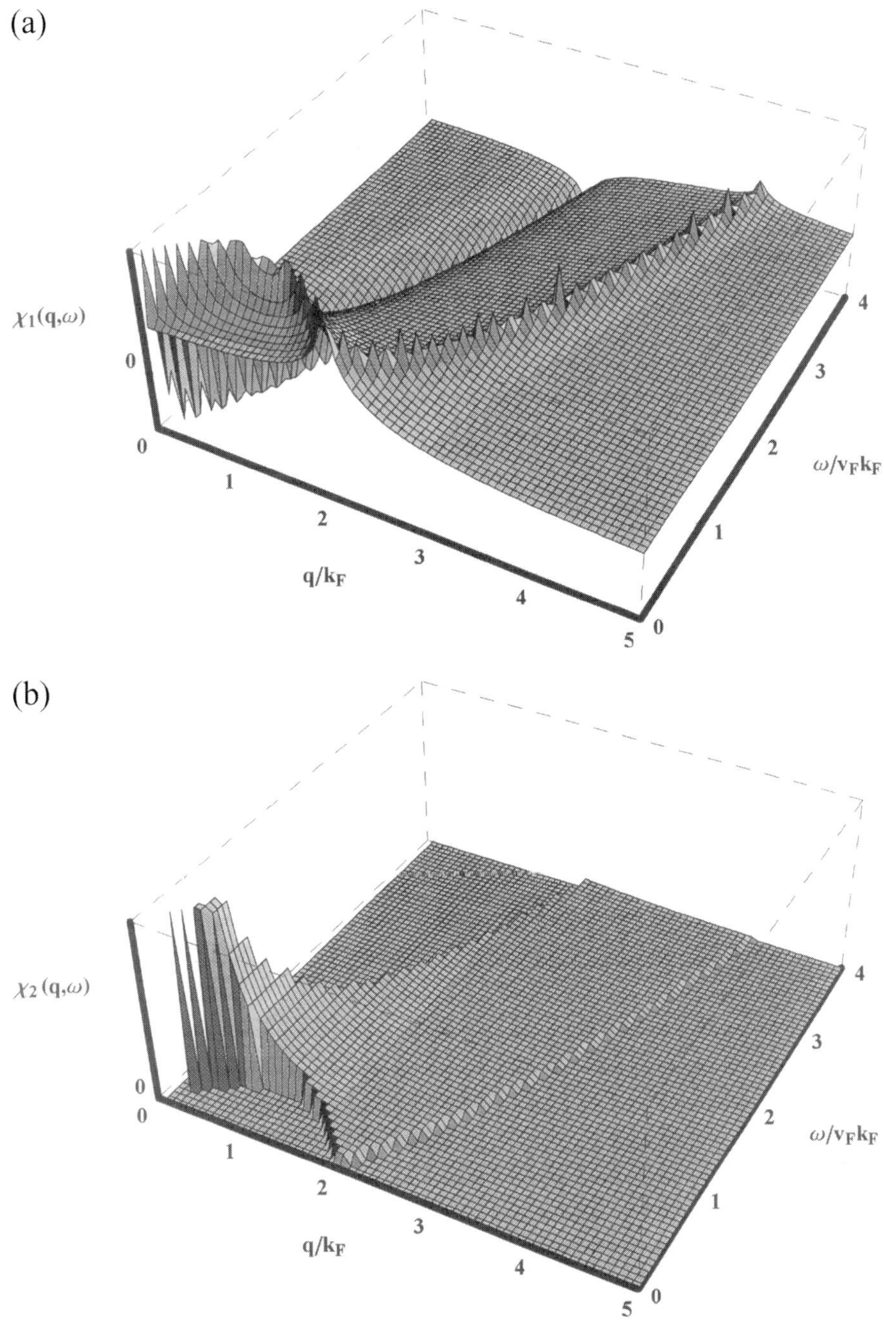

Fig. F.3. Frequency and wavevector dependence of (a) the real and (b) the imaginary parts of the Lindhard response function $\hat{\chi}(q,\omega)$ of a free-electron gas at $T=0$ in one dimension as calculated using Eqs (F.17). The peaks are artifacts of the numerical procedure.

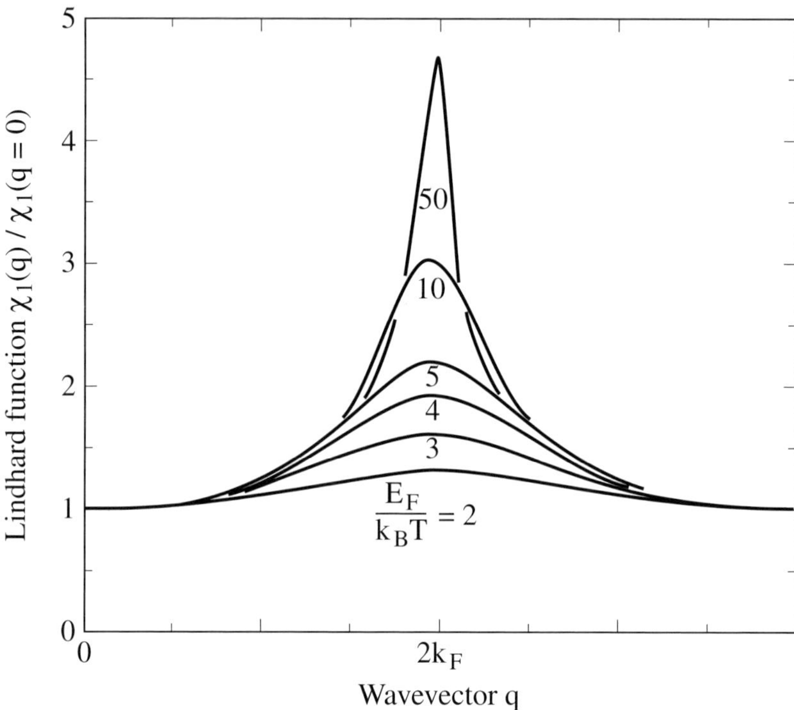

Fig. F.4. One-dimensional response function $\chi_1(q)$ at finite temperatures as a function of wavevector normalized to the $\mathbf{q} = 0$ value. The temperatures indicated are normalized to the Fermi energy $\mathcal{E}_F$.

in the one-dimensional case. It has a pronounced singularity at $q = 2k_F$, which reflects that all states with the momentum difference $2k_F$ have the same energy:

$$\chi_1(q, 0) = -\frac{e^2 k_F}{q} D(\mathcal{E}_F) \ln \left| \frac{q + 2k_F}{q - 2k_F} \right| \tag{F.18}$$

assuming a linear dispersion relation around $\mathcal{E}_F$; this approximation is only valid near the Fermi wavevector k_F. For small q values $\chi_1^{1D}(q)$ is given by the Thomas–Fermi approximation Eq. (5.4.19). In contrast to a three-dimensional electron gas the response function diverges at $q = 2k_F$ in one dimension. For finite temperatures the divergency of $\chi_1^{1D}(q)$ at $q = 2k_F$ decreases and the peak broadens as shown in Fig. F.4. According to the mean field theory we find

$$\chi_1^{1D}(2k_F, T) = -e^2 D(\mathcal{E}_F) \ln \frac{1.14\mathcal{E}_F}{k_B T} \quad . \tag{F.19}$$

F.2.3 Dielectric constant

In one dimension the equation for the dielectric constant is given by

$$\hat{\epsilon}(q,\omega) = 1 + \frac{4e^2 k_F}{\hbar q^3 v_F}\left[\mathrm{Ln}\left\{\frac{\frac{q}{2k_F} - \frac{\omega+i/\tau}{qv_F} + 1}{\frac{q}{2k_F} - \frac{\omega+i/\tau}{qv_F} - 1}\right\} + \mathrm{Ln}\left\{\frac{\frac{q}{2k_F} + \frac{\omega+i/\tau}{qv_F} + 1}{\frac{q}{2k_F} + \frac{\omega+i/\tau}{qv_F} - 1}\right\}\right] .$$

(F.20)

Using Eq. (3.2.7) the real and imaginary parts of the dielectric constant are

$$\epsilon_1(q,\omega) = 1 + \frac{4e^2 k_F}{\hbar q^3 v_F}\left[\ln\left|\frac{\frac{q}{2k_F} - \frac{\omega}{qv_F} + 1}{\frac{q}{2k_F} - \frac{\omega}{qv_F} - 1}\right| + \ln\left|\frac{\frac{q}{2k_F} + \frac{\omega}{qv_F} + 1}{\frac{q}{2k_F} + \frac{\omega}{qv_F} - 1}\right|\right] , \quad \text{(F.21a)}$$

$$\epsilon_2(\mathbf{q},\omega) = \begin{cases} 0 & \frac{\omega}{qv_F} + \frac{q}{2k_F} < 1 \\ \frac{4\pi e^2 k_F}{\hbar q^3 v_F} & \left|\frac{\omega}{qv_F} - \frac{q}{2k_F}\right| < 1 < \frac{\omega}{qv_F} + \frac{q}{2k_F} \\ 0 & \left|\frac{\omega}{qv_F} - \frac{q}{2k_F}\right| > 1 \end{cases} \quad \text{(F.21b)}$$

The real part of the dielectric constant is plotted in Fig. F.5, together with the imaginary part. In the static case ($\omega \to 0$), the real part of the dielectric constant simplifies to

$$\epsilon_1(q,0) = 1 + \frac{8e^2 m}{\hbar^2 q^3}\ln\left|\frac{q + 2k_F}{q - 2k_F}\right| . \quad \text{(F.22)}$$

If the wavevector q is comparable to the Fermi wavevector, Eq. (F.20) tends to

$$\hat{\epsilon}(q,\omega) = 1 - \frac{8e^2 v_F/\hbar}{(\omega + i/\tau)^2 - \frac{q^4 v_F^2}{4k_F^2}} , \quad \text{(F.23)}$$

which is similar to the value from the semiclassical treatment for large k_F.

In Eq. (5.4.10) we arrived at the three-dimensional dielectric constant in the $\omega \to 0$ limit, i.e. the well known Thomas–Fermi dielectric response function. In one dimension the dielectric constant in the limit $\omega \ll qv_F$, or $\omega\tau \gg 1$, is given by

$$\epsilon_1(q,\omega) \approx 1 - \frac{4\pi N e^2}{mq^2 v_F^2} . \quad \text{(F.24)}$$

Interestingly, the difference between one and three dimensions is just a factor 2/3. We also arrive at this factor during the discussion of one-dimensional semiconductors in Section 6.3.1.

F.2.4 Single-particle and collective excitations

In one dimension, the single-particle excitation spectrum is significantly different from the three- and two-dimensional cases. For $0 < q < 2k_F$, $\Delta\mathcal{E}_{\min}$ is no longer

(a)

(b)

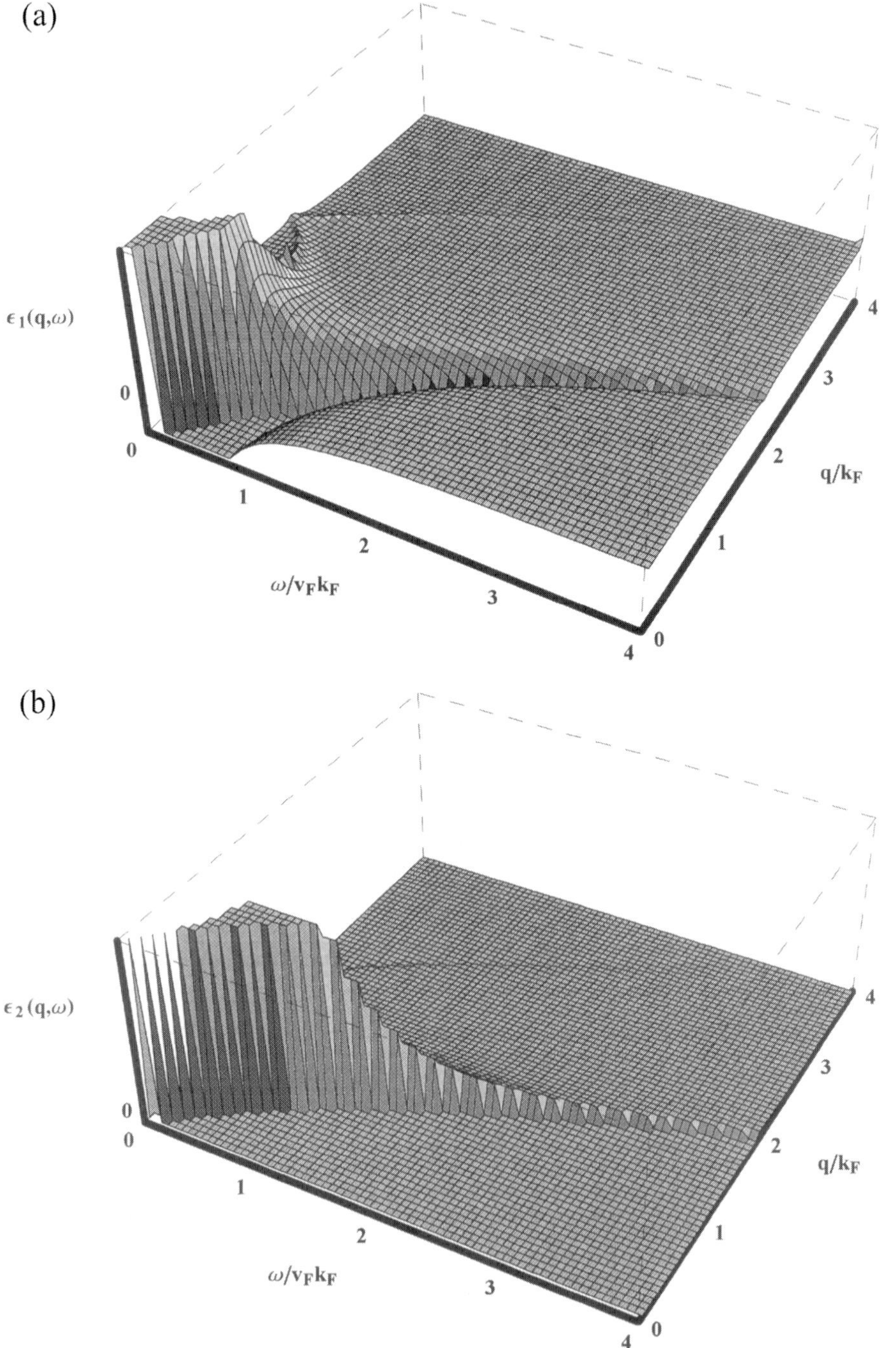

Fig. F.5. Frequency and wavevector dependence of the (a) real and (b) imaginary parts of the dielectric constant $\hat{\epsilon}(q, \omega)$ of a free-electron gas at $T = 0$ in one dimension after Eqs (F.21).

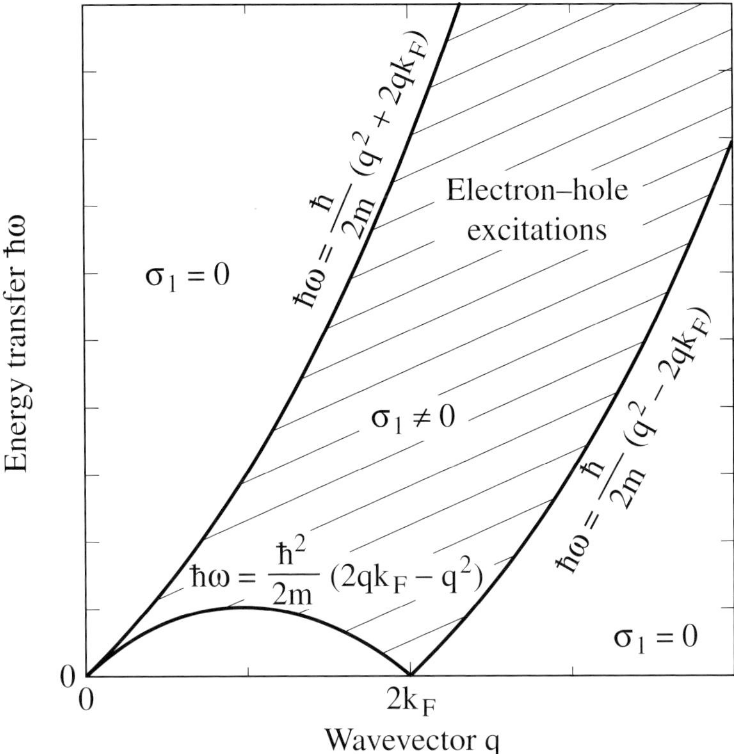

Fig. F.6. Energy spectrum of the excitations shown as a function of momentum for an electron gas in one dimension. The shaded area indicates the pair excitations possible. Note, at low energies only excitations with momentum transfer $q = 0$ and $q = 2k_F$ are possible.

zero. The only zero energy transitions occur at $q = 0$ and $2k_F$. Between these two values, we have

$$\Delta\mathcal{E}_{\min}(q) = \frac{\hbar^2}{2m}(2qk_F - q^2) = \left| \frac{\hbar^2}{2m}(q^2 - 2qk_F) \right| \quad . \tag{F.25}$$

Fig. F.6 shows this one-dimensional excitation spectrum.

Because of the divergency of the absorption at the boundary (shown in Fig. F.3b, for example), in one-dimensional metals collective excitations are only possible outside the continuum of single-particle excitations indicated by the hatched area in Fig. F.6. The plasma frequency has a dispersion

$$\omega_{\mathrm{p}}(q) = \frac{2Ne^2}{\epsilon_s ma^2} |\ln\{qa\}|^{1/2} qa + \mathcal{O}(q^2) \quad , \tag{F.26}$$

which is linear in first approximation.

References

[And82] T. Ando, A.B. Fowler, and F. Stern, *Rev. Mod. Phys.* **54**, 437 (1982)

[Cza82] A. Czachar, A. Holas, S.R. Sharma, and K.S. Singwi, *Phys. Rev. B* **25**, 2144 (1982)

[Hau94] H. Haug and S.W. Koch, *Quantum Theory of the Optical and Electronic Properties of Semiconductors*, 3rd edition (World Scientific, Singapore, 1994)

[Lee83] J. Lee and H.N. Spector, *J. Appl. Phys.* **54**, 6989 (1983)

[Mar95] N.H. March and M.P. Tosi, *Adv. Phys.* **44**, 299 (1995)

Further reading

[Hau96] H. Haug and A.-P. Jauho, *Quantum Kinetics in Transport and Optics of Semiconductors*, Springer Series in Solid State Sciences **123** (Springer-Verlag, Berlin, 1996)

[Hee79] A.J. Heeger, *Charge-Density Wave Phenomena in One-Dimensional Metals*, in: *Highly Conducting One-Dimensional Solids*, edited by J.T. Devreese, R.P. Evrard, and V.E. von Doren (Plenum Press, New York, 1979)

[Lie97] A. Liebsch, *Electronic Excitations at Metal Surfaces* (Plenum Press, New York, 1997)

[Ste67] F. Stern, *Phys. Rev. Lett.* **18**, 546 (1967)

[Voi94] J. Voit, *Rep. Prog. Phys.* **57**, 977 (1994)

Appendix G

Important constants and units

Throughout the book we have used cgs, or Gaussian, units. Without doubt, SI units are more convenient when equating numbers and analyzing experimental results. Therefore we provide conversion tables for commonly used quantities. For a discussion of problems of units and the conversion between Gaussian (cgs) and rational SI units (mks), see for example [Bec64, Jac75].

Table G.1. *Conversion table.*

In order to convert the equations in cgs units into those in the SI system, the relevant symbols have to be replaced by the corresponding one on the right hand side of the table.

Quantity	Gaussian (cgs) systems	SI (mks) system
Speed of light	c	$c = \dfrac{1}{(\epsilon_0\mu_0)^{1/2}}$
Electric field	$\mathbf{E}$	$(4\pi\epsilon_0)^{1/2}\,\mathbf{E}$
Electric displacement	$\mathbf{D}$	$\left(\dfrac{4\pi}{\epsilon_0}\right)^{1/2}\mathbf{D}$
Scalar potential	ϕ	$(4\pi\epsilon_0)^{1/2}\,\phi$
Charge density	ρ	$\dfrac{1}{(4\pi\epsilon_0)^{1/2}}\rho$
Electric polarization	$\mathbf{P}$	$\dfrac{1}{(4\pi\epsilon_0)^{1/2}}\mathbf{P}$
Current density	$\mathbf{J}$	$\dfrac{1}{(4\pi\epsilon_0)^{1/2}}\mathbf{J}$
Dielectric constant	$\hat{\epsilon}$	$\dfrac{\hat{\epsilon}}{\epsilon_0}$
Conductivity	$\hat{\sigma}$	$\dfrac{\hat{\sigma}}{4\pi\epsilon_0}$
Magnetic field	$\mathbf{H}$	$(4\pi\mu_0)^{1/2}\,\mathbf{H}$
Magnetic induction	$\mathbf{B}$	$\left(\dfrac{4\pi}{\mu_0}\right)^{1/2}\mathbf{B}$
Vector potential	$\mathbf{A}$	$\left(\dfrac{4\pi}{\mu_0}\right)^{1/2}\mathbf{A}$
Magnetization	$\mathbf{M}$	$\left(\dfrac{4\pi}{\mu_0}\right)^{1/2}\mathbf{M}$
Permeability	$\hat{\mu}$	$\dfrac{\hat{\mu}}{\mu_0}$
Impedance	$\hat{Z}$	$4\pi\epsilon_0\hat{Z}$
Poynting vector	$\mathbf{S}$	$\dfrac{c}{4\pi}\mathbf{S}$
Energy density	u	$4\pi u$

Note: In the case of the electric field $\mathbf{E}$ we must remember to replace $1/4\pi$ by ϵ_0.

For convenience, in Table G.2 we also list some fundamental physical constants used throughout the book. These values are only given in SI units since quantitative values are no longer given in cgs units such as erg, dyne, etc. However, some

Table G.2. *The fundamental physical constants.*

Quantity	Value in SI units
Speed of light in a vacuum	$c = 2.997\,924\,58 \times 10^8 \text{ m s}^{-1}$
Permittivity of free space	$\epsilon_0 = 8.8542 \times 10^{-12} \text{ A s V}^{-1} \text{ m}^{-1}$
Permeability of free space	$\mu_0 = 1.2566 \times 10^{-6} \text{ V s A}^{-1} \text{ m}^{-1}$
Elementary charge	$e = 1.602\,189 \times 10^{-19} \text{ A s}$
Mass of electron	$m = 9.010\,953 \times 10^{-31} \text{ kg}$
Mass of proton	$m_\mathrm{p} = 1.672\,61 \times 10^{-27} \text{ kg}$
Mass of neutron	$m_\mathrm{n} = 1.674\,82 \times 10^{-27} \text{ kg}$
Planck constant	$h = 6.626\,176 \times 10^{-34} \text{ J s}$
	$\hbar = h/2\pi = 1.054\,589 \times 10^{-34} \text{ J s}$
Avogadro constant	$N_\mathrm{A} = 6.022\,05 \times 10^{23} \text{ mol}^{-1}$
Boltzmann constant	$k_\mathrm{B} = 1.380\,66 \times 10^{-23} \text{ J K}^{-1}$
Bohr radius	$r_\mathrm{B} = \frac{\hbar^2}{me^2} = 5.291\,77 \times 10^{-11} \text{ m}$
Rydberg constant	$Ry = \frac{me^4}{2\hbar^2} = 2.179\,91 \times 10^{-18} \text{ J}$

Table G.3. *Table of units.*

Quantity	cgs	SI	SI	cgs
Conductivity σ	s^{-1}	$1.1 \times 10^{-12} \ \Omega^{-1} \text{ cm}^{-1}$	$\Omega^{-1} \text{ cm}^{-1}$	$9 \times 10^{11} \text{ s}^{-1}$
Magnetic field $\mathbf{H}$	Oe	$\frac{10^3}{4\pi} \text{ A m}^{-1}$	A m^{-1}	$4\pi \times 10^{-3} \text{ Oe}$
Magnetic induction $\mathbf{B}$	G	10^{-4} T	T	10^4 G
Magnetic susceptibility χ_m	$\text{emu cm}^{-3} \text{ Oe}^{-1}$	4π	1	$\frac{1}{4\pi} \text{ cm}^3 \text{ Oe emu}^{-1}$
Pressure p	bar	10^5 Pa	Pa	10^{-5} bar
				$7.53 \times 10^{-3} \text{ torr}$

Note: Oe = oersted; G = gauss; T = tesla; Pa = pascal. 1 emu = 1 G cm^3; we also found the susceptibility per mass and per mole.

quantities still occur in different units, like the magnetic field. In Table G.3 we give only those which are important in our context. Most important are certainly the units of energy and its equivalent such as temperature and frequency; the corresponding conversion is listed in Table G.4, which can also be found in Chapter 8.

Table G.4. *Conversion table.*

	$\mathcal{E}$ (J)	$\mathcal{E}$ (eV)	T (K)	f (Hz)	ν (cm^{-1})	λ (m)
$\mathcal{E}$ (J)	1	6.242×10^{18}	7.244×10^{22}	1.509×10^{33}	5.035×10^{22}	1.986×10^{33}
$\mathcal{E}$ (eV)	1.602×10^{-19}	1	11 605	2.418×10^{14}	8066	1.240×10^{-6}
T (K)	1.381×10^{-23}	8.617×10^{-5}	1	2.084×10^{10}	0.671	0.0149
f (Hz)	6.626×10^{-34}	4.136×10^{-15}	4.799×10^{-11}	1	3.336×10^{-11}	2.998×10^{8}
ν (cm^{-1})	1.986×10^{-23}	1.240×10^{-4}	1.490	2.998×10^{10}	1	1×10^{-2}

Note: The conversion between energy $\mathcal{E}$, temperature T, frequency f, wavenumber $\nu = 1/\lambda$, and the corresponding wavelength λ is performed by using the basic physical constants: the elementary charge $e = 1.602 \times 10^{-19}$ A s, the Boltzmann constant $k_B = 1.381 \times 10^{-23}$ J K^{-1}, the speed of light in a vacuum $c = 2.998 \times 10^{8}$ m s^{-1}, and Planck's constant $\hbar = 1.055 \times 10^{-34}$ J s. (Note that λ is inversely proportional while all other scales are proportional to the frequency.)

References

[Bec64] R. Becker, *Electromagnetic Fields and Interaction* (Bluisdell Publisher, New York, 1964)

[Jac75] J. D. Jackson, *Classical Electrodynamics*, 2nd edition (John Wiley & Sons, New York, 1975); 3rd edition (John Wiley & Sons, New York, 1998)

Further reading

[Bir34] R.T. Birge, *Am. Phys. Teacher* **2**, 41 (1934); **3**, 102 and 171 (1934)

Index

cavity *see* resonant cavity
cavity ring-down technique 270, 287
CeAl$_3$ *324*
center frequency *see* Lorentz model *and* resonant
    structure
cgs units *see* units
(CH)$_x$ *362*, *365–6*
Chambers formula **112**, 189, 313, **430**, 431, 442
    see also non-local response *and* skin effect,
        anomalous
characteristic impedance *see* impedance
charge carrier concentration N 85, 93, **106**, 120, 138,
    140, 179, 302, 358, **424**
charge conservation 11, 78
charge density ρ 10, **15**, 53, 78
    external ρ_{ext} 16, 52–3
    induced ρ_{ind} 52–3
    polarizational ρ_{pol} 16
    total ρ_{total} 16
charge density wave (CDW) *174*, 175, 387, 389, 392
    commensurate 175, 362
    incommensurate 175
    optical properties 197, *202*, *389*
circuit representation **218**, *219*, *231*, **236**, *237*, 241,
    288
clean limit **186**, 187, 199, 372, 381, 385, **443**
clystron 211
coaxial cable 217, 273
coherence 211, **212**, 245, 261–2, 281
coherence factor **182**, 185, 197, 375, 394
    case 1 **183**, 198, *389*, *393*
    case 2 **183**, 191, 375
coherence length
    ξ 186, 379, *382*, 442
    l_c 212
coherence temperature T_{coh} 323, 384
Cole–Cole relaxation 391–2
collective excitation **131**, 171, 187, **197**, 198, **309**,
    310–11, 371, **387**, *389*, 390, *392*, 428, 459
conductivity
    complex $\hat{\sigma}$ **18**, *23*, 61, 208, 322
    density wave 182, 196, **197**, *199*, *200*, *202*, 387–8,
        389, 392–4
    imaginary part σ_2 18, *23*, 61
    longitudinal component $\hat{\sigma}^{\text{L}}$ 55, **78**, 79, 120
        $\mathbf{q}$ dependence *123*, **130**, 132
    metal **95**, *96*, 303–5, 322, *324*, 453
    real part σ_1 16, 18, *23*, 61, 304
    semiconductor *141*, 153–4, 157, 167, *168*, 350,
        351, *359*, 363, *364*
    superconductor 181, **191**, *192–3*, *259*, 375, *376*,
        379, *380*, *419*
    transverse component $\hat{\sigma}^{\text{T}}$ 55, 73, **76**, 79, 88, **89**
        $\mathbf{q}$ dependence 115, **118**, *119*, 133
contact 209, 247
continuity equation 11, 78, 179
convolution 399–400
Cooper channel 175
Cooper pair 173, 176, **179**, 383
correlation length *see* coherence length
Coulomb gap 331

Coulomb gauge 11, 48
Coulomb glass 4, 331
Coulomb repulsion U 361–2
    screened μ^* 383
Coulomb's law **10**, 13, 16–17, 48, 446–7, 453
critical point 154, 345–6, *347*, 348, 366
Cu *314*
CuO *355*
current–current correlation function 72, **105**
current density $\mathbf{J}$ 10, **16**, 93, 109, 179, 208, 254, 430,
    435, 439
    bound $\mathbf{J}_{\text{bound}}$ 16
    conduction $\mathbf{J}_{\text{cond}}$ 16
    diamagnetic $\mathbf{J}_{\text{d}}$ 73, 80, 89–90, 439
    displacement $\mathbf{J}_{\text{bound}}$ 16
    external $\mathbf{J}_{\text{ext}}$ 16
    longitudinal component $\mathbf{J}^{\text{L}}$ 48, 51, 79
    paramagnetic $\mathbf{J}_{\text{p}}$ 73, 80, 439
    superconducting 179, **180**, 189, 439, 441
    total $\mathbf{J}_{\text{total}}$ 16
    transverse component $\mathbf{J}^{\text{T}}$ 48, 51, 73, 76–7, 79, **89**
cutoff energy $\mathcal{E}_{\text{c}}$ 177–8, 325
cutoff frequency 222–3, 229, 325
cutoff wavelength *see* cutoff frequency
cyclotron resonance 316

dc conductivity σ_{dc} **93**, 270, 284, **302**, 306, 323, 326,
    330–1, **342**, 388
de Haas–van Alphen effect 316
Debye–Hückel screening length *see* screening length
Debye relaxation 255, 391
Debye temperature θ_{D} 383
density of states
    electronic (DOS) D_l **154**, 161, 177, 184, 319, 328,
        331, 340, **344**, *345*, 367, 446, 454
    phonon $F(\omega_{\text{P}})$ 327, 383, *384*
    see also joint density of states
density wave (DW) 173, 387
    commensurate 175, 362, 394
    incommensurate 175, 387, 394
    see also collective excitation; internal deformation;
        pinning; *and* single-particle gap
depolarization 233
detector 212, *213*
diamond *349*
dielectric constant
    complex $\hat{\epsilon}$ **17**, 18, *23*, 62, 65, 86, 250, 279, 340
    density wave 390, 392, 394
    glass 255, 392
    imaginary part ϵ_2 **17**, *23*, 62, 87, 90, *341*, *347*, *348*
    longitudinal component $\hat{\epsilon}^{\text{L}}$ 49–54
    metal 95–6, *97*, 303–4
    one dimension *157*, *158–9*, *347*, **457**, *458*
    $\mathbf{q}$ dependence 123, **127**, *128*, 130, 131, 169, 452,
        457
    real part ϵ_1 **16**, *23*, 61, 66–7, 87, 90, *140*, 303, *306*,
        335, *341*, *343*, 390
    semiconductor **138**, 139–40, *142*, **153**, 154, *157*,
        158, 161–2, *168*, *347–8*, *367*
    static **67**, 155, 340, *349*, 354, 390, 394–5, 447, 452,
        457

plasma frequency (*cont.*)
  dispersion *117*, **131**, 132, *310*, 453, 459
  longitudinal plasma frequency ω_L 51, 310
  transverse plasma frequency ω_T 51
plasma waves 38, *309–10*, 453
Poisson's equation **11**, 48, 85, 120, 446
polarimetry *see* ellipsometry
polarizability $\hat{\alpha}$ 65, **67**, 138, 140
polarization **P 15**, 16, 52, 131, 137, 140, 251–2, *255*, *391*
polarization of electric field vector 12, *14*, 31, *32*, 262, 275–6, 278, *280*
polyacetylene *see* $(CH)_x$
potential *see* scalar potential *and* vector potential
potential energy *V* 78, 85
power absorption *P* **26**, 27, 74–5, 233, 269, 424–5
  broken symmetry ground state 184–5, 191
  resonant structure 237
  semiconductor 152
  transmission line 220, 226
Poynting vector **S** 15, **27**
prism spectrometer 248
propagation constant $\hat{\gamma}$ 219
propylene carbonate 254, *255*
pseudopotential method 344, 345
pulse technique *see* time domain spectroscopy
pyroelectric detector *213*

quality factor *Q* **235**, 236–7, 240, 270, 286, 289, 293, 420
quantum Monte Carlo (QMC) method 363
quantum phase transition 4
quarter wave plate 279, *280*

radiation detector *see* detector
radiation source *see* source
Raman scattering 1, 33, 215
random phase approximation (RPA) 53
Rayleigh criterion 264, 400
reflection coefficient $\hat{r}$ **34**, 35, *39–41*, 44, 63, 69, 232, 408, 416
reflection regime 146
reflectivity *R* **36**, 44, *276*, 303, *341*, 377, *393*
  extrapolation 246, 266
  film R_F 410, *412–13*, *415*
  Hagen–Rubens behavior *99*, **103**, 246, 307, *308*, 414
  measurement 275–6, 278
  metal *99*
  plasma edge *99*, **104**, 304, 312, 318, 340, *361*, 415
  semiconductor *144*, *158*, *168*, *361*
  spin density wave *201*, *393*
  superconductor *194*, *378*
refractive index
  complex $\hat{N}$ **21**, *23*, 29, 63, *98*, *143*, 304
  imaginary part *see* extinction coefficient
  real part *n* **21**, 22, *23*, 63, 69, *98*, *143*, 285
relaxation rate **93**, 95, 100, 105, 108, **139**, 186, 208, 302, 306–7, 310, **319**, *433*
  frequency dependent 302, 322, *324*, 326, 427
  renormalized **322**, 325, 384, 387

relaxation regime 100, **103**, 303, 313, 328, *433*, 434, *435*
relaxation time approximation 93, 105, 108, 139
resistivity *see* dc conductivity
resolution 248, 253, *263*, 264
resonant cavity *214*, *236*, 238, *239*, **290**, *291*, 375
resonant structure 217, **234**, *236*, *237*, 269, 286
  coaxial line 238
  dielectric 238
  discrete elements *see* *RLC* circuit
  frequency ω_0 **235**, 236–7, *237*, 239, 242, 289, *292*
  open *see* Fabry–Perot resonator
  perturbation 241, 290
  stripline *214*, *236*, 238
  waveguide *see* resonant cavity
  width Γ 235–7, *237*, 240, 242, *292*
  see also quality factor
response
  longitudinal response 47, 52, 81, 120, 223
  static limit 120
  transverse response 47, 115
response function 1, **18**, 56, **57**, *64*, 71, 81, 208, 251, 257
rigid-band approximation 120
RLC circuit 218, 236, *237*, *288*
Rydberg series 163–6, 355–7

sampling theorem 253, 264, 400
scalar potential Φ **10**, 13, 54, 78, 84
scaling regime 330
scattering matrix *S* 232, 248
scattering of waves 217, 230
scattering rate *see* relaxation rate
screening effect 120, 209, 445
screening parameter
  classical 121
  Debye Hückel λ_{DH} 121, 446
  Thomas–Fermi λ 121
  two dimensions 446
second quantization 79, 173
selfconsistent field approximation (SCF) 53
semiconductor 136, 303, 339
Shubnikov–de Haas effect 316
Si *312*, *349*, 350, *351*
Si:B 335
Si:P 331, 335, 357–8, *359*, 360
SI units *see* units
single-particle excitation 115, **117**, 130, 171, 173, 187, 197, 309–11, 371, **388**, 390, 393, 428, 452, 459
single-particle gap Δ **177**, *178*, 208, 275, 362, 365, **373**, 375, 379, *382*, *389–90*, 393, 394
  anisotropic 371–2, 385–6
  see also bandgap
skin depth δ **25**, 44, *114*, 180, 209, 233, 273, 275, 313, 431
skin effect **25**, 99, 313, 381, *433*, 434, *435*
  anomalous 113, **312**, 314, 381, **429**, *432*, *433*, 434, *435*
Sn *382*